T4-AJW-068

Automotive Lubricants Reference Book

A.J. Caines
and
R.F. Haycock

Published by:
Society of Automotive Engineers, Inc.
400 Commonwealth Drive
Warrendale, PA 15096-0001
U.S.A.
Phone: (412) 776-4841
Fax: (412) 776-5760

Library of Congress Cataloging-in-Publication Data

Caines, A. J. (Arthur J.), 1932-
Automotive lubricants reference book / A.J. Caines and R.F. Haycock.
p. cm.
Includes bibliographical references and index.
ISBN 1-56091-525-0 (hardcover)
F. (Roger F.) II. Title.
TL153.5.C35 1996
629.25'5--dc20 96-22048
CIP

ISBN 1-56091-525-0

SAE Order No. R-145

Preface

When asked by the SAE to write a companion volume on lubricants to their Automotive Fuels Reference Book, we asked ourselves a series of questions:

> Who are the expected readers?
> What depth of coverage is needed for each topic?
> Do we have the experience and knowledge to author such a book?

Discussion with the SAE publishing division enabled satisfactory answers to be made to these and other questions, and we hope the resulting volume will be a useful contribution to the rather limited literature on the topic.

First, the book is not intended for those who are already experts in the field of automotive lubrication. Prospective readers will be automotive engineers who want to understand more about the properties and composition of lubricants used in the machinery they are concerned with, or scientists or engineers in the petroleum industry who have worked in another field and need a primer on lubricants for the specialized area of automotive applications.

The depth of coverage is limited by the number of topics we felt needed to be included in the one volume, and is broadly in line with that of the Automotive Fuels Reference Book. Our treatment of the subject matter is essentially qualitative and the use of mathematics is minimal. As we have found in the initial reviews of the text, those who are expert in any of the topics we cover will feel our coverage is too simplistic for their own area, but may well feel other areas are dealt with adequately. Our advice to them is not to linger over the area they already know.

There may be surprise at the authorship by two Englishmen with similar career paths within one corporation, albeit a major international one. We hope that we can highlight complexities which it may be difficult to appreciate from a North American viewpoint. Also, through our U.S. connections, we have had very strong links with North American lubricant activities while being able to relate directly to European attitudes and developments. Colleagues in the Asia-Pacific region have helped with Japanese and other developments from that area.

Both of us have worked in both the petroleum and chemical sectors of Exxon Corporation, mainly in areas where lubricant testing was an important factor. However, the views expressed are our own and not necessarily those of our past or present employer. AJC had close contacts with the U.S. military, particularly in the 1970s when they were key actors in the improvements of oil quality, and later worked closely with the national petroleum companies of Africa, the Middle East, and Eastern (now called Central) Europe. RFH has had close contacts with European approval organizations, spent 10 years in liaison with the motor industry, acquiring a good understanding of lubricant problems as seen by that industry, and is currently the chairman of the British Technical Council of the Motor and Petroleum Industries. Both of us have seen motor oils develop from uncompounded monograde oils to the sophisticated part or wholly synthetic wide energy conserving multigrades of today.

You will notice that we have frequently taken a historical approach in our presentation of the various topics. This is deliberate, because oil qualities have evolved steadily over the last half century, with no major discontinuities, and we believe an understanding of how this evolution took place is necessary to understand oil quality today.

We give considerable attention to specifications and oil approval systems. The desire for products to have the widest coverage of equipment and geographical areas usually conflicts with the ever more complex needs of specific markets. Specifications often have short lives, but the organizations that develop them last much longer and are key actors in the lubricants business.

Few existing books give much discussion of the relationship between test methods and formulation technology, and we have tried to cover this adequately. While the fundamentals of formulation are presented, detailed formulations to meet specific quality levels precisely cannot be given. Such formulations are proprietary to formulating companies, whether oil company or additive supplier, and each formulator has different components and expertise to utilize in producing an oil to meet a given quality. Furthermore, the optimum means of meeting a given quality level can change quite rapidly, requiring running changes within the life of a specification, and so a precise formulation often has a short life.

Much of the book is given over to engine oils, but other automotive lubricants such as transmission oils and greases are considered. We have also included a brief section on the key features of the industrial oils which may be used in the production of automotive equipment, in the belief that this will be of interest and useful to those working in the automotive industry. Our coverage of the blending, handling and usage of lubricants is probably unique in a book of this type, and our attention to health and environmental aspects of lubricants reflects the growing concerns in these areas.

We have collected in the appendices some of the less readily digestible material which we felt should be included to make the book a work of reference, and hope we have succeeded in keeping the main sections both interesting and easily readable. We have also included a glossary, a list of relevant acronyms, and for those who studied chemistry a long time ago, a primer in petroleum chemistry.

A.J. Caines
R.F. Haycock
May 1996

Acknowledgments

To Exxon Corporation, for providing us with a total of over 60 years of experience in this or closely related spheres of activity.

To its affiliates, Esso Petroleum Co. Ltd. and Exxon Chemical Ltd., and particularly their operations at the Esso Research Centre site at Abingdon, U.K., where advice, expertise, graphics and library facilities were made readily available to us by their respective managements.

To Exxon Chemical for allocation of a portion of R.F. Haycock's working time.

To the colleagues and ex-colleagues who allowed us to pick their brains and advised us of major errors and omissions in draft versions, but particularly to:

> Mary Allen, Karen Ball, Alain Bouffet, Roy Cole, Monty Crook, Jim Eagan, Gordon Farnsworth, Derek Fraser, Penny Hughes, Charlie Keller, Mike Kingsland, Eric Lewis, Owen Marsh, George de Montlaur, Jim Newcombe, Joe Noles, Bob Northover, Richard Phillips, Renee Roper, Larry Smith, John Smythe, Nigel Tilling, Malcolm Waddoups, Skip Watts, Tony White, Bob Wilkinson, Tony Yates

To industry contacts who also gave very helpful advice including:

> Dennis Groh, Norman Hunstad, Eric Johnson, Richard Kabel, Mike McMillan, Roy Smith

and especially to John May and Max Gairing for significant review of draft material.

Plus the reviewers engaged by SAE who were generally kind, and provided many helpful suggestions and additions.

To the secretaries and word processing operators who have contributed significantly in development and finalization of the work and suffered innumerable changes with good grace: Georgina Denton, Elizabeth Roberts, Zoë Ludwig, Valerie Oglesby, and particularly Kate Baker.

Contents

Chapter 1—Introduction and Fundamentals 1
1.1 Lubricants in History 1
1.2 Functions of a Lubricant 4
1.3 Approval of Lubricants for Use 6
1.4 Friction and Wear, Lubrication and Tribology 11
1.4.1 The Mechanics of Friction 11
1.4.2 Dry Friction 15
1.4.3 Lubricated Sliding 16
1.4.4 Boundary Lubrication 19
1.4.5 Extreme Pressure Conditions 20
1.4.6 Elasto-Hydrodynamic Lubrication 21
1.5 Solid and Grease Lubricants 22
1.5.1 Solid Lubricants 22
1.5.2 Greases 26
1.6 Unlubricated Conditions 27
1.7 Lubrication Requirements of Different Systems 29
1.7.1 Simple Systems 29
1.7.2 Internal-Combustion Engines 34
1.7.3 Special Systems 46

Chapter 2—Constituents of Modern Lubricants 49
2.1 Base Stocks 49
2.1.1 Conventionally Refined Petroleum Base Stocks 51
2.1.2 Modern Conversion Processes 59
2.1.3 Reclaimed Base Stocks 62
2.1.4 Other Types of Base Stocks 63
2.2 Additives 69

Chapter 3—Crankcase Oil Testing 97
3.1 Introduction 97
3.2 Laboratory Bench Tests 99

3.2.1 Tests for Physical Properties ... 99
3.2.2 Chemical Tests and Properties ... 116
3.3 Performance Testing ... 124
3.3.1 Bench Performance Tests ... 124
3.3.2 Laboratory Engine Tests ... 127
3.3.3 Field Testing ... 133
3.4 Precision and Accuracy; Testing Statistics of Automotive Lubricants ... 139
3.4.1 Basic Statistical Principles ... 139
3.4.2 Laboratory Tests on Petroleum Products ... 148
3.4.3 Engine Tests ... 153
3.4.4 Reference Lubricants ... 158
3.4.5 Reference Fuels ... 162
3.5 Tests on Used Oils ... 164
3.5.1 Tests for Evaluation of Oil Condition ... 164
3.5.2 Testing for Equipment Condition ... 170

Chapter 4—Crankcase Oil Quality Levels and Formulations ... 175
4.1 Evolution of Quality Levels ... 175
4.1.1 Gasoline Engine Oils ... 179
4.1.2 Diesel Engine Oils ... 186
4.1.3 Multi-Purpose Gasoline/Diesel Oils ... 189
4.1.4 Super Tractor Universal Oils (STUO) ... 194
4.2 Formulating a Crankcase Oil ... 196
4.2.1 Choice of Base Stocks ... 197
4.2.2 Choice of Viscosity Modifier ... 199
4.2.3 Developing the Performance Package ... 200
4.2.4 Evaluating and Finalizing a Formulation ... 203
4.3 Specialized Crankcase Oils ... 206

Chapter 5—Practical Experiences with Lubricant Problems ... 211
5.1 Problems of Use of Inappropriate Lubricants ... 212
5.2 Lubricant/Design Interactions ... 214
5.3 Inadequate Test Procedures ... 227
5.4 New Marketing Initiatives ... 228

Chapter 6—Performance Levels, Classification, Specification and Approval of Engine Lubricants 231
6.1 Definitions 231
6.2 Performance Measurement 233
6.2.1 Performance Parameters 233
6.2.2 Performance Requirements for Gasoline Engine Oils .. 236
6.2.3 Diesel (Commercial Oil) Tests 239
6.2.4 Problems with Engine Test Procedures 243
6.3 The Organizations Involved and Their Roles 248
6.3.1 The United States 248
6.3.2 Europe 259
6.3.3 Japan 264
6.3.4 Other Countries 265
6.3.5 The International Scene 266
6.4 General Comments 267
6.4.1 Advantages and Disadvantages of Establishing a New Quality Level 268
6.4.2 Motor Industry and User Quality Level Philosophies ... 271

Chapter 7—Other Lubricants for Road Vehicles 275
7.1 Gear Oils 275
7.1.1 Introduction 275
7.1.2 Additives 279
7.1.3 Automotive Gear Oil Formulation 281
7.1.4 Gear Oil Testing 284
7.1.5 Gear Oil Specifications and Quality Levels 287
7.1.6 Limited-Slip Differentials 290
7.2 Automatic Transmission Fluids (ATF) 291
7.2.1 Development of Automatic Transmissions 292
7.2.2 Characteristics of a Conventional Automatic Transmission 294
7.2.3 Requirements of an Automatic Transmission Fluid 299
7.2.4 ATF Testing 302
7.2.5 ATF Formulation 304
7.2.6 ATF Approvals and Specifications 306
7.2.7 Tractor Hydraulic Fluids 308

7.3 Greases ... 309
7.3.1 Introduction ... 309
7.3.2 Characteristics of Common Greases ... 313
7.3.3 Grease Manufacture ... 315
7.3.4 Grease Testing ... 316
7.3.5 Use of Grease in Motor Vehicles ... 321

Chapter 8—Other Specialized Oils of Interest ... 325
8.1 Two-Stroke Oils ... 325
8.1.1 Automobile Engines ... 325
8.1.2 Mopeds, Motor Scooters and Lawnmowers ... 327
8.1.3 Chainsaws ... 328
8.1.4 Motorcycles ... 329
8.1.5 Outboard Motors ... 330
8.1.6 Two-Stroke Oil Tests and Specifications ... 331
8.1.7 General-Purpose Two-Stroke Oils ... 334
8.2 Gas Turbine Oils ... 335
8.3 Railroad Oils ... 337
8.4 Hydraulic Oils ... 340
8.5 Air-Conditioner (Refrigerator) Lubricants ... 342
8.6 Industrial Lubricants in Automobile Plants ... 343

Chapter 9—Blending, Storage, Purchase and Use ... 353
9.1 Deciding on Oil Composition ... 354
9.2 Purchasing the Components ... 357
9.2.1 Component Specifications ... 357
9.2.2 Checking Incoming Materials ... 360
9.3 Oil Blending ... 364
9.3.1 Batch Blending ... 364
9.3.2 Automated and In-line Blending ... 369
9.4 Quality Control ... 370
9.5 External Monitoring Schemes ... 372
9.6 Storage of Lubricants ... 373
9.6.1 Bulk Storage in Tanks ... 373
9.6.2 Barrel Storage ... 374
9.6.3 Cans and Small Packages ... 375

9.7 Purchasing Lubricants ... 375
9.7.1 Quality Considerations ... 376
9.8 Oil Use for Small Users ... 377
9.9 Use of Lubricants in Large Plants ... 378
9.10 Complaints and Trouble-Shooting ... 380
9.10.1 Complaint Procedure ... 380
9.10.2 Laboratory Examination of Samples ... 381
9.10.3 The Usual Causes of Complaints ... 382

Chapter 10—Safety, Health, and the Environment ... 385
10.1 Introduction ... 385
10.2 Notification Laws for New Substances ... 386
10.3 Classification and Labeling ... 387
10.4 Toxicology of Lubricants ... 392
10.4.1 Base Stocks ... 392
10.4.2 Additives ... 393
10.4.3 Unused Lubricants ... 394
10.4.4 Used Lubricants ... 396
10.5 Biodegradability ... 397
10.6 Lubricant Effects on Automotive Emissions ... 398
10.7 Disposal of Used Lubricants ... 400
10.8 Transportation ... 402
10.9 Marketing Aspects ... 403

Chapter 11—The Future ... 407
11.1 The Influences for Change ... 407
11.1.1 The End User ... 408
11.1.2 The Oil Companies ... 409
11.1.3 The Vehicle Manufacturer ... 410
11.1.4 The Technical Societies ... 411
11.1.5 Environmental Pressures ... 412
11.2 Predicting the Future ... 414
11.3 Changes to Existing Types of Formulation ... 416
11.3.1 Alternative Base Stocks ... 416
11.3.2 Additive Technology ... 418
11.4 External Factors Influencing Oil Quality ... 419
11.4.1 New Hardware ... 419

11.4.2 Hardware Problems 421
11.4.3 Demands of Add-on Devices 422
11.4.4 Alternative Fuels 423
11.4.5 Emissions Effects 423
11.4.6 Safety, Health and Environment 424
11.4.7 Oil Supply and Consumer Buying Habits 425
11.5 Developments in Testing, Classifications, and Approvals 426
11.5.1 Test Costs 427
11.5.2 Quality Approval Procedures 429
11.5.3 Oil Quality Development 430
11.6 Future Crankcase Oils 431
11.7 Other Automotive Lubricants 432
11.8 Conclusions 433

APPENDICES

1. Glossary 437
2. Common Automotive Acronyms 489
3. Basic Petroleum Chemistry 493
4. The S.I. System of Units 501
5. Engine Oil Tests (SAE J304) 507
6. Precision of Laboratory Tests 517
7. Engine Oil Performance Classifications (SAE J183 plus J2227) ... 521
8. Approximate Engine and Rig-Test Prices - 1994 551
9. Engine Oil Viscosity Classification (SAE J300) 557
10. Crankcase Lubricant Specifications
    (a) ILSAC GF-1 567
    (b) CCMC 577
    (c) ACEA 581
11. Gear Oil Classifications (SAE J306 plus J308) 589
12. ATF Specifications and Approvals 599
13. Automotive Greases (SAE J310) 609
14. Classifications and Specifications for Two-Stroke Oils 625
15. Properties of Engine Oils (SAE J357) 635

Index 649

Chapter 1

Introduction and Fundamentals

1.1 Lubricants in History

A *lubricant* can be defined as a substance introduced between two surfaces in relative motion in order to reduce the friction between them. It is not known precisely when lubricants were first deliberately and consciously used, but various forms of primitive bearing were known in the Middle East several thousands of years B.C. It is reasonable to assume that if the concept of a bearing had been developed then the use of a lubricant with that bearing was highly likely even if only water. A Mesopotamian potter's wheel dating from 4000 B.C. contained a primitive bearing with traces of a bituminous substance adhering to it. This suggests the use of a lubricant originating from surface petroleum deposits in the area. By 3000 B.C., wheeled chariots were in extensive use in the Middle East, although few traces of lubricant materials have been found associated with remnants of such vehicles. A notable exception of a somewhat later period is a well-preserved Egyptian chariot of 1400 B.C. with definite traces of both chalk and animal fat in the wheel hub, suggesting a primitive grease had been in use.

Egyptian murals dating to about 2000 B.C. show statues being dragged along the ground with liquids being poured ahead of a transporting sledge, presumably as a lubricant. There has been much speculation as to whether these liquids were water, natural oils, a type of liquid grease, or even blood. D. Dowson, in his excellent book *The History of Tribology*(1), suggests these statues were in fact pulled along balks of timber lubricated by water, and he actually reconciles the expected frictional resistance for such a system against the magnitude of the slave power depicted in the hieroglyphs.

The Greek and Roman civilizations produced many devices based on the wheel including lathes, pulleys, gears, crane mechanisms, etc. From the remains of a Roman ship which was recovered in the 1930s[2] we know that the principle of ball and roller bearings was understood at this time. Pliny in the First Century A.D. listed the known lubricants which could be used in machinery of the era, these being principally animal fats and vegetable oils. This remained the situation up to the Industrial Revolution, with olive oil being common in southern Europe and oil derived from various seeds such as rape (colza) and linseed being more commonly applied in the north and west of Europe. Petroleum would have been used in those places where it was available as surface seepages, notably in Russia and the Middle East.

In Britain the Industrial Revolution started around 1760 and lasted for about eighty years, during which time the development of large-scale machinery based on iron and steel was achieved, the steam engine was invented, and the concept of the railway developed with self-propelled steam locomotives. To lubricate all this machinery, animal oils such as sperm oil (from the sperm whale) and neatsfoot oil (from animal hooves) were added to the existing lubricants, while palm oil and ground-nut oil were imported to supplement the locally available vegetable oils. During the period mineral oils obtained from distillation of coal or shales also became available as the residue when a light illuminating grade of petroleum had been distilled off. Graphite (black lead) and talc also came into use as solid lubricants for sliding surfaces.

Greases were developed originally by combining soda and animal oils. Later, lime was also used, and solid lubricants added to the greases provided further anti-friction properties.

Early use of mineral oils involved distilled coal or shale residues as described above. However, in the 1850s, small quantities of petroleum oil began to be produced in the U.S., Canada, Russia, and Romania. From the 1880s, petroleum was produced in quantity from wells in the U.S., and the modern petroleum industry was born. Liquid petroleum has to be distilled and "fractionated" into a range of products in order to be fully exploited, and the heavier of these can find use as lubricating oils. It was soon discovered that by distilling under reduced pressure, so called *vacuum distillation*, fractions could be separated without the heavier products oxidizing and deteriorating. This is because the boiling point of the fractions is reduced as the pressure is lowered, and lower temperatures are sufficient to separate the mixture. By the 1920s superior lubricants were being produced by vacuum distillation, and some of these fractions were being combined with soaps to form greases.

Additives to improve performance of petroleum-based oils were developed and saw increasing use in the 1930s. Initially seen as ways to improve the physical properties of the lubricants, the ability to control the deterioration of the oil itself became more and more important as the use of the internal-combustion engine grew. This led to the development of the so-called "detergent" lubricant additives which both reduced the oil oxidation and reduced the formation of deposits in engines. These were increasingly used in diesel engines from the 1940s, but only began to be used significantly in gasoline engines a decade later. Sludge-reducing additives were developed around this time and were used in gasoline engines from the 1960s, and in diesel engines from 1970. Modern lubricants are now highly specialized and complex products, and later in this book we will discuss their technical requirements and how these are met by combinations of base stocks and additives. Additives of various types now form a considerable proportion of the more sophisticated oils, of which crankcase lubricants are the largest and best-known example. A discussion of the principal additive types and their modes of operation will be included in Chapter 2, while in Chapters 4, 7 and 8 we will discuss how base stocks and additives are combined to meet the technical requirements of modern lubricants.

Conventional petroleum base stocks have become less able to satisfy the most severe modern demands, particularly for high-temperature performance, and significant additions of synthetic or specially refined petroleum stocks are now common in passenger car oils. In the future, for the most demanding applications, conventionally fractionated and refined base stocks may have to be replaced almost entirely by synthetic stocks or base stocks produced from petroleum in new ways.

1.2 Functions of a Lubricant

The basic functions of a lubricant are of course to *reduce friction* and *prevent wear*. In practice, lubricants are called upon to fulfill other functions, some of which are equally vital to the operation of the equipment in which they are employed. The automobile and engine manufacturer Mercedes-Benz has listed more than 40 properties required from engine oils.[(3)] Specialized lubricants such as hydraulic or transmission oils will add other properties for our consideration, while solid and semi-solid products such as greases have more restricted functions and properties measured in special ways.

The desired properties can have positive aspects (the oil shall prevent wear, etc.) or negative aspects (the oil shall not corrode the engine parts, etc.). The ability of an oil to meet any particular requirement is usually a matter of degree, rather than an absolute fact, and therefore questions of testing and test limits and the acceptance of limitations of both lubricants and machine have to be taken into account.

A simplified list of the positive properties particularly applicable to a motor oil can be given as follows:

1. *Friction reduction*—This reduces the energy requirements to operate the mechanism and reduces local heat generation.
2. *Wear reduction*—An obvious need for keeping the equipment operating for a longer period and in an efficient manner.
3. *Cooling*—In an engine the lubricant is an initial heat transfer agent between some parts heated by combustion (e.g., pistons), and the heat dissipating systems (sump, cooling jacket, etc.). In addition, and in other systems, the lubricant dissipates heat generated by friction or the mechanical work performed.
4. *Anti-corrosion*—Either from its own degradation or by combustion contamination (see Section 1.7.2) the oil could become acidic and corrode metals. Moist environments and lack of use can also cause rusting of ferrous components. The lubricant should counter all of these effects.
5. *Cleaning Action*—The oil should prevent fouling of mechanical parts from its own degradation products or from combustion contamination. Deposits, classified usually by descriptive terms

such as "solid carbon," "varnish" or "sludge," can interfere with the correct and efficient operation of the equipment. In extreme cases, piston rings may become stuck, and oil passages blocked, if the oil does not prevent these effects. Deposit prevention and the dispersion of contaminants are included under this heading.

6. *Sealing*—The oil should assist in forming the seals between pistons and cylinders (piston to rings, and rings to cylinder walls).

In addition to providing the above functions on a continuing and economic basis, a lubricant has to have certain properties which are dictated by the equipment in which it is used. There are necessary compromises between antagonistic requirements, some listed as negative limitations, as summarized below:

An oil should not:

1. *Have too low a viscosity.* This will allow metal-to-metal contact and subsequent wear, and can increase oil leakage.
2. *Have too high a viscosity.* This will waste power and, in the case of engines, cause starting difficulties.
3. *Have too low a viscosity index.* This means that it must not thin down too much when hot (or thicken up too much when cold).
4. *Be too volatile.* High volatility will appear as a loss of oil (high oil consumption) from the boiling away of the lighter constituents, and it has been said to also cause deposits.
5. *Foam unduly in service.* If an oil foams, this can result in loss of the lubricating properties of the oil, and/or loss of the oil itself from the engine.
6. *Be unstable to oxidation or chemical attack.* Engine oils in particular are subject to high temperatures and contamination by acids and other chemicals. The oils must be resistant to these to preserve their beneficial properties.
7. *Attack emission systems components, coatings or seals.* Catalytic converter performance can be degraded by unstable oils or unsuitable additive treatment. Some equipment contains paints or coatings and most have elastomeric sealing components. None of these should be seriously degraded by the oil.
8. *Produce deposits from residues.* If an oil decomposes on hot metal components (e.g., in the ring zone), it produces oxidation products

which polymerize to form a yellow or brown layer known either as "varnish" or "lacquer." This can build up and further carbonize to "solid carbon." Either type of deposit can prevent movement of parts which should be free to move (e.g., the piston rings). Apart from not producing deposits on moving parts in an engine, the lubricant should also not cause significant deposits in the combustion chambers which would lead to pre-ignition.

9. *Be unduly toxic, or of unpleasant odor.* Required for the comfort and health of the user.
10. *Be unduly costly.* This is often a real restraint, not because high-cost oils are not worthwhile in terms of engine operating economics, but because competition between suppliers limits the price which can be charged to the user, and hence the acceptable ingredient cost.

These requirements are studied in greater depth in subsequent sections.

1.3 Approval of Lubricants for Use

Having discussed in broad terms the properties required in a lubricant, and before we consider in detail the tests which can be performed to measure such properties, let us consider how you can determine if a given lubricant is suitable for your purpose. Of course, in the simplest case you can take the word of the supplier who might state "This oil is suitable for all modern automobiles." Such a bold statement is in fact unlikely to be found, and the supplier will more likely refer to different oil specifications which he claims the oil will meet. Reference to the vehicle handbook or operator's manual will indicate the specification requirements of suitable oil and you can see if they match up. If so, you can go ahead on the basis that the supplier's claims are valid. In effect, you approve your own use of the oil.

An easier approach is if the vehicle manufacturer has approved certain oils or types of oil and lists these in the manual. Practices differ in different countries, and in some cases only one brand and grade of oil will be officially approved, in others several competing brands of similar qualities will be listed, and in others lubricants approved by some official body and carrying a certification mark will be specified. In the U.S., and possibly in other countries in the future, the API "doughnut" and the new API (formerly ILSAC) "starburst" certification mark represent such "seals of quality" (see Figs. 80 and 81).

Approvals can therefore be given by the lubricant user, the vehicle (or engine) manufacturer, or some independent body. In Chapter 6 we will look at the organizations and procedures employed in some detail, but some historical background at this point will provide a useful perspective for the next sections.

The approving of lubricants for use was initially done on a very simple empirical or trial-and-error basis, simply seeing if the equipment would operate for a satisfactory length of time on the oil being evaluated. However, in the 1930s there was an upsurge of studies into friction and lubrication, and the development of tests which could distinguish one lubricant from another in several different ways. The importance of viscosity was soon realized, as well as the need to be aware of properties such as solidification or pour point, and acidity or corrosiveness. Gradually a pattern of product development arose which consisted of control of fundamental properties such as viscosity, pour point and acidity, together with other properties measured in laboratory gadgets which tried to simulate operating conditions, or in other words in "rig-tests." Final approval for use of a lubricant was usually based on practical experience obtained by testing the lubricant in the equipment and service for which it was designed. This field service experience would also be used to set oil change intervals.

A significant change took place at the end of the 1930s, which introduced the concept of testing oils in real engines, but in the laboratory. In the mid-1930s the Caterpillar Tractor Company had produced a range of heavy-duty tractor and earthmoving equipment with sophisticated high-power diesel engines. Under the typical severe service conditions for which they were designed it was found that the pistons rapidly became carbonized and the rings stuck so that efficient operation lasted for only a short period. It was found that metal soaps dissolved in the oil were effective in alleviating this carbonization, and thus the first "detergent" additive was developed. This was aluminum dinaphthenate. The problems of screening new and differing lubricants for detergency quality by field testing soon became apparent, and Caterpillar embarked upon the development of a laboratory engine test which would permit rapid screening of the new oils. They were joined by the U.S. military and U.S. navy laboratories. The first engine test was ready for use in 1940 and immediately became part of a lubricant specification employed for screening oils for both Caterpillar and other severe diesel engines.

Development of products and means of testing and approving them in the laboratory proceeded rapidly from this time, but was given even greater acceleration during World War II. The wartime needs of military and aviation equipment, and the difficulties of providing adequate servicing routines, rendered it imperative to set at least minimum standards for the quality of lubricants being purchased for the armed forces. The U.S. army introduced its specifications 2-104 for heavy-duty detergent oils in 1941, to be elaborated in 2-104B in 1943. Also in 1943, the U.S. navy (whose great interest was in operating diesel-engined submarines without breakdown problems) issued their 14-0-13 specification. The U.S. forces and other military bodies around the world continued to develop specifications and to be leaders in setting standards of performance until recently. In Europe, for cost reasons, parallel engine tests were developed in smaller, locally available engines, initially to provide approximately equivalent tests to those used in the U.S., but later used to develop different European specifications. The large size of today's lubricant testing industry, particularly that part which relates to testing in engines run on laboratory test beds, arose initially from the boost given in wartime by the need to screen lubricants.

The military approval of lubricants represents a case where the user of the equipment is setting requirements and standards for their quality, rather than that of the manufacturer of the equipment. For a long time military authorities around the world, and particularly the U.S. military, set the standard by which lubricants were judged. The underlying philosophies behind these standards were not always uniform. For example, the U.S. military, using the expertise of equipment manufacturers on their committees, developed a specification and approval system which was available to oil companies in most non-Comecon countries and which provided standards and lubricant approvals to rationalize quality levels around the world. In Europe, the British and other military authorities worked with the oil companies and developed their own tests and quality philosophies, and had smaller-scale approval systems in their areas of influence. During the 1980s the role of military bodies declined and today the major specification activities are conducted by equipment manufacturer groups in the U.S., Europe and Japan. Engine manufacturer involvement has grown steadily and today they are much more significant in setting oil quality standards than any military body.

From the beginning, it was clear that many engine manufacturers had specific requirements for their engines due to special metallurgy or design considerations, and they therefore wished to specify test requirements which were unique to their equipment. Thus we had the situation where test requirements of a specialized nature were added by individual manufacturers to more general requirements typical of the industry as a whole. This proliferation of test requirements led to difficulties in producing oils of general applicability which could be used in a mixed fleet of engines of different types and from different manufacturers. It also increased both the costs of the lubricants developed and of the test programs required. Attempts to rationalize the situation on initially national but later international lines are still going on with moderate success, but the formulation of a general lubricant, for example for the automotive car market, still requires considerable ingenuity and a balancing of often conflicting properties and requirements.

Whether the results of laboratory engine or rig-tests are very meaningful has long been a subject of debate. These tests are run under so-called "accelerated conditions." In other words, factors such as engine temperature, power output, valve spring loading, or fuel composition are made artificially severe in order that the test duration can be shortened. In addition to this, in some cases the results are compared on a scale which covers only the beginning of the deterioration process, with "failure" being arbitrarily described as a level of deposition or wear which in service would pose no problems. This policy shortens testing time and cost significantly and prevents expensive catastrophic engine failures, but the correlation between early deterioration and service failures is often weak.

In the early days of engine testing it was usual to run different engine tests to measure different key properties such as piston cleanliness (oil detergency), sludging (dispersancy), wear and corrosion. Today a wide variety of tests is still required for most specifications but sponsoring manufacturers tend to want all such properties measured in each engine. This may have merit, but can lead to formulation difficulties if different engines have incompatible needs. Due to the expense of both developing new test equipment and then installing this in the many testing laboratories around the world, there is also a prevailing tendency to keep existing test procedures with perhaps minor changes to the operating conditions or fuel characteristics, even though the equipment in real terms represents obsolete technology. These problems will be discussed more fully in later chapters, but it must be said at this stage that the two alternatives of either running only simple physical and chemical tests on a lubricant, or alternatively, subjecting it to real-life service conditions over a period which may be extended to several years, both carry overwhelming disadvantages compared to some form of accelerated mechanical testing.

Another subject for later discussion is whether or not equipment manufacturers (particularly the designers and developers of new automotive engines) should design their equipment to use a standard lubricant which is already available in the marketplace, or whether, as can happen at the moment, they should design the equipment for the best mechanical performance and if necessary require new high-quality oil to be developed to ensure reliability. The oil companies sometimes see this as relying on them to solve problems with new equipment, but certain engine manufacturers feel that requirements for improved oil qualities are a necessary part of progress. Most engine manufacturers adopt an intermediate position and often take an active part in improving oil quality to enable engineering advances to be made. Some, however, have a policy of designing "robust" engines which will operate successfully on a worldwide range of lubricants of differing qualities, accepting that reliability must not be sacrificed to ultimate performance in a global market.

In the U.S., engine manufacturers have usually tried to design with contemporary lubricant quality in mind, but continuous demand for improved emissions control has led to increases in severity of operation which have given field problems and the need to improve lubricant quality.

In Japan the "robust" philosophy predominates in relation to export markets, although there are stringent requirements for the local single company approval of preferred ("Genuine") oils.

In Europe, a wide variety of attitudes can be detected, including the most extreme ones. In France an approval regime operates with the single chosen supplier being expected to give strong technical support to the Original Equipment Manufacturer (OEM). In Germany very comprehensive specifications are set with difficult targets and these are amended to cope with field problems.

These problems of different philosophies and attitudes in different countries, together with others related to the variety and complexity of oil approval schemes, are now being addressed and it is hoped that eventually new methods of testing, new ways of specifying oil quality, and new international systems of approval will come into being, which will be to the benefit of equipment manufacturers, the oil companies, and particularly the end users of the equipment.

1.4 Friction and Wear, Lubrication and Tribology

Before considering the testing and formulation of lubricants in detail, we need to review the fundamentals of lubrication and explain some of the terminology. This section is a historical and mainly qualitative account of the development of the basic concepts.

1.4.1 The Mechanics of Friction

The concept of friction is generally well understood, even the sound of the word (from the Latin *frictus*, to rub) giving some indication of its meaning. Friction can be either beneficial or can produce problems. We rely on friction between our shoes and the ground when we walk, or even stand still. Early man learned to produce fire by the frictional heat generated by rubbing pieces of wood together, and we use only an improved version of this process when we strike a match or use a flint cigarette lighter. Production of music often requires frictional effects, as does painting. Many day-to-day activities would be very difficult if things did not stay where they were put, under the influence of friction. We rely on friction to keep a car on the road, and also to provide the braking forces and the drive through the clutch. In engineering fabrication we rely on friction to hold clamped parts together and to retain bearings and other "interference-fit" parts in position.

In the present context, however, we shall regard friction as an undesirable property arising from the surface roughness of two bodies in close contact and relative motion. The friction forces absorb power and give rise to wear, which ultimately will lead to a machine becoming unserviceable.

To minimize the friction between surfaces we can introduce a lubricant. The word comes from a Latin root, meaning slippery, and again the concept is quite familiar. The degree to which lubrication can reduce friction depends not only on the choice of the lubricant, but also on the particular situation and circumstances. The science relating friction and wear with lubrication is now called *tribology*. This word was coined in 1965 by a committee which had been set up to advise the British Government on the need for more investment in the study and teaching of lubrication science. Their report(4) suggested that very large financial savings would result from improvements in lubrication practice, and from the teaching of and research into lubrication science. They considered that the word lubrication had a low-technology image, and proposed the use of the word tribology for lubrication science. Coming from the Greek word *tribos*, meaning rubbing, this is obviously concerned with friction, but is taken to mean the science and technology of friction reduction and hence the prevention of wear.

Our understanding of friction has taken a long time to develop. Although Aristotle around 500 B.C. recognized the existence of frictional forces, the first one to study friction scientifically was Leonardo de Vinci in 1470. He measured the forces to move blocks over horizontal and inclined surfaces, and to rotate simple axle rods (Figure 1).

He found that greasy substances, fine powders, and small rollers all substantially reduced frictional forces. He pronounced that the force of friction was proportional to the load on the moving surface and independent of

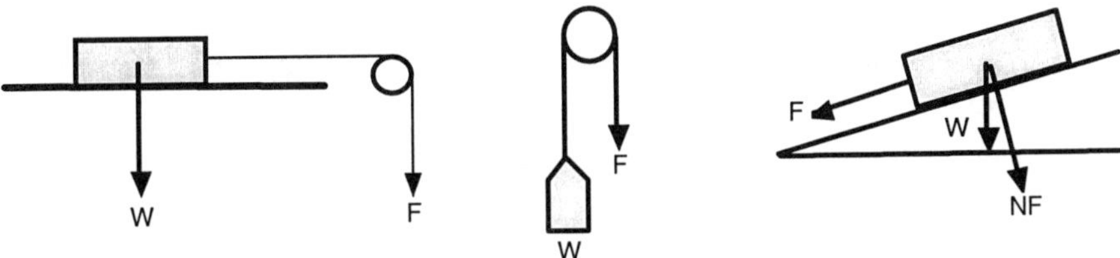

Fig. 1 Leonardo's friction experiments.

the area of contact. For smooth and polished surfaces he found the frictional resistance was one-quarter of the weight of the object. (In today's terms the coefficient of friction, μ, was 0.25. This is a quite reasonable value for wood or iron blocks on a wood substrate.) Leonardo appears to have correctly distinguished between the load on the surfaces and the weight of the objects, which were usually blocks. If we look at Figure 2, the supporting surface is at an angle θ to the horizontal, and the normal force NF (the load) is the weight W times cosθ. It is the surface load and not the weight which is important.

After Leonardo, Guillaume Amontons was one of the next to study friction seriously. His apparatus (Figure 3) used springs to apply the forces, and he studied the sliding of copper, iron, lead and wooden objects when lubricated with pork fat. Reporting to the French Academy in 1699[(5)] he stated that the force to move all these objects was the same, namely one-third of the applied load. (The consistency and magnitude of his results was probably largely due to his choice of lubricant!)

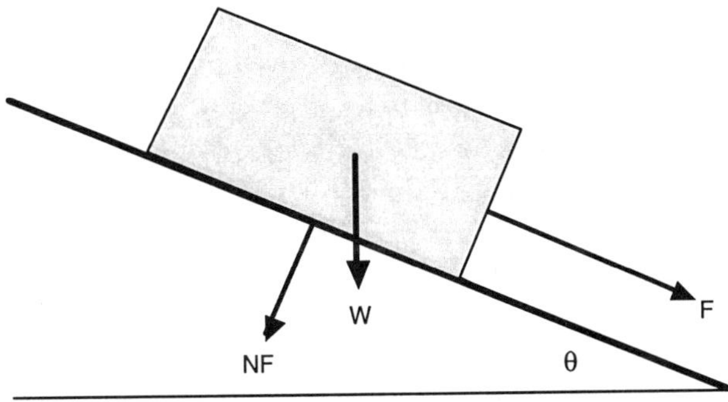

Fig. 2 Block on inclined plane.

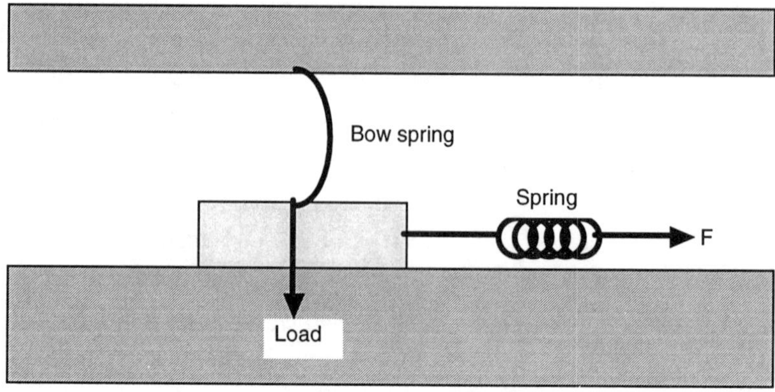

Fig. 3 Amontons' apparatus.

Amontons' work was extended and elaborated by scientists such as de la Hire[6] and Desaguliers[7], and particularly by Coulomb. Coulomb was a military engineer who devoted as much of his spare time as possible to scientific studies. In 1781 he won an honorary prize at the French Academy of Sciences for his work on measurement of friction for many substances and with many lubricants.[8] Like Amontons and many others, he believed friction arose from the interlocking of asperities (high spots) on the sliding surfaces, and there was no actual adhesion between the surfaces. This view tended to prevail until 1938, when Bowden and Tabor showed that adhesion was a most important component of the friction of solids.[9]

Other important contributions were made by Newton and Hooke. Newton studied fluid flow, and came close to defining viscosity by stating that the resistance to flow was proportional to the velocity of movement of the fluid.[10] So called "Newtonian" fluids obey this dictum. Hooke is famous for his studies of elasticity, and applied this to the studies of rolling friction.[11] He stated that a surface which is deformed by a roller but which recovers elastically after this has passed does not create a great resistance to rolling, but a surface which does not fully recover (is non-elastic) provides severe resistance. He believed there was a contribution from adhesion to the frictional force.

Euler in 1750 wrote papers on friction[12,13], and was probably the first to distinguish specifically between *static* and *dynamic* friction. If we refer again to Figure 2, increasing the angle θ until the block starts to slide gives a measure of the static coefficient of friction. The normal force is Wcosθ and the force F making the block slide is Wsinθ. The friction coefficient is F divided by NF, or tanθ. However, if θ is unchanged, the block will accelerate, indicating that the friction is less when it is in motion than when static. This concept of static and dynamic friction is still useful, but in practice there may be close to a continuous transition from one state to the other, factors such as the presence of some traces of lubricant being critical to sliding behavior, and some motion taking place in actual measurements of so-called static friction.

The complexity of the situation existing between sliding surfaces, and the modern understanding of the processes involved, is covered in the next sections.

1.4.2 Dry Friction

Let us consider two ordinary planar objects with no specific lubricant between them, although surface films, including contaminants and oxidation products may be present.

Figure 4 represents a close-up view of the two surfaces in contact.

No surfaces are completely smooth, and machined surfaces will show a series of saw tooth patterns arising from the machining process. Highly polished or electroplated surfaces may show smaller or more rounded

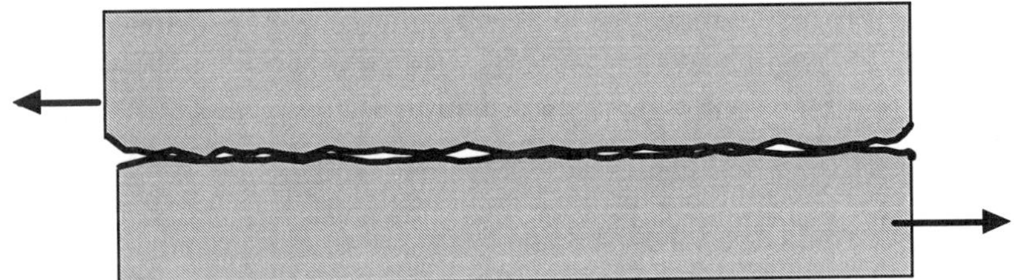

Fig. 4 Unlubricated sliding.

asperities, but completely smooth surfaces are virtually unknown. In addition to the small-scale irregularities, surfaces are very seldom planar and therefore on the larger scale some areas will be in closer contact than others.

If the two surfaces are moved relative to each other the asperities will collide and we can imagine four possibilities:

1. The two surfaces may separate and ride over each other as proposed by Amontons and Coulomb.
2. The two surfaces may deform, either temporarily or permanently, permitting sliding against the force of deformation.
3. The asperities on one surface may knock off those of the other.
4. The asperities may weld together and be pulled out of either one surface or both.

We can see that in the last two cases material is lost from the surface. This then becomes wear debris, and promotes further wear by acting as an abrasive between the two surfaces. The surfaces themselves may have the larger irregularities removed, but on a smaller scale will become both rougher and chemically active. We shall return to the concept of dry friction again later, considering the use of dry solid lubricants in Section 1.5.1, and the effect of removal of inherent films from the surfaces in Section 1.6.

1.4.3 Lubricated Sliding

Let us now look at the introduction of a lubricant, particularly a liquid lubricant, between the two surfaces. Figure 5 shows the lubricant separating the surfaces completely and the force required to move one against the other is merely that to move the lubricant, and the friction is much lower.

To maintain such a continuous film of lubricant, for example in a bearing, we require several things:

1. The lubricant must have a high enough viscosity not to be squeezed out by the forces pushing the surfaces together. Viscosity (or thickness of consistency) is the most important characteristic of a lubricant. Low-viscosity (thin) fluids like water offer little

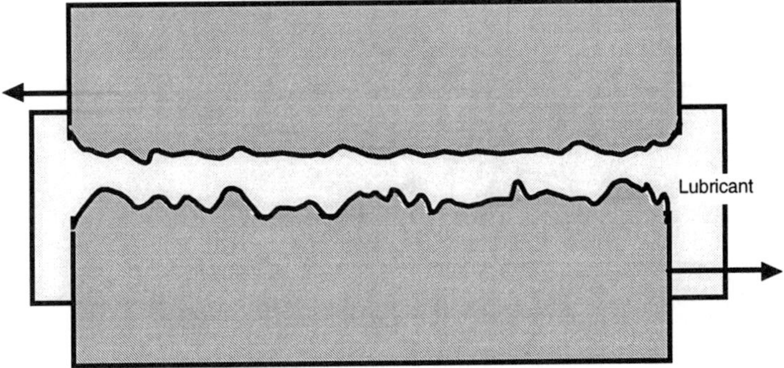

Fig. 5 Fluid lubrication.

resistance to movement. High-viscosity (thick) fluids like molasses offer considerable resistance. Lubricating oils are intermediate in viscosity.

2. The lubricant is confined, or channelled, into the gap between the surfaces. This means that a bearing and its lubricant supply must be properly designed.
3. The pressure resulting from the force between the two surfaces is kept at a level at which the above two factors can maintain the film. In other words, the bearing must not be overloaded.

In practical bearings, whether sliding linear bearings or rotating journals, it is arranged that the surfaces are at a slight angle so that a wedge of oil is formed. In the case of a journal (cylindrical) bearing, this arises naturally from the clearance being made asymmetric by the loading on the bearing. The pressure of the oil in the wedge (and hence the load it can sustain between the surfaces) is then influenced not only by the viscosity of the lubricant but also by the rate of sliding between the two surfaces, higher relative speeds generating higher pressures within the oil wedge. This concept that high sliding speeds promote good lubrication is very important, and in the case of reciprocating motion (where at the end of travel the speed becomes zero) is critical. The concept of the wedge in a linear bearing is depicted in a simple form in Figure 6, and the pressures generated in a journal are depicted in Figure 7. Lubrication under such conditions is known as *hydrodynamic lubrication*.

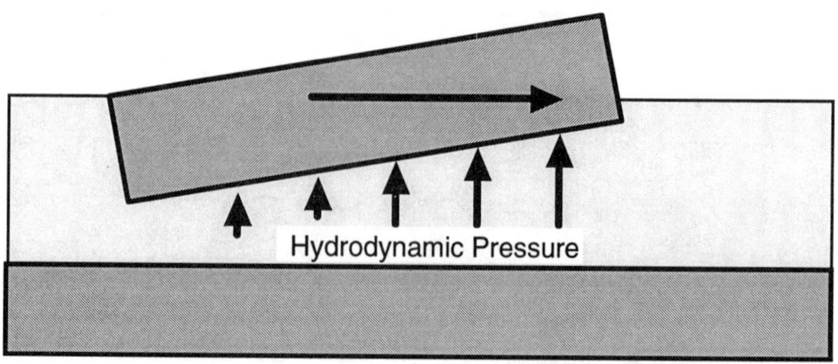

Fig. 6 Lubricant supporting a sliding wedge.

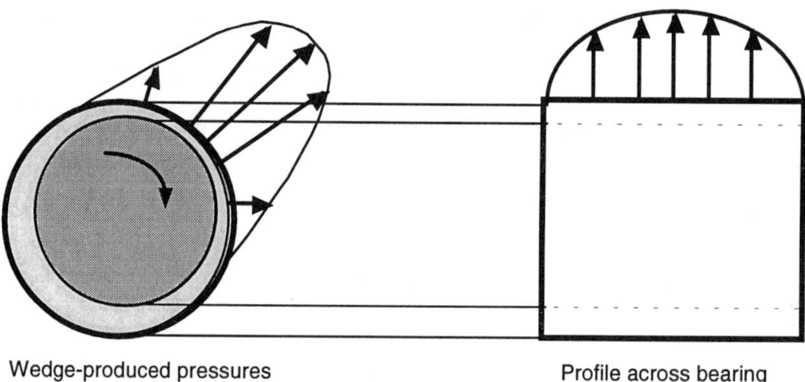

Fig. 7 Pressures in a journal bearing.

The pressure profile in a journal bearing was first discovered by Beauchamp Tower and reported to the Institute of Mechanical Engineers in London in two papers in 1883 and 1885.[14] He had first drilled a bearing in order to supply more lubricant when to his surprise the oil spurted out under pressure. Subsequently he drilled a series of holes and determined the pressure profile. Tower's work greatly interested Osborne Reynolds who developed the now famous equation which bears his name, a complex mathematical relationship between the thickness of the oil film, sliding velocity, the load, the oil pressure and the viscosity of the lubricant. He reported his studies to The Royal Society of London in 1886.[15]

1.4.4 Boundary Lubrication

If the lubricant film is incomplete for any reason (low oil viscosity, poor oil supply, excessive loading, poor geometry, etc.), then the surfaces will start to touch. Friction increases as the degree of touching increases, and adhesion and welding come into play. Temperatures rise, the oil thins out, and the surfaces eventually seize together. The condition where the surfaces are just touching, but lubricant is supporting most of the load, is called *boundary lubrication.* At an early stage in the study of friction and lubricity, it was found that natural fats and oils were good lubricants and helped prevent friction and wear. They can also be used as additives in petroleum oils to improve lubrication under boundary conditions. Natural oils are mainly glyceryl esters of long-chain organic acids, and the acids, synthetic substitutes, or related alcohols are all effective as so-called "lubricity" improvers. The mechanism is that the polar constituents of these molecules attach themselves to the metal surfaces, while the organic chains adsorb a layer of oil which is held on the surface. This is illustrated in Figure 8.

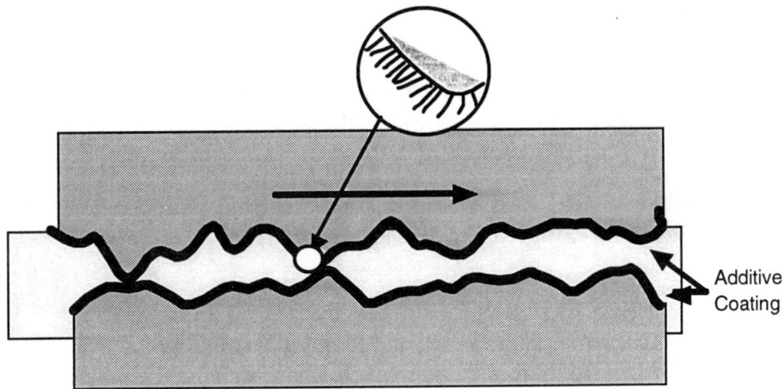

Fig. 8 Boundary lubrication.

These additives are more correctly called *boundary lubrication additives*, and are very effective at minimizing friction and wear provided that the temperature is not high enough to cause either decomposition of the molecules or desorption from the surface. The most potent types are effective down to zero sliding speed (no oil wedge), where they significantly reduce the coefficient of static friction. In this role the designation of "friction modifier" is usually applied.

1.4.5 Extreme Pressure Conditions

If the temperatures are too high, the viscosity of the oil too low, or the load is too great, then the asperities on the two surfaces will come into a more severe degree of contact. In these circumstances another type of additive known as an *extreme pressure additive* can be effective. The term "extreme pressure" is an old one, originating in the conditions found in highly loaded gears where the phenomenon is most frequently found. Modern references often consider it to be the ultimate stage of boundary lubrication, but we prefer to distinguish between the two because of the differences in severity which mean different additive treatments are (or can be) used.

When the asperities strike each other and undergo deformation, great heat is generated at the contact points and this can be used to trigger a chemical reaction with certain additives present in the oil. Elements such as sulfur, phosphorus and chlorine in the additives react with the metal under these conditions of local heating to form metallic compounds such as sulfides, phosphides and chlorides. These compounds have relatively weak crystal structures which readily shear apart to allow sliding and thereby prevent welding of the metal and tearing of the surface (Figure 9).

As their name suggests these compounds permit high loads to be carried by the lubricant, and while there is metal loss by chemical reaction (chemical wear), this takes place only at the asperities and the surface tends to be-

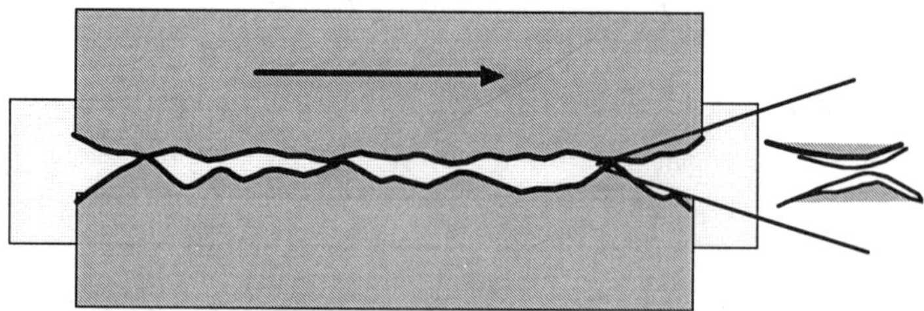

Fig. 9 Extreme pressure action at contact points.

come smoother in service. Provided the loading is not increased, the requirement for the extreme pressure effect of the additive will become less and less as the asperities are reduced in height and both the chemical wear of the metal and the depletion of the additive may well become minimal. The surfaces are then described as being "run-in."

1.4.6 Elasto-Hydrodynamic Lubrication

So far we have discussed four types of lubrication:

1. Dry lubrication
2. Full fluid lubrication
3. Boundary lubrication
4. Extreme pressure lubrication

The discussion of these four types of lubrication has been deliberately simplistic. They have considered the cases where one surface is sliding against another, with or without a perfect lubricating film between them.

In the real world a fifth type of lubrication regime is common, namely *elasto-hydrodynamic lubrication.*[(16)] This is concerned with rolling friction, and, although its existence could have been forecast from Hooke's work, it was not fully appreciated until the 1950s. It probably occurs to a greater or lesser degree in most systems which have moderate or heavy intermittent loading of components, such as gears and ball or roller bearings. Elasto-hydrodynamic lubrication is characterized by the occurrence of two phenomena:

1. The metal deforms elastically under load (i.e., it recovers the original profile exactly when the load is released). This effectively results in an increase in the apparent contact area between the opposing metal surfaces, and a reduction of the load per unit of bearing area.
2. The high pressures generated in the oil film lead to an effective increase in its viscosity. Fluids in general have positive viscosity/pressure coefficients and for mineral oil at very high pressures the increase is large.

The first of these effects increases the surface area of the thin film of oil present between the opposed surfaces, and thereby reduces the stress within the oil. The second effect reduces the tendency for this oil to be squeezed out by the pressure between the surfaces. Elasto-hydrodynamic theory explains many of the cases in which wear is found to be less than expected if only classical hydrodynamic/boundary/extreme pressure theory is considered. Gear mechanisms and ball and roller bearings are systems where the phenomenon is now considered to be very important, but any system where intermittent loads are concentrated in small localized areas is likely to have a wear control contribution from this lubrication mode.

Obviously, loading cannot be increased indefinitely without an unmodified oil film breaking down, and when it does boundary and extreme pressure lubrication additives may be useful. The significance of elasto-hydrodynamic lubrication is that the need for such additives, normally used in sliding conditions, may be less than simple theory would predict. It follows that lubrication and wear characteristics need to be measured practically in operating machinery, rather than calculated from simple parameters such as lubricant viscosity and bearing clearances.

1.5 Solid and Grease Lubricants

These require separate consideration, because of their different physical nature.

1.5.1 Solid Lubricants

If no lubricating liquid (or an inadequate one) is used to separate surfaces in relative motion, then there are several ways in which wear can be reduced. Generally, a smooth surface finish is beneficial, although clean smooth surfaces can weld together and a grooved or roughened surface may retain traces of lubricant more readily than a smooth one.

Chemical or heat treatment of the metal can cause metallurgical changes, increasing the surface hardness or perhaps reducing the tendency for the surface to flake or spall. As these do not comprise generation of an actual lubricant, we have considered them to be outside the scope of this book. Some surface treatments, however, do produce surface layers which can be considered to amount to lubricants, although the distinction between surface protection and lubrication is somewhat blurred.

Initially, we propose to discuss solid materials which are clearly recognizable as lubricants, and which do not form part of the surfaces which require lubrication.

The use of such specific solid lubricants provides dry lubrication superior to that arising from the innate surface films which exist on metal surfaces in air. Early dry lubricants were talc, graphite, and molybdenum disulfide, all of which occur as natural minerals and which can be used directly after a degree of cleaning and processing. Their structure is layer-like series of platelets which adhere loosely to the surfaces and slide readily over each other to reduce friction. For maximum effect, purity and particle size are very important and great care is exercised in selecting raw materials and in process control.

A more recent solid is the fluorinated polymer polytetrafluoroethylene or PTFE ("Fluon™," "Teflon™," etc.). This thermally stable, soft plastic has a low coefficient of self friction and again coats the surfaces to allow them to slide easily over each other.

Initially the purified and powdered solid lubricant was simply rubbed onto the sliding surfaces and replenished as necessary. Such simple application was more successful on materials such as wood (used quite frequently in early machinery and particularly in textile manufacture) than it was on shiny metal surfaces where it was easily swept away by the motion of the mechanism. One solution was to incorporate the material into a thick oil or a fatty substance, thus providing a composite lubricant where the medium (oil or fat) provided adhesion and initial lubrication, and the solid provided resistance to seizure under heavy loads. Such mixtures are not greases in the conventional sense (see next section). At lower concentrations of solid lubricant in oil the solid can act as a type of EP additive, although of limited application.

A high solids mixture is used for some applications where equipment is subjected to severe heating. In such applications the oil or fat may carbonize to a matrix in which the solid lubricant is embedded, and which can provide short-term lubrication. An example would be for the withdrawal of trolleys from a furnace or kiln.

This concept of embedded solids has been extended to lower temperatures and long-life lubrication by incorporating the solid lubricant in a plastic or resin which is molded or machined into bearing material. Plastics themselves do not usually have a particularly low coefficient of friction either for plastic on plastic or metal on plastic, unless a fluid or dry lubricant is also present. Exceptions include nylon and PTFE, which can be used as bearing materials or can be incorporated into oils or greases as solid lubricant additives. Plastics in bulk form can be used only for relatively lightly loaded situations, due to their deformability and their low melting points.

Some metals are used as solid lubricants, particularly for exceptional circumstances and in extreme environments. The metals can be as powders usually in paste form, or as an applied surface coating. The most usual metal is lead as a flashing (thin surface coating), but other metals such as copper, antimony and indium have been shown to be useful. Such metals have to have low shear strengths and high deformability, but even then surface films of oxides or sulfides may need to be formed for them to be effective. For very soft metals under short-term conditions it is often found that the coefficient of friction reduces with applied load. This can be attributed to the formation of a more-perfect film of (metal) lubricant.

Phosphating of steel parts, at one time frequently applied to gears, provides a measure of EP performance somewhat analogous to the use of phosphorus-based additives. During the running-in process that it is intended to facilitate, the coating is rapidly lost from the high spots and opinions differ as to its efficacy. Nitriding and case hardening, on the other hand, are examples of metallurgical processes producing surface changes which are not in themselves of the nature of lubricants.

The same distinction really applies to ceramics, whether used as surface coatings or massively in the form of ceramic components. However, the early promise shown by ceramics as a means of reducing friction is worth some discussion, although their application to date has been very limited.

Until the 1980s, they had not found a significant place in automotive engine engineering. However, at that time pressures to improve fuel economy by reducing internal friction and reducing vehicle weight caused the automotive industry to examine the potential which may be offered by ceramics such as silicon nitride, silicon carbide, Sialon and zirconia. Japanese manufacturers were at the forefront of this activity and at least one manufacturer (Isuzu) produced a prototype engine entirely from ceramic components. In the U.K., an industry consortium on Ceramic Applications in Reciprocating Engines (CARE) sponsored by the British Government, published its findings in 1990.

Early work(17,18,19) gave conflicting evidence as to the benefits which might accrue from ceramics. Values of coefficient of friction from 0.06 to 0.8 had been claimed for ceramic running against ceramic. Possible reasons for these discrepancies could include surface contamination and neglect of the effect of surface finish upon friction. The valve train and cylinder liner/piston ring interface which operated under mixed and boundary lubrication(20) were felt to be areas where use of ceramics might significantly reduce frictional losses.

Published data(21) suggest that it may be particularly dangerous to make generalizations from limited experiments. While some results indicate reduced friction compared to conventional metallurgy, others gave unacceptable levels of wear and friction. Monolithic components behave differently from coated ones. Catastrophic levels of wear and friction have been seen in the absence of lubricant.

In 1995, there are relatively few examples of ceramics being used for tribological purposes in commercial engines. They find application in turbocharger rotors and one European manufacturer (Peugeot) is using silicon nitride as a valve train component.

In general, solid lubricants are used only where geometry or environment precludes the use of a full film of a liquid or oily lubricant, but the occasions when these provide the best or only solution to a lubrication problem are increasing all the time as we seek to use mechanisms in extreme environments. The most difficult environment is probably space, because of the vacuum and temperature extremes, and here solid lubricants are frequently used.

1.5.2 Greases

Greases may sometimes appear to be solid although they can also be quite fluid. We would like to define them as lubricating oils held in position by a gel structure. These gels are usually based on soaps (e.g., sodium, calcium, or lithium stearates), but can also be formed by certain minerals (such as bentonite) or other solids which have a particular crystalline structure able to both absorb oil and bind it into a gel-like form. The nature, quantity and structure of the gelling agent can be chosen to provide greases of varying consistency (solidity).

Greases are superior to the fats used in earlier times because they have the ability to retain the oil in place up to a relatively high but distinct temperature, known as the “drop point” at which the grease appears to melt. Fats consist of mixtures of (mainly glyceryl) esters which melt individually over a range of temperatures, commencing at quite moderate levels. Fats are also more chemically unstable, being susceptible to oxidation and to the formation of acidic and odorous by-products.

As well as oil and the gelling agent, greases can incorporate solid lubricants. Graphite or “Moly” greases are well known, and a copper-containing grease is marketed for high-temperature uses. PTFE-containing blends are also appearing. The oil constituent of greases can contain other additives, including, for example, extreme pressure agents, and hence greases are very versatile if space or geometry prevents the application of full fluid lubrication. The formulation and properties of the most common types of grease are discussed in Section 7.3.

1.6 Unlubricated Conditions

So far we have been considering mainly the most usual situation where sliding surfaces (usually of metal) are lubricated by a special agent (usually an oily liquid). For completeness and of relevance when particularly hostile environments are to be encountered, we need to look briefly at so-called unlubricated conditions, the extreme of the dry lubrication regime we started with.

When solid surfaces rub together under load, the friction is neither infinite nor predictable. A key factor is the presence or absence of surface films. Materials can be classified broadly in three types: crystalline solids, metals, and plastics. When like rubs against like, specific frictional characteristics are found, but when unlike materials are in contact, various different effects can be seen, dependent always on the geometry of the contacts. Provided the loading is light, or well spread to give low pressures, in many cases friction is not excessive and sliding can take place for some time without seizure.

However, when such sliding contacts take place under vacuum, different results again are found. Generally, friction and surface damage increase rapidly and go on increasing as the materials abrade, and as they are held longer under a vacuum. Metal surfaces will eventually weld together completely. What is happening here is that films on the surfaces which can act as primitive lubricants are being removed by the effects of outgassing under vacuum, and then by abrasion.

The surface of normal metals in air can be represented as in Figure 10.

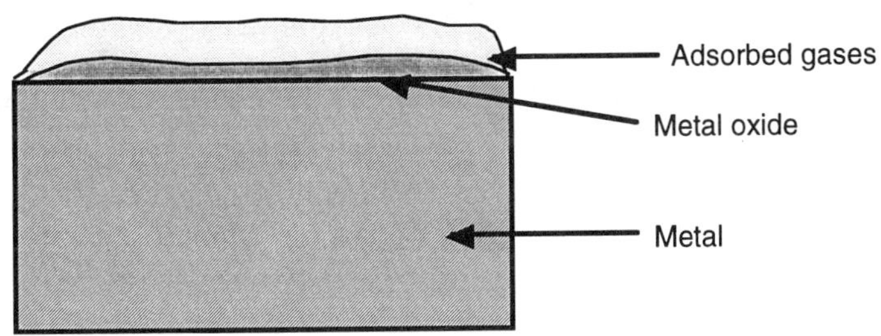

Fig. 10 Metal surface in air.

The metal is covered with a layer of oxide (and possibly traces of other compounds) from reaction with the air, and onto this are adsorbed atmospheric gases such as water vapor, carbon dioxide, oxygen, etc. The adsorbed layer acts as a lubricant under light loads, but as loading increases the oxide layer will permit sliding until it is worn away and true metal-to-metal contact occurs, resulting in surface damage and possible welding of the two surfaces. Outgassing under vacuum removes the adsorbed layer to a greater or lesser degree, and also prevents exposed metal from re-oxidizing by depriving it of oxygen from the air (Figure 11).

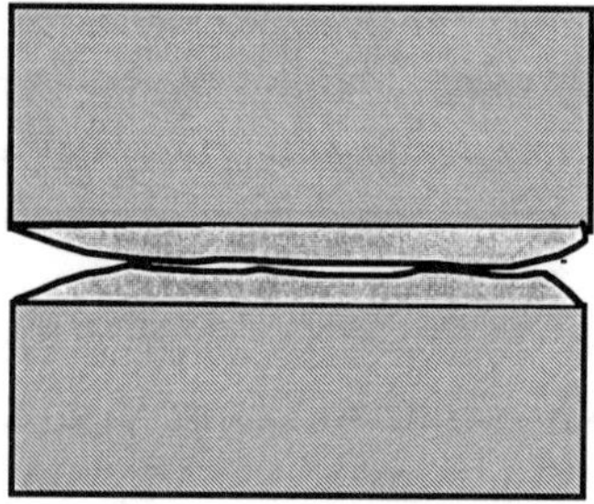

Fig. 11 Surfaces in vacuum.

Metals can exist in essentially amorphous form or as crystals. Many non-metallic crystals such as diamond and ruby have also been used in bearings, and studied in friction experiments. With crystals, surface adsorbed films also exist, but when outgassed and true crystal-to-crystal contact takes place it is then found that the orientation of the crystal planes is important. Diamond will self-weld in certain orientations and not in others. Hexagonal crystal cobalt with basal planes in contact shows a low adhesivity and a moderate coefficient of friction, while other orientations and contacts with other substances can demonstrate very high friction.

For further reading on this topic the early monographs by Bowden and Young(22) and Bowden and Tabor(23) on "The Friction and Lubrication of Solids" are of interest, and a good summary was given in the *Standard Handbook of Lubrication Engineering* by O'Connor and Boyd published in 1968(24), and sponsored by the American Society of Lubrication Engineers (ASLE).

1.7 Lubrication Requirements of Different Systems

1.7.1 Simple Systems

In this section we are looking in a little more detail at the sliding systems and journal bearings discussed earlier in general terms. To achieve hydrodynamic lubrication for the relative motion of two surfaces, an element of play (freedom of movement) must be provided in the design to enable the surfaces to separate and form the oil wedge which is sustained by continuing motion (Figure 12).

The degree of separation of the two surfaces is affected by the applied load and the viscosity of the lubricant as well as by the bearing design, but in general it is stable if these factors are kept constant. An exception is in journals if the viscosity, loading, speed and the clearances are mismatched, when bearing "whirl" can occur with the center of the journal describing small circles.

There is an optimum viscosity for a given situation which provides the minimum friction. It is obvious that in hydrodynamic lubrication the fluid

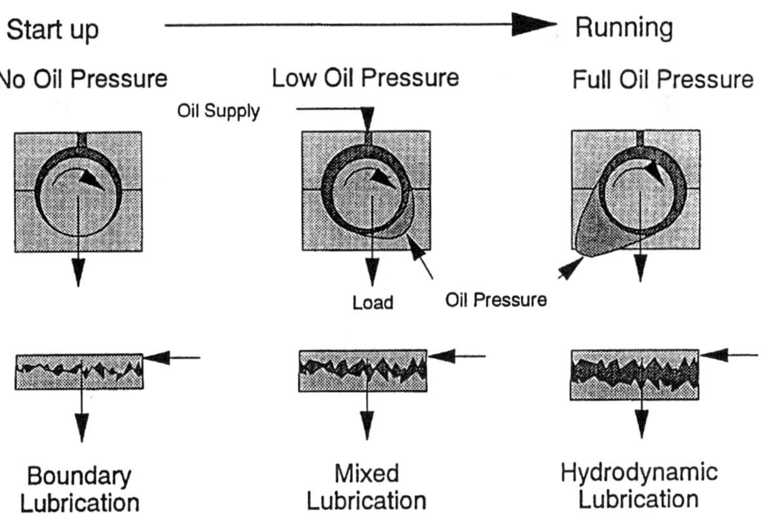

Fig. 12 Wedge generation in a bearing. (Source: Esso)

frictional forces will increase with fluid viscosity, and that the consequent drag will be proportional to the area of the bearing. This viscous drag also increases with bearing speed, as the motion of the surfaces performs increased work on the oil film.

In the case of boundary lubrication conditions, however, these relationships are inverted, because increased speed, bearing area, and increased viscosity tend to promote the regeneration of a true fluid film with consequent reduction in friction (which in boundary lubrication comes mainly from surface contact and not from the internal friction in the lubricant).

In either case it can be said that the friction is a function of the viscosity of the lubricant, the area of the bearing, and the relative speed of the two bearing surfaces. Expressing this mathematically we can say for a journal bearing:

$$F = (f)\ ZNA$$

where F is the frictional drag imposed by the bearing
Z is the oil viscosity
N is the journal speed (e.g., in rpm)
A is the load-carrying area of the bearing
(f) is a mathematical symbol indicating that there is some relationship between the two sides of the equation.

The ratio of the friction to the applied load L is called the coefficient of friction, μ. Thus:

$$\mu = \frac{F}{L}$$

The load can be expressed in terms of a pressure P which is exerted on the bearing area:

$$P = \frac{L}{A}$$

Using these relationships we can express the first equation for the coefficient of friction as follows:

$$\mu = (f)\frac{ZN}{P}$$

This is a well-known relationship for the friction in a bearing, but the nature of (f) is not defined. This may be derived from measurements in an actual bearing which are plotted to produce a curve as in Figure 13. This type of curve is often called a *Stribeck curve*, after a German professor who presented it in a paper to the Berlin Research Institute in 1902.[25] It applies equally to both rotating journal bearings and linear sliding surfaces, although in the common case of reciprocating motion there are rapid speed changes and reversals.

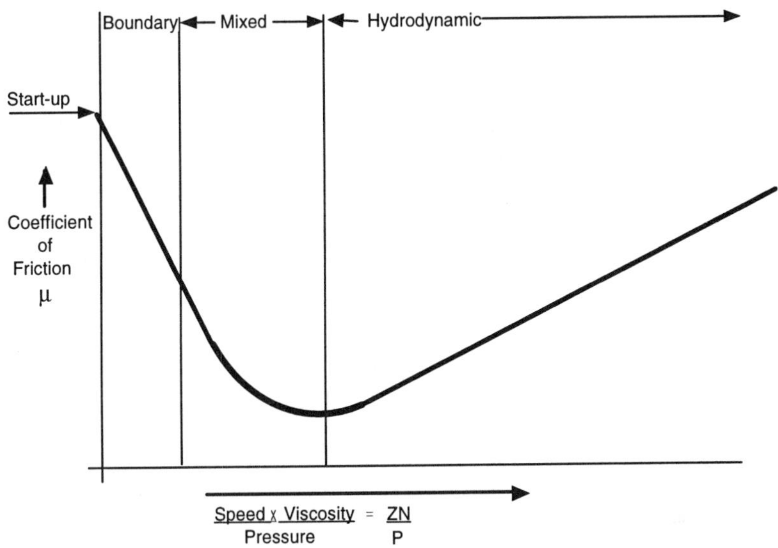

Fig. 13 Friction regimes in a bearing.

The precise shape of the curve will depend on the actual bearing design and dimensions, but we can see there is an area of minimum friction, which represents the transition between boundary lubrication and full fluid lubrication. As lubricant viscosity and bearing speed increase the friction rises to well above the optimum level. At low values of viscosity and speed the friction rises sharply in the area of boundary lubrication. When the speed is zero the coefficient of friction is a maximum and is effectively the coefficient of static friction. This can be measured by applying sufficient force to the bearing to just produce the commencement of movement. As we can see from the diagram the coefficient of friction then continues to fall until the minimum or optimum value is reached at which time full fluid lubrication has taken over. It is clear from this diagram that lubricant viscosity has to be matched to bearing speed to minimize friction.

At low levels of sliding or rotational movement the high levels of boundary friction which exists may be erratic and give rise to noise or chatter, regardless of whether the speed is actually increasing or decreasing. Systems that exhibit this phenomenon when speed is decreasing include oil-immersed brakes and clutches, while machine tool slideways exhibit a similar so-called "stick-slip" phenomenon at both the starting and the stopping of motion. To overcome such problems, special boundary lubricant additives are employed ("friction modifiers") which help to produce smooth transitions between rest and relative motion.

The time the bearing surfaces (or two sliding components) spend at rest is also an important factor, for residual oil between the two surfaces takes some time to leak away or be positively expelled by the load which is forcing the surfaces together. As an example we can consider the reciprocating motion of the piston rings in an engine. At normal revolutions a lubricant film is retained between the rings and the cylinder walls during the time the piston moves through a rest position and reverses its direction of motion. However, when the engine has been stopped for awhile, the oil will be forced out of the gap between the rings and the cylinder walls and initial lubrication will be very much under boundary or even unlubricated conditions when the engine is restarted.

From the above we can see that viscosity is the principal controllable property for obtaining satisfactory lubrication of simple mechanisms, or of the rotational and sliding motions contained within more complex machinery. If the apparatus operates continuously and is maintained within a relatively narrow temperature range, then the oil can be chosen to give a satisfactory viscosity at that temperature. However, the viscosity of a petroleum lubricant varies quite markedly with temperature, so that if equipment is in intermittent use and there is consequently a range of temperatures at which it has to operate, then the various viscosities of the lubricant corresponding to the operating temperature profile must be suitable for its lubrication, if not always optimum. An oil where the rate of change of viscosity with temperature is minimized should be chosen to prevent the onset of boundary lubrication as it heats up, while not giving excessive viscous drag when cold. Such an oil is said to have a high *Viscosity Index.* Motor oils of this type are known as "Multigrade Oils" as they meet both low- and high-temperature viscosity requirements in various oil specifications (see Section 6.3.1).

In addition to correct viscosity, use of boundary lubricant additives may be required for start-up conditions, for slow reciprocating motion, or if the oil is likely to become too thin at peak operating temperatures to provide satisfactory lubrication. Typically, organic acids and esters have been used, and in some situations solid lubricants such as molybdenum disulfide may also be effective. The possible use of other additives, such as those to control oil oxidation, will be discussed in Section 2.2 on oil additives. Such additives are more for prolonging oil life than for improving basic lubrication.

In the above paragraphs we have referred to hydrodynamic lubrication where the oil film is created and maintained by the relative motion of the surfaces, whether this be linear or rotational motion. There is, however, another way of ensuring the presence of a lubricating film, and this is to inject oil between the moving surfaces under sufficient pressure to maintain an adequate oil film. This is known as *hydrostatic lubrication*, which for certain applications can provide low friction and virtually wear-free bearings. However, to be successful, good sealing of the bearings is required, which introduces an additional source of friction and leads to complications in both the design and maintenance of the equipment. The oil needs to be supplied at a constant pressure and to maintain a uniform viscosity as far as possible, making the system more suitable for large, continuously operated machinery where temperatures are constant. As the oil film is maintained by external pressure, the relative motion of the two surfaces is unimportant, and hydrostatic bearings can be used for stop-start devices such as machinery slideways. Friction, wear, and stick-slip problems are eliminated, but the precision of the system and its complexity makes such equipment expensive.

1.7.2 Internal-Combustion Engines

Engines such as the steam engine, which are powered by externally generated pressure, can be considered to be relatively simple devices whose lubrication is governed mainly by the considerations discussed above. However, when the fuel is burned internally within the engine, lubrication becomes a much more difficult process because combustion products and residues may contaminate the lubricant.

The type and quality of the fuel used is therefore important, as is the type of combustion cycle employed. The power output in relation to engine size, which is related to combustion efficiency, will influence the peak temperatures the lubricant must withstand, and these are much greater than for simple machinery or external-combustion engines. Figure 14 shows the key components of a typical internal-combustion engine.

Spark-ignition engines, also known as Otto-cycle engines, use a volatile fuel such as gasoline which is vaporized/atomized into a stream of intake air, either in a carburetor or by a fuel injection nozzle. The mixture of air and fuel is compressed by the piston(s) rising in the cylinder(s) and ignited

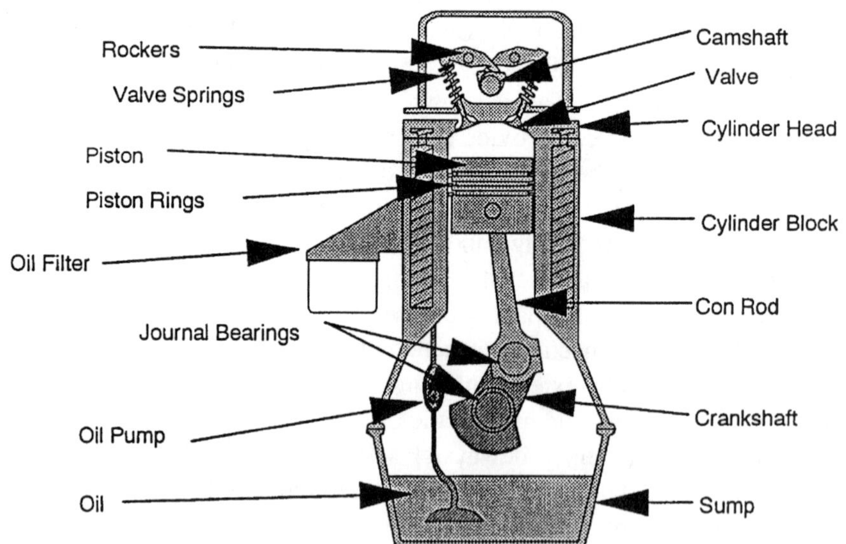

Fig. 14 Sectional view of engine.

by a spark plug in the combustion chamber. The high pressure generated by the burning gases drives down the piston and produces usable power. For simplicity we will refer to these as gasoline engines, although use of other liquid fuels such as the lower alcohols and also gaseous fuels is possible.

A conventional diesel engine runs on a relatively involatile fuel which is injected at very high pressure into the combustion chamber after a charge of intake air has been compressed and thereby raised to a high temperature by the ascending piston. The fuel self-ignites on mixing with the hot air, and power is produced. Diesel fuel is required to self-ignite readily in the combustion chamber, the opposite of the requirement for gasoline where self-ignition (before the spark) would result in engine damage. Diesel-type engines are generically known as compression-ignition engines.

In a four-stroke engine each operation takes place once in every four piston movements (two up and two down) (Figure 15).

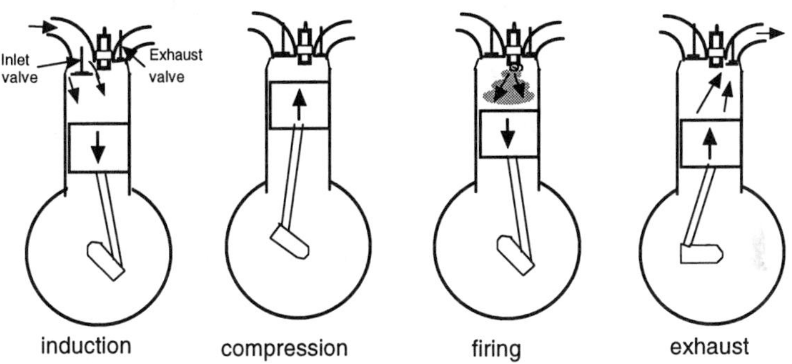

Fig. 15 The four-stroke engine cycle.

In a two-stroke cycle the operations (induction, compression, firing, exhaust) are arranged to overlap so only two piston movements or one crankshaft rotation are needed to complete the cycle (Figure 16).

In the two-stroke engine the intake air enters under pressure either from an external compressor or from crankcase pressure generated by the descending piston. Most two-strokes use ports which are covered and uncovered by the piston movement to control the gas flow, but some designs use

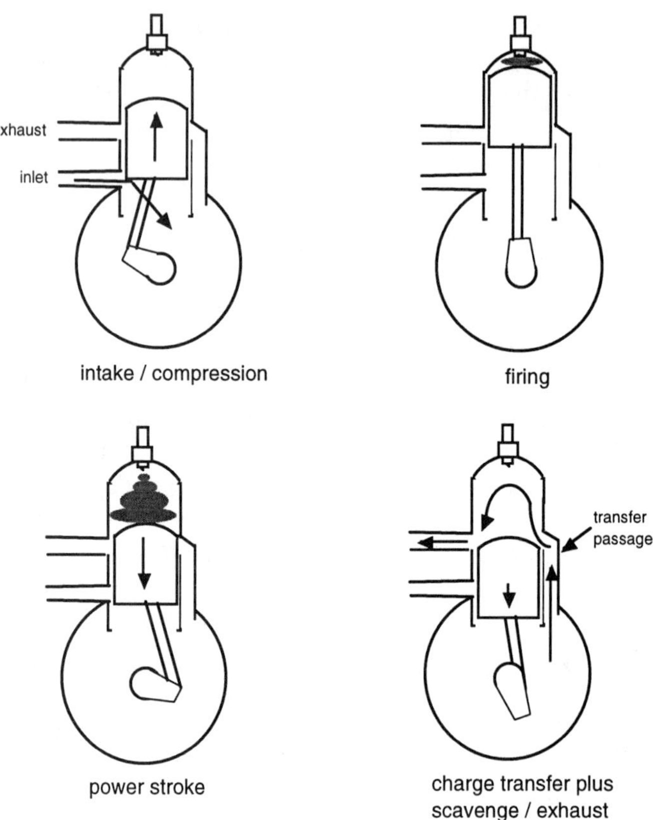

Fig. 16 The two-stroke cycle.

conventional poppet valves. Four-stroke engines now virtually all use multiple poppet valves, although sleeve valves were used in both aviation and passenger car engines at one time, and some rotary four-stroke engines used ports. Both diesel and gasoline engines can be of either four-stroke or two-stroke design.

As well as ignition quality (octane rating for gasoline and cetane rating for diesel fuel), the degree of refining and the presence of harmful impurities or beneficial additives are of importance in relation to how the fuel burns and the effect this may have on the lubricant.

Gases (methane, LPG, etc.) can be used to fuel gasoline-type engines with little modification and to produce very clean combustion. Gases can also be burned in multi-fuel "diesel" engines although for gases considerable differences from a normal diesel system are required, including a continuous source of ignition (hot wire, etc.). Most engines, however, burn liquid fuels. Once started on conventional fuel, diesel engines will run on many fuels, including methanol.

The majority of combustion products are evacuated via the exhaust system, but a significant proportion leaks past the piston rings, contaminating the engine oil and causing other adverse effects (Figure 17).

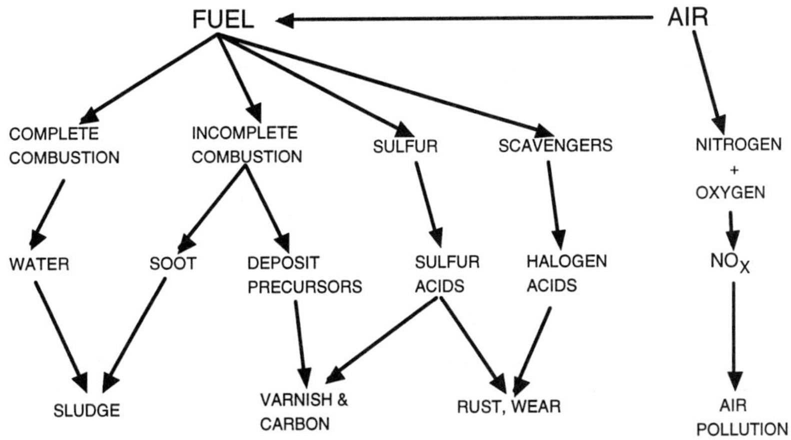

Fig. 17 Fuel combustion and crankcase contamination.

This leakage of the products of combustion is known as *blowby*, and is particularly significant in the small mass-produced automobile engines of popular cars. These are designed down to a size, and a price, and assembly-line production means that the tolerances cannot continuously be held as tightly as when parts are hand-fitted, although development of part-matching techniques has improved the average quality of fit. The direct connection of the piston to the crankshaft via a connecting rod and wrist-pin is a problem, especially in short-stroke engines, because it imposes side loadings on the piston and impairs ring sealing, therefore tending to increase the amount of blowby gases passing into the crankcase.

Modern positive crankcase ventilation (PCV) systems recirculate a proportion of the blowby via the induction system to the combustion chambers, but a considerable amount is trapped by the large volume of oil spray in the crankcase and is absorbed by the oil.

Generated by the combustion process and appearing in the blowby gases are carbon dioxide, water, acidic components, and some hydrocarbon residues. From the excess of intake air, the high combustion temperatures also produce some nitrogen oxides.

The carbon dioxide and the water arise from combustion of the hydrocarbon fuel:

$$C_7H_{16} + 11O_2 \rightarrow 7CO_2 + 8H_2O$$

Heptane

or,

$$C_7H_8 + 9O_2 \rightarrow 7CO_2 + 4H_2O$$

Toluene

It can be seen from these chemical equations that large quantities of water are produced by the combustion of hydrocarbons. Of course, when the engine is at normal operating temperature most of the water remains in the vaporized form and passes out of the exhaust system, but the water vapor in the blowby gases passes into the relatively cooler crankcase and condenses to liquid form. It can settle on various parts of the engine mechanism and cause rusting, or it can mix with the lubricant to form sludge. The coolest part of an engine is usually under the rocker cover, and condensed water frequently collects there and produces sludge (Figure 18).

Addition of dispersing additives to the lubricant results in the water mixing intimately with the oil to prevent these effects, while at the same time carrying some of the water to the hotter parts of the engine where it can be flashed off and vented via the PCV system. An equilibrium is set up whereby the proportion of water mixed into the crankcase oil depends very much on the prevailing engine temperature, as well as the overall engine design.

The carbon dioxide produced, while it has the characteristics of a very weak acid, can essentially be regarded as harmless to engines and lubri-

Fig. 18 Production of rocker cover sludge.

cants. However, fuels contain impurities which can give rise to more serious problems from acid formation. Crude petroleum contains significant quantities of sulfur, not all of which is removed in the refining processes. Particularly in the case of diesel fuel, residual sulfur in the fuel burns in the combustion chamber to produce sulfurous and sulfuric acids:

$$2S + 2O_2 \rightarrow 2SO_2$$
$$2SO_2 + O_2 \rightarrow 2SO_3$$

Then,

$$SO_2 + H_2O \rightarrow H_2SO_3 \text{ (sulfurous acid)}$$
$$SO_3 + H_2O \rightarrow H_2SO_4 \text{ (sulfuric acid)}$$

While gasoline often has a lower sulfur content than diesel fuel, in the case of leaded gasoline the scavengers used to reduce the build-up of lead salts in the combustion chambers are compounds such as ethylene dichloride and ethylene dibromide. Excess of these compounds reacts in the combustion chamber to produce complex chlorine and bromine oxy-acids and some hydrochloric and hydrobromic acids.

All of these acids can produce corrosion and corrosive wear in the engine, and they can also act as catalysts both for the degradation of the oil and also for the formation of gums and varnishes. These are sticky or lacquer-like deposits which can prevent free movement of parts of the engine, and arise from the partial combustion of some of the fuel, which produces reactive hydrocarbon substances known as "deposit precursors" and which appear in the blowby. These, together with other products from the degradation of the lubricant, polymerize in the presence of acids to form sticky deposits (gums), which in the case of hot engine parts such as the piston skirts can be baked to form brown or yellow varnishes. Continuous varnish formation will ultimately lead to production of hard deposits of solid carbon (Figure 19).

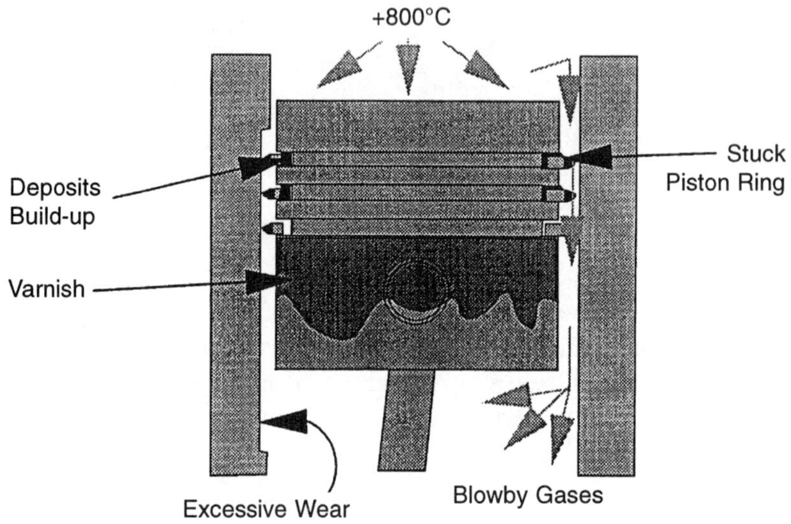

Fig. 19 Piston deposits and stuck rings.

Alkaline additives are introduced into the oil to neutralize these acidic compounds and prevent their harmful effects on engine and lubricant. The dramatic increases in engine life in the last few decades is primarily due to the introduction of these additives first into diesel engine lubricants and then into passenger car motor oils. The introduction of lead-free gasoline in many areas and proposed reductions in the maximum permissible sulfur content of diesel fuels will greatly reduce acid formation and lead to new types of oil formulation and probably further increases in oil and engine lifetimes.

A further contaminant, which appears in greater quantity in the diesel engine, is fuel soot. This arises from the incomplete combustion of fuel in the cooler parts of the combustion chamber, and, as for the other contaminants, a proportion appears in the blowby. Before the advent of dispersant additives, soot and water in the lubricant produced large quantities of grey/ black sludge in the crankcase and in the oil passages of an engine, resulting in lubrication problems if the oil was not changed frequently. Soot can also adhere to varnish deposits and accelerate the build-up of carbon, if the varnish-forming tendencies are not sufficiently controlled.

In summary, a modern automotive oil not only has to be stable at high temperatures and provide suitable viscosity for the moving parts throughout a wide range of temperatures, but also has to counter the effects of these various contaminants. It will include additives that disperse water, soot and other constituents of the blowby and carry these around in the lubricant, which remains homogeneous. Other additives in the oil react with and neutralize the various acidic contaminants which would otherwise give rise to rusting, corrosion or deposit formation problems.

1.7.2.1 Medium and Large Diesel Engines

In the preceding section we discussed the relatively small engines found in automobiles, trucks, and buses. Engines in these vehicles are compact and operate at high speed, this being possible because of the relatively low inertial loading on the components due their small size and light weight. In general, as engines increase in size and power output, there is a corresponding reduction in the maximum engine speed which can be achieved. In the following paragraphs we will review various other types of engines which, because of design or operating differences, have different or additional requirements to the more numerous road-vehicle types. Both larger and smaller types are considered, and some are included for their interesting features even if not strictly within the confines of the automobile industry.

Medium-sized, medium-speed diesel engines are employed in railroad locomotives, earthmovers, in oil drilling operations, in small-scale and stand-by power generation, and as auxiliary engines for shipboard use. Larger engines are used in power generation and particularly for marine propulsion. In all these applications the running costs and particularly the cost of fuel is a major consideration. The tendency is therefore to use lower-cost, lower-quality fuels and to utilize highly detergent lubricants to overcome the deficiencies of these fuels in terms of their adverse effects on the engines. Low-quality fuels tend to have high sulfur contents, and the lubricant must therefore have high alkalinity in order to neutralize the acids generated by combustion of such fuels.

The following will give a review of some of the other important requirements of these various classes of engines.

1.7.2.2 Railroad Locomotives

Diesel railroad locomotives are normally powered by engines specifically designed for this purpose, combining a high power output with restricted space requirements. (Differentiation is made here between the separate locomotive for hauling unpowered stock, and diesel railcars or self-powered coaches, which typically employ a multiplicity of converted bus engines for motive power.) For high-powered locomotives heat dissipation can be a problem, and, taken with the lower fuel quality which is often used, lubricants are required to protect the engine against piston groove deposits which lead to ring sticking, and against corrosive wear. The oil must also be protected against deterioration by adequate anti-oxidant treatment. Some engines cannot tolerate zinc-containing oils. Railroad oils are discussed further in Section 8.3.

1.7.2.3 Marine Engines

For auxiliary marine applications (winches, power generation, etc.) trunk piston engines (i.e., of conventional design) are used, as they are for similar land-based uses. These are very similar to those in larger off-road equipment (earthmovers, etc.) and can indeed consist of larger versions of such engines, perhaps employing sixteen cylinders instead of a more usual six.

In the big marine propulsion diesel, however, which is most frequently of two-stroke design, the interesting feature is the use of a cross-head. This is shown diagramatically in Figure 20.

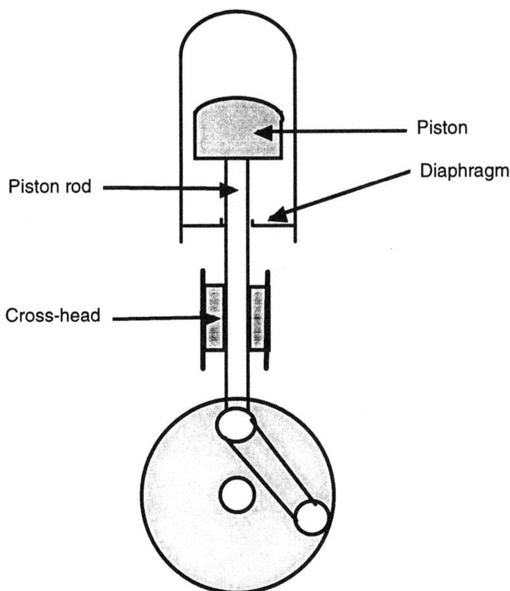

Fig. 20 Schematic of cross-head engine.

Instead of pivoting the connecting-rod at the piston, this is rigidly bolted to the piston and moves linearly with the cross-head. This mechanism will be very familiar to devotees of the old steam railroad engine, the cross-head being where the linear motion is converted to an angled drive to the crank-shaft, rather than this taking place at the base of the piston itself. Moving the pivoting point from the piston to a cross-head has several advantages. First, side forces on the piston are eliminated, piston-ring wear reduced, and piston-ring-to-cylinder contact is improved and piston sealing more consistent over the life of the engine. However, the big advantage is that with a linear connecting-rod motion it is possible to seal the crankcase from the blowby and other contamination coming from the combustion chamber and the piston-ring areas.

This is done by a suitable sealing gland. In large marine propulsion engines the ability to separate the engine effectively into two parts is particularly useful. Low-cost residual fuel of high sulfur content is often used, and to counteract the acid attack and deposit formation that this would otherwise cause in the piston-ring area, a very alkaline cylinder oil is injected at ring level. The piston-rod gland seals off the blowby, unburned fuel, and cylinder oil residues from the crankcase oil. This can then be a non- or mildly detergent type, designed specifically for optimum crankshaft and cross-head lubrication, and having a long service life. Separation of the combustion and blowby zones from the crankcase is possible only when such a cross-head design is used, and in the marine case is fundamental in permitting the use of cheap but poor-quality fuel. It is, however, perhaps surprising that the advantages this confers, together with the more linear forces seen by the piston, have not been included in medium-sized engines of lower power outputs than for the marine case. It could be argued that a miniaturized medium-speed version of a marine propulsion engine would have some very attractive features for truck and off-road diesel applications, but we are unaware of any such designs being produced. The arguments against such an engine relate to the size-complexity-cost of such a design versus a conventional diesel of similar power output.

1.7.2.4 Two-Stroke Engines

The two-stroke cycle is not limited to small utility engines and outboards. As mentioned above a large number of marine propulsion units are two-strokes, and very successful Detroit Diesel truck engines and the General Motors EMD railroad engines are two-stroke diesels. Such larger types employ positive pressure-charging from an external blower and use valves as well as or instead of porting. In Europe, earlier successful two-stroke truck and railroad engines have generally been succeeded by conventional four-stroke designs.

Theoretically the doubled frequency of firing strokes should make two-stroke engines hot-running and prone to oil oxidation, but the relatively low-speed operation in most applications enables heat to be extracted satisfactorily. In the case of the Detroit Diesel, this has a fairly high level of oil consumption and fresh additive is continuously supplied with replenished oil.

Two-stroke engines have been used for passenger cars (Saab, Trabant, etc.) but were abandoned on grounds of air pollution and noise. These were similar to the "2-T" engines discussed below. Modern designs have been developed which offer high efficiency and low emissions, and it is possible that the two-stroke may reappear in motor vehicles to a significant extent.

1.7.2.5 "2-T"

This expression, possibly originating from the French "deux-temps" or the German "Zwei-Takt" (for two-stroke), is frequently used to denote the small two-stroke, normally aspirated gasoline engines used in garden machinery, mopeds and motorbikes, and for many outboard motors. Sizes of these engines range from below 50cc for single-cylinder engines used for gardening applications, to several litres for multi-cylinder engines.

This type of two-stroke engine is characterized by the absence of normal valve gear, although there may be automatic reed valves. Air is drawn into the closed crankcase by the upward movement of the piston, then passed to the combustion chamber by means of a transfer port and passageway as the piston commences its downward stroke. At the top of each stroke, the transfer port is closed off by the rising piston and the gas is compressed prior to ignition. There is no lubrication sump as such, the oil passing through the engine on a once-through basis, eventually either being burned, or passing out with the exhaust gases.

Traditionally the oil has been supplied in a mixture with the gasoline. The ratio of gasoline to oil generally now ranges from 25:1 to 100:1 depending on the application and oil quality, although early types used down to 6:1. More recently for high-performance motorcycles (particularly those of Japanese origin) and some outboard motors it has become the practice to inject oil separately into the engine, metering the quantity according to speed and load.

The quality needs for 2-T oils depend to a large extent on the application, but the key requirements are for good lubricity, piston cleanliness, low deposits especially in the exhaust system, low smoke emission, and, more recently, biodegradability. These are discussed in more detail in Section 8.1.

1.7.3 Special Systems

Gears, gas turbines, hydraulic systems, and industrial machines have requirements which are specific to the individual application, but more straightforward than the complex requirements of the internal-combustion engine. Details of their requirements and the applicable technology are discussed in Chapters 7 and 8.

References

1. Dowson, D., The History of Tribology, Longman, 1979.
2. Ucelli, G., Le Navi Di Nemi, Liberia Dello Strato, Rome, 1950.
3. Mercedes-Benz Betriebsstoff - Vorschriften, *Specifications for Service Products* (Fuels, Lubricants, etc.), p.221.
4. Jost, H.P. *et al.*, "Lubrication (Tribology)," Report to Education and Science Ministry, UK, HMSO, London, 1965.
5. Amontons, G., De la Résistance Coussie Dans Les Machines, Histoire Acad. Royale, Paris, 1699.
6. de la Hire, P., Sur les frottements des machines, Histoire de la Academie Royale, Paris, 128-34, 1706.
7. Desaguliers, J.T., *Phil. Trans. R. Soc. London*,VII, 100 ,1725.
8. Coulomb, E., "Theorie des Machines Simples," *Mem. Math. Phys.*, X 161, Paris, 1785.
9. Bowden, F.P., and Tabor, D., "The area of contact between stationary and between moving surfaces," *Proc. R. Soc. London*, A169, 391, 1939.
10. Newton, J., Philosophiae Naturalis Principia Mathematica, 4, 268, 1687.
11. Gunter, R.T., "The Life and Work of Robert Hooke," *Early Science in Oxford*, Vols.6,7, Oxford, 1930.
12. Euler, L., "Sur le Frottement des Corps Solides," *Mem. Acad. Sci. Berl.*, No.4, 122-32, 1748.
13. Euler, L., "Sur la Diminution de la Resistance du Frottement," *Mem. Acad. Sci. Berl.*, No.4, 133-48, 1748.
14. Tower, B., *Proc. Inst. Mech. Eng.*, 622-659, 1883/4 and 29-35, 1884.
15. Reynolds, O., *Philos. Trans. Roy. Soc. London*, 177, 157-253, 1886.

16. Dowson D., Higginson, G., Elastohydrodynamic Lubrication, Pergamon Press, London, 1966.
17. Dalal, H.M., Chiu, Y.P., and Rabinowitz, E., ASLE Trans., 18211, 1975.
18. Page, T., and Adawaye, O.O., *Proc. Br. Ceram. Soc.*, 26, 193, 1978.
19. Fischer, T.E., and Tomizawa, H., *Wear*, 105, 29-45 1985.
20. Ball, W.F., Jackson, N.S., Pilley, A.D., Porter B.C., SAE Paper No. 860418, Society of Automotive Engineers, Warrendale, Pa., 1986.
21. Bovington, C.H., *et al.*, "Tribology of Ceramics in Valve Train & Piston Ring/ Cylinder Liner Applications," CARE Consortium Conference, 20 June 1990.
22. Bowden, F.B., Young, J.E., *Nature*, London, 164, 1089, 1949.
23. Bowden, F.B., Tabor, D., "The Friction & Lubrication of Solids," OUP Monographs, Clarendon Press, 1950, 1964.
24. O'Connor, J.J., and Boyd, J., Standard Handbook of Lubrication Engineering, McGraw-Hill, 1968.
25. Stribeck, R., "Die Wesentlichen Eigenschaften der Gleit-und Rollenlager," *Z. Ver. dt. Ing*, 46, 1341-1348, 1432-1438, 1463-1470, 1902.

Further Reading

26. Schilling, A., Automobile Engine Lubrication, Vol. 2
27. Clauss, F.J., Solid Lubricants and Self-lubricating Solids, Academic Press, 1972.
28. Neale, M.J. (Ed), Tribology Handbook, Butterworths, 1973.
29. *Proceedings of First European Tribology Conference 1973*, Inst.Mech.Engineers, 1975.
30. Cameron, A., and McEttles, C.M., Basic Lubrication Theory, 3rd Edn., John Wiley, 1981.
31. Gohar, R., Elastohydrodynamics, John Wiley, 1988.
32. Bosser, E.R. (Ed.), CRC Handbook of Lubrication, Vol II, CRC Press, 1984.
33. Leeds-Lyons Symposia on Tribology, 1986, 1987, and subsequently.
34. Buckley, D.H., "Friction, Wear, and Lubrication in Vacuum," NASA SP-277, Washington, DC, 1971.

Chapter 2

Constituents of Modern Lubricants

2.1 Base Stocks

The basic fluid that constitutes the major part of a lubricating oil is almost invariably made from a mixture of two or more components, generally referred to as the *base stocks*. These can be petroleum-derived lubricating oil, but increasingly quantities of other chemical fluids are being employed as partial or whole substitutes for mineral oils. Most such chemical fluids are readily miscible with petroleum fractions, and this is a requirement for use in lubricants for general sale, in order to avoid problems from the use of different types and brands of oil in admixture.

The past decade of the 1980s has seen the introduction of significant quantities of petroleum-derived base stocks produced by new processes. These are frequently called "unconventionally refined base oils" but "modern petroleum base oils" would also be an apt description. Conventional refining merely extracts existing suitable material out of crude petroleum, and base stock yield and quality is therefore heavily dependent on the type and quality of the crude oil. The new processes are catalytic conversion processes, specifically hydrocracking, hydro-isomerization, and catalytic dewaxing, and base oil quality is less dependent on crude oil or feedstock quality and more dependent on specific operating conditions within the plant. The economics of such plants are dependent on how they are integrated into a refinery complex, and on the assumptions made about feedstock and by-product values.

Use of the term "synthetic" to describe base stocks and finished products has been the subject of some controversy in recent years. SAE J357 defines synthetic base stocks as chemical products and includes all products directly derived from petroleum as "refined petroleum base stocks" (see Appendix 15).

However, the term "synthetic" has high marketing significance, describing oils having generally a higher level of performance, and distinction in chemical terms has become blurred by the availability of base stocks derived from petroleum feedstocks by conversion processes. Since most base stocks, be they purely hydrocarbon, esters or other chemical types, are ultimately derived from petroleum feedstocks, and the term "synthesis" covers a wide range of chemical processes, it has proved impossible to find an agreed definition of the term "synthetic" base stock. It is now generally accepted that the term may be used in marketing products which have been formulated to deliver the levels of performance which can be achieved only by use of judiciously chosen base stocks from a variety of sources, including severe reforming and cracking processes at a petroleum refinery.

The base oil market is approximately static on a total volume basis, but many old smaller plants have been closed and the shortfall largely made up by the building of new conversion plants, although some de-bottlenecking of conventional refineries has also taken place. Some improvements in the quality of conventionally refined stocks are being made, for example by improving fractionation and so reducing the content of volatile components in the lighter cuts. In volume terms the conventionally refined base stocks still predominate in all markets and most applications, but in the passenger car motor oil sector the percentage of conversion-processed stocks along with other chemical fluids has risen dramatically and this will continue. (They will also be used in specialized industrial oil areas.)

Mention will also be made of base stocks which arise from the reclaiming of used (waste) oils. Conservation pressures will demand more and more lubricant reclamation, and while not without problems, lubricant base stocks can be recycled, if at some cost.

We will now consider these different types of base stock in more detail in the following sections.

Plate 1. Base oil manufacturing plant.

2.1.1 Conventionally Refined Petroleum Base Stocks

Lubricating oils can be produced from a wide variety of crude oils, and taking into account the many variations in available processes, there are many different qualities of petroleum lubricants available. Crude oils and the conventional lubricating oils refined from them are generally classified as either *paraffinic*, *naphthenic*, or *intermediate*. The names relate to the relative preponderance of either paraffins (straight or branched hydrocarbon chains) or naphthenes (cycloparaffins) in their composition, both types also containing some aromatics (alkyl benzenes and multi-ring aromatics). These chemical structures are illustrated diagramatically in Figure 21.

Each angle in these structures represents a carbon atom, and typical lubricating oil base stocks have between twenty and forty carbon atoms in each molecule. Within each type of structure many variations are possible, and literally millions of different molecular types make up the typical lubricating oil. A brief introduction to petroleum chemistry is given in Appendix 3.

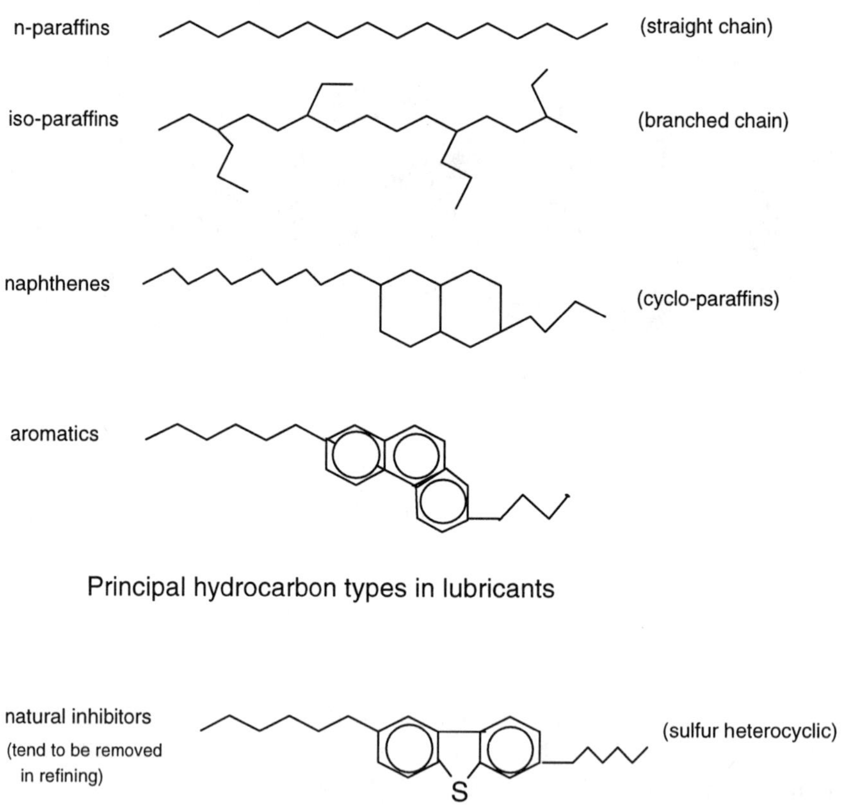

Fig. 21 Lubricating oil constituents.

Long chain, higher-molecular-weight paraffins (i.e., those that boil at higher temperatures and have more atoms in each molecule) are solids at low temperatures, and when a raw paraffinic oil is cooled these separate out as wax and the oil eventually gels to a solid. To enable a paraffinic oil to flow at low temperatures, the heaviest wax has to be removed and perhaps some additive introduced to modify the way the rest of the wax crystallizes (see Section 2.2).

Naphthenic crudes and the lubricating oils extracted from them contain little or no wax, and remain liquid down to low temperatures (i.e., they have low pour points). However, they have a high viscosity-temperature coefficient, thinning down considerably when heated. (The rate of change of viscosity with temperature is normally indicated by a number known as the Viscosity Index, an arbitrary number developed in the 1930s to classify base stocks. A Pennsylvanian paraffinic stock of good viscosity-temperature characteristics was given a Viscosity Index [or V.I.] of 100, and a U.S. Gulf stock of poor viscosity-temperature characteristics a value of zero. The V.I. of a base stock was interpolated between these two limits by comparing the viscosities at 100°F when the reference and the test oils had identical viscosities at 210°F. More recently, the V.I. has been calculated from viscosities at 40°C and 100°C, and the scale has been extended beyond 100. Naphthenic base stocks have low V.I. and paraffinic stocks a high V.I. of 90+.)

It is important to realize that the terms naphthenic and paraffinic indicate only a tendency in the composition, and extracted base stocks are a mixture of all types of compound.

Naphthenic stocks have higher densities and greater solvency powers than paraffinic stocks, and on thermal decomposition tend to form softer carbon deposits, a reason for their preference in the era of uninhibited diesel lubricants, when carbon formation was accepted as a routine phenomenon and engine overhauls were consequently frequent. Today they are hardly used in crankcase oils, and are produced in smaller and smaller quantities for certain specialized industrial oils.

2.1.1.1 Conventional Refining Processes

Distillation

Crude oil is a wide-boiling mixture of hydrocarbons, from light gases to heavy asphaltic material. To perform the initial separation of the basic products, atmospheric distillation (i.e., at normal pressure) is employed. This is illustrated diagramatically in Figure 22.

The heavy residual fraction is dark in color and contains the major impurities and considerable oxidized and degraded material. It can be used directly as fuel oil, or further distilled either to provide feed for a catalytic cracker or for lubricating oil production. To prevent further degradation of the fraction, this second distillation is done under vacuum, which has the

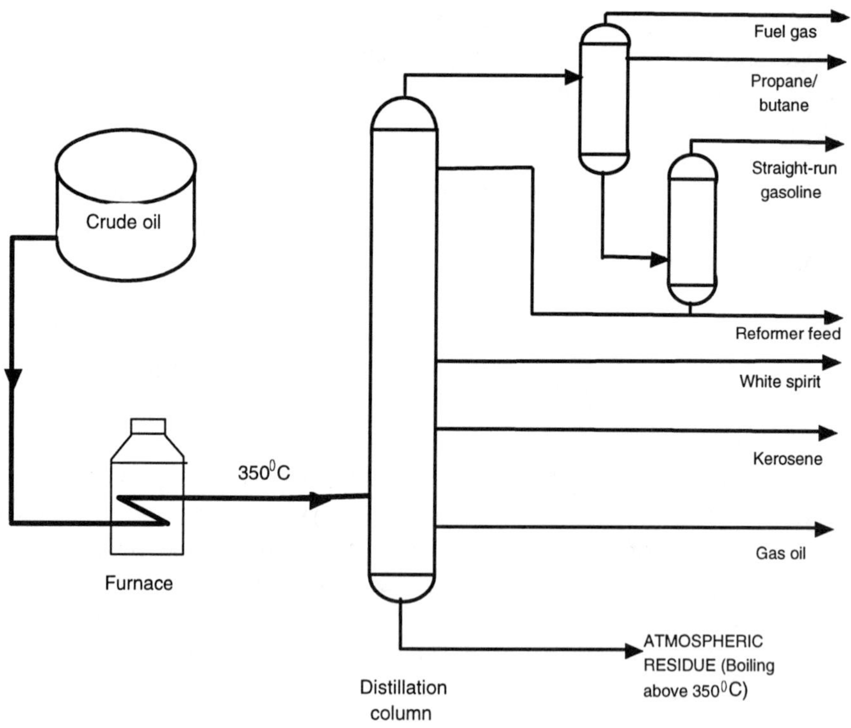

Fig. 22 Atmospheric distillation of crude oil.

effect of lowering the temperature at which the material boils and is thereby distilled. Vacuum fractionation yields the primary lubricating oil streams with a range of viscosities (Figure 23).[(1)]

Traditionally the fractionation of lubricant streams has been by the so-called "tray-type" of distillation column containing relatively few horizontal baffles to collect condensate at different levels and temperatures and cause it to mix with the rising vapor. Such columns have a large throughput capacity, but do not provide very good fractionation. Some of these older columns are now being replaced or supplemented by packed columns, which can produce narrower cut fractions more able to meet the volatility requirements in modern specifications.

The heavy concentrated residue from vacuum distillation is the feed for the so-called "bright stocks," heavy fractions used in gear oils and other viscous lubricants. To remove asphalt and impurities the residuum is mixed with propane, in which aromatic residues are insoluble. These and

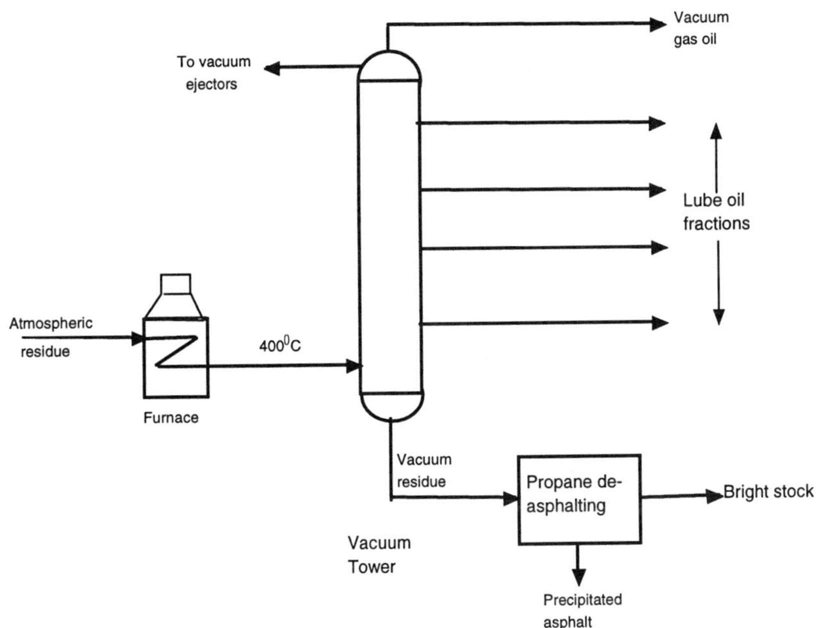

Fig. 23 Vacuum distillation of atmospheric residue.

other impurities are precipitated and the asphalt (bitumen) recovered. The oil is steam-distilled, stripped of solvent, and passed through to the extraction units.

Solvent Extraction

This process removes most of the aromatics and some of the naphthenes, and increases both V.I. and base stock stability.(2) It has replaced earlier acid-treating processes with their hazardous oleum treatment and waste disposal problems.(3) The solvent (phenol, furfural, or more recently N-methyl-2-pyrrolidone [NMP]) is passed in a counter-current against the lube feedstock stream, and is recycled after stripping from the aromatic extract and the raffinate (the refined or extracted oil). This is shown in Figure 24.

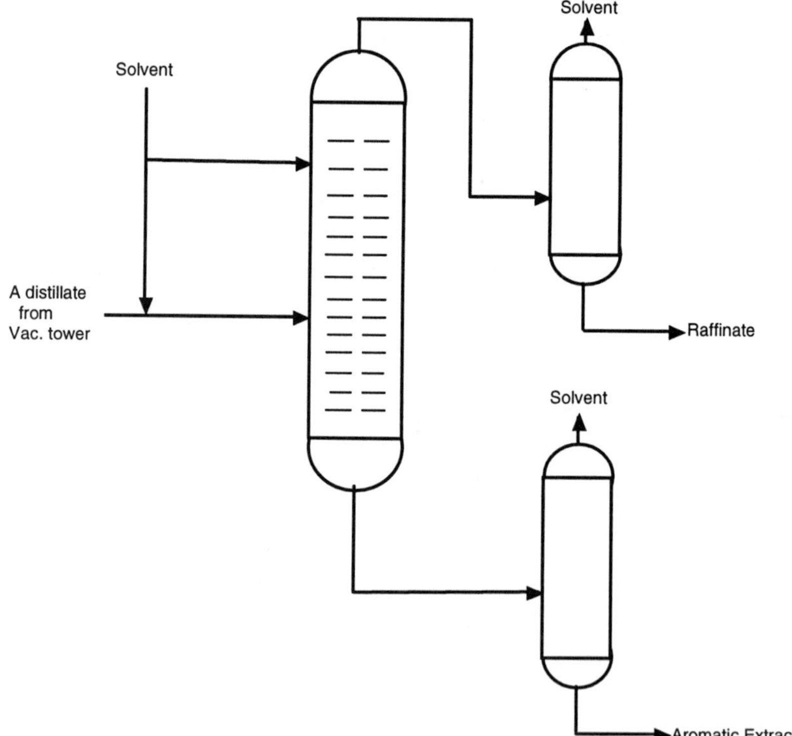

Fig. 24 Solvent extraction of distillates.

Finishing Processes

The solvent extraction process does not remove all the reactive and unstable material from the base stock, and without a finishing process the oil would soon darken and precipitate sludge, especially when exposed to light.

An early finishing process was clay treatment, whereby the oil was mixed with "fuller's earth," which adsorbed reactive aromatic and unstable molecules, and was then filtered off. The contaminated clay residues presented a waste disposal problem and there were also large product losses. Hydrofinishing is now the normal process, whereby the raffinate is passed through a heated reactor packed with a catalyst (typically nickel and molybdenum oxides on silica and alumina) and hydrogen is passed in under pressure (Figure 25). Hydrogenation reactions convert the unstable compounds into stable ones. Aromatics, for example, are converted into naphthenes. Hydrofinishing does not substantially reduce the product yield, it increases the V.I., and removes some of the sulfur compounds (which are converted to hydrogen sulfide) and other trace materials. The

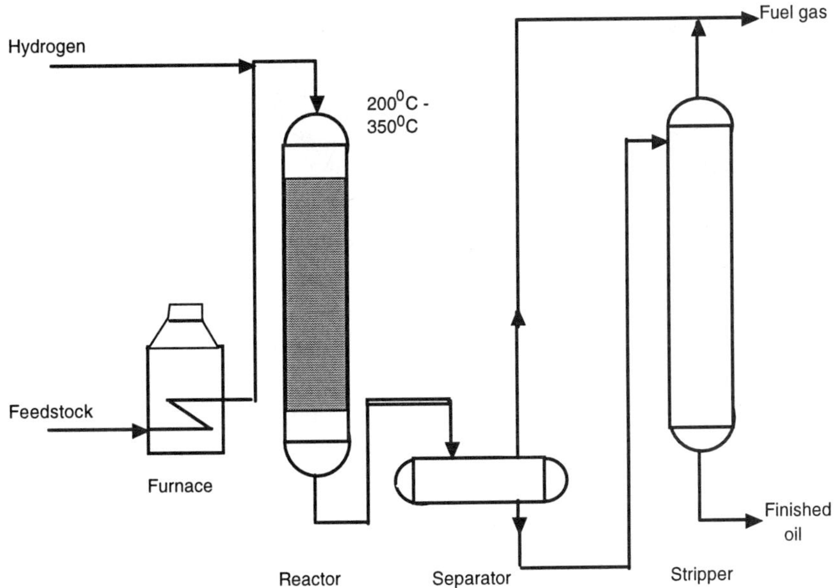

Fig. 25 Hydrofinishing of base oil stocks.

extent of these effects is very dependent on the processing conditions — catalysts, temperature, pressure, etc.

While these reactions result in good base stocks from the point of view of general molecular structure, hydrogen processing has the disadvantage of removing the so-called "natural inhibitors" from the base stocks. These comprise certain sulfur compounds and a proportion of the aromatics.(4) Hydrofinishing was initially introduced when motor oils had relatively low additive contents, and this depletion of natural inhibitors was found to be a disadvantage. The level of hydrotreating was therefore restricted, or extra inhibitors were added. These could be either recycled extracts added to the base stock, or specific anti-oxidant supplements to the additive package. With today's highly formulated motor oils, the contribution of "natural inhibitors" to their overall stability is very small and severe hydrotreating is widely adopted.

Dewaxing

Unnecessary for most naphthenic stocks, the need to remove wax from paraffinic stocks to improve their low-temperature flow properties imposes major cost and product loss penalties. The raffinate to be dewaxed is mixed with a solvent in which the heaviest wax is insoluble at low temperatures. Propane was an early solvent (and still must be used to dewax bright stocks), but most plants now use a ketone solvent such as methyl ethyl ketone or methyl isobutyl ketone or a mixture of ketone (typically methyl ethyl ketone) and toluene.(5) A mixture of the chlorinated hydrocarbons dichloroethane and methylene chloride (Di-Me) has been used with some success,(6) but fears of ozone depletion from leaks of these chemicals are likely to prevent wide-scale adoption in the future.

The chosen solvents and the raffinate are mixed together and refrigerated to a temperature at which enough of the wax crystallizes out and can be removed by a rotary filter. The solvent is stripped from both the oil and the wax portions and recycled (Figure 26).

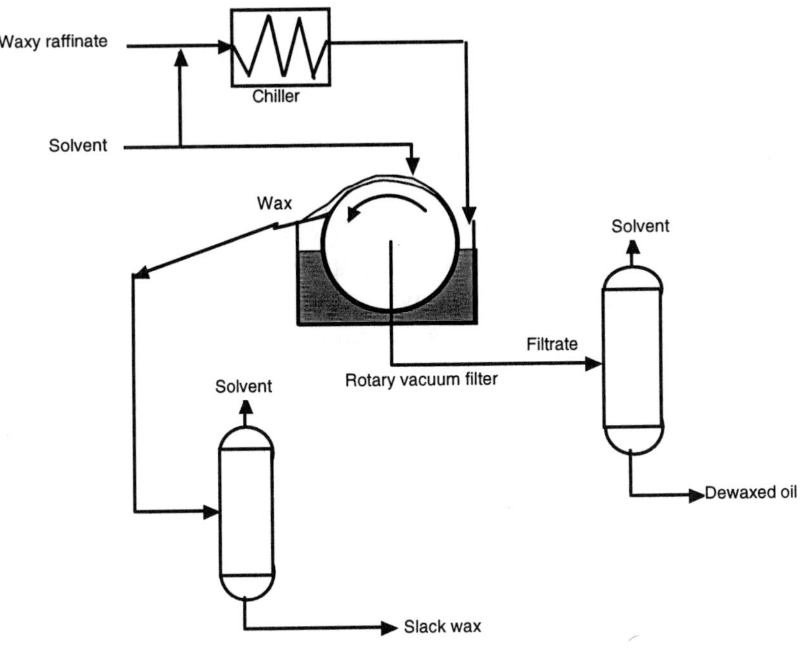

Fig. 26 Solvent dewaxing of raffinates.

Oil is allowed to "sweat" out of the "slack wax" removed from the filter, leaving paraffin wax from the distillate streams and microcrystalline wax from the bright stock. If pure waxes are needed for sale, rather than as feedstocks for other refinery units, they can be solvent-washed to remove residual oil.

Catalytic dewaxing, discussed in Section 2.1.2.1 below, can also be employed after conventional extraction processes, although it is more usually associated with conversion (hydrocracking) base oil production.

2.1.2 Modern Conversion Processes

The hydrofinishing process described in 2.1.1.1 is normally a relatively mild process which results in some saturation of aromatics and other unstable compounds but does not significantly crack molecules with a resulting loss of viscosity, or cause other major molecular changes. By increasing the severity (temperatures up to 420°C and pressures up to 4000 psig or 28 MPa), the molecules in the oil can be modified in several ways to produce a high-quality base stock, and one which is less dependent on the composition of the original crude oil. At the highest severity the process is known as *hydrocracking*, and in this case the feedstock can be slack wax, fuel oil, or other heavy feed locally available, and high-quality base stocks along with valuable fuel by-products will result. In particular, hydrocracking of slack wax (from a conventional dewaxing plant) can produce very high V.I. oils of up to 150 V.I.[7]

A lubricant hydrocracker used to improve the quality of streams from a conventional lube stock fractionator will produce yields from 40% to 70% on feedstock, but if special base oils are to be produced in a high severity lube or fuel hydrocracker the yield of such base stocks may be as little as 5% of feedstock, with co-products being of sufficient value to justify this low yield of the lubricant fraction. In such cases, where the lubricant is in effect a by-product in a larger operation, there will be a need for distillation, extraction, dewaxing, and probably hydrofinishing steps before the stream is suitable for base stock use.

The hydrocracker product approaches the characteristics of synthetic polyalphaolefin stock (see below), but an even closer approximation comes from a process usually known as wax hydro-isomerization. This uses wax as feedstock, operates at somewhat lower temperatures, and uses a special catalyst to produce highly paraffinic stocks in high yield. Cracking is reduced, and the conditions favor reforming of the molecules to more useful types.

The principal reactions taking place in hydrogen reforming and hydrocracking are as follows:

(a) SATURATION
  Polyaromatics → Polynaphthenes
  (higher V.I.; lower pour)

(b) RING-OPENING
  Polynaphthenes → mononaphthenes
  (higher V.I.; lower pour)

(c) REFORMING (Isomerization)
n-paraffins → branched paraffins
(lower pour; some V.I. loss)
(d) CRACKING
High M.W. → Medium M.W.
(lower viscosity and boiling point)
(e) DESULFURIZATION
Sulfur compounds → H_2S
(loss of natural inhibitors)
(f) DENITROGENATION
Removal of nitrogen- and oxygen-containing heterocyclic compounds
(improved stability)

2.1.2.1 Catalytic Dewaxing

This process has some similarities to the hydrocracking processes, but the catalyst takes the form of a highly porous molecular sieve. Straight chain or lightly branched paraffins (alkanes) enter the pores of the sieve, are trapped, and cracked to lighter products. Long side chains of complex molecules can also become trapped and removed, thus some useful components such as alkyl aromatics can be lost.

Catalytic dewaxing is a highly specific and efficient process for removal of the straight chain hydrocarbons. The resultant base oil tends to be lower V.I. than solvent dewaxed stocks, where temperature of wax separation rather than molecular shape determines the portion of product which is removed. However, as most straight chain material is removed in catalytic dewaxing, overall low-temperature properties are superior to solvent extracted stocks and they are highly suitable for low-temperature (e.g., Arctic duty) oils. In such cases, low-temperature fluidity is much more important than conventional V.I.

Catalytic dewaxing is expected to displace solvent dewaxing for light motor oil stocks produced by conventional refining. The value of wax as a by-product and feedstock, and the continuing need for large quantities of conventional base stocks for other applications, will, however, ensure that conventional processing remains in use for the foreseeable future.

With the possibility of base stocks being produced by new processes or new combinations of processes, the API has proposed a new classification system which focuses on the paraffinicity and sulfur content of the stocks. The present proposal (January 1995, API Publication 1509) is for classifications as follows:

	Saturates	Sulfur	V.I.
Group 1	below 90%	over 0.03%	80 to 120
Group 2	90% and over	0.03% and less	80 to 120
Group 3	90% and over	0.03% and less	over 120
Group 4	(polyalphaolefins)		
Group 5	(others)		

Conventionally refined stocks would fall into Group 1, while moderately hydrocracked stocks including those catalytically dewaxed would fall into group 2. Cracked wax products would be in Group 3. Polyalphaolefins are given a special category Group 4, and all other stocks are Group 5. The API groupings are used to define which engine tests must be performed when substituting base stocks from different groups in finished oils carrying API licenses.

2.1.3 Reclaimed Base Stocks

During the time of the petroleum shortages in the late 1970s, several processes were developed to produce high-quality base stocks by reprocessing used oils.[8] Some plants were commissioned and operated for a time but were generally uneconomic and, with the return of low crude oil prices, have generally closed down. High costs were associated with the collection of used oils, and the processes were quite sophisticated, having to re-refine a cocktail of used motor oils, industrial lubricants, and cutting oils. The collection process was always a problem, although considerable success can be achieved, as for example in Germany where over half of used oil is collected. Germany remains in the lead for used oil collection, but the reason today is environmental rather than economic. Much used oil is burned as fuel at the present time.

Reclaiming may have a resurgence as environmental regulations make it more and more difficult to dispose of used oil. Used oil can either be incinerated or reprocessed, but dumping will be prohibited. Incineration provides an energy source, but is wasteful of a useful commodity and could add to air pollution.

Refined oils can satisfactorily replace petroleum base stocks provided the re-refining process has removed all contaminants. In fact, re-refined oil may have superior oxidation characteristics to virgin stocks, because easily oxidized compounds will have been reacted during use and then removed during reprocessing.

Simple filtration to remove gross contamination does not amount to re-refining. Oils collected for reprocessing or disposal usually contain mixtures of several types of product, and will contain soot, sludge, water, salts and dirt as insoluble contaminants, and fuel residues, oxidized material and additive residues as dissolved material. Vacuum distillation is virtually essential to refractionate the lubricants, and product is finished by clay or hydrogen finishing processes, or a combination. Processes differ mainly in the way the gross and solid contaminants are removed initially. Processes which have proven to work technically include IFP acid/clay,(9) IFP propane precipitation, Berc alcohol/Ketone precipitation, KTI thin film evaporator,(10) Phillips ammonium sulfate, and the Matthys Garap continuous process. The economics of all processes depend critically on the acquisition cost of the waste oil and the disposal costs of sludges and residues. Environmental pressures, with taxation or subsidy implications if required, may eventually cause a considerable amount of reprocessing capacity to be brought on stream.

2.1.4 Other Types of Base Stocks

Until the arrival of the jet aircraft engine in the late 1940s, petroleum-based lubricants performed adequately the services demanded of them, any minor shortcomings being accepted because of their low cost. For the majority of applications today, a cost/benefit analysis still indicates that petroleum base stocks which are extracted from crude oil have an advantage over chemical types which have to be specially synthesized. Hydrocarbon and other chemical fluids can, however, be tailored to meet critical requirements, and can provide superior performance over longer periods in cases where a lubricant is heavily stressed, or has to meet restrictive physical requirements. Often being constituted from single types of molecules and usually of restricted molecular range, they can provide combinations of properties (e.g., good low-temperature fluidity with good load carrying and low volatility) that the broad-cut complex mixture of a petroleum stock cannot attain. Sometimes fluids will be selected because they burn more cleanly or possess greater solvency than petroleum stocks, both properties which contribute to lower engine deposits when used in crankcase oils.[11, 12, 13]

Other types of base stocks are described below:

1. <u>Synthesized Hydrocarbons</u>
   The synthesis of petroleum substitutes from carbon monoxide and hydrogen via the Fischer-Tropsch process was commercialized in Germany in 1936, and provided that country's main source of fuel and lubricating oils during World War II. The process has also been operated in the U.S., and until recently has been a prime source of petroleum-type material for South Africa during the period of crude oil embargos. Fischer-Tropsch material has a broad compositional spectrum and is best regarded as substitute petroleum. Of greater interest from the formulator's viewpoint are synthesized hydrocarbons which have restricted compositional ranges and thereby benefit from the absence of both very light or very heavy material in their composition.

2. <u>Olefin Oligomers</u>
   These are polymers of moderate molecular weight, made from olefins derived from petroleum cracking. Polybutenes (from C_4 olefins) were developed in the 1930s, and higher-molecular-weight material was the basis of the first viscosity improver (see Section 2.2). Lower-molecular-weight versions have been used as lubricant thickening agents and synthetic oils for special purposes (e.g., electrical insulants).

Polyalphaolefins are products obtained by polymerizing higher olefins (typically C_{10} to C_{14}), which are usually obtained from wax cracking (Figure 27).

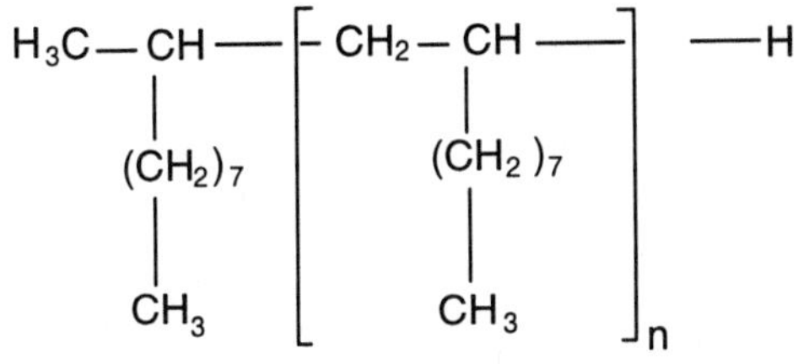

$$H_3C-\underset{\underset{CH_3}{|}}{\underset{(CH_2)_7}{\underset{|}{CH}}}-\left[CH_2-\underset{\underset{CH_3}{|}}{\underset{(CH_2)_7}{\underset{|}{CH}}}-\right]_n-H$$

Fig. 27 Structure of a polyalphaolefin.

For higher stability they are often hydrogenated to saturate any residual unsaturation, when they can be considered as very stable and very pure paraffin hydrocarbons. They have been widely used in the late 1980s as low-viscosity blending agents for multigrade oils, where their contribution to volatility is much less than that of a similar viscosity conventional petroleum stock.

3. Alkylated Aromatics
   These are principally the alkyl benzenes, which are commonly synthesized as intermediates in the production of synthetic sulfonic acids (Figure 28). The alkyl chain is typically made from tetrapropylene.

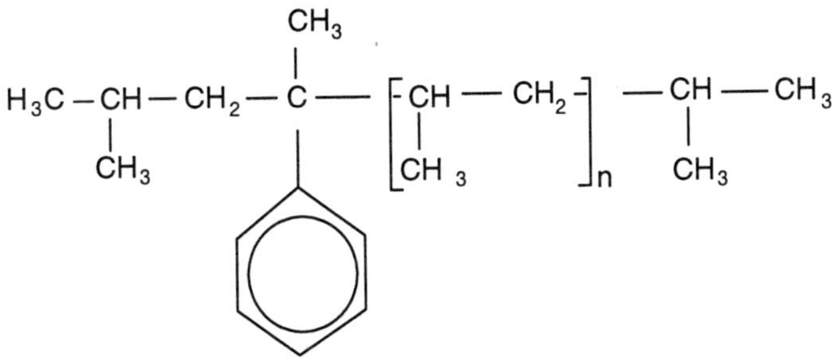

$$H_3C-\underset{\underset{CH_3}{|}}{CH}-CH_2-\overset{\overset{CH_3}{|}}{\underset{\underset{C_6H_5}{|}}{C}}-\left[\underset{\underset{CH_3}{|}}{CH}-CH_2-\right]_n-\underset{\underset{CH_3}{|}}{CH}-CH_3$$

Fig. 28 An alkyl benzene, derived from polypropylene.

Their low pour points and high solvency have made the alkyl benzenes useful in certain industrial oil applications (particularly in refrigerator oils), and also in two-stroke lubricants.

4. Organic Esters
Diesters (dibasic acid esters) formed the basis of the first jet engine lubricants, and were the components of most of the "semi-synthetic" motor oils marketed in the 1970s and 1980s. Esters are the product of combining organic acids with alcohols, and diesters are made from diacids such as adipic and azelaic acids with mono alcohols (C_8 to C_{12} chain length). These compounds are shown diagramatically in Figure 29.

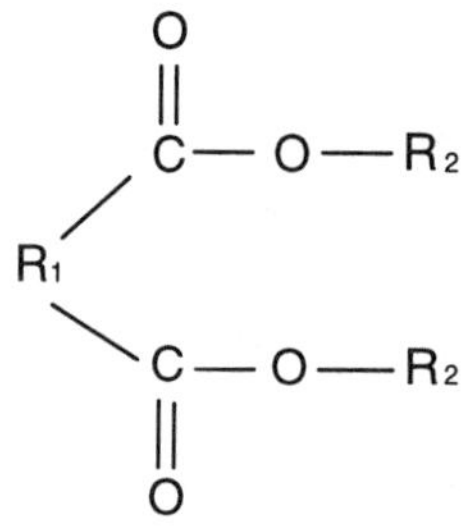

Fig. 29 The diester from $R_1(COOH)_2$ and R_2OH.

Phthalate esters are diesters based on the benzenoid structure of phthalic acid (1,2,benzene dicarboxylates). C_{36} dimer acid esters have been used for special applications. Triesters exist as the trimellitates (1,2,4,benzene tricarboxylates).

Polyol esters represent an inverse chemistry in which mono acids of 5 to 10 carbon chain lengths are combined with polyhydric alcohols such as glycols or the neopentyl alcohols. They tend to have more compact shapes than the diester molecules, providing greater thermal and mechanical stability than the typically long molecules of the diesters. Neopentyl esters are particularly stable, because their central carbon atom, around which four others are grouped, has no hydrogen atom attached. The structure of a neopentyl ester is depicted in Figure 30.

Neopentyl esters are expected to find considerable future use in compressor and refrigeration lubricants.

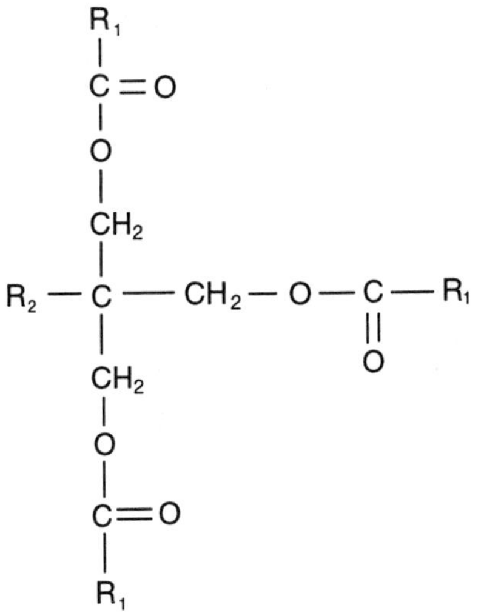

```
           R1
           |
           C=O
           |
           O
           |
           CH2
           |
    R2 —   C  — CH2 — O — C — R1
           |              ||
           CH2            O
           |
           O
           |
           C=O
           |
           R1
```

Fig. 30 Structure of a neopentyl ester.

Borate esters, from a combination of alcohols with boric acid, are displacing polyglycols in automotive brake fluids.

5. Other Fluids
   The most important of these are:

   - Polyglycols
     Also known as polyalkylene glycols (PAGs) and polyethers, these have been used extensively in industrial oils, and formed the basis of the early synthetic types of automotive brake fluids. Their polar nature confers good lubricity properties. They have good viscosity-temperature and low-temperature performance characteristics, but are available in a rather limited range of viscosities.

     They are manufactured by adding ethylene oxide and/or propylene oxide to a starting compound containing oxygen, normally an alcohol, glycol, or ether. This is done at moderate pressure and temperature in the presence of a catalyst. The structure is shown in Figure 31.

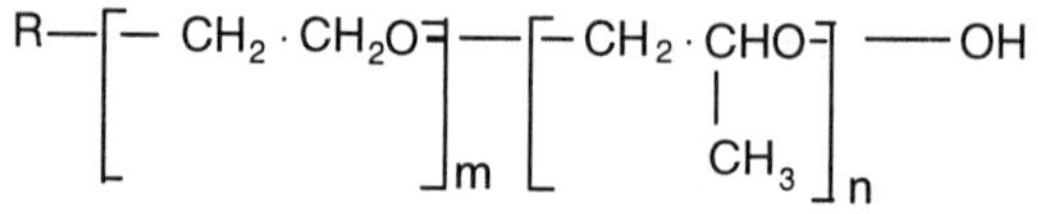

$$R-\left[-CH_2\cdot CH_2O-\right]_m-\left[-CH_2\cdot \underset{\displaystyle CH_3}{\underset{|}{CHO}}-\right]_n-OH$$

Fig. 31 A polyglycol based on ethylene and propylene oxides.

Types containing ethylene oxide are water soluble and have high V.I. They therefore find application in fire-resistant hydraulic fluids. A high propylene oxide content confers low pour point properties and water insolubility, and made them useful for brake fluids over many years.

- Silicones
  These are specialized fluids based on chains of alternating silicon and oxygen atoms with side chains appended. They can be cross-linked to form rubbery solids, and are also used as specialty lubricants in instruments, etc. High-molecular-weight versions are used in petroleum crankcase oils as foam suppressants. They are not normally miscible with hydrocarbon oils.[15]

- Phosphate Esters
  Declining in use because of their toxicity, these products formed the basis of fire-resistant hydraulic fluids for many years. (They are being replaced in such applications by water/glycol/polyglycol mixtures.) They are manufactured by reacting substituted phenols and/or cresols with phosphoric acid.

- Halogenated Hydrocarbons (e.g., Chlorofluorocarbons)
  The compounds based on methane and ethane have enjoyed extensive use as refrigerants, but high-molecular-weight versions can be useful lubricants. The halogen content of such fluids provides substantial EP and anti-wear properties in fluids of intrinsic low viscosity. Fears for the depletion of the ozone layer in the stratosphere by escape of chlorocarbons and chlorofluorocarbons into the atmosphere pose severe problems in the case of refrigerants, but should not be a problem for the less-volatile lubricant grades. However, the presence of chlorine in lubricants is now being

challenged on the grounds that on combustion dioxins may be formed. This could be either in an engine or through disposal by incineration. It appears unlikely that use of this type of synthetic will increase.

2.2 Additives

Oil additives are those materials which are added in small amounts to a lubricant to improve its properties. Small amounts of specialized base stocks, whether petroleum or synthetic, are not normally considered as additives but as blending components. Modern oils, and particularly motor oils, contain considerable quantities of additives as well as several base stocks. A premium crankcase oil may contain as much as 20% additive material, although it should be noted that this does not mean 20% of actual chemical materials, because the materials sold as additives are normally oil solutions of the active ingredients, which can represent as little as 5% of the total "additive" for some particular types.

Interestingly, the first additives could be considered to be mineral oil additions to natural vegetable or animal oils which were employed to extend the quantity and render these natural lubricants somewhat cheaper. When mineral oils began to be used in major quantities as lubricants, the situation was reversed and the first additives for mineral oils were in fact natural vegetable and animal oils which provided "lubricity" to the mineral oils, which were considered to be lacking in this rather ill-defined property. Today, we would interpret this as adding polar molecules to the mineral oil to enhance its boundary lubrication performance.

The first synthetic chemical additives were developed in the early 1930s by the Standard Oil Development Laboratories (now Exxon Research) to improve the physical properties of lubricants. Other types of additives followed from various suppliers, and by the end of the 1930s a significant industry had developed which received a considerable boost by the demands for improved lubricants brought about by World War II. The simplest way to consider the development of the various types of additives is

Plate 2. Additive manufacturing plant.

to deal with them separately under the classifications of functional types into which they are normally divided.

Pour Point Depressants

The first branded synthetic additive was Paraflow which was commercialized in 1932 as a means of lowering the temperature at which a mineral oil congealed. This temperature is called the *pour point*, and such additives are called *pour point depressants*. It was realized very early on that very little actual separation of wax from the oil could stop it from flowing, because the wax crystallized in large plates which linked together to form a retaining network preventing oil flow. Various petroleum extracts had been tried as a means of reducing the separation of wax, including the addition of some different waxy type material, but these had very limited success. Paraflow was a combination of the type of wax found in typical lubricating oils combined chemically with naphthalene. When the oil is cooled and wax commences to precipitate, the waxy constituent of the additive crystallizes together with the wax from the base stock, but the bulky naphthalene constituent interrupts the crystal surfaces and prevents continued crystal growth. The wax is therefore precipitated in small crystals and does not form an interlocking structure, so the oil is able to continue to flow despite the wax floating within it. It is important to realize that as a pour-depressed lubricant or base stock is cooled, at a certain temperature (the *cloud point*) wax still precipitates out from the oil and is visible as turbidity, which increases as the oil is further cooled, until the pour point is finally reached. Pour point depressants can produce very large reductions in the pour point temperature of lubricants, but some types are more effective in certain base stocks than others. Little or no effect is seen on the cloud point, and they are of course completely ineffective in naphthenic or synthetic base stocks which contain no wax.

Exxon Paraflow was followed shortly afterwards by Santopour from Monsanto, and during World War II the Acryloid polymers were developed by Rohm and Haas. These were long chain polymers of alkyl acrylates or methacrylates derived from higher alcohols, and the majority of pour depressants in use today are various polyester materials which are now made by a wide selection of manufacturers. The mechanism of operation remains the same, however, namely co-crystallization with the wax in the base stock but interference with crystal growth and the prevention of agglomeration.[16]

Viscosity Modifiers

The viscosity temperature characteristics of base stocks were first defined in terms of a Viscosity Index by Dean and Davis in 1929.[17] With a measurable property to target, interest was soon aroused in the possibility of improving the Viscosity Index (V.I.) of base stocks. Initially, thickening thinner base stocks with colloidal soaps was tried but was not very effective. Exxon's second pioneering additive was a polymer made from refinery butenes (unsaturated C_4 molecules). This polybutene was called Paratone and permitted the production of lubricants of 120 V.I., sufficient to allow year-round operation for an oil in suitable climates. The term "Viscosity Index Improver" was applied to such additives for many years, but today the expression "Viscosity Modifier" is preferred.

Viscosity modifiers increase the V.I. because they are more soluble in the base stock at high temperatures than at low temperatures. At high temperatures the polymer chains are said to be solvated, which means that they are surrounded by lubricant base stock molecules and extend into the oil, interfering with its flow and thereby increasing the viscosity considerably. At low temperatures, however, the polymer is less solvated and tends to be attracted to itself rather than the base stock molecules, and therefore forms small coils or clumps of polymer which interfere less with the flow of the oil (Figure 32). While the oil itself has thickened due to the reduced temperature, the added thickening from the polymer is less than that found at higher temperatures.

For a high-V.I. multigrade oil the concentration of polymer in the base stock may be from 0.5% to 2%, although it will usually be supplied by the additive company as oil solution. This may contain anywhere from 6% to 60% of polymer, necessitating higher oil treatments of the purchased additive.

Fig. 32 Effect of viscosity modifier polymers in oil (a) at low temperatures and (b) at high temperatures.

It must be pointed out here that the improvement of viscosity/temperature properties produced by viscosity modifiers, and the ability to manufacture so-called multigrade oils, is related very much to the test methods used to measure the viscosity, particularly the low-temperature viscosity. Various new ways of measuring low-temperature flow have been introduced over the years, and products differ in the way they perform in the various tests.

For such reasons polybutene is not now widely used as a viscosity modifier for motor oils, although a successor olefin co-polymer based on ethylene and propylene enjoys considerable success. Polymethacrylates were developed for V.I. improvement in the late 1930s, to be followed by polyacrylates and polyfumarate co-polymers. Many of these polymers were modified by special chemical side chains to make them multi-functional, e.g., providing them with pour depressancy or dispersancy properties as well as viscosity improvement. Other polymers have been introduced subsequently, notably polyalkylstyrenes in 1948, and in the 1970s a new generation of polymers including ethylene-propylene co-polymers, polyisoprene, polybutadiene, and styrene-butadiene co-polymer. New polymers continue to be introduced from time to time. The structures of some of the key polymer types are shown in Figure 33.

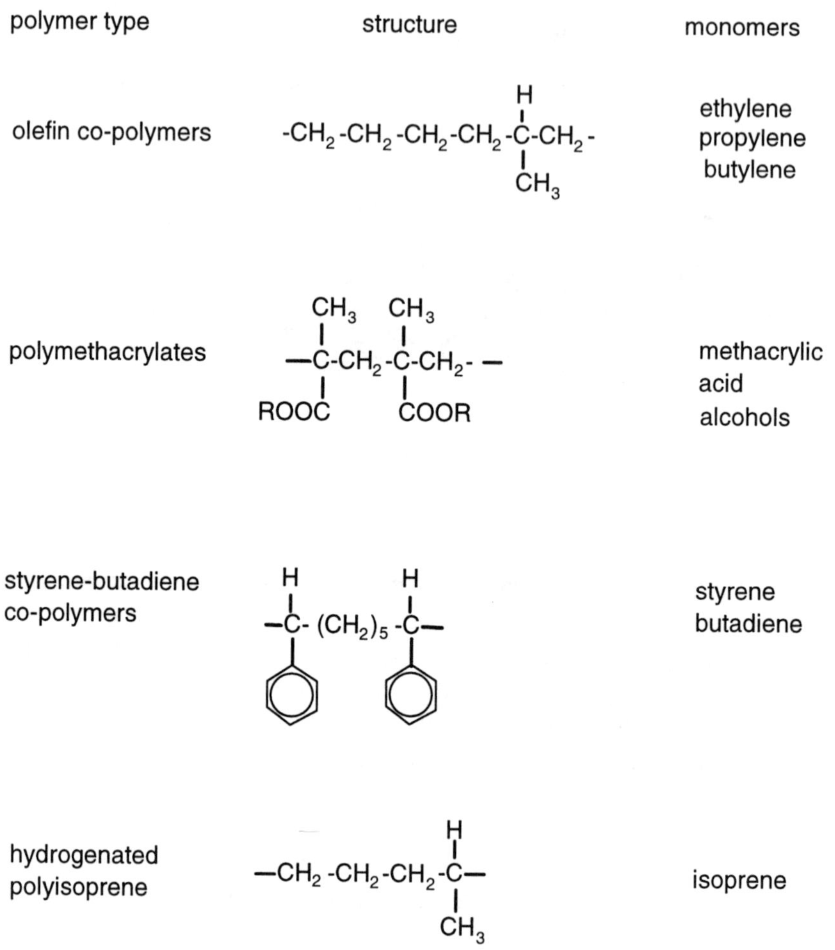

polymer type	structure	monomers
olefin co-polymers	$-CH_2-CH_2-CH_2-CH_2-C(H)(CH_3)-CH_2-$	ethylene propylene butylene
polymethacrylates	$-C(CH_3)(COOR)-CH_2-C(CH_3)(COOR)-CH_2- -$	methacrylic acid alcohols
styrene-butadiene co-polymers	$-C(H)(C_6H_5)-(CH_2)_5-C(H)(C_6H_5)-$	styrene butadiene
hydrogenated polyisoprene	$-CH_2-CH_2-CH_2-C(H)(CH_3)-$	isoprene

Fig. 33 Structure of some viscosity modifier polymers.

The search for new polymers in the late 1960s and early 1970s was sparked by the interest in developing highly potent stable molecules which would not cause deposit problems in diesel engines due to their decomposition at high temperatures. The polyester products suffered because they tended to require high active ingredient treat levels (i.e., they were not particularly potent molecules) and were somewhat chemically unstable. In addition there was a reduction in the temperature at which the low temperature performance was measured, and the concept of V.I. for classifying motor oils was dropped (although it has remained useful for categorizing base stocks and industrial oils). The new polymers were not as efficient as V.I. improvers, but were good high-temperature thickeners, more thermally stable and with good low-temperature properties. From this time, the expression "V.I. Improvers" was replaced by "Viscosity Improver" or "Viscosity Modifier," the Viscosity Index no longer being of significance in motor oils.

While oil-thickening potency versus molecular treat level is very important for controlling diesel engine deposits, the shear stability of the resultant oil is an equally important parameter. Thickening potency is highest for simple straight chain molecules which contain the minimum side chains for conferring oil solubility. Unfortunately, long straight chains are liable to break in two under strong shearing action, so a compromise between diesel debit (minimized by potent long straight chains) and shear stability (maximized by short chains) has to be made when selecting a polymer as a viscosity improver. One interesting way of attempting to make this compromise is the adoption by Shell of a star-shaped aggregation of polyisoprene molecules, which has reasonable potency and relatively short multiple side chains. Most other varieties of the new polymers balance chain length versus potency, with different chain lengths being available for different uses.

As well as permitting year-round use for one grade of oil, polymer-improved oils have been found over the years to reduce engine friction (thereby improving fuel consumption) and also, provided the base stock is not unduly volatile, to improve the oil consumption compared to an equivalent non-multigrade oil.

Anti-Oxidants

In the presence of air, oil oxidizes at high temperatures, darkens, becomes acid, and produces sludge. Unrefined or mildly refined base stock contains certain natural inhibitors or anti-oxidants which are progressively removed as the severity of refining is increased.

In the early days of petroleum refining, base stocks were acid treated to produce high-quality material for specialized uses such as for turbine oils and electrical oils. These heavily refined stocks were found to be very subject to oxidation, and in the 1920s work in Paris and at M.I.T. showed that certain phenolic and/or amine products could inhibit the oxidation of such oils. These types of compounds still find use today, although many different varieties of anti-oxidants have been developed and marketed since their early use in the specialized oils.

Early motor oils had little or no refining and so retained their natural inhibitors. In addition, motor oil was changed frequently because it was very cheap and also because it became rapidly contaminated with combustion residues. As the interest in high V.I. oils increased, motor oils were more severely extracted, and particularly after hydrofinishing was introduced, were often severely hydrogen treated to increase the V.I. These processes removed most of the natural inhibitors and the base stocks had a reduced oxidation stability. With a desire to increase the oil change period for motor oils, the need for oxidation inhibition in motor oils became imperative.

Developments in other areas had already resulted in compounds which improved other properties but which also provided anti-oxidant capability at the same time. Some early premium grades of motor oil contained small additions of ZDDP (see Anti-Wear Additives below) to provide some anti-wear and also some anti-oxidant properties. Diesel oils were also utilizing more and more detergent additives, many of which contained sulfurized components which provided anti-oxidant capability of a sufficient level. Detergent-inhibitors later became common in gasoline engine oils, and it is only in the last fifteen years or so with increasing oil drain intervals and more severe oxidation requirements that it has become necessary to add specific oxidation inhibitors to motor oils to supplement the inhibition coming from other types of additives. Amines and substituted phenols are typically employed for use as such supplementary anti-oxidants.

The use of traces of copper as an anti-oxidant is of some interest. Copper has long been considered a pro-oxidant for lubricating oil and in many situations this remains true. However, in motor oils and particularly in association with other motor oil additives, a few hundred ppm or less of copper has been shown to have remarkable anti-oxidant properties. The use of this low-cost inhibitor is very attractive to formulators who have access to this technology. One disadvantage is that when used oil analysis for trace metals is being performed, allowance has to be made for the copper introduced as anti-oxidant.[18]

Detergents and Detergent-Inhibitors

The definition of these types of additives is somewhat vague, but we would like to define a detergent additive as one whose primary function is to minimize deposits in the hot parts of an engine, particularly on the piston and in the piston ring grooves. A detergent-inhibitor is an additive which functions as a detergent but also contains components that provide inhibition against oil oxidation and bearing corrosion. Detergents, particularly the sulfonates, also have some ability to keep soot and other contaminants dispersed in the oil. The mechanisms of deposit formation and prevention were discussed in Section 1.7.2.

In 1930 Rosenbaum patented metal ricinoleates as additives for lubricating oil to promote engine cleanliness. At this time engine power ratings were relatively modest and no great use was seen for such additives. However, in the mid-1930s, the Caterpillar Tractor Company introduced a new range of medium-speed diesel engines for their earthmoving equipment. These were high-output engines that immediately ran into problems of piston deposits and ring sticking. After a few hundred hours of operation, engines lost power and had to be disassembled and pistons rings freed in the grooves. This was done by hand cleaning in baths of oil or solvent, and it is said that some users found cleaning to be easier if "detergents" in the form of metal soaps were added to the cleaning oil. (This may or may not be the origin of the term detergent in relation to promoting piston cleanliness, but it is now clear that lubricating oil detergent additives do little to remove existing deposits in an engine, although these may be lost by mechanical means. Detergents as we now know them serve not to clean away deposits, but to prevent them from forming in the first place by combating combustion-derived contaminants.)

With an unacceptable problem on their hands, Caterpillar sought the help of the research division of Standard Oil of California (SOCAL) and, in 1935, commenced marketing for use in their tractors their own brand of lubricating oil which contained aluminum dinaphthenate as a detergent. This detergent oil relieved the problem considerably, and the concept of using detergent additives in oils for diesel engines immediately took off. Most oil companies and some chemical companies joined a race to produce better detergents, one small oil blender called Lubri-graph enjoying such success that they eventually became the largest producer of detergent additives, while remaining outside the circle of the major oil companies. This company's name today is Lubrizol.

SOCAL, whose additives division later became Oronite, followed up their initial naphthenate chemistry with calcium phenyl stearates and calcium cetyl phenates, while Union Oil and Lubrizol produced calcium dichloro-stearates. (Stearate chemistry was later abandoned due to bearing corrosion problems.) Exxon discovered the utility of simple phenates in 1938, being followed by developments from Union Oil, SOCAL, and Mobil. Standard Oil of Indiana, later Amoco, developed phosphonates in the early 1940s to be followed a decade later by thiophosphonates from Union Oil, Lubrizol, and Exxon. Shell developed their unique salicylate chemistry in 1955 and engaged Lubrizol to manufacture large quantities on their behalf. Salicylates share some of the characteristics of both phenates and naphthenates. Overbased versions (see below) of these different chemical types were developed in due course.

In the 1950s there was a tendency to produce complex detergent-inhibitors of uncertain composition by combining several of these compounds, sometimes with different metals, and overbasing and sulfurizing the lot to finish up with a potent "witch's brew" of very strong inhibiting properties. These mixtures, frequently with a large thiophosphonate content, were excellent detergent additives for gasoline or mild diesel engines but tended to lack the thermal stability required for the more severe diesel tests. With increasing interest in multi-purpose oils suitable for both gasoline and severe diesel engines their use has steadily declined and is now very low.

It is worth noting here that many additives, even if of essentially simple composition (one type of compound), possess multi-functionality. That is to say the same chemical can have more than one effect in a lubricant. Thus compounds which combine viscosity modification with pour depressancy, detergency with anti-oxidant properties, anti-wear with anti-oxidant properties, and dispersancy with viscosity modification are all well known.

In today's formulations the most common detergent-inhibitors are over-based sulfurized phenates, or their close relations, the salicylates. These have medium/high basicity, and the phenate structure provides anti-oxidant properties. Additional anti-oxidant performance and bearing corrosion inhibition is given by the sulfur content. The structure of a metal phenate is illustrated in Figure 34, and of a salicylate in Figure 35.

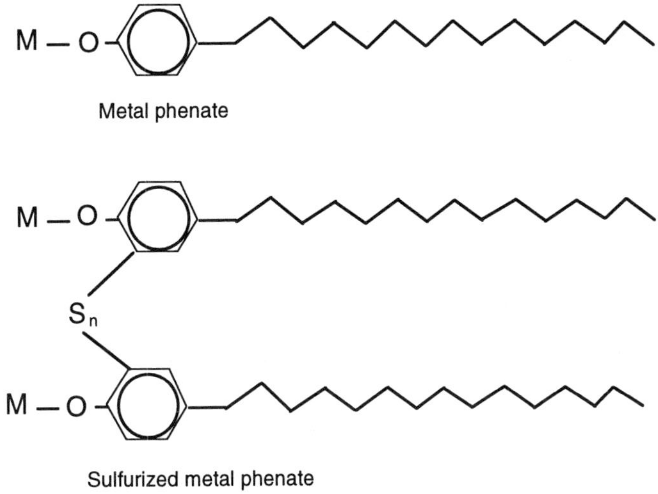

Metal phenate

Sulfurized metal phenate

Fig. 34 Structure of (a) metal phenate, (b) sulfurized phenate.

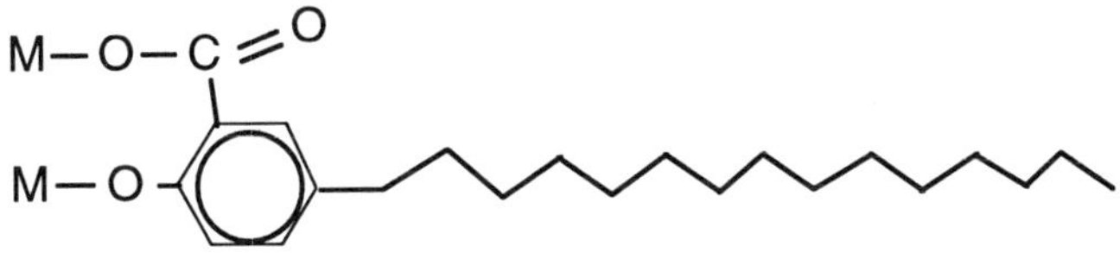

Fig. 35 Structure of a metal salicylate.

These diagrams and those that follow are intended to give an idea of the structures of the organic radicals concerned. The metal atoms, shown as M in each case, may be monovalent (e.g., sodium) or more likely divalent (e.g., calcium or magnesium). It is important to realize that the bonding (see Appendix 3) between the organic radicals and the metal is largely ionic, so that a radical is not permanently associated with a given metal atom, but can migrate and exchange with others. Furthermore, in the case of a divalent metal, the second valency may be associated with another radical of the same type, a radical of a different type, or linked to inorganic radicals such as hydroxide or carbonate.

The picture is therefore complex for a simple monobasic radical such as a phenate or sulfonate (see below), but the sulfurized phenate and the salicylate are effectively dibasic. They offer increased possibilities for association with metal atoms, but in addition also frequently give rise to the situation where only one of the functional phenolic or acidic groups is effectively reacted with metal. In practice, therefore, it is not possible to indicate single chemical structures for detergent additives, and they are defined in terms of the ratios of the metal to organic and inorganic radicals, on an overall basis and without considering how they are bonded together.

Not all detergents have to be inhibitors, however, and the oil companies realized by the mid-1940s that a by-product of the acid refining of lubricating oils (sodium sulfonate) could be used to produce calcium and barium sulfonates which were effective detergents. These tended to promote oil oxidation rather than reduce it, but were excellent for suspending soot and decomposition residues in the oil. Many of the sulfonates produced had a so-called "alkaline reserve" which implied that they contained excess alkali which could be used to neutralize combustion-generated acids, particularly sulfuric acid in the case of diesel engines. Later the technique of "overbasing" sulfonates and phenates to yield very high alkali reserves (expressed as base numbers of 150-400 T.B.N.) were developed by key manufacturers such as Shell, Lubrizol, Union Oil, Mobil, Texaco and Exxon. Overbased additives are excellent for neutralizing combustion acids, thereby preventing metal corrosion (and particularly cylinder bore corrosive wear) and preventing acid-catalyzed oil degradation and degradation product polymerization. Sulfonates are also good anti-rust additives.

The structure of a sulfonate is illustrated in Figure 36, while Figure 37 illustrates overbasing.

$M{-}SO_3$

Fig. 36 Structure of a metal sulfonate.

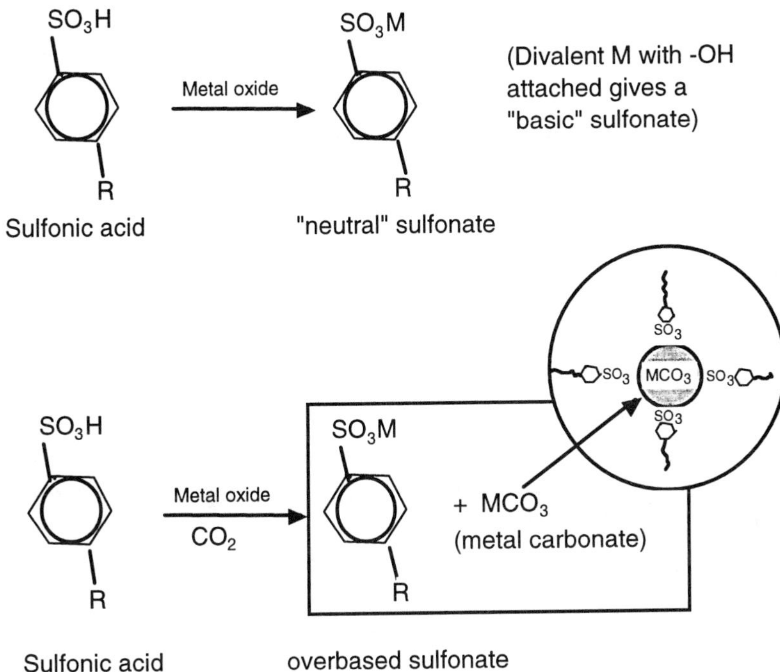

Fig. 37 Overbasing of detergent-inhibitor with metal carbonate.

Barium compounds were initially considered to be superior detergents to their calcium analogs. However, a barium detergent produces a greater weight of ash when burned than its calcium or magnesium analogs, and this, coupled with toxicity fears in Europe, has led to a major decline in the use of barium additives since the 1970s. Magnesium additives, particularly the highly overbased 400 T.B.N. magnesium sulfonate, have become increasingly popular in the last 20 years. Originally conceived as an additive for boiler fuels, magnesium sulfonate has been widely adopted in crankcase oils because of its high T.B.N. and its relatively low ash content. Magnesium phenates followed magnesium sulfonates into the armory of available detergent additives in the mid-1970s. Sodium-based additives (particularly thiophosphonates and sulfonates) have been used as detergents. They are very effective, but because of their high surface activity, care has to be taken when formulating with such materials, and they are not widely used.

Dispersants

It was recognized early on that one of the functions of a detergent additive was to suspend fuel soot in the oil, to prevent it from agglomerating and depositing on the hotter parts of the engine. Many soaps, and especially the sulfonates, also suspend and solubilize water of combustion and reduce sludge formation. A gasoline engine usually runs cooler than a diesel engine and also less efficiently, and the oil therefore tends to have higher levels of water contamination. When the detergents developed for diesel engines were initially tried in gasoline engines they had little effect on sludge control, and also caused problems of valve train cam and tappet wear. This latter problem was eventually solved by the use of anti-wear additives (see Anti-Wear Additives), but the sludge problem required the development of a new class of additives capable of dealing with so-called "cold sludge." These are the ashless dispersants whose prime function was to disperse water and render it harmless, but which are now also used in diesel engines and can also disperse other contaminants and act as mild detergents.

In 1952 Du Pont patented various nitrogen-containing polymers of methacrylates and co-polymers of methacrylates and fumarates as additives for controlling gasoline engine sludge.[20] They were followed by Rohm and Haas with different methacrylates and Exxon with fumarates, and other companies with similar types of polymeric compounds. These additives were multi-functional and not only suspended water but also acted as viscosity modifiers[21] and in some cases also as pour depressants. These early polymers were not very thermally stable, and new materials, based on polyisobutylene reacted with maleic anhydride and neutralized with polyamines or polyalcohols, were developed by Shell, Oronite, Lubrizol and others. These dispersants were more efficient as dispersants than the earlier polyesters,[22] but were also capable of being developed to levels of thermal stability where they could be used in diesel engines. This became common in the 1960s for the lower-output diesel engines, and by the 1970s high-power supercharged diesels were using large doses of the more thermally stable ashless dispersants both for sludge suspension and as ashless detergents. The "succinimides" based on polyamines became the dominant type. The concept of multi-purpose oils which can satisfactorily lubricate both gasoline and diesel engines was born at this time, when it became possible to formulate an oil to meet the then highest standards of diesel performance and the highest standards of gasoline performance at the same time. Alternative types of dispersant based on polyhydroxy compounds (such as pentaerythritol) or amino-hydroxy compounds are still sometimes used and are in effect polyisobutenyl succinate esters. These function in the same way as the purely nitrogen-based materials, having the polar head based on oxygen or oxygen plus nitrogen rather than on nitrogen alone. They were fashionable for a time and were favored by some formulators as at least part of the dispersant treatment. Their use is, however, much less than that of the polyamine-based types, in which must be included various levels of amine-to-acid ratios as well as the strict ratio for 100% succinimides, a cyclized structure.

Various other ways of coupling a hydrocarbon chain to a polar entity have been tried, of which the most successful has been by the Mannich formaldehyde process. Different additive suppliers have developed their own preferred chemistry and optimized their products in ways which leave them free of patent domination and able to formulate successfully. A diagrammatic illustration of dispersant structure and mode of action is shown in Figure 38.

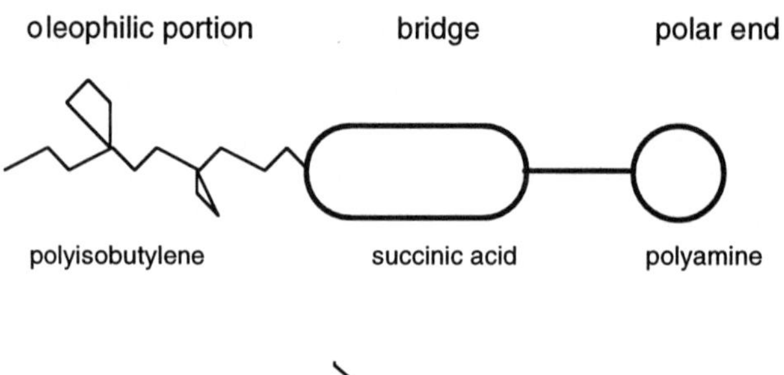

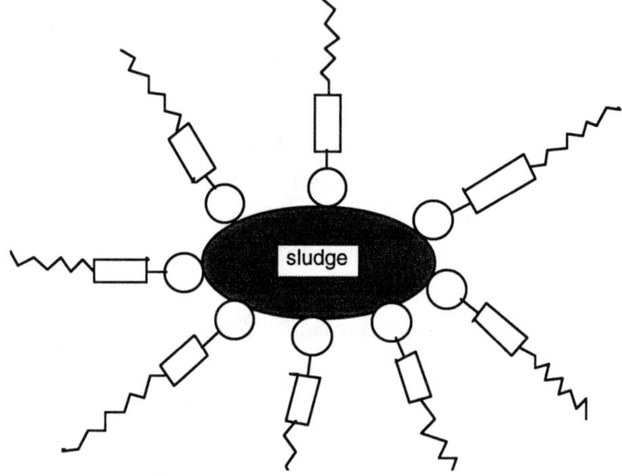

Fig. 38 Dispersants and their action.

Detergent Formulations

From the above it will be seen that the lubricant formulator has a wide choice of possible compounds which will reduce deposits and prevent corrosion and wear. For commercial and patent reasons every formulator will not have access to every type of additive, but the following table shows that a wide choice exists:

1. To inhibit oxidation (usually providing bearing corrosion protection as well)

    Phenates (especially metal phenates)
    Sulfurized phenates
    Phenolic compounds

Amines
Salicylates
Phosphonates
Thiophosphates (incl ZDDP)
Sulfurized oils and many sulfur compounds
Carbamates
Copper compounds

2. To prevent acid corrosion and acid catalysis

"Overbased" additives	- by neutralization
Sulfonates	- by sequestration/solubilization
Dispersants	- by sequestration/solubilization
Neutral phenates/ Other salts of weak acids	- exchange of strong acids for weak

3. To suspend acids and fuel residues (including water)
Metal soaps (especially sulfonates)
Dispersants (e.g., succinimides)

Mixtures of several of these compounds will be required for a complete detergent formulation and the choice of which types to use will depend on the specific specification requirements for the oil and the most economic way of meeting them for a given formulator. The economics as seen by the formulator will involve the required treat level of each additive versus its intrinsic chemical costs, whether or not he has patent rights, and whether or not he manufactures the components in-house. Some typical choices will be discussed in Chapter 4 on the formulation of lubricants. Other additives, as discussed below, will be required to complete the formulation, and these may have beneficial, harmful, or neutral effects on the detergent performance.

Anti-Wear Additives

The boundary lubricant additives referred to in the previous sections (usually natural acids or their esters) provided sufficient anti-wear activity in the early days of non-detergent gasoline engine oils, while diesel engines were more massively built and operated at lower speeds and so did not initially exhibit any wear problems. Bearings were initially made of the soft conformable babbit material (a tin/copper/antimony alloy) which was relatively inert chemically and had the capacity to absorb small amounts of foreign material. However, as engine power outputs grew, babbit ceased to be strong enough to bear the increased loadings on the bearings, and harder bearings of cadmium/silver, cadmium/nickel and copper/lead were developed. These materials were strong but were not as chemically inert as babbit, and were attacked by the acids generated from oil oxidation. They were also unable to absorb into the bearing surface foreign material such as grit and wear debris, and in consequence improvements in filtration were developed. In the 1930s organic acid inhibitors, bearing corrosion inhibitors and various anti-wear agents were developed to protect these bearings, many of the compounds being multi-functional and providing protection of bearings against both corrosive and mechanical wear. These compounds included organic phosphates, sulfurized sperm oil, dithiophosphates, dithiocarbonates, and culminated in 1941 with Lubrizol's development of zinc dialkyldithiophosphates (ZDDP). Initially used at low concentrations (0.1% to 0.25%) as a bearing passivator and oil anti-oxidant, ZDDP was soon found to be remarkably effective as an anti-wear additive. The anti-wear activity extends from boundary lubrication up to true EP activity for heavily loaded steel-on-steel sliding mechanisms. The chemical structure is shown in Figure 39.

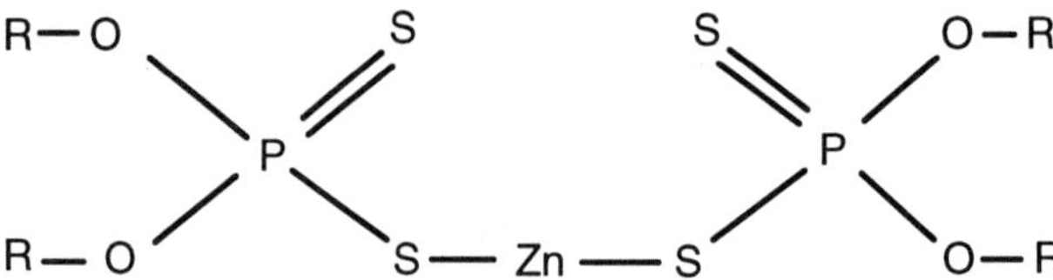

Fig. 39 Structure of ZDDP.

In a gasoline engine the valve gear is heavily stressed due to the high engine speeds producing high sliding speeds between cams and tappets, which are basically poorly lubricated. For systems without hydraulic backlash take-up (the norm in Europe until the 1990s), high impact loads also resulted from the reaction between cam-follower (tappet), the pushrod, and the rocker. Impacts in this chain leading from the cam to the valve stem were increased in severity as valve spring loadings increased, and also mechanical resonances in this cumbersome system tended to produce high instantaneous loads at the contacting surfaces. For the early non-detergent gasoline oils, small additions of additives such as ZDDP were sufficient to provide anti-wear protection for these various parts, which are often called the "valve train." However, when attempts were made to introduce detergent additives into gasoline engines, or to use diesel lubricants containing detergents in gasoline engines, there were many failures with heavy wear resulting particularly in the cam and tappets. The initial reactions were that these detergents were either chemically attacking the metal, or that their apparently colloidal metal compounds were actually abrading the surfaces. This is now known not to be the case, and the effect is due to the highly surface-active nature of the detergents which causes them to compete strongly for possession of the metal surfaces with boundary layer and anti-wear additives or natural lubricity compounds in the oil. As the majority of detergents do not themselves have significant anti-wear capability, the surfaces become relatively unprotected and wear takes place where loadings are heavy. To overcome this the concentration of ZDDP, or other anti-wear additives, must be increased substantially in order for it to compete successfully with the detergent and obtain some measure of occupation of the metal surfaces.

Today ZDDP is the predominant anti-wear additive used in crankcase oils, although it is in fact a class of additive rather than one particular chemical. The solubilizing groups which enable the metal dithiophosphate to be soluble in oil can either be alkyl (straight or branched chains) or aryl (aromatic rings). The anti-wear activity (or rather the sensitivity of the additive to commence giving anti-wear protection) varies inversely with the thermal stability of the particular structure. This increases with carbon number and in the order secondary alkyl (the least stable and the most potent) through primary alkyl to aryl types (the most stable but least potent). Diesel engines run considerably hotter in the ring zone than gasoline engines, and ZDDP decomposition tends to produce lacquer in this area. On the other hand, diesel engines because of their design and metallurgy tend to have less wear problems than gasoline engines, so that for a simple diesel oil a more stable but less potent type of ZDDP can be tolerated. However, when formulating multi-purpose oils for use in gasoline engines, high-speed passenger car diesel engines, and larger diesel engines, it is necessary to select carefully between the possible ZDDP types available, and sometimes to use balanced mixtures of two or more types. In some countries restrictions on lubricant phosphorus content, caused by concern for exhaust catalyst poisoning, can limit the level of ZDDP which can be used.

The "extreme pressure" additives used in lubricants such as gear oils are also in essence anti-wear additives, although frequently treated as a separate category. The term "extreme pressure" was coined by the SAE in the 1920s to reflect the heavy loadings on gear teeth, particularly in hypoid back axles. These were developed by Packard in 1926, and by the 1930s were in wide-scale use after suitable lubricants had been developed. Lead soaps were initially tried as additives but required additional sulfur- and/or chlorine-containing materials to control the wear sufficiently. Some of these compounds proved corrosive to bearing or bushing materials and required careful selection and possibly special inhibition. Sulfur/chlorine/lead additives persisted for many years, but new compounds were developed based on sulfur and phosphorus which were good anti-scuff and anti-wear agents yet were thermally stable and did not promote corrosion of components or rusting of steel cases in moist conditions. Although not now used significantly, ZDDP featured in some of these additive mixtures in the 1950s and the 1960s. Lubrizol and Monsanto were companies particularly active in the area of gear oil additives, but most petroleum additive companies developed some formulations. Various EP additives were developed in the work on gear oils which were later used in other areas such as in industrial and metal-working oils. Gear oil formulation is discussed in more detail in Section 7.1.

Friction Modifiers

Originally a minor class of additives, these have recently come into more prominence owing to the desire to improve the fuel economy of vehicles, and the discovery that such additives can make a small but significant contribution to this. There are two types of friction modifier:

1. *Friction reducers*—These include the boundary lubricant additives and extensions of this technology to provide increased "slipperi-

ness" and lower friction coefficients. Typical materials would be long chain molecules with highly polar groups to anchor the molecule to the metal surfaces. The best products result in lower heat generation and therefore less wasted energy, hence their application as fuel economy additives. Related materials have also been employed for many years to reduce the coefficient of static friction relative to dynamic friction in automatic transmission fluids. This results in smoother gear changes but can result in excessive slip and wear of the various clutches if taken to extremes.

2. *Friction enhancers*—In contrast to the requirements of General Motors, Ford's automatic transmissions (until the mid-1970s) required a fast lock-up to avoid excessive clutch slip and consequent wear. This was provided by the requirement for high static coefficient of friction, and permitted both longer fluid life and the development of more compact transmissions. (The somewhat harsh gear changes that resulted, and the development of universal automatic transmission fluid technology, have recently led to modifications to Ford's original requirements.) Additives to enhance static and low sliding-speed friction include detergents, sulfides, and other types which are attracted to the rubbing surfaces but provide little lubricity in themselves. By carefully balancing the combinations of anti-wear and friction-reducing additives with those that promote high coefficients of static friction, it is possible to formulate transmission fluids that provide a positive lock-up when the clutches are applied but one which is not too harsh. It is important that these properties are retained over a long service life for the lubricant and just not for the initial properties of new fluid, so a careful selection of possible additives must be made.

Pro-friction or "traction" additives are also required for certain other applications, ranging from the wheels of railroad locomotives to precision instruments. In general the mechanism is the same, namely providing additives to cover the metal surfaces without providing a high degree of lubricity.

Rust and Corrosion Inhibitors

As well as the inhibition of bearing corrosion which was discussed under the heading of detergent inhibitors, it is also necessary to protect an engine against rusting in the areas where water tends to collect, for example under the rocker-cover and in the sump. Other metals which may be found in an engine (aluminum, magnesium, copper alloys, etc.) may also tend to suffer from corrosion under conditions of excess moisture or acidic by-products of combustion. Detergent additives which neutralize acids and detergent or dispersant additives which keep water in suspension will help considerably in overcoming these problems, but if necessary other booster additives may be added. These may often be surface active materials with a very high affinity for metal surfaces and which allow hydrocarbon molecules close to the surface while preventing the ready access of water which would cause corrosion (Figures 40 and 41). Copper alloy corrosion is countered by typical bearing corrosion inhibitors or specialized additives which coat the alloy surfaces with a thin layer of inert and strongly adhering reaction product. Triazoles and thiodiazoles are two families of modern general-purpose corrosion inhibitors.

For rust inhibition of steel parts (such as automobile panels) in storage or transit, the traditional recipe of a mixture of lanolin (degras) and sodium sulfonates is being replaced by special surface active compounds dissolved in oil, which again coat the material and prevent moisture from having

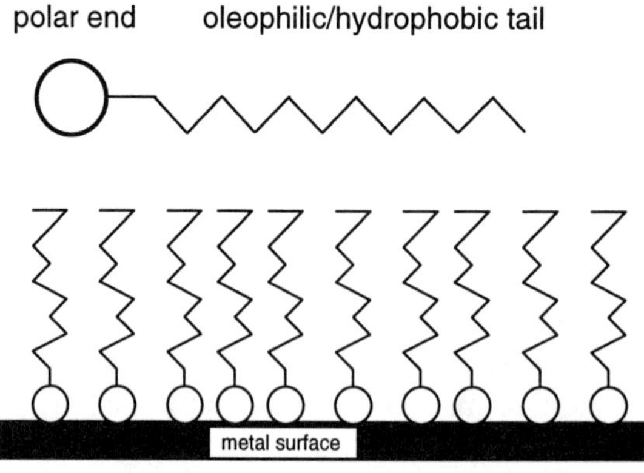

Fig. 40 Action of a rust inhibitor.

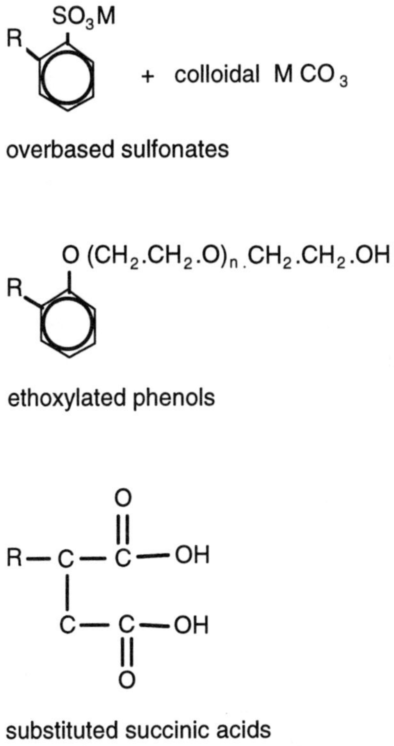

R–C₆H₄–SO_3M + colloidal M CO_3

overbased sulfonates

R–C₆H₄–O $(CH_2.CH_2.O)_n.CH_2.CH_2.OH$

ethoxylated phenols

R—C—C(=O)—OH
|
C—C(=O)—OH

substituted succinic acids

Fig. 41 Typical rust inhibitors.

access to the surface. Some of these materials are based on oxidized wax, providing another reason for the retention of conventional dewaxing capacity in base oil plants.

Emulsifiers

Dispersant additives effectively emulsify relatively small amounts (a few percent) of water from combustion into the lubricating oil to prevent it from producing sludge. For lubricating emulsions, such as are used for metal-working and increasingly for hydraulic oils, more comparable proportions of water and oil are required. Emulsions can either be of oil-in-water or of water-in-oil (invert emulsions). Oil-in-water emulsions tend to be used when the primary requirement is for cooling (as in metal turning operations), while water-in-oil emulsions tend to be used when the need for lubricity is much greater. Contamination or bad working practices can lead to the emulsion breaking, or one type of emulsion reverting into another with possibly serious consequences.

Many types of chemicals can be used as emulsifiers but the common characteristic is that they have one part of the molecule that is oil soluble linked to another part that is water soluble (Figure 38). The type of emulsion formed when mixing oil, water and an emulsifier depends on the precise chemistry, including such factors as the polarity of the molecule and its ionization, and the length of the oil-soluble hydrocarbon chain. Sodium sulfonates are the most widely used industrial oil emulsifier. For special applications non-ionic emulsifiers such as ethoxylated alcohols may be used.

Demulsifiers

These are chemicals which have the property of breaking an emulsion, separating it into water and oily layers. The largest outlet is for breaking crude oil emulsions often produced in the oil field. A future large outlet may be in the precipitation of oil from metal-working emulsions for conservation reasons, but there is a tiny outlet in the motor oil additive business which is of interest.

The key role of dispersants is to emulsify small quantities of water into the bulk of the engine oil. However, in engines running under cold conditions quite large amounts of water can condense in the cold rocker-cover area and the breathing tubes associated with it. Dispersants may well in this case permit the formation of oil-in-water emulsions which can have the appearance of a dense white sludge. Emulsion sludge can partially block the air ventilation system, adversely affect emissions, and is generally regarded as undesirable, although few if any cases of actual lubrication failure have been reported due to its presence.

There are two ways of overcoming the problem. It has been found that by adding small quantities of auxiliary emulsifiers to the crankcase oil the formation of this invert emulsion can be prevented. These emulsifiers are in many ways similar to but distinct from the conventional dispersants. They are sometimes known as demulsifiers because their presence prevents the formation of this emulsion sludge, and if already present it will slowly disappear when such materials are present in the crankcase oil. Another approach is to use a chemical to break the oil-in-water emulsion into its constituent parts, with the water passing into the crankcase where the high level of agitation will allow it to be emulsified by the dispersants. Such additives are true demulsifiers and do not appear to have deleterious effects on the dispersancy at the concentrations used.

Anti-Foam Additives

Many lubrication systems, and especially that of a traditional wet sump engine where the crankshaft beats air and oil together, can lead to the production of a layer of foam on the surface of the lubricant, particularly in the presence of detergents or other surface-active additives. For crankcase oils, very small quantities of silicone oil added to the lubricant can reduce the foam remarkably. Silicone oil is essentially insoluble in mineral oil and globules of silicone on a microscopic scale help to keep the oil film around the air bubbles sufficiently thick to permit rapid oil drainage and consequent collapse of the foam. In the production of many detergent additives silicone is added to prevent foam in processing, and this appears in the final product. It may not be necessary to add further silicone to the finished motor oil, but this can be done if necessary. It is important to note that silicone anti-foam must be very finely dispersed in oil or it will either agglomerate or settle out, or both.

One disadvantage of silicone oil is that it tends to promote air entrainment, that is, the retention within the oil of a large number of small air bubbles. In motor oils a small amount of entrainment is not usually harmful, but in some industrial oils it is most undesirable. In these cases anti-foam additives usually take the form of ester-based polymers which are used in larger amounts than silicone, but have less serious air entrainment properties. In oils with very simple formulations no anti-foam additive may be necessary.

Other Additives

Minor and special additives employed in certain types of formulations will be mentioned in Chapters 4, 7 and 8 which deal with formulations and requirements of different lubricants.

Interchangeability of Additives

From the early developments in the 1930s and 1940s there has been a very large research and development effort by the oil and additives companies in the area of crankcase oil technology. This effort can be characterized as following three principal directions:

1. Understanding the mechanisms of oil degradation and deposit formation in engines.

2. Development of new chemical materials which alleviate the above, or which improve the physical properties of lubricants.

3. Development of alternative materials or processes which avoid patent coverage obtained by rival companies, but produce equivalent beneficial effects.

The last of these has considerable significance. With certain exceptions of commodity-type additives (some ZDDPs are examples) the products of one company are not fully interchangeable with similar products from another. Thus interchanging sulfonates, phenates, or dispersants from different suppliers is unlikely to produce identical performance in all areas, even if the superficial inspection data (metal content, TBN, etc.) are the same. A formulation is therefore essentially fixed at the time of development, and the oil blender is tied to the original additive supplier(s) if major retesting and approvals are to be avoided.

References

1. Pritchard, J.J., *Oil & Gas Jnl.*, 76 (26) 152-158, 1978.
2. *Hydrocarbon Processing*, 43 (9) 201. 212, 215, 1964.
   Hydrocarbon Processing, 55 (9) 209, 1976.
   Hydrocarbon Processing, 57 (9) 113-209, 1978.
3. Sager, F., Palmquist, F., *Pet. Refiner*, 31 (6) 139-145, 1952.
4. Burn, A.J., and Greig, G., "Optimum Aromaticity in Lubricant Oil Oxidation," *Journal Inst. Pet.*, November 1972.
5. Bushness, J.D., and Eagen, J.F., *Oil & Gas Jnl.*, 73 (42) 80-84, 1975.
6. *Hydrocarbon Processing*, 61 (9) 101-208, 1978.
7. Bull, S. and Marmin, A., "Lube Manufacture by Severe Hydrotreatment," World Petroleum Congress, Bucharest, 1979.
8. *Proceedings of International Conference on Waste Oil Recovery and Re-use*, 3, 1978, copyright Am. Pet. Re-refiners, Washington, D.C.
9. Quang, D.V. *et al.*, *Hydrocarbon Processing*, 57 (9), 157, 1978.
10. Goosens, A.G., *et al.*, "Erdöl, Kohle, Erdgas," *Petrochem.*, 29, 419, 1976.
11. Zisman, W.A., Murphy, C.M., Advances in Petroleum Chemistry and Refining, Vol II, Chap. 2, Kobe K.A. and McKetta, J.J. (Eds.), Interscience Publ., NY, 1959.
12. Gunderson, R.C., and Hart, A.W., Synthetic Lubricants, Reinhold Publ. Corp., NY, 1959.
13. O'Connor, J.J., and Boyd, J., Standard Handbook of Lubrication Engineering, Chap. 11, McGraw-Hill, NY, 1968.
14. Smith, K.W., Starr, W.C., and Chen, N.Y. "A new process for dewaxing lube base stocks," API 45 mid-year refining meeting, May 1980.
15. Meals, R.N., and Lewis, F.H., Silicones, Reinhold Publ. Corp., NY, 1959.
16. Sanin, P.J., *Review Inst. Pet.*, 16, 468, 1961.
17. Dean, *et al.*, "Viscosity Index of Lubricating Oil," *Ind. & Eng. Chemistry*, 32, 102, 1940.
18. "Copper Compounds as Antioxidants," *PARAMINS Post*, Issue 4-2, p.1.

19. "Lubrizol—Fifty Years of Improving Performance," The Lubrizol Corporation, 1978.
20. Du Pont, *Ind. & Eng. Chemistry*, 46.(6), 17A, 1954.
21. Willis, "The Control of Low Temperature Sludge in Passenger Car Engines," SAE Paper No. 550293, Society of Automotive Engineers, Warrendale, Pa., 1955.
22. Crail, I., *et al.*, "The Effect of Polymeric Dispersants on Engine Sludge," *J. Inst. Pet.*, 1963.

Further Reading

23. Smalheer, C.V., and Smith, R.K., Lubricant Additives, The Lezius-Hiles Co., Cleveland Ohio, 1967.
24. Klamann, D., Lubricants and Related Products, Chapter 9, Verlag Chemie, 1984.
25. *Proceedings of International Colloquia on Additives for Lubricants*, Esslingen, 1986/88/90.
26. Modern Petroleum Technology, Institute of Petroleum, John Wiley, ISBN 0471 262765, Sec.26, "Lubricating Oils," 963-1007.
27. Mortier, R.M., and Orszulik, S.T., Chemistry and Technology of Lubricants, (Castrol) Blackie/VCH Publications Inc., N.Y.
28. Copan, W.G., and Haycock, R.F., "Lubricant Additives and the Environment," CEC/93/SP02, CEC Symposium, May 1993. Republished by the Technical Committee of Petroleum Manufacturers in Europe as ATC Document 49.

Note:

A complex and voluminous patent literature exists for petroleum additives which we have not cited. Study of this is recommended for those with a serious interest in this area. The commercial patent search organizations can be asked to provide listings by topic, but the complexity cannot be exaggerated.

Chapter 3

Crankcase Oil Testing

3.1 Introduction

In the preceding chapters we have discussed various oil properties or characteristics (such as viscosity, wear prevention, and ability to keep machinery operating satisfactorily) in a rather abstract manner, without saying how these properties are measured. This chapter will indicate how key properties are usually determined without going into the fine details of test equipment design or operation. These can be found in standard national and international test method books, of which the most well-known are the volumes published by ASTM.[(1)] In many cases we will quote the relevant ASTM test method designation to assist those who may need more than a summary of how tests are performed. Details of apparatus design, where it can be purchased, operating procedures, and information on the precision of the test will be found in the ASTM methods.

The reasons for testing an oil (or any other material) are essentially to predict how it will perform in service.[(2)] Measurement of physical properties like viscosity provide initial information on its suitability, and therefore strictly could be considered as measures of performance. However, it has become the practice in the oil and additive industries to use the expression "performance testing" for those tests conducted in some form of operating machinery and which measure properties such as wear prevention and engine cleanliness. This can be full-scale machinery or a laboratory device of reduced scale or complexity.[(3)] The latter type is usually referred to as a "rig-test."

We propose to stay with this narrower definition of performance testing and will divide tests into three categories as follows:

1. Tests for physical properties (density, viscosity, volatility, etc.)
2. Tests for chemical properties or composition (elemental content, hydrocarbon types, etc.)
   Note: Many analytical techniques for chemical composition use physical methods such as spectroscopy and chromatography rather than wet chemical analysis.(4,5,6)
3. Tests for performance (anti-wear properties, corrosion prevention, detergency, etc.). These can be further divided into:
   - Special laboratory apparatus or "rigs"
   - Test bed evaluation in real but selected equipment
   - Field tests

Tests can also be divided into empirical tests giving results expressed in arbitrary units, and tests that measure fundamental or absolute properties. The results from empirical tests depend on the design of the apparatus as much as the properties of the material under test, and consistent results are obtained only by standardization of the apparatus and its method of operation. The various types of flash point apparatus are typical examples of empirical tests, and so are engine tests. Most petroleum testing was originally of an empirical nature, but now methods are designed to enable results to be expressed in fundamental units whenever possible. Of course the use of fundamental units does not of itself provide more information about the properties of a lubricant than a purely empirical measurement of the same property, viscosity in Saybolt seconds being just as useful as viscosity in milliPascal seconds. The important factor is the availability of comparative data from many sources in the same units to enable international assessments and comparisons to be made more readily. The use of fundamental units increases the possibilities for this, and the "SI" system of units is now being generally adopted. Notes on the SI system are given in Appendix 4.

In the sections below, test methods, mainly relevant to crankcase oils, are discussed in a rather arbitrary order based on importance in an automotive context. Fundamental principles are reviewed with an indication of the procedures. ASTM method numbers are given in most cases. Some less frequently used tests are not described here but may be found mentioned briefly in the glossary (Appendix 1).

Special tests for other types of lubricant will be covered in Chapters 7 or 8.

The testing of used oils will be covered separately in Section 3.5 because the reasons for testing used oils are different from those for testing new oils. There are special tests involved, and the interpretation of the results of used oil analysis requires background experience and data, and an understanding of the processes involved in the lubricant's use.

3.2 Laboratory Bench Tests

3.2.1 Tests for Physical Properties

3.2.1.1 Viscosity[7,8]

A fluid flowing over a stationary surface does not move as a whole at the same velocity (Figure 42). There is a continuous variation of flow velocity from effectively zero at the containing surface to a maximum at the distance farthest from it.

When a fluid flows down a pipe there is a velocity gradient across the pipe with fluid essentially stationary at the pipe walls and moving with maximum velocity in the center of the pipe (Figure 43). The force required to sustain the flow increases with rate of flow and narrowness of pipe.

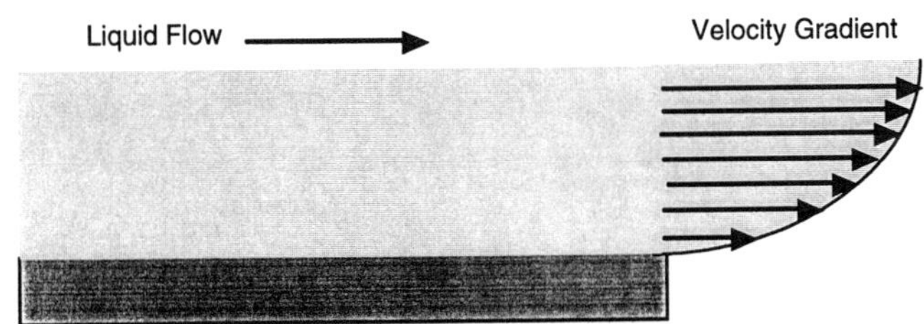

Fig. 42 Flow of liquid over a surface.

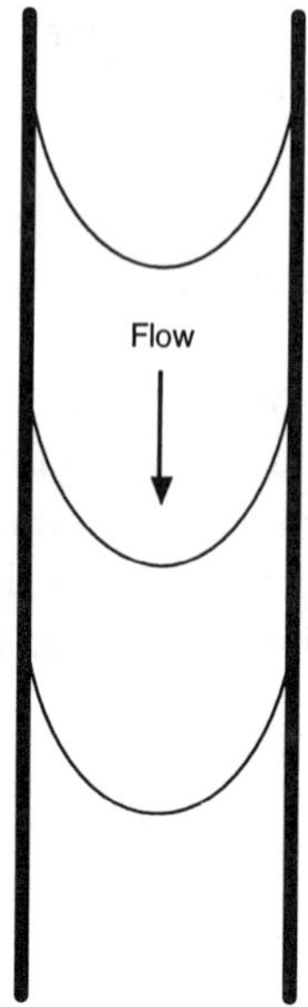

Fig. 43 Differential flow rates.

One can envisage a series of layers, or rather annuli, of the fluid which are in relative motion, with a shearing effect between the different layers. The narrower the pipe and the faster the flow rate, the greater the *shear rate* will be. High shear rates are found in fast flow down narrow tubes, and also around rapidly rotating spindles immersed in fluid.

The viscosity of a fluid is in essence its resistance to flow, which can be defined as the force required to move a given layer of fluid past another layer at a given speed and at a standard separation (Figure 44).

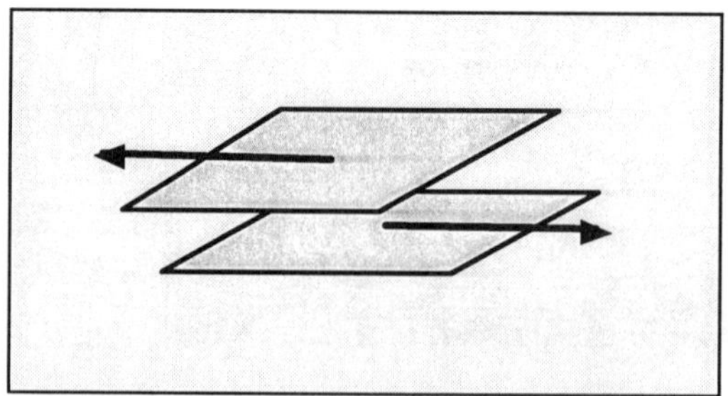

Fig. 44 Laminar flow and viscosity.

The traditional unit is the poise, which was originally defined as the force in dynes to move 1 cm^2 of surface in the fluid past a parallel 1 cm^2 at a distance of 1 cm within the fluid, and at a speed of 1 cm/second. This is the dynamic viscosity, and for petroleum products a more convenient unit is the centipoise (1/100 poise), which is now expressed as 1 milliPascal second in SI units; 1 cP = 1 mPa·s. When the viscosity of a fluid is measured by allowing it to fall under its own weight, a related unit, the centistoke, is more useful; in SI units this is expressed as mm^2/second. This is the kinematic viscosity, and is numerically equal to the dynamic viscosity divided by the density of the fluid.

The velocity of flow produces a shear rate in the liquid, dependent on geometry. The force acting upon the oil causing it to flow is called the *shear stress*. This is related to the shear rate and viscosity, and the relation can be expressed as:

$$\text{shear stress} = \text{viscosity} \times \text{shear rate}$$

or,

$$\text{viscosity} = \frac{\text{shear stress}}{\text{shear rate}}$$

For so-called "Newtonian" fluids the viscosity measured is not affected by the shear rate. For certain oils, however, particularly multigrade motor oils containing polymeric additives, the viscosity depends on the shear rate, normally being lower at high shear rates than a Newtonian fluid of the same low shear viscosity (Figure 45). This reduced viscosity at high shear rates is a temporary and reversible phenomenon, and is not to be confused with the permanent loss of viscosity occurring in multigrade oils when polymer molecules are broken up by high-energy shearing forces in the equipment. For a non-Newtonian fluid an apparent viscosity can be measured at a given shear rate. This is constant for a Newtonian fluid but depends on the shear rate for a non-Newtonian fluid.

For the normal measurement of viscosity (for viscosity classification of oils, etc.) low-shear instruments are used. Early empirical instruments such as the Saybolt, Redwood, and Engler viscometers consisted of a cylinder with a plugged hole at the bottom. The temperature was controlled using an annular water bath, and the time in seconds for a given

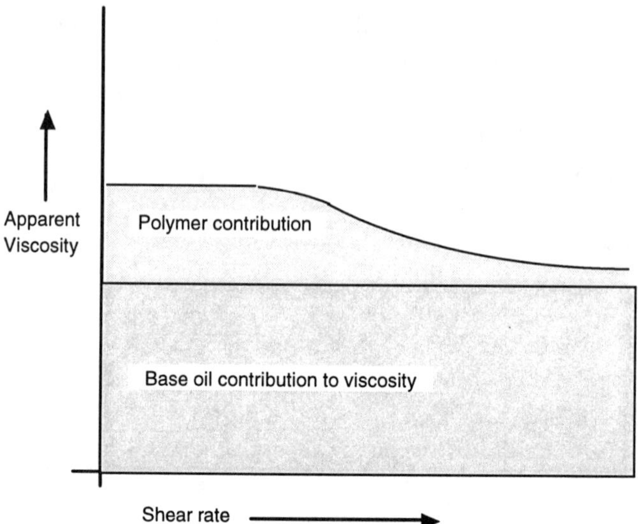

Fig. 45 Temporary viscosity loss at high shear rates.

volume of oil to flow through the hole when the plug was removed was a measure of the viscosity (Figure 46).

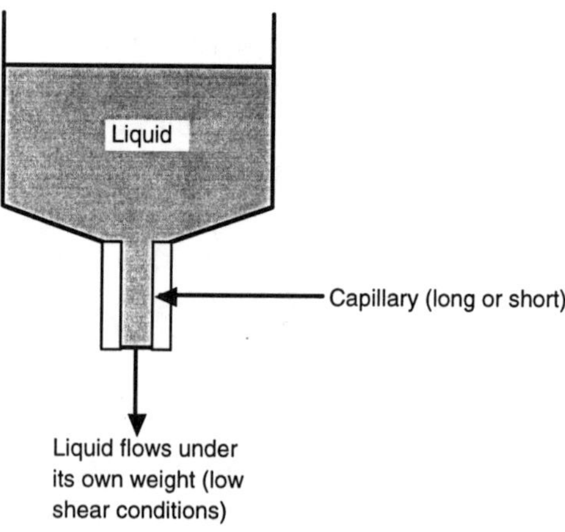

Fig. 46 Kinematic viscosity utilizes gravity.

The more accurate modern method is to use a capillary tube suspended in a temperature-controlled bath. A bulb contains the fluid which flows out under its own weight through a capillary (Figure 47). The time of flow again gives a measure of the viscosity, but instead of using a series of accurately made identical viscometers all giving the same efflux time for the same viscosity, the tubes are individually calibrated against fluids of known viscosity and the efflux times directly converted to fundamental units by a simple calibration constant.

This type of viscometer is normally used at temperatures of 40°C and 100°C, the temperatures from which the V.I. can be read off from tables (see ASTM D 2270). 100°C viscosities are also used to define the summer grades of motor oil in the SAE viscosity classification described more fully in Section 6.3.1 and given in full in Appendix 9.

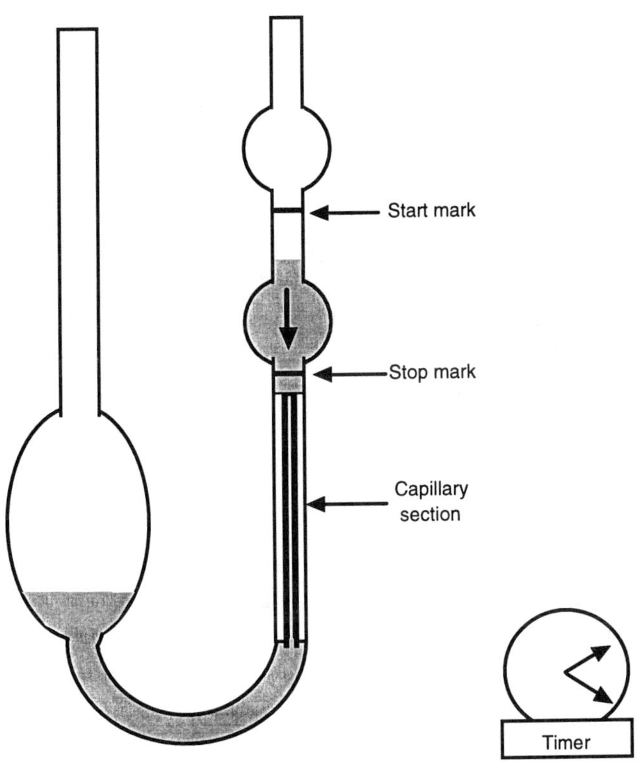

Fig. 47 U-tube determination of kinematic viscosity.

Rotational viscometers measure the dynamic viscosity, utilizing the viscous drag on a rotor immersed in oil and measuring the torque required to rotate it at a given speed, or the speed achieved for a given torque (Figure 48).

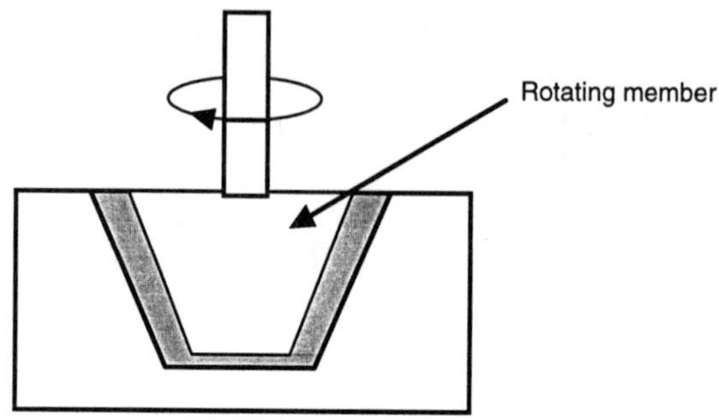

Fig. 48 Dynamic viscosity uses mechanical force.

Important examples for the automotive field include the Brookfield Viscometer, Cold Cranking Simulator (CCS), the Mini-Rotary Viscometer (MRV), all for low-temperature use, and the Tapered Bearing Simulator (TBS) used for measuring high shear viscosities at 150°C.

The Brookfield is a rapid, direct-reading instrument employing an assortment of rotary spindles and variable speed settings (very low to medium) to provide a wide range of measurements. It is immersed in the test liquid, the speed of rotation selected, and the torque reading is a measure of the apparent viscosity. The Brookfield is normally used at low temperatures. The standard method is ASTM D 2983, and method D 5133 describes means of finding the minimum temperature for flow at very low shear rates (Figure 49).

Between 1957 and 1967 a viscometer with a very specific use was developed. This was the Cold Cranking Simulator (CCS) whose purpose was to define the winter grades of motor oil in a way that correlated with engine crankability, rather than by use of the V.I. system. It uses a non-cylindrical rotor enclosed closely by a stator (Figure 50) and was originally operated at a single temperature, namely 0°F or –18°C.

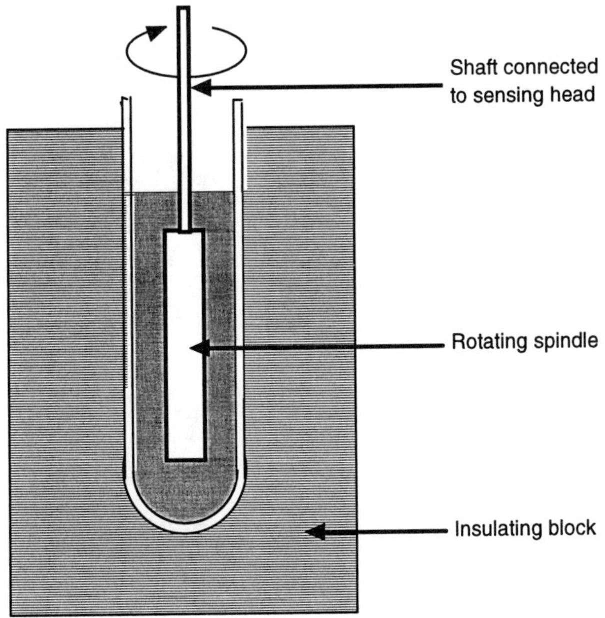

Fig. 49 The Brookfield viscometer.

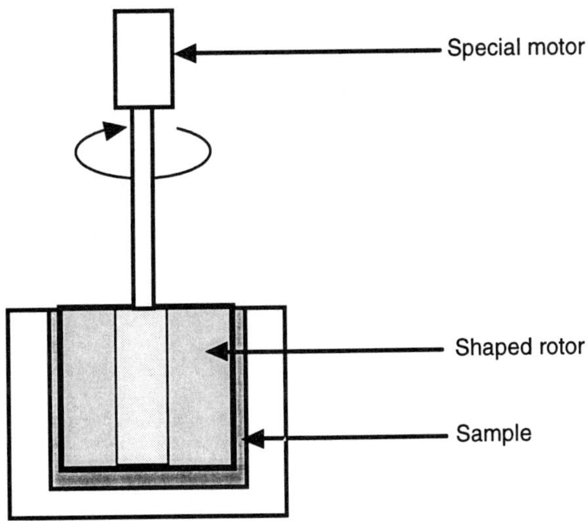

Fig. 50 The Cold Cranking Simulator.

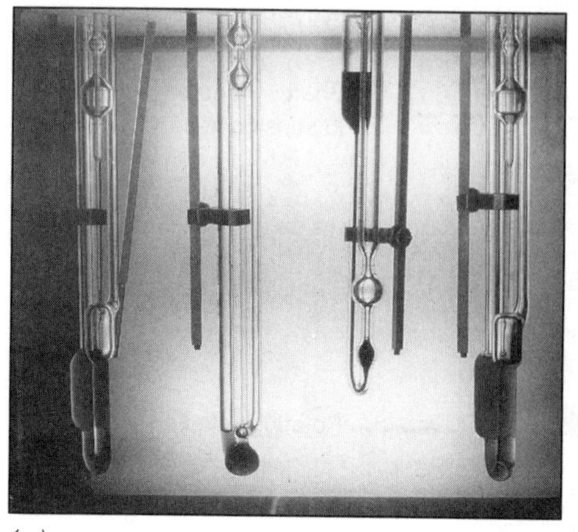

(a)

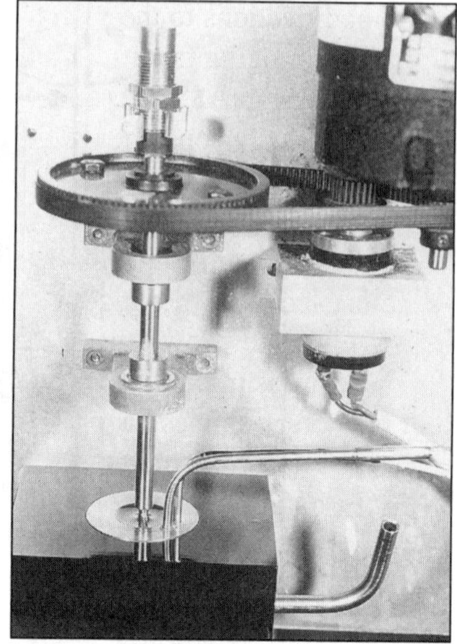

(b)

(c)

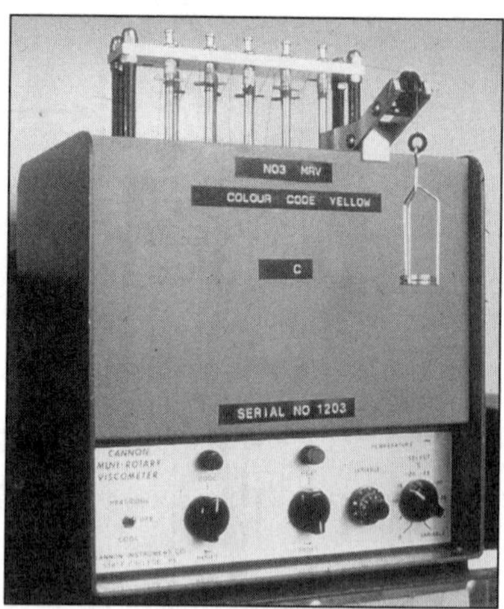

(d)

Plate 3. Viscosity and cold flow measurement. a) viscometer U-tubes, b) cold cranking simulator, c) autopour apparatus, d) mini-rotary viscometer apparatus.

Recent modifications to the viscosity classification system now require test temperatures ranging from –5°C for 25W oils to –30°C for 0W oils. The method is given in ASTM D 5293. A particular type of electric motor applies a relatively constant torque to the rotor, and the speed of rotation is related to the viscosity by calibration with standard oils.

In the 1980s problems of oil pumpability in cases where cranking was satisfactory led to further modifications of the viscosity classification, and the adoption of new low-temperature viscometers. Low-temperature, low-shear viscosity is important for predicting the possibility of "air binding" in motor oils after vehicles have stood at low temperatures for a considerable period. The non-Newtonian motor oil can gel to a semi-solid and fail to flow to the oil pump inlet when the engine is started. The oil pump then pumps air instead of oil to the engine and both the pump and other engine parts can be rapidly damaged. Even if "air binding" does not take place, an oil can be so viscous after standing at low temperatures that the rate of pumping oil to sensitive bearings and rockers may be inadequate, and again engine damage can result. The Brookfield method ASTM D 5133 is believed to correlate with these problems and it is recommended that this test is performed on new oil formulations. It is, however, time-consuming and does not readily permit tests on large numbers of samples.

After a run of severe winters in N. America and Europe in the early 1980s with many engine damage problems, it was necessary to devise a test which correlated with the problems, was easy to perform, and permitted testing of multiple samples. The Mini-Rotary Viscometer (MRV) was developed as an inexpensive solution. Simple spindles are immersed in the oil, and strings are wound around the tops of the spindles. These pass over pulleys, and weights are attached to provide the turning force (Figure 51). The stepwise addition of small weights provides a measure of the yield stress, the minimum force required to break the oil's gel structure and to cause the spindle to turn. When a larger fixed weight is used, the speed of rotation of the spindle gives a measure of the viscosity of the fluid. Variations have been made to the detailed pre-test oil conditioning technique to improve correlation with field problems. (See relevant classifications and specifications.) The SAE J300 viscosity classification now requires testing in the MRV to the TP-1 procedure for winter classifications. The ASTM test method is D 4684.

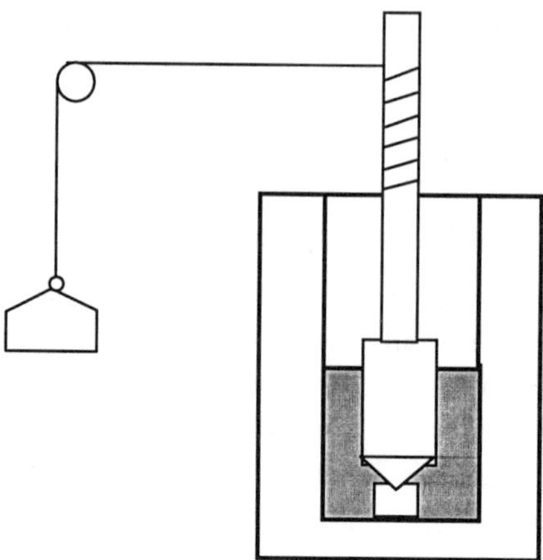

Fig. 51 The Mini-Rotary Viscometer.

Engine speeds and engine oil temperatures have been increasing steadily over the past decades. In Europe, extended high-speed motorway driving has become common, while in the U.S. the downsizing of engines has been important. Globally, air-smoothing of vehicles has led to less direct cooling of the oil pan and of the engine in general. Engine manufacturers have consequently become more concerned over the high-temperature/high-shear (HTHS) viscosity of motor oils, and requirements have been introduced into motor oil specifications. Fine bore capillary viscometers can be used for such measurements (ASTM D 4624), but rotary direct reading instruments are preferred. The Tapered Bearing Simulator (TBS, ASTM D 4683) utilizes a conical journal bearing which is brought into close proximity to generate high shear conditions. The torque on the lower member is measured when the top component is rotated at a fixed speed. A similar European instrument is the Ravenfield Viscometer (ASTM D 4741) (Figure 52).

The following table summarizes the principal types of viscometers used for automotive lubricant measurements.

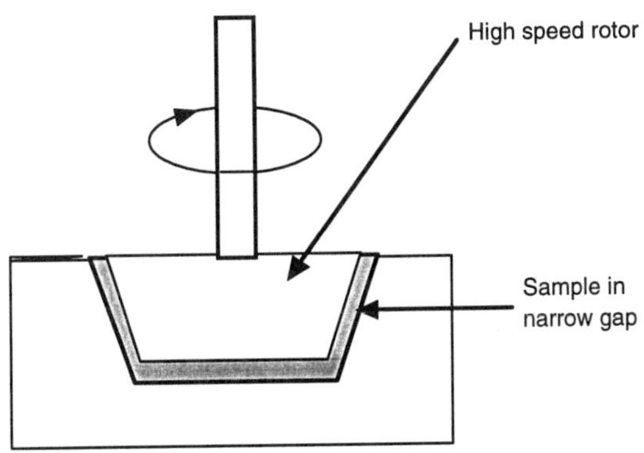

Fig. 52 High-temperature/high-shear viscosity.

TYPE	DESIGNATION	SHEAR RATE	TEMPERATURE	INDICATES	ASTM METHOD
Capillary		Low	40°C , 100°C	SAE oil grade	D 445
Capillary		Very high	150°C	Bearing protection	D 4624
Rotational	Scanning Brookfield	Very low	–10°C to –40°C	Low-temperature pumpability	D 5133
Rotational	Brookfield	Low to med.	–10°C to +40°C	Gear oil viscosity	D 2983
Rotational	MRV	Low	–10°C to –35°C	Low-temperature pumpability	D 4684
Rotational	CCS	High	–5°C to –30°C	Engine crankability	D 2602
Rotational	TBS or Ravenfield	Very high	150°C	Bearing protection	D 4683/ D 4741

3.2.1.2 "Shear Stability" of Multigrade Oils

As discussed earlier and illustrated in Fig. 45, non-Newtonian fluids exhibit a lower viscosity when subjected to high shear rates than Newtonian fluids having the same viscosity at low shear. The available energy causes the polymer coils to elongate and squeeze through narrow orifices without the polymer itself being changed. As the stress is removed the polymer recovers its original configuration and the level of viscosity at low shear remains unchanged. This type of reversible viscosity loss is often called "temporary viscosity loss," and is seen in tests such as the Tapered Bearing Simulator discussed above.

In high-temperature and/or high-energy situations, and particularly where localized oil shear is much higher than that of the average within the volume of oil, the polymer will undergo a permanent change by rupture of the molecular chains (Figure 53). Oil flow which induces extension of the

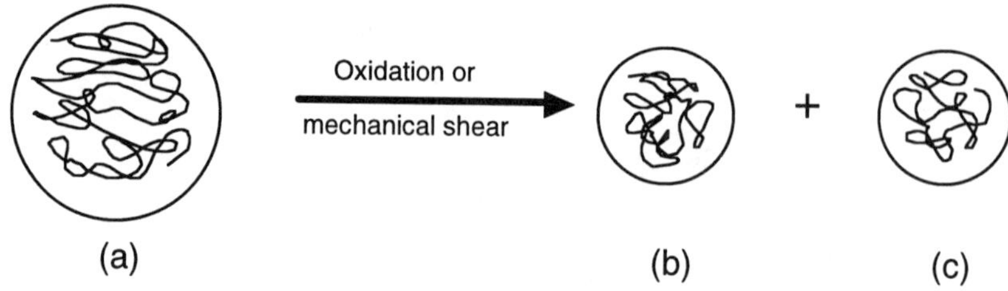

Fig. 53 Polymer rupture gives viscosity loss.

polymer molecules, and cavitation phenomena, can also cause polymer breakdown.

When a long molecule breaks into two shorter molecules, the overall thickening effect is reduced and a permanent loss in the oil viscosity results. The longer the polymer chain, the higher the oil temperature, the greater the shearing action and the greater the tendency for rupture to occur. The greatest shearing effect occurs in areas of thin film lubrication at high sliding speeds particularly in oil pumps, gearboxes, valve gear, and piston ring zones.

The level of permanent viscosity loss depends on the molecular weight and structure of the polymer used to thicken the oil. Different samples of polymers can therefore be assigned a *Shear Stability Index* (SSI) from which the performance, in terms of in-service viscosity loss, can be estimated. Figure 54 illustrates the concept of Shear Stability Index(9) graphically.

Various types of apparatus have been used to measure Shear Stability Index, including motored gearboxes, sonically agitated systems (e.g., as in ASTM D 2603), and various systems of contra-rotating discs/paddles. The most common technique, however, is to use a diesel injector pump to spray the oil through the diesel injector nozzle, recycling the oil so that it undergoes many passes through the nozzle. In the nozzle it is subjected to high-speed thin film changes of direction, and is then atomized. Various diesel

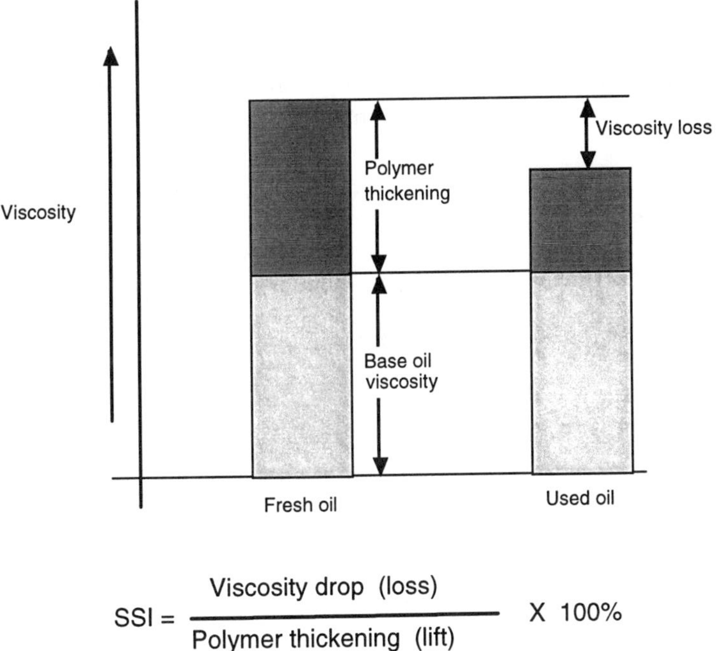

Fig. 54 Illustration of shear stability index.

injector rigs and test methods are in use, but in Europe the Kurt Orbahn, named after the apparatus manufacturer, uses a Bosch injector pump and is specified for most multigrade oil specification testing. In the U.S. a similar method is described in ASTM D 3945. Used oil viscosity loss in the L-38 engine test is also used as an indication of shear stability by the U.S. military, and in the API SH engine oil classification.

The rate of bulk oil viscosity loss in an engine will depend on the sump capacity relative to the volume undergoing shear breakdown, as well as on the SSI of the polymer used.

3.2.1.3 Pour Point (ASTM D 97)

For many years the pour point was the sole test used to assess low-temperature fluidity, and while it is now backed up by the MRV for motor oils and the Brookfield for gear oils, it still remains the criterion for most other oils. The apparatus is simple and inexpensive, but the results are entirely empirical and arbitrary. A small quantity of oil is placed in a glass cylinder with a thermometer monitoring its temperature (Figure 55).

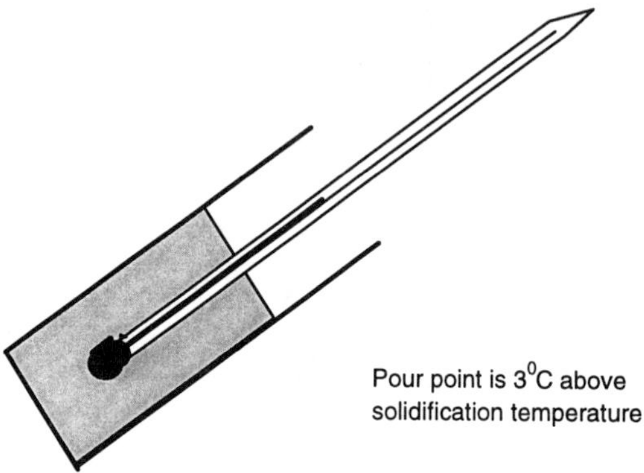

Fig. 55 Determination of pour point.

The oil is cooled rapidly through a series of cooling baths and around the expected pour point it is removed for examination every 3°C. This consists of tilting the cylinder and watching for movement of the surface of the oil. The pour point is said to be at the last 3°C interval at which the oil moves. Below the pour point a sufficient network of precipitated wax has formed to cause gelation. Gelation is affected by the rate of cooling, the temperature profile during cooling, and any disturbance to the oil. It is therefore important to follow the procedures precisely, and to ensure that no sources of vibration are close to the apparatus.

For laboratories handling large numbers of samples, an automatic pour point apparatus is available which provides automated cooling and measures the pour point by means of a small rotational viscometer. The apparatus is very convenient and normally can be correlated very well with the manual method. Care should be taken, however, with unknown or unusual samples, as the two tests are fundamentally different in the way the pour point is measured.

The Stable Pour Point is a test which was devised by the U.S. military to avoid potentially over-optimistic results in relation to low-temperature fluidity which the normal pour point test can sometimes produce. In this test the oil is cooled more slowly over a period of one day, and is then taken through a complex heating and re-cooling cycle over a further 5 days before the pour point is measured in a period of slow cooling (Figure 56). This type of test is now being superseded by the new low-shear/low-temperature viscometric tests for motor oils.

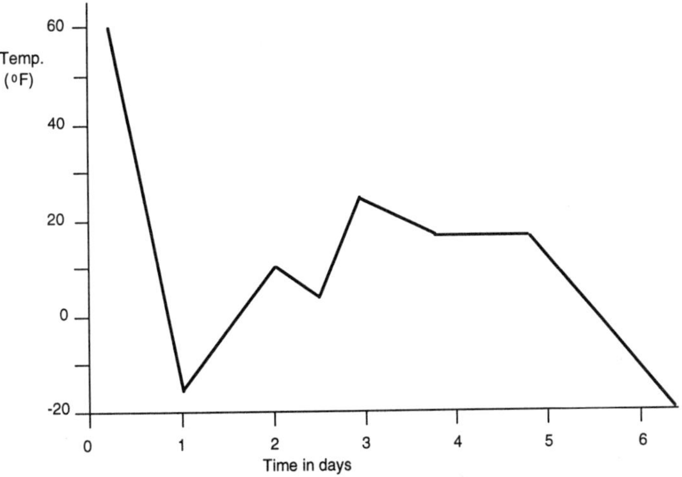

Fig. 56 Temperature cycle for pour stability test.

3.2.1.4 Flash Point

Petroleum products contain a spectrum of molecular species, and when they are heated the more volatile fractions collect in the air space above the liquid. If a sufficient quantity collects it can be ignited when a flame is applied to the mixture of vapor and air. The temperature at which this occurs can be measured in a special apparatus but the results will be very dependent on its geometry. If the air space is normally closed off, as is the case in the Pensky-Martens flash point tester (ASTM D 93), the vapors will collect faster and the flash point will be lower than in cases where the cup of oil is open to the atmosphere, as for the Cleveland Open Cup tester

(ASTM D 92). These and other flash point tests are entirely empirical, and there is no correlation between different models. The Pensky-Martens flash tester has been successfully automated by close simulation of the manual procedure (ASTM D 3828, the "Setaflash" apparatus).

Flash point tests can give some indication of the width of the distillation cut for base stocks, and for finished oils it can reveal contamination by other more volatile materials such as fuel products. Additives present in a motor oil can either lower or raise the flash point and the test has no great significance in relation to finished motor oil quality. Flash points of automotive lubricants are in a range where fire hazards are not significantly flash point dependent.

3.2.1.5 Volatility

The volatility of a base stock is governed by its distillation characteristics, and is expressed as the percentage vaporized at a given temperature. At a given measurement temperature a light narrow cut stock may appear no more volatile then a broadly cut heavier stock, or a so-called *dumbbell blend*, and some type of distillation is preferred to single-point measurements if the product is to be fully analyzed. The impact of oil volatility on engine performance and apparent oil consumption will depend on the individual engine design, the climate in which it operates, and the driving regime. There is no simple and direct relationship between measured volatility and perceived field performance.

Volatility of motor oils is, however, becoming of increasing concern, for high volatility can lead to oil loss by evaporation, air pollution increase, and an increase in piston deposits. In Europe small engine sizes and high road speeds lead to high oil temperatures, and the engine manufacturers are specifying motor oil volatility limits. The test most commonly used is the Noack test (DIN 51581), in which the sample is heated to 250°C and a stream of air carries away the volatile components. The Air Jet test (ASTM D 972) has been used for similar purposes in North America. Vacuum distillation or gas chromatographic simulated distillation provides more information, but is more difficult to use for specification setting.

The European vehicle manufacturers, represented by ACEA, the primary governors of European motor oil quality, specify Noack volatility tests for all viscosity grades. Some of the U.S. car manufacturers apply similar restrictions to SAE 10W grades, especially to factory fill oils.

Volatility restrictions on motor oils promote the use of synthetic components, or unconventionally refined petroleum stocks, because these have narrower distillation curves than conventional base oils.

3.2.1.6 Foaming Tendency and Stability (ASTM D 892)

Considerable aeration of oil takes place within an engine, particularly when the crankshaft is partially immersed in the oil. A foam of air bubbles forms upon the surface of the oil and if this does not collapse as soon as it forms, it will build up and may result in oil loss from the dipstick, oil filler, or other orifices. The build-up of foam may also cause the oil level in the crankcase to drop until the oil pump receives a compressible mixture of air and oil foam instead of pure oil. The oil pump itself may be damaged and bearings and other forced-lubrication parts may suffer from oil starvation. The oil/air foam also will not support heavy loads as well as a fully liquid oil. In a dry sump engine (with a remote oil reservoir) the situation may be even more critical because the oil pump may suck air or foam into the oil, and all lubrication in this case is by direct oil feed with no splash application. In motor oils the use of detergent and dispersant additives has greatly increased the tendency for foam formation, unless special precautions are taken to limit the foaming tendency.

The standard method for measuring and specifying foaming tendency and stability is ASTM D 892, in which air is bubbled through a porous stone inside a cylinder containing oil. The test can be conducted at various temperatures (usually 24°C and 93.5°C) and the volume of foam above the liquid level is measured while the air is blowing, and at a set interval (1-10 minutes) after the air is turned off. There are wide differences in the settling times and foam levels specified by various approval authorities, although most require zero foam after settling. What constitutes zero foam is a matter of interpretation, and the design of the apparatus makes the foam level while blowing very variable. Despite the limitations of the method and attempts to devise superior foam tests methods,[10] ASTM D 892 remains the only significant test found in specifications. At the time of writing, a "Sequence 4" run under high-temperature conditions (150°C) with a metal diffuser is under study within ASTM. The objective is to simulate the more severe conditions encountered in modern engines and, for example, in hydraulic valve lifters.

3.2.1.7 Density

This is not a significant property for finished lubricants where density is heavily influenced by base stock choice and the additive content. For base oils, density gives an indication of the paraffin content (low density) versus the naphthenic content (high density). Test methods used include hydrometers (ASTM D 1298), proprietary gravity balances, and density bottles or pyknometers.

3.2.2 Chemical Tests and Properties

3.2.2.1 Neutralization Number, Acidity, Base Number (ASTM D 664, D 2896, D 4739)

The acidity or alkalinity of an oil can be measured by running into it an amount of either alkali or acid, respectively, that will neutralize the acid or alkali present. This is known as titrating the acid or alkali content. As agents of known strength are used, and the "end point" of neutralization is indicated either by a color change indicator or a pH meter, the amount of acid or alkali in the sample can be measured quantitatively and accurately. The titration is performed in a special solvent mixture in which both the oil and the aqueous reagents (alkali or acid) are soluble.

Acidity may be caused by oxidation of the lubricant, by contamination with acids from fuel combustion, or from some additive components. Alkalinity is derived from detergent additives and new crankcase oils today are strongly alkaline.

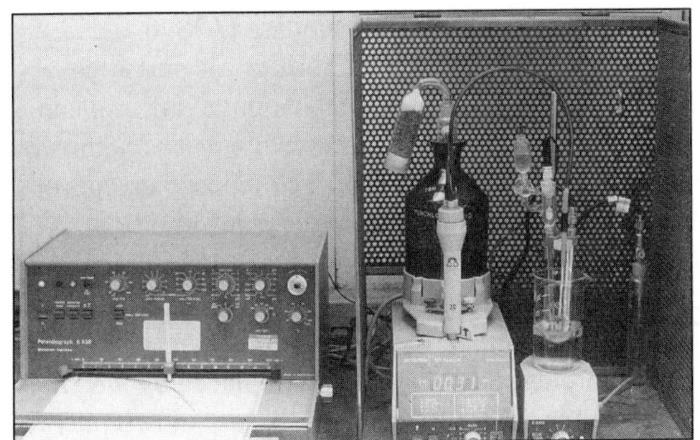

(a)

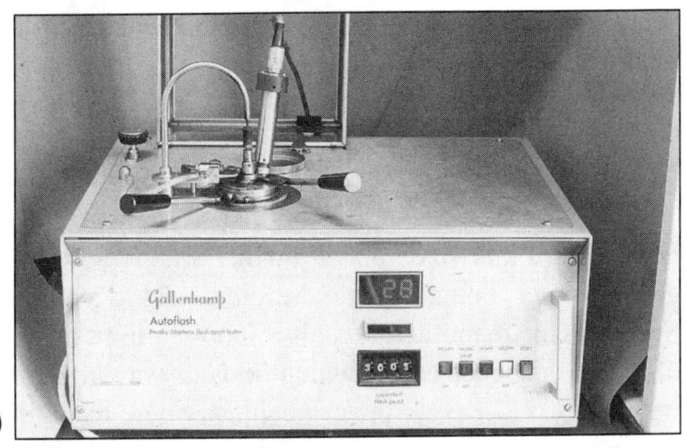

(b)

(c)

Plate 4. Routine laboratory tests. a) Titrimeter, b) Autoflash, c) X-ray analyzer.

The two ASTM test methods for measuring alkalinity are D 2896 and D 4739. D 2896 is usually specified by approval authorities. It measures mildly alkaline as well as strongly alkaline constituents in the oil, and can be said to provide a better measure of the total additive content of a known new oil. It can be argued that weak alkalinity is not particularly useful for acid neutralization and prevention of corrosion and deposit formation, so that for specifying oil alkalinity in relation to fuel sulfur content the preferable test would be one that measures strong alkalinity alone. Originally, ASTM D 664 was considered satisfactory, but this relatively simple method suffered from poor reproducibility and is now replaced by ASTM D 4739. This method measures both strong and weak bases, and employs potentiometric titration at a very slow rate (90 s per 0.1 mL reagent added). The slow addition rate makes the method suitable for used oil analysis, where neutralization is often slow.

3.2.2.2 Sulfated Ash (ASTM D 874)

For an oil of known composition this test gives a good check on the detergent additive content. It is therefore useful as a control test in oil blending. Some engine manufacturers also impose a limit on the sulfated ash content of motor oils, because oil which finds its way into the combustion chamber will be burned leaving the ash behind, where it could cause pre-ignition. Ash contributes to crown land deposits above the piston rings, and can lead to valve leakage and subsequent seat burning. Recent work has shown that ash from lubricating oil can contribute significantly to diesel engine particulate emissions.

A sample of oil is burned in a silica dish and sulfuric acid is added to the resulting ash to convert the various metal oxides into their sulfates. The residue is then heated under carefully controlled conditions until the weight of the ash is constant. The intention is to convert any zinc sulfate (which is present from anti-wear additives in the oil) back into zinc oxide. For the results to have any significance it is important that this stage is correctly performed. The test is still not completely quantitative, however, for phosphorus compounds can produce refractory pyrophosphates giving higher and variable results, and since the introduction of magnesium as a common detergent additive there is even more variability in the results because magnesium sulfate is partially but not repeatably converted to magnesium oxide during the heating process. However, the test still provides a reasonable degree of quality control over additive metal content. If it is to be used for this purpose the following conversion factors for the weight of metal present in the oil to the expected sulfated ash content can be used:

Estimation of Sulfated Ash from Metal Content:

Metal:	Multiply metal % by:
Barium	1.7
Calcium	3.4
Magnesium	4.5
Sodium	3.1
Zinc	1.25

Sulfur, Phosphorus, Chlorine and Boron contents may be ignored. Calculated values from the above table should be checked against carefully made oil blends whenever possible.

3.2.2.3 Elemental Analysis

This is required for the control of additive concentrates and is preferred to sulfated ash for measuring the concentration of additive present in finished lubricants. The old techniques of wet chemical precipitation, filtering, and weighing have been superseded except for reference purposes by modern analytical equipment which uses comparative instrumental techniques. These relate the test sample readings to those given by standard calibration material. Some methods can use oil as received, in others it needs to be diluted with a solvent, while some require more complex sample preparation techniques. A brief review of the most important analytical techniques is given below:

<u>Flame Photometry</u> (ASTM D 3340)

At one time this was a general-purpose technique used for many elemental analyses in chemical laboratories, but now its use is largely confined to older laboratories and for the rapid measurement of sodium and lithium contents found in greases and some lubricating oils. The sample can be diluted with solvent, or for greater sensitivity can be ashed and dissolved in water. In either case the liquid is aspirated into a flame. Characteristic wavelengths of light are emitted in the flame depending on the elements present, the light is dispersed by a spectrometer, and the relevant wavelengths are measured electronically. The intensity of emission from the sample is compared with standards of known elemental concentration and the concentrations in the sample thereby estimated.

Atomic Absorption Spectroscopy (ASTM D 4628)

This technique has largely replaced flame photometry for routine analysis in many petroleum laboratories. A monochromatic beam of light characteristic of the element to be measured is produced by a hollow-cathode lamp containing the element. This light is passed through an intense flame, such as from an acetylene/nitrous oxide burner, and the sample in solvent solution is aspirated into the flame. Atoms of the element being measured are produced in the flame and absorb the characteristic radiation, and the drop in its intensity can be measured and related to the elemental content of the sample using calibration standards. Elements routinely measured by atomic absorption include barium, calcium, magnesium and zinc.

Emission Spectroscopy

Flame photometry is a form of emission spectroscopy, but the term is usually applied to more complex apparatus employing an electric arc or a plasma to excite the spectral lines from the sample, rather than a flame. The higher-energy excitation source gives higher sensitivity and produces more useful spectral lines for the purpose of analysis, which can help to overcome problems of interferences between elements.

Older equipment used an arc struck between graphite electrodes, and with a rotating graphite disc dipping into the sample as the lower electrode, an oil could be analyzed rapidly without any sample preparation. Direct-reading instruments used photomultipliers to detect the emitted spectral lines and permitted analysis of many different elements to be completed in a few minutes. At one time the method was primarily used for the measurements of wear metals in used oils (see Section 3.5).

Inductively Coupled Plasma Emission Spectroscopy (ASTM D 4951)

Usually known as ICP, this is a more recent technique of emission spectroscopy, tending to be less sensitive than arc systems for some elements but normally giving greater accuracy. A stream of ionizable gas such as argon is excited by a powerful radio-frequency coil and is converted into a plasma. The sample in a solvent is sprayed into the plasma and characteristic spectral lines are emitted. The method has moderate sensitivity for barium and phosphorus but more for calcium, magnesium, sulfur, and zinc. It can also be used for trace metals in used oil analysis, when the ASTM method is D 5185.

X-Ray Fluorescence Spectroscopy (ASTM D 4927)

This has become a very widely used analytical technique of high speed and good precision. Sample preparation is not normally needed, the oil being placed in a small cell with a thin plastic film window at the bottom. This is attached to a vacuum chamber and the sample is irradiated through the window with powerful x-rays. The sample emits secondary x-rays whose wavelengths are characteristic of the elements present. These are analyzed with a crystal diffraction spectrometer and the intensity at selected wavelengths is measured with an x-ray counter (e.g., a Geiger counter). The method is sensitive for most elements, but is unsuitable for those lighter than silicon and therefore excludes sodium and magnesium. The precision of the method is relatively good and can sometimes be further improved by extending the counting time (the time taken to measure the emitted x-rays). There are considerable inter-element interference effects, and correct calibration is essential for accurate analysis.

Note on Precision and Accuracy of Chemical Analysis of Petroleum Products

Surveyors and engineers will find it difficult to appreciate the comparatively low levels of precision found in chemical analysis of petroleum products compared to physical measurements within their own disciplines. There are many reasons for this, but the most important are listed below:

(i) Precise basic standards can be defined for most measurements of length or distance (a bar of metal or the wavelength of light, etc.). Concentration standards are, however, much more difficult to set up.

(ii) Where concentration standards can be produced, for example by adding weighed amounts of a pure substance to a diluent, such samples often have limited application.

(iii) This is because in most cases the standard samples need to be quite similar to the unknown samples for analysis. In spectrographic analytical methods the different elements interfere mutually with each other's output signals, and the base matrix can also have a considerable influence.

(iv) For classical wet chemical analysis, precipitated material can be lost, or false precipitates of other interfering elements can be unknowingly produced.

(v) Spectroscopic devices have considerable "noise" in both the excitation and detection areas which is significant in relation to the magnitude of the signals being measured and limits the sensitivity at low concentration levels.

(vi) Petroleum samples are normally liquids which are not completely stable either chemically or physically.

Repeatability of spectrographic methods for lubricating oil analysis lies in the range of 2% to 5% of the mean value, while estimates of reproducibility from cooperative test programs show values ranging from around 4% to 20%. This does not mean that two laboratories working together (e.g., a supplier and a customer) cannot come to considerably closer agreement for a specific analysis. The problems of setting meaningful specifications for both additives and lubricating oils will be discussed further in Section 9.1. Accuracy depends on the quality of the reference or calibration standards used. Precision and accuracy are discussed in Section 3.4, with elemental analyses specifically covered in 3.4.2.

3.2.2.4 Infrared "Fingerprinting" of Oils

With the availability of simple inexpensive infrared absorption spectrophotomers it is becoming increasingly common for purchasers of oils or additives to demand a reference sample of the material to be supplied, and to take an infrared spectrum of this sample. It is unlikely to provide much detailed information on the composition of the sample, but can be used as a "fingerprint" to identify the material.

With moderate interpretative skills, the spectrum can be used to identify which of several known additives or oils is under test, and if the composition of these has been changed. However, both base oils and additives can show changes in their absorption spectra, depending on processing conditions or raw material sources, without there being necessarily a change in quality or performance. Use of "fingerprinting" can persuade suppliers to discuss such changes with customers rather than supply without declaration of a change. Some purchasers will reject deliveries of oil or additives if the infrared spectrum of the new product does not match that of a retained reference sample.

Spectral changes are always worthy of discussion with a supplier, with whom there needs to be an understanding of possible and permissible variations. On the other hand, significant quality changes could take place and not show up in an infrared spectrum.

3.2.2.5 Identification of Unknown Materials and Structural Analysis

This type of activity is normally the concern of the research laboratory, rather than the routine oil testing laboratory. Methods used to identify chemical types in unknown samples include gas and liquid chromatography, infrared absorption, mass spectroscopy, and nuclear magnetic resonance. Details of these and other useful techniques will be found in analytical texts and periodicals.

3.3 Performance Testing

The ultimate test for performance quality is service use over an extended period of time. Some reduction of the consequent delay in evaluation may be obtained by conducting controlled field tests, but such tests have to be carefully designed and the control over procedures quite rigorous in order to arrive at significant results. Field testing can be lengthy, and also expensive if special (e.g., pre-measured) units are used for testing. Field testing is discussed under Section 3.3.3 below.

To provide faster and less-expensive quality assessments, laboratory tests are generally used at least in the development phase of new products. For convenience laboratory tests can be divided into two types:

1. Bench performance tests.
2. Engines or other equipment run on a test bed.

Of these, bench performance tests give the quickest evaluation, but may not represent field conditions very well.

3.3.1 Bench Performance Tests

Oxidation Tests

Many simple bench tests for oil stability have been developed over the years, with hot oil being contacted with air or oxygen, and various parameters being measured to assess degree of oxidation. With the increasingly high inhibitor level in crankcase oils these are now mainly used in the testing of base oils and some industrial oils; crankcase oil oxidation tests are mainly engine tests.

Shear Stability Tests

These have already been discussed under viscosity measurements. The sonic shear tester was in vogue for some time, but diesel injector rigs are now most common.

(a)

(b)

(c)

(d)

Plate 5. Laboratory performance tests: a) Noack volatility, b) shear breakdown, c) FZG gear test, d) CEC cam and tappet rig.

In Europe, electrically motored engines, usually with integral gearboxes, were used for some time to evaluate viscosity loss in particular equipment. The tests were not representative of over-the-road conditions as the oils were not subjected to engine heat or contamination. The simpler diesel injector rigs were found to produce equivalent results more quickly and economically.

Cam and Tappet Rigs

These are electrically driven arrangements of cams and tappets which permit evaluation of anti-wear performance of different oils against different metallurgies. The most well-known is the European CEC Cam and Tappet Rig which permits different valve gear components to be mounted and lubricated with various oils, used or unused. Some engine manufacturers have had simpler equipment using their own hardware in the past. This type of apparatus is used more for research and problem-solving than for routine oil evaluation. A weakness of rig-tests of this type running on fresh oils is that no account can be taken of the interaction of anti-wear additives with contaminants, oxidation products and especially soot formed during combustion. This concern is particularly relevant to diesel engine oils.

In Japan an electrically driven Toyota 3A engine has been used to evaluate valve train wear in gasoline engines and has been found to correlate well with fueled MS IIIE and MS VE tests.

Corrosion Tests

The most commonly used test is the copper strip corrosion test (ASTM D 130) in which a polished copper strip is immersed in the sample and heated, and its color and condition after the test is compared with a bank of standards. Over the years manufacturers of large engines have required various bench corrosion tests to be performed on samples of their bearing metals before approving oils. Typical tests would be the silver corrosion test of EMD (The Electro-Motive Division of General Motors, builders of railroad engines) and the Mirrlees corrosion test.

The most frequently performed bench rust test is ASTM D 665, which was originally to measure the rust-prevention characteristics of steam turbine oils but is occasionally used for automotive products. Steel cylinders are immersed in a stirred mixture of oil and water and the rusting measured after 24 hours.

Tribological Testers

These are small testing machines to evaluate the anti-wear and extreme-pressure characteristics of lubricants. They are normally applied to crankcase oils only if the oil is for multi-purpose use, for example if it is to double as a gearbox oil or to be used in non-engine applications in tractors or other equipment. Purveyors of proprietary additives for older engines have frequently used various types of tribological testers, often of unique design to demonstrate anti-wear performance for their products. The correlation of such devices with the real wear problems in an operating engine is at least questionable. Details of standardized tribological testers are included under Gear Oil testing in Section 7.1.

3.3.2 Laboratory Engine Tests

In this type of test a representative engine is mounted on a test bed and provided with fuel, a cooling system, and a means of absorbing the power produced (a dynamometer or "brake"). The engine can range from a full-sized, multi-cylinder commercial engine to a small, single-cylinder engine developed specifically for laboratory testing. It will be run to a specified procedure, continuously or intermittently, and with programmed variations in speed and load. The conditions of operation (temperature, speed, load, etc.) may well be more severe than would be the case in service use, in order to reduce the testing time. After completion of the tests the engine will usually be completely dismantled, the parts rated and/or measured for deposits and wear, and the oil submitted for inspection tests.(11)

For industry-wide tests, and particularly for those used in setting performance specifications, there will be a supervisory body to lay down procedures and arrange for reference oils and laboratory check samples to be

Plate 6. Caterpillar engine test stand.

available. In some aspects of engine performance, fuel quality is an important parameter and this will be carefully specified, and for most tests a stock of fuel may have to be laid down and replenished from time to time. Engines and parts will have to be carefully selected and controlled to ensure consistency, and supplies must be made available even if the test engine is no longer used in production vehicles. Organization of continuing and consistent parts supply is one of the most difficult tasks of overseeing authorities.

The size of engine used for laboratory testing has shown considerable increase over the last few years; full-size large truck engines are now commonly run on test beds. Computer control is now widely used for engine testing, and all of the organization, supplies, control, and rating of engine parts render it both a manpower-intensive and expensive operation.

Military organizations, engine manufacturers, oil companies, and additive companies are all active in test method development, and the latter two in the development of lubricants using engine test evaluation. Over the last 25 years a large independent engine test industry has grown, running tests on a contractual basis for those companies or laboratories which do not have the resources or desire to conduct the tests themselves, or have not installed a particular test in their laboratories. There are so many different tests that no one laboratory can run them all.

Early on, engine manufacturers had run automotive engines on test beds, both for purposes of engine development and for proving tests. The initiation of laboratory engine tests for oil-quality evaluation started with the need of Caterpillar to evaluate oils for their tractors, which was discussed in Section 2.2 on additives. Their single-cylinder test engine, the L-1, was developed in the late 1930s and was adopted by the U.S. military in the first specification to require engine test performance measurement. In 1942 a full-sized Chevrolet gasoline engine was used on a test bed in the L-4 test to evaluate oil oxidation and thickening. Throughout the 1940s and 1950s a large number of test methods were developed and used in specifications. Most test methods were developed around existing commercial engines, using the smaller types where possible. In Europe, low-cost test methods were developed using the small, single-cylinder Petter engines originally designed for dump-truck and concrete-mixer use. In the U.S., a single-cylinder laboratory test engine was developed as cooperative effort to replace the Chevrolet L-4 and remains current today as the Labeco L-38 oxidation and oil thickening test.

In Japan, local engines have been used to produce oil evaluation tests which parallel the U.S. Sequence tests, with some success.[12]

Some research laboratories have developed flexible single-cylinder engines that may be adapted to meet the configuration of commercial multi-cylinder engines. The AVL engine is a good example, and is discussed in Section 6.2.

However, in general, development of engines specifically for test purposes has proved time-consuming and expensive, and there have tended to be doubts on their relevance to the real world. Hence the tendency has grown to use commercial engines of increasing size and complexity as they have evolved. Large engines require larger test beds and dynamometers, are thirsty for fuel, and require more time for rating and rebuilding than smaller types, and therefore the costs of engine test evaluation are escalating rapidly. Details of current engine tests will be covered in Chapter 6 and Appendix 5.

Requirements for Successful Engine Testing

Development of new engine tests is a very time-consuming operation, and maintenance of valid testing in established engine testing laboratories is both expensive and manpower-intensive. In the paragraphs below we will attempt to set out the primary considerations which must be taken into account.

When a new test is developed, it needs to be relevant to quality needs in the field and therefore should reflect as far as possible the normal operating conditions, metallurgy, and fuel quality which are found in service. The latter is becoming a matter of concern in many countries where new vehicles use only unleaded gasoline while most of the oil qualification tests continue to use leaded fuel. A similar discrepancy is anticipated for diesel

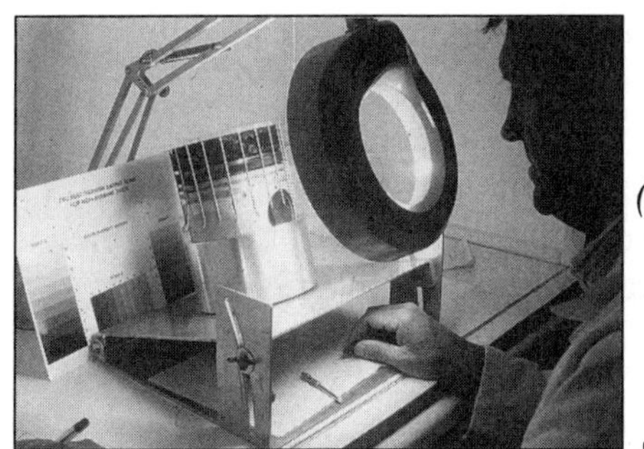

(a)

(b)

(c)

(d)

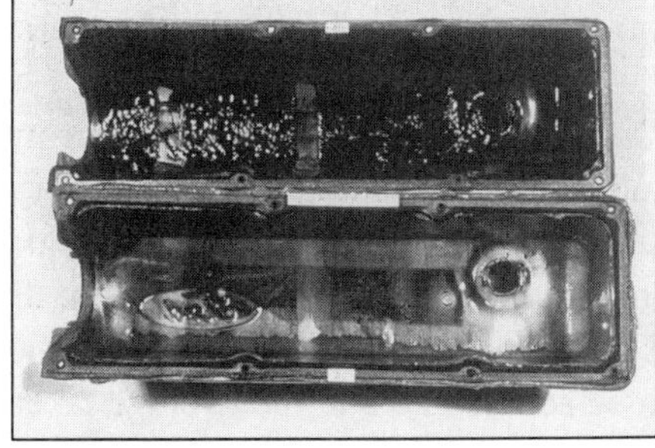

Plate 7. Engine parts and rating. a) Rating booth, b) pass/fail Caterpillar pistons, c) PSA TU3M hot test pistons, (d) pass /fail Sequence V rocker covers.

vehicles as a consequence of legislation for severe limitation of sulfur content of diesel fuel.

Attempts to shorten test time by making conditions very severe are common, but become of doubtful merit if outside the extremes of foreseeable operating conditions. A test must correlate with available field service data to be meaningful.

A test needs to be as repeatable as possible so that a given oil tested at different times in one laboratory produces the same effective result, and reproducible so that different laboratories will agree on the results. To achieve this, tight control of part quality is vital, but so are consistent engine-building techniques, operating practices, and constant fuel quality.

A test is no good if it is very repeatable but does not discriminate. In other words, it must clearly distinguish between oils of different qualities so that when a passing level is defined, oils under test can form a distribution of clear passes, marginal passes, marginal failures, and clear failures. Several proposed new engine tests have been abandoned because discrimination was inadequate.

After a test is run the engine is disassembled and the parts rated. Disassembly and part washing procedures must be standardized, and rating methods carefully spelled out for each test. Skilled and consistent rating comes with experience and training, and, for the more important tests, worldwide rating meetings are organized where selected raters can discuss rating practices and participate in rating exercises.

Rating of wear usually involves measurement by a prescribed technique and is easier than deposit rating. For piston lacquer deposits and sludge rating, special "rating booths" should be used with standard backgrounds and specified types of illumination. For circular objects such as pistons it is usual to use a graticule to divide the part into segments, each of which is rated separately, and the average taken. Rating is normally based on scales of 1 to 10 for different levels of deposit, defined by photographs, and different parts or areas are multiplied by different weighting factors in order to arrive at a correctly balanced assessment.

During test development, passing limits will be established, usually based on test results of oils known from field experience to be borderline satisfactory. For controlling the test after release to testing laboratories, reference oils will be needed to define passing, borderline passing, and failing results. These may or may not be the same as the development oils. Reference oils serve several purposes:

(i) They permit an engine laboratory to judge if they are obtaining acceptable results from their tests.
(ii) They permit, via "blind" testing, a controlling or approval authority to judge if a laboratory's results can be considered valid.
(iii) For some tests and some specifications it is possible to judge pass/fail criteria in relation to a reference oil. This can be in terms of difference between a candidate oil test result and a recent reference oil test result, or the reference results can be used to adjust the initial rating of the test oil. (These rating adjustments are likely to be increasingly used in the future as the so-called "statistical testing" criteria are increasingly used, where all tests results must be considered.)

Because of the problems, engine testing is never as repeatable or reproducible as we would wish. New efforts are underway to improve the situation. Control of parts and fuel quality is essential, along with clear operating procedures; but perhaps the most important thing is to ensure that the most appropriate hardware to measure the test parameter is chosen in the first place (see Sections 6.2.4 and 6.4.1).

Some comments on engine testing statistics are included in Section 3.4.3.

3.3.3 Field Testing

It might be expected that field testing would be the most meaningful performance evaluation to which a lubricant could be subjected, but the truth is that it is surprisingly difficult to obtain worthwhile results from field tests. To do so, a great deal of time and effort has to be expended, and a good field test is not cheap to run. With care, however, valuable results can be obtained, and some field testing is always desirable when new technology has been developed in order to be sure that the standard laboratory tests have adequately screened the new approach.

Field tests generally fall into one of three distinct types: uncontrolled tests, controlled tests, and caravan testing. Each of these categories will be explained and discussed separately.

3.3.3.1 Uncontrolled Testing

This is what many people think of in terms of field testing. A fleet of vehicles chosen indiscriminately from available sources has the oil changed to that under test, and carries on with normal operations. Sometimes the collection of vehicles is split and two oils are compared, one of which may be a standard commercial oil or the oil normally used by the vehicle. Typical sources of vehicles are the personal automobiles of employees, or the haulage trucks and tankers of an oil company or additive supplier. Testing may well continue for over twelve months, and while it continues watch is kept for any unusual signs of engine distress or other operational problems. A score is kept of a number of problems for the test oil versus the reference oil or against some indication of previous experience.

When failures do occur it is very difficult to decide if these are due to the test lubricant, the individual engine and its previous history, or the driving regime to which it has been subjected. If no failures are found then there is a feeling of confidence about the quality of the test oil, but the severity of the test in relation to possibly more extreme service is often not adequately debated.

The best that can be said for this type of testing is that if the oil were to be seriously deficient in some important property (which is, in fact, unlikely) then there is a chance this would show up in this type of testing, provided it was sufficiently extensive. Uncontrolled tests should be regarded as only supportive, rather than definitive of oil quality.

3.3.3.2 Controlled Field Tests

These are tests where at the beginning of the test period the condition of the engine used in the test is known exactly, and test conditions are controlled and monitored. The test is normally commenced with brand new or reconditioned engines. If measurements of wear are to be a feature of the test, then each engine must be stripped down and measured before the test commences. If deposits only are to be assessed, then this stage is not necessary and the engines can be installed without this extra step. At the end of the test the engines are removed from the vehicles, stripped down, and completely rated for deposits, wear, and general condition. In some cases they may have an intermediate inspection during the course of the test in which case they may be partially stripped in order to evaluate how the test is progressing. During the course of the test the condition of the lubricant will be carefully monitored.

When a test is being considered, the precise objectives must be carefully defined. Is the primary concern to measure piston deposits, engine wear, oil consumption, general oil condition, or to study some other criterion? Are average or mixed types of service conditions being considered, or very severe operations? Having decided on the objectives of the test, a suitable fleet or fleets of vehicles must be found. The absolute minimum number of vehicles to be run on each oil in the test is three, but this can easily prove insufficient and a realistic minimum is probably better set at five or six. The vehicles in the test need to be on similar types of duties and cover approximately the same mileages during the course of a year. Typical fleets for automobiles would be limousine or taxi companies, while for heavy-duty oil evaluation a haulage (trucking) company would be a normal choice. Fleet owners can be quite cooperative if offered new engines for their vehicles and free supplies of oil!

A typical arrangement would be for the organizer of the field test to provide the new engines for the vehicles, and in return for this valuable hardware the fleet owner undertakes to run the field test in accordance with agreed guidelines, to make trucks or cars available for intermediate and final assessments, and to provide oil samples as necessary. He will also have to carry a slight risk that there may be engine problems, but the test organizer can protect against excessive downtime by keeping a spare engine available.

As well as supplying and fitting new engines, the test organizer will also have to arrange for continuing supplies of test oil, reference oils and filters, and make careful arrangements with the fleet operator to ensure that the correct oil is used at all times in each vehicle. We regard it as particularly important that some of the vehicles are run on a well-defined reference oil. There is often an inclination to save money by omitting the reference oil but as field test results are usually "good," it is then impossible to tell whether the new oil is actually "better" or "worse" than any other. If the fleet operator is to provide oil samples in between oil changes, then details of sampling procedure must be agreed upon, and suitable receptacles and equipment for sampling must be provided. Details of duties, lubricant consumption, mileages, and any special situations must be recorded by the driver or the fleet operator, and the test organizer should provide paperwork in the nature of forms requiring the minimum effort to complete.

Such controlled field tests are often run by an oil or additive company in cooperation with an equipment manufacturer. This would apply particularly if one of the purposes of the testing is to obtain OEM approval, as for example for "Volvo Drain" approval.

In such cases the OEM will insist on defining the details (including severity) of the test, will often help to find a suitable fleet operator, and will send representatives to examine the engine condition at test completion, and compare the results of test and reference oils.

Even if an approval is not involved, the OEM can provide valuable advice on technical areas of special interest, and will normally be highly interested in the test results.

The length of time needed to run controlled tests varies with the purpose. In Europe, for light-duty vehicles such as taxis, a typical period might be 80,000 km (50,000 miles). By this time any significant developments will usually have become apparent. For heavy-duty trucks, it is more usual to run for a longer period of up to 320,000 km (200,000 miles) which may often take two or three years. This would typically be the mileage up to a first major overhaul. In the U.S., with longer typical journeys, test mileages can be considerably greater.

Such long tests have inherent problems. Apart from the risk of accidents, there are risks that the lubricant technology may have been made obsolete for some reason, such as a change in a critical specification, or the fleet operator may have lost a contract to deliver goods to a particular location, resulting in a change to test severity. Careful choice of fleet is very important, but the test sponsor must be sensitive to changing circumstances, and from time to time may need to abort a test early or modify the objectives. Maintaining good will with the fleet operator is always important.

At the beginning of the test, as the engines are new or reconditioned, it will be necessary to have an initial running-in period. This can be on the test oil or possibly on a special running-in oil, and can serve as a trial of the test procedure. It is important that this is carefully controlled and that all vehicles begin the test proper after the same amount of running-in.

It is quite common for one or two of the vehicles on each test oil to be lost from the program for such reasons as accidents, gasket leaks, or non-oil-related breakdown. These engines are best eliminated from the test, which is why it is considered that three vehicles are not really sufficient for a meaningful trial.

Having participated in one fleet trial many fleet owners are more than willing to continue to partake in such exercises for the benefits they bring. Others, on the other hand, find that the time and trouble involved in ensuring that the correct oils are used and that records are kept make them unwilling to participate again.

3.3.3.3 Caravan Tests

This type of test is the most difficult to set up and the most expensive to run, but can give relatively precise results in a short period of time. The value and meaning of the results depends on the conditions and duration chosen, but in general such tests are extremely useful. Typical objectives of such tests would be to measure parameters such as fuel economy with different lubricants, or emission levels, but in principle a wide range of evaluations is possible.

The concept is that a fleet of identical vehicles traces out a standardized route with specified speeds and timings, and this route is repeated over and over again until the desired mileage has been accumulated. Usually a mixture of slow urban driving in traffic with some fast highway sections will be specified, but this depends on the test objectives. As far as possible all vehicles are required to undergo the same driving conditions, and therefore they often will follow each other in a "caravan" of vehicles. Sometimes vehicles are driven at high speed on test tracks to accumulate mileage quickly. Such high-speed testing alone is suitable for measuring high-temperature performance but is otherwise generally a poor predictor of real marketplace conditions.

Hired drivers and hired vehicles with dummy loads are normally used, with new or reconditioned engines being in place at the start of the test. In some cases such as short-duration fuel-economy testing the reconditioning may consist only of sufficient flushing to remove all traces of the previous lubricant. The driving of the vehicles is very tedious and is usually continuous until the test mileage is achieved, and so several drivers per vehicle may be needed. Drivers may be rotated between vehicles in order to remove any driver effects. Wind speed and acceleration effects may be important for cases such as fuel consumption measurements, and therefore it may be desirable to also rotate the positions of each vehicle in the caravan.

Referencing against results with a standard lubricant is an essential part of most caravan tests. The reference oil can be in some vehicles with test oils in others, or for short-duration tests the various oils can be run sequentially in the same engine. In this case a test could theoretically be run with a single vehicle provided it consistently followed the prescribed route, but it would be usual to average the results from several vehicles.

Due to its relatively short duration, a caravan test will not demonstrate effects which depend on aging of oils or engine deposits, but yields comparative data between oils in a precise and rapid manner.

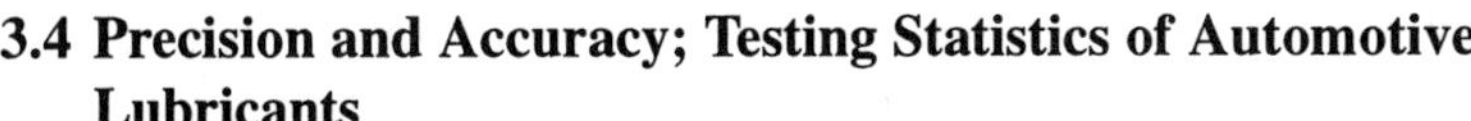

3.4 Precision and Accuracy; Testing Statistics of Automotive Lubricants

3.4.1 Basic Statistical Principles

In this section we want to give a brief introduction to the use of statistics as applied to matters of testing and measurement for those who have little or no familiarity with statistical principles. We hope our simplistic approach will not give too much offense to professional statisticians while setting out the usefulness and limitations of statistical methods to the uninitiated. An important aim is to distinguish between *precision* and *accuracy* of measurements. Statistical methods are more concerned with precision than with accuracy. By *precision* we mean how closely different measurements of a property of a sample approach each other, and therefore how closely any single result is likely to be to the average result of such measurements. By *accuracy* we mean how closely a result, or the average of a series of results, approaches the *true value* of the property. This true value is at least difficult and sometimes impossible to measure absolutely, although it may be possible to state that it lies within a certain range. In some cases, for convenience, it may be *defined* in terms of primary or secondary standard (or reference) materials, and results from the test under investigation assessed by comparison with such nominal values.

A basic concept in the statistical approach is that of *random variation*. In other words, no matter how much care is taken in making a measurement, residual errors will exist which arise from the instrument used, the technique employed, the observations made or some other influencing factor. We are talking about random errors to either side of some mean value, and not consistent errors due to some basic fault in instrument or technique. These random variations will give rise to a *distribution* of results about a *mean value*. The spread of the results about the mean value gives an indication of the precision of the method used, a narrower spread indicating a higher level of precision.

In the absence of other information, the mean result will be taken as the best estimate of the true value, but in fact there is initially no means of knowing how accurate the result really is. This can be assessed by calibration/comparison with standards of known value. Consistent errors give rise to inaccuracy, and will be present in both the measurement of the test and standard samples. If the instrument is badly calibrated, if the technique is faulty (for example by measuring at the wrong temperature), or if there is a consistent calculation error, then the measurements could be precise but inaccurate. Using a series of standard samples and calculating the result for the unknown test sample from the results of the standard samples is the usual way of overcoming inaccuracy. If the test sample lies within the range of the standard samples the process of calculation is called *interpolation*. If, less desirably, the test sample lies outside the range of the standards, then the result has to be *extrapolated.*

We also need to look at the resolution of the instrument or measurement system used. If we measure the length of a small metal bar with a ruler calibrated in millimetres, we may obtain say 10 equal results of 18 millimetres, which looks like good precision because the spread is zero. If a micrometer is used, the spread of results might be between 17.853 and 17.847 mm, or 0.0335% of the mean. The more accurate micrometer is showing a spread of results when no spread was found with the ruler. The result with the ruler is perhaps better expressed as 18.0 ± 0.5 mm, giving a spread of 5.6% of the mean. The micrometer is therefore more precise, although at first the ruler appeared to offer more consistent results. Again, we must point out that the micrometer result is not necessarily more accurate than the ruler. It may have been damaged, not used correctly, or perhaps it was calibrated in inches and an incorrect conversion factor was used. There is no doubt, however, that it is more precise, and if we calibrate it it is possible to check and control the degree of accuracy within certain limits.

First, however, let us look at the spread of test results in more statistical terms. If an object is set up as a reference and measured say at weekly intervals by an operator who is checking an instrument, then the results will show a spread, although some results may occur several times. We can plot the results on a frequency diagram (Figure 57). The results below are rounded to the nearest 0.01:

<u>Results</u>:	
	9.99
	10.02
	10.00
	10.01
	10.01
	9.98
	10.00
	9.98
	10.00
	10.02

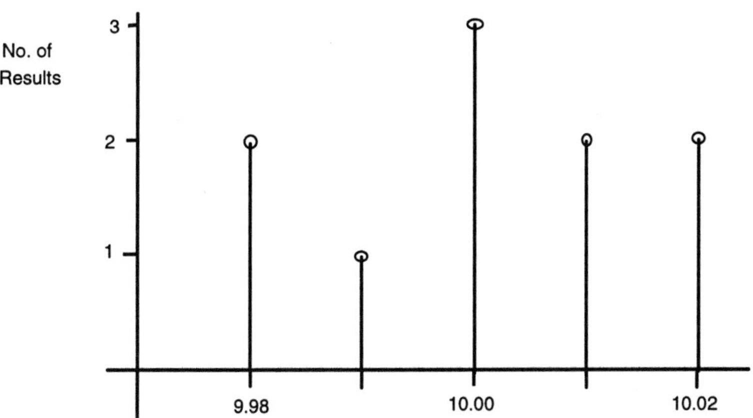

Fig. 57 Frequency of test results.

The precise nature of the measurements and the units used are not important, but from this plot we can see that the mean result is close to 10.0 and the spread is ± 0.02. If we want to assess the likelihood of results falling outside this range, we need more results. If 100 results are plotted, and we do not round them so much, the chart would look more like Figure 58.

Another way to plot the data is to use a *histogram* in which results are not rounded at all but grouped into narrow ranges or cells, and the number of results in each cell is plotted (Figure 59).

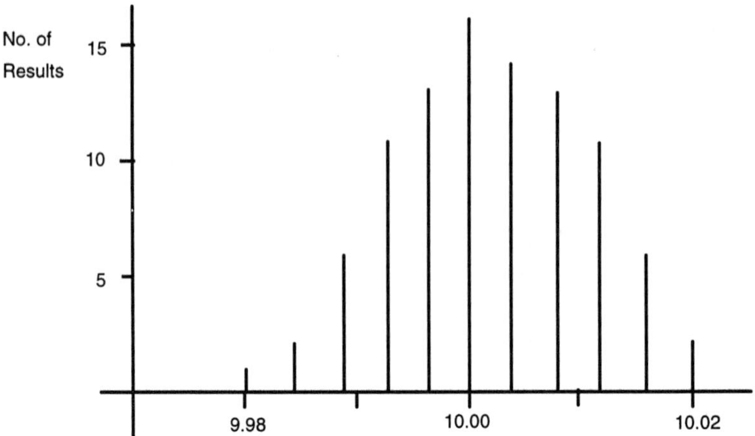

Fig. 58 Effect of more results—less rounding.

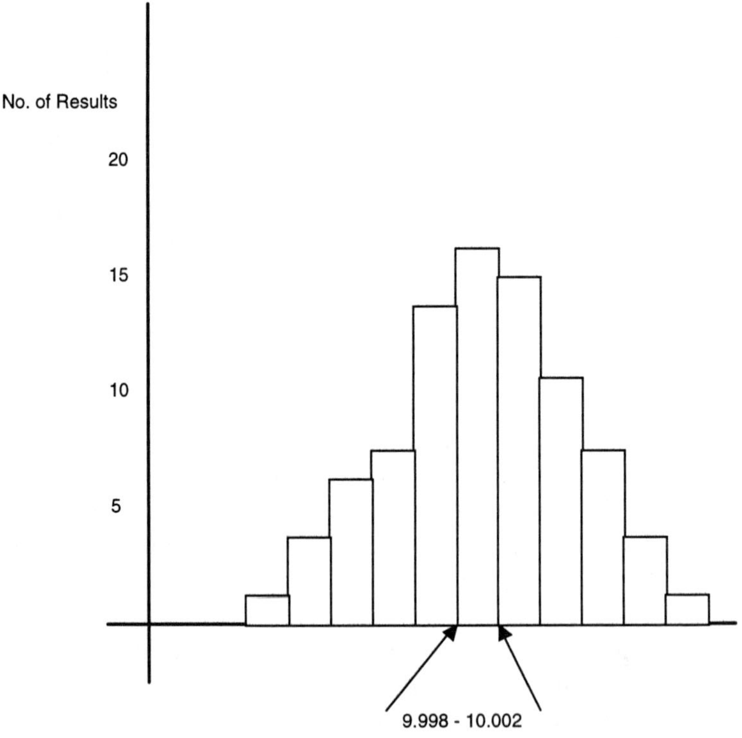

Fig. 59 Histogram of grouped results.

Provided we are dealing with random variation, the more results we have the closer the plot approaches a smooth curve, which is often well-approximated by the so-called *Gaussian Curve* or *normal distribution* (Figure 60).

The degree of spread of the bell-shaped curve depends on the precision of the measurements, and is known as the *dispersion* of the population (of results) (Figure 61).

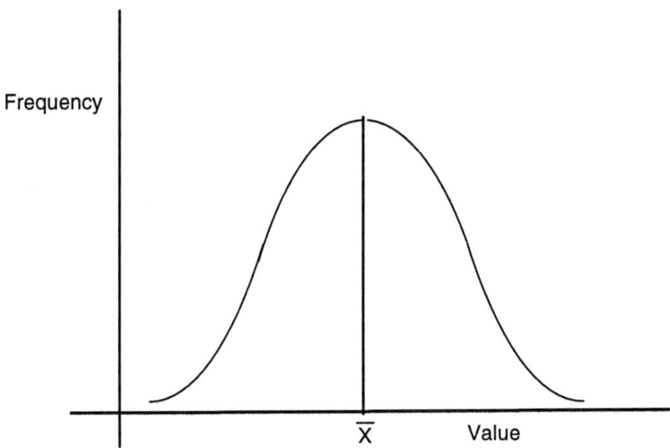

Fig. 60 Normal distribution curve.

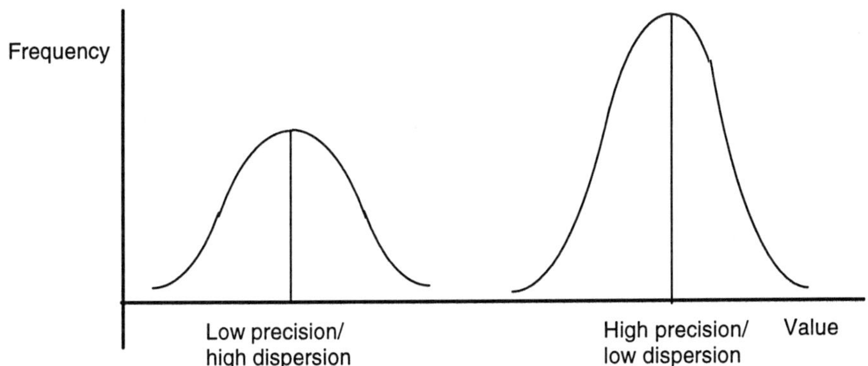

Fig. 61 Dispersion of results.

Dispersion can be expressed in terms of the *range* of the results (highest minus lowest), or in terms of the *standard deviation*. This is calculated from the difference between each result and the mean or average result. The average is normally designated as $\overline{x}$ ("x bar").

Standard Deviation,

$$s = \sqrt{\frac{(x_1 - \overline{x})^2 + (x_2 - \overline{x})^2 + (x_3 - \overline{x})^2 + (x_4 - \overline{x})^2 + (x_5 - \overline{x})^2 \ldots + (x_n - \overline{x})^2}{n - 1}}$$

$$\text{or,}\quad s = \sqrt{\frac{\sum_{i=1}^{n} (x_i - \overline{x})^2}{n - 1}}$$

A related property is the population standard deviation or *root-mean-square deviation*, σ (sigma), which is often in fact used as the standard deviation. It is calculated on the total population of results, whereas estimates of standard deviation, s, can be made on a representative sample.

$$\sigma = \sqrt{\frac{\sum (x_i - \overline{x})^2}{n}}$$

$$\text{or,}\quad \sigma = s\sqrt{\frac{n - 1}{n}}$$

(s and σ are essentially the same if n is large)

Two other expressions frequently used are *variance*, equal to s^2, which is a measure of dispersion which is simply additive, so the dispersion of two distributions can be added together. This is useful if the data result from different sources and need to be consolidated. The other expression is the *coefficient of variation*, the standard deviation divided by the mean:

$$\upsilon = \frac{s}{\overline{x}}$$

This is a pure (or dimensionless) number, whereas the standard deviation is expressed in the units of measurement.

The relationship between the normal distribution curve and σ (or standard deviation(s)) is shown in Figure 62.

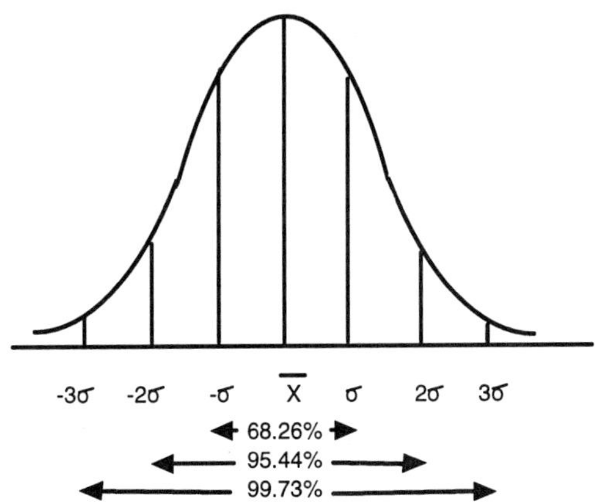

Fig. 62 Proportions of distribution related to standard deviation multiples.

Therefore if the population of results fits a normal distribution and we can estimate σ, then the proportion of results falling outside given limits can be calculated, or, conversely, the chances of obtaining a particular result. Useful estimates of s and σ can be made from very few results (or "degrees of freedom"), but the quality of the estimate will improve as the number of results is increased.

So far, we have assumed that the distribution of results about the mean is symmetrical. This is not always the case, and the distribution can be *skewed* (Figure 63).

The mean result $\overline{x}$ is no longer at the center of the range, and this type of distribution is not so easy to use. Where possible, the data is *transformed* (for example, by taking logarithms of the results) to convert the data into a *normal distribution.*

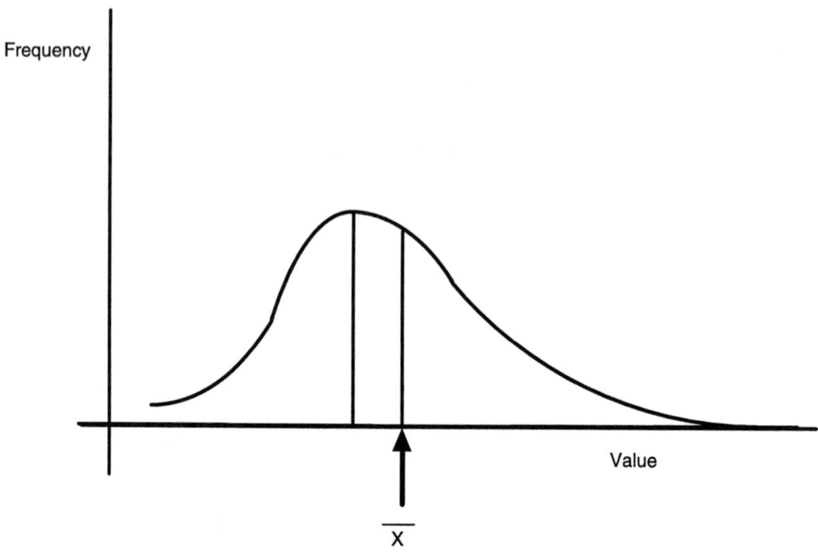

Fig. 63 A skewed distribution.

3.4.1.1 Complex Situations

In our example, we considered one operator measuring a single object with a single test instrument. If we introduce a second operator and/or a second means of measurement, then we find two overlapping distributions which probably will not have the same mean (Figure 64).

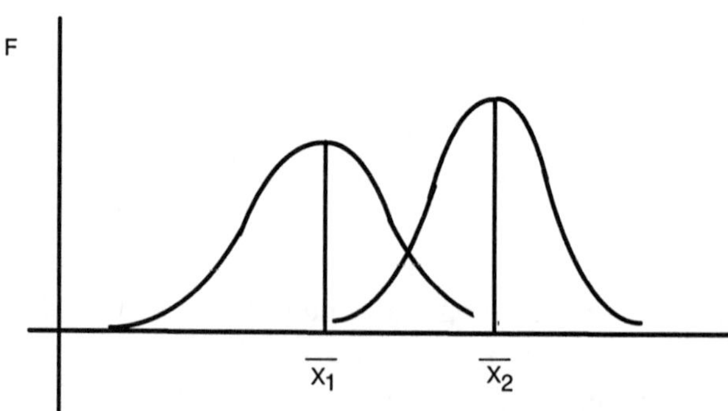

Fig. 64 Two related distributions.

With several operators, several measurement techniques, or other sources of variation, then a new wider distribution may emerge as the sum of the individual distributions (Figure 65).

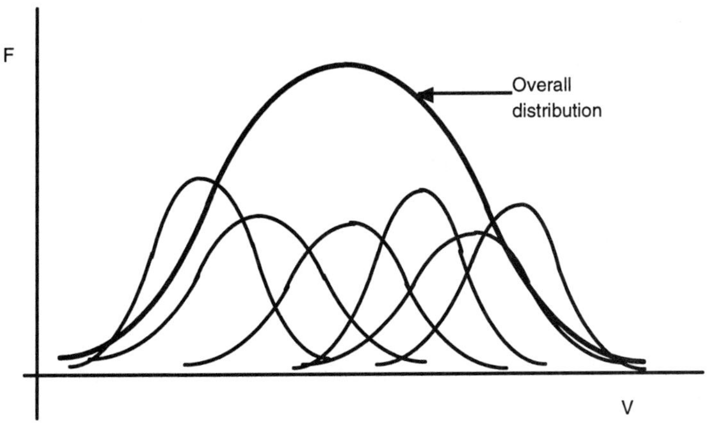

Fig. 65 Multiple related distributions.

It is often necessary to analyze a wide and complex distribution for the different contributing factors involved. In our example, if there are many different objects to be measured (for example, from a production line) then the overall spread would be expected to increase significantly. Separating production variation from testing variation is a common and important exercise. In general, assignable causes of variation should be sought in order to simplify a mixture of distributions and attempt to find the major sources of difference.

3.4.1.2 Accuracy

It might be thought that the more operators and measuring techniques employed, the closer to the true value the overall mean would be. This could only be true if bias errors in some measurements cancelled those in the opposite direction in others. As explained earlier, to assess accuracy, calibration is performed with reference samples of similar type to those to be measured, and whose properties are known to a high degree of accuracy.

These may be made to be of certain values, or may be measured by the most precise and accurate methods available, on an exceptional rather than routine basis. (There will always remain some doubt as to the exact values of the properties measured.) These reference samples (or in some cases a single sample) are tested by the routine method(s), and the results compared with the declared values of each sample. The spread of results (standard deviation) can be used to see how many results need to be averaged to meet the required degree of precision. If the means do not agree with the declared values, then the testing can be investigated to try to determine the source of error. In certain cases, the difference between the declared value of a standard and the mean of its test results can be used to apply a correction to the measurements. It is, of course, essential to ensure that the merit of the declared value of the chosen reference sample is greater than that of the merit of the values obtained by routine testing. It is not always possible to be sure of this, and sometimes for operational convenience values of reference samples are declared as standards when their true accuracy or merit is not known. While this is undesirable, it is a way out of the problem of consistent differences between two or more testers (e.g., analytical laboratories) which arises from use of different standards.

3.4.2 Laboratory Tests on Petroleum Products

The results of petroleum tests are generally much less precise than physical measurements of length, both for general properties and in particular in relation to the chemical composition. The reasons for the poor precision of the latter are discussed in Section 3.4.1 above, but the situation causes problems in connection with quality control, and particularly in agreeing oil specifications between suppliers and end users, or in connection with additive specifications between additive supplier and oil blender.

Thc ASTM and other national testing or petroleum organizations publish precision data on the various tests used in petroleum laboratories. These data are arrived at from lengthy cooperative testing programs in which samples are carefully prepared and circulated to participating laboratories who test them according to laid-down procedures. The results are analyzed statistically, and values of *repeatability* and *reproducibility* are calculated.

Repeatability is defined as: "A quantitative expression of the random error associated with a single operator in a given laboratory, obtaining repetitive results with the same apparatus under constant operating conditions on identical test material. It is defined as the difference between two such results at the 95% confidence level."

Put simply, this means that a single test operative should not find a difference greater than the repeatability value between any pair of duplicate results, in 19 cases out of 20.

Reproducibility brings other laboratories and locations into the picture, but the method should be the same for each, and samples are maintained as near identical as possible. It is defined as: "A quantitative expression of the random error associated with operators in different laboratories, each obtaining a single result on an identical test sample when applying the same method. It is defined as the difference between two such single and independent results at the 95% confidence level."

Values for reproducibility tend to be several times those for repeatability (see Appendix 6 for the values for common tests). Apart from highlighting the difficulties of comparative testing, their significance for routine laboratory testing is considered to be less than the values for repeatability. Single test results are compared, whereas in analysis or testing for physical properties duplicate results would normally be obtained and averaged. Again, different laboratories may be more or less familiar with the specific technique being correlated, and some would not normally regard it as their preferred method. Thus reproducibility values may not refer to the optimum testing regime for each laboratory, and not reflect the best comparison of results which could be obtained without restraint.

Repeatability and reproducibility are related to the standard deviations of the sets (populations) of results obtained from cooperative programs, grouped in terms of single operator or different operators. The exact relation depends on the number of degrees of freedom (related to the number of results) and is 2.77 s for an infinite number of results rising to 2.8 s for 100 results and 3.15 s for only 10 results.

The significance of either parameter in relation to a normal distribution is shown in Figure 66.

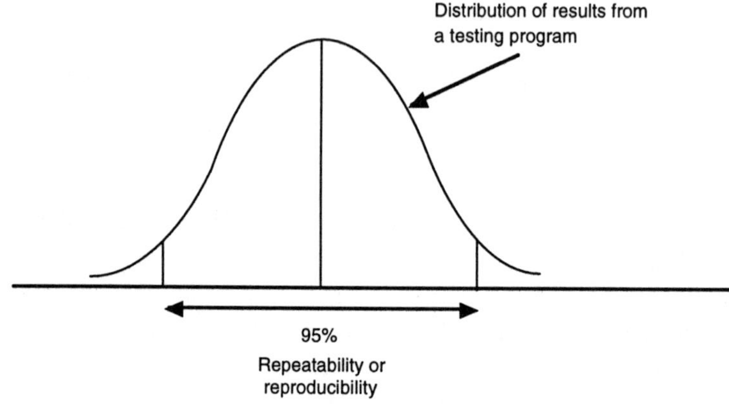

Fig. 66 Test precision measurement.

3.4.2.1 Setting Specification Limits

While published values of repeatability and reproducibility are of some interest, they are not particularly helpful in connection with specification setting. For a manufacturing specification we need to consider the degree to which manufacture can be controlled and the precision which the plant laboratory can achieve. A period of testing on early production samples will indicate where limits could be set. The standard deviation of all results on acceptable production can be used to set limits of $\pm$4 s, at which 99.99% of future similar product will test "in-grade."

Note that the overall standard deviation can be considered the result of two components, the production variation and the testing precision. These can be added if variance V is used, equal to s^2.

$$V_{Total} = V_1 + V_2 \text{ or } s_{Total} = \sqrt{\frac{s_1^2 + s_2^2}{2}}$$

Such production specification limits are all very well if the producer has a well-controlled process and the purchasers accept his product without question. If, however, the purchaser wishes to specify the quality limits he will accept against his own testing, then things become much more complicated. If the two parties can agree on the acceptable variation of the product, problems of testing tolerances still have to be resolved. The testing Variance (or standard deviation) has to reflect the fact that two laboratories, at least two operators, and possibly different test methods are involved. For reasons outlined above, the published reproducibility values are not much help, particularly for chemical analysis where they would frequently suggest unacceptably wide specification limits. What has to be done is for the two parties' laboratories to get together to discuss testing methods and agree on one or more reference samples of stated composition. A joint testing program is run on the reference and other samples and the between-laboratories precision is calculated. A specification can then be set of at least ± 3 s overall, when there is approximately a 95% chance of results on on-grade product meeting the specification.

If the purchaser finds the width of this specification excessive, then negotiations must take place, and the supplier must see if he can reduce his manufacturing variation. The two laboratories must also try to reduce the testing Variance. It is no solution, and will only lead to problems, if the specification is set at less than ± 3 s of the determined overall variation.

An interesting expression of this, which has been written-up by the Ford Motor Company, is that of Process and Machine Capability. These employ the same concept, with Machine Capability applying to manufacture of parts to meet a given specification, and Process Capability applying to a complete process, which can be taken to include several stages of manufacture, and by extension, the testing process as well. The specification width is related to the machine or process variability as shown in Figure 67.

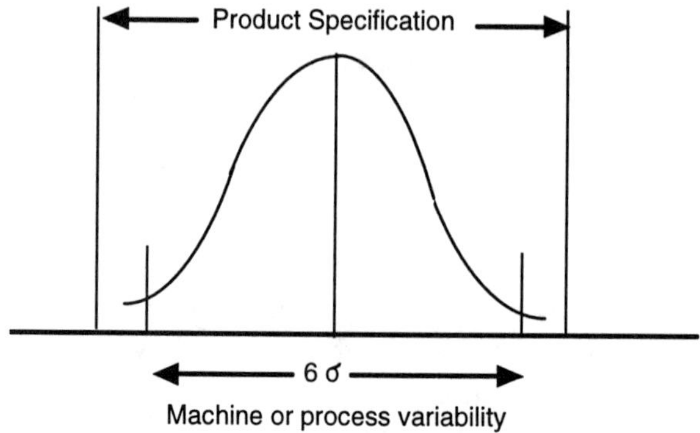

Fig. 67 Machine or process capability relates specification width to achieved precision.

If the standard deviation of the machine or process is measured as s, then

$$\text{Machine Capability, } C_m = \frac{\text{Specification Width}}{6s}$$

$$\text{Process Capability, } C_p = \frac{\text{Specification Width}}{6s}$$

Thus, if the specification happens to be at ± 3 s, then C_m or $C_p = 1.0$.

If the specification is at ± 4 s, then C_m or $C_p = 1.33$.

However, if production is not centered on the mid-point of the specification, then the effective capabilities C_{mk} and C_{pk} are used, based on the mean x of the results (Figure 68).

C_{mk} or C_{pk} is the minimum of $\dfrac{\text{Spec. Max} - x}{3s}$ and $\dfrac{x - \text{Spec. Min}}{3s}$.

A penalty is therefore imposed if production is not at mid-specification.

It is essential that both the theoretical and effective capabilities are kept above a value of 1.0, or there will be many production rejects.

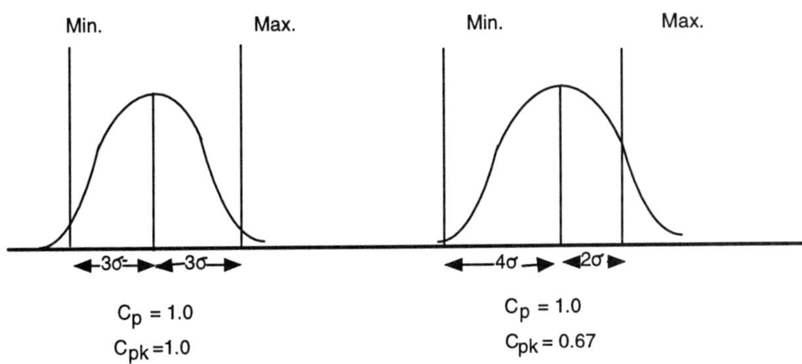

Fig. 68 Effective capability falls if production is not centered on specification mid-point.

3.4.3 Engine Tests

The basic principles of statistics as applied to engine testing are no different from those applied in other areas, but the perspective is a little different and the area has special problems. These special problems arise particularly because measurement is not at all precise, and because testing costs are orders of magnitude greater than for simple laboratory bench tests.

Single parameters such as mass, length or volume can be measured with a standard deviation of better than one percent and often better than (less than) 0.1%. Standard laboratory bench tests on lubricants will typically have standard deviations in the range 1-10% with the simpler tests such as measurement of kinematic viscosity being at the low end of the range and more complex tests such as rig-tests being at the other extreme.

Engine tests will have standard deviations in the range 5-30% but can be even worse than 30% for some parameters in some tests.

The usual approach to using a test of inherently poor precision is to run replicate tests, but when engine tests may cost in excess of $50,000 each, no tester can afford to pay to achieve an exceedingly high level of confidence in the result. There is, however, no alternative to the use of engine tests if formulators are to have confidence that oils will perform acceptably in the real world.

The concepts of repeatability and reproducibility become blurred when engine testing is considered. An engine test runs usually for many hours, or even weeks, and during this time may be tended by many operators. Sometimes an engine block is used many times with test components such as bearings, valves, or pistons and rings changed for every test. Sometimes the complete engine may be replaced. Within a single laboratory there may be only one test cell or "stand" capable of taking a particular engine test or there may be many. Test variation will be greater if many stands are used than if the same one is always employed.

Sometimes a laboratory may employ only one rater to measure test performance. In others many may be used. Again a greater variance will occur if many raters are used.

For historical reasons relating to the nature of the market, U.S. testing tends generally to consider engine testing statistics in terms of individual test stands, whereas in Europe the laboratory is the unit which defines repeatability, rather than the stand. The U.S. has a relatively small number of standard engine tests run in large numbers of cells in a few laboratories. In Europe there are more types of test and more laboratories, but it is common for laboratories to have only one test stand for a particular engine type.

An engine test method has to be carefully devised so that for minimum cost and time the test can identify oils which are satisfactory and reject those that are not. The test usually has to correlate with field performance or it is useless, and before a test is devised it is essential to find a range of oils which produce both satisfactory and unsatisfactory results in the field. The test in the laboratory then has to rank the oils in the same order and show sufficient *discrimination* to separate the satisfactory from the unsatisfactory. This will partly depend on the test *precision* but not only on this. It is possible that the test will initially give very similar results on a wide range of oil qualities, in which case it will need to be modified. Figures 69 and 70 illustrate the effects of poor precision and of poor discrimination on differentiating oils.

In the examples in Figure 69, it is difficult to distinguish A from B, for many of the possible results which could be obtained could apply to either oil. In Figure 70 the combination of good precision and good discrimination enable the quality of the two oils to be easily distinguished.

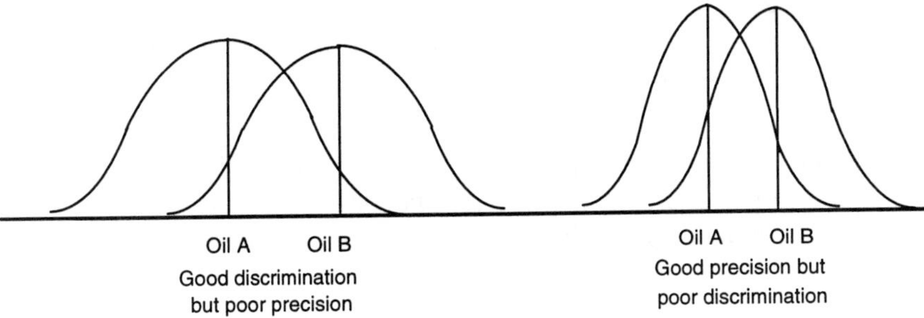

Fig. 69 Oils A and B are not easily distinguished.

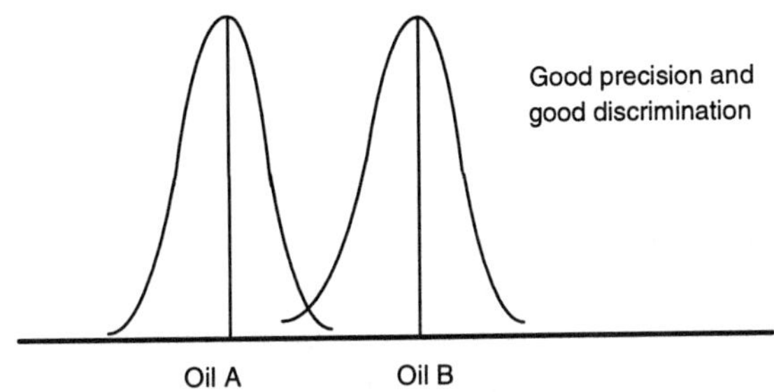

Fig. 70 Oils A and B are clearly different.

In developing a test method, precision is measured by repeatability and reproducibility defined in the same way as for other test methods subject to the limitations mentioned earlier. The discrimination is also defined in relation to 95% confidence limits, in other words, the difference between two results (of a particular rated test parameter) which can be distinguished to 95% confidence. Normally when a test is set up, reference oils are developed with the following qualities in relation to the required standard:

Failing
Borderline Pass/Fail
Passing
and sometimes, Excellent Pass

The discrimination should be such that at normal levels of precision these qualities can be distinguished by the test. A test may well have good discrimination for one property (e.g., piston deposits) but not for another (e.g., cam wear), so that it is important that the test is used to judge the qualities where it discriminates well, and not those where it does not. This has not always been the case in the past.

Given that a satisfactory test and limits have been set up, the attitude to the limits and test precision has in the past depended on the reason for testing. Until recently it was permissible to test an oil for an indefinite number of times, and a single pass among many failures was acceptable to most of the various approval authorities (with notable exceptions). This practice enabled an oil formulator, at least in theory, to balance the cost of reformulation or extra additive against the cost of additional testing to achieve a passing result. This technique of a "bounced" pass was assisted by poor test precision, so there was little incentive for the formulator to improve it. This is illustrated in Figures 71 and 72.

It is uncertain to what extent this tactic was employed as a deliberate means of obtaining economic formulations, but it is undoubtedly true that when a formulation was close to finalization, and considered adequate for its purpose by an expert formulator, then repeat testing to obtain the final passing results was routinely practiced. Approval authorities who permitted repeat testing were fully aware of the situation, although not all mem-

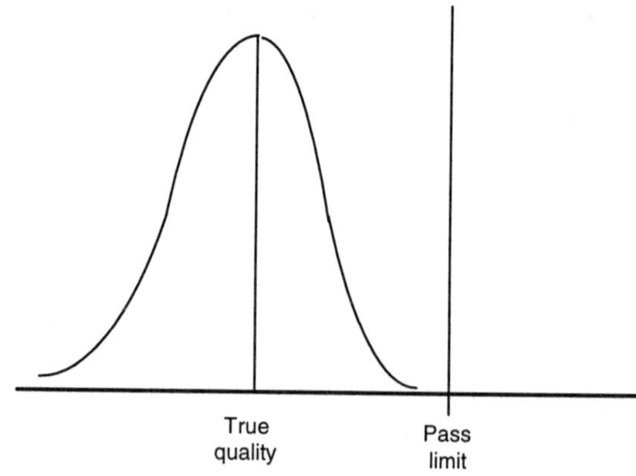

Fig. 71 No chance of a passing result.

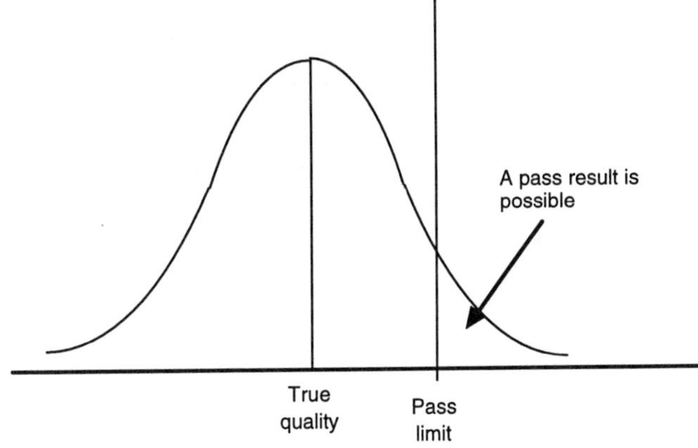

Fig. 72 The possibility of a "bounced" pass.

bers of their review committees approved. It was allowed to continue on the basis that passing limits in some cases were set at such a severe level that any pass justified acceptance of an oil. The U.S. military propounded this philosophy in face-to-face discussions with one of us (A.J.C.) in the early 1970s. The passenger car OEMs on the review panel were never happy with the situation, however, and felt that having set realistic passing levels for tests in their engines they finished up with oils of lower quality than intended. It should be said that the "bounced pass" was generally more associated with diesel deposit testing than with gasoline engine tests, and frequently involved the Caterpillar diesel tests where failing results were often indistinguishable from passes except to a skilled rater.

This situation has long been deemed unsatisfactory by many, and particularly the OEMs. Some European manufacturers in effect demanded "first time passes," while in the U.S. the emergence of the "Tripartite" approval body resulted in pass limits being set at levels equivalent to satisfactory field performance: levels which some OEMs still considered too low. Led by the additive industry on both sides of the Atlantic, new testing rules have been introduced through the CMA Code of Practice (see Section 6.3.1) and adopted by the API and some key approval authorities. These are very similar to those which have always applied to approvals for the British Ministry of Defence, and require tests to be registered before running, and only one test to be normally allowed. This can be described as the "pass first time" rule. For borderline results and where there are grounds for expecting the oil to be of suitable quality despite an initial fail, repeat testing is permitted and an oil will be accepted if the average result of the tests (including the initial failure) amounts to a passing result. One test result may be discarded if three or more results are available. This is the so-called "statistical testing" scheme, which is only statistical insofar as it recognizes the possibility of variability of test results. (The British scheme differed in that two passing results were required to nullify an initial failure, and the results were not averaged.)

As a result of these new rules, we can expect a significant increase in the overall level of oil quality, although for new tests it may be that passing standard will be set somewhat lower than if the multiple testing concept was still permitted.

3.4.4 Reference Lubricants

There are particular difficulties in finding suitable reference oils for engine tests, partly because engine tests often measure several properties at once, and partly because there is usually a need for correlation with field performance which is not usually required with other test methods.

In an ideal world, several reference oils will be available which have shown good correlation with the field and which give good discrimination in the test. Such a situation is illustrated in Figure 73.

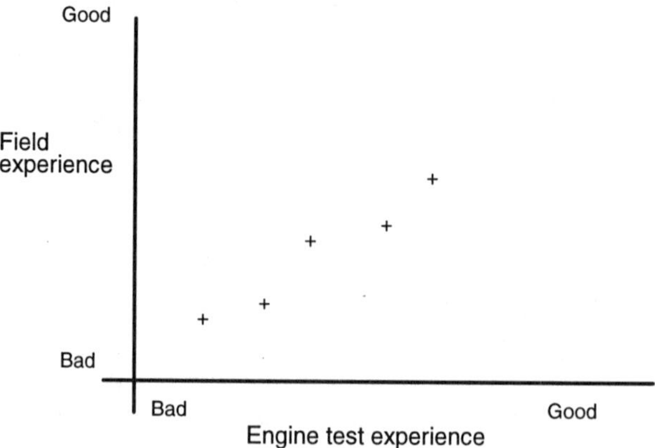

Fig. 73 An ideal set of reference oils.

The reality is that a good spread of results may possibly be obtained with specially selected calibration oils, but that some of these (good and bad) may have been chosen to give a certain performance in the test and may not be suitable for commercial use in the service being evaluated. Oils with genuine field experience often give results within a narrow range, reflecting the fact that they were probably developed to meet the same broad targets but some met these better than others. The situation when oils with field experience are used for calibration is illustrated in Figure 74.

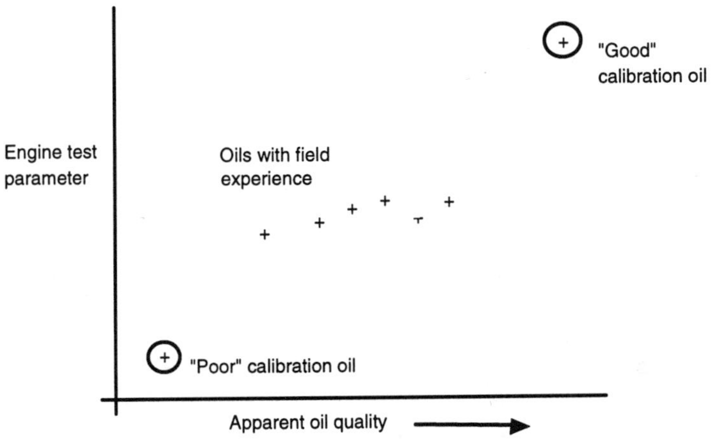

Fig. 74 Problems with specific test parameters.

Current thinking among lubricant engine testing experts in the U.S. is that oils should not be accepted as industry reference oils unless they do have field experience. This approach has undoubted merits but can have the debit that a test may be of doubtful benefit for the development of future higher-quality oils because the response is unknown. Figure 75 attempts to show the problem of extrapolating higher field performance from a series of calibration oils which are not too dissimilar.

It has also been proposed that a good reference oil for any one test should have at least acceptable performance in all others in a specification. This is another way of saying that it must give adequate field performance if so used and makes a lot of sense for ensuring that realistic specification limits are set, but may be somewhat restrictive for development of the test itself.

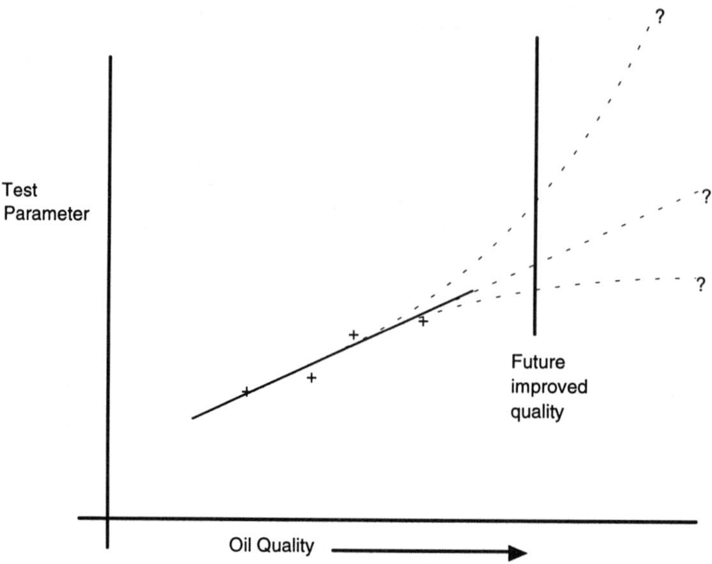

Fig. 75 Problem of extrapolation to a higher-quality level.

Another difficulty arises where tests may measure more than one parameter, for example, both engine deposits and wear. An oil may perhaps have excellent credentials based on field experience to be a good reference oil for one test quality, but may not necessarily be an appropriate candidate for measuring the other quality.

In the U.S., reference lubricants for standard industry tests are stocked by the ASTM. For laboratory results to be accepted for API licensing, the laboratory must run a series of reference oil tests. Some of these will be "blind," i.e., the oil is known to be a reference oil but the nature of the oil is not disclosed. Some will be "double-blind," i.e., the oil will be submitted to the laboratory as if it were a "candidate oil" for approval purposes. Laboratories which report results for the oil outside the tolerance band built up from industry experience will have the test stand declared "non-approved" and will have to take corrective action.

In Europe, the CEC (see Section 5.3.2) provides reference lubricants and runs "round-robins" at intervals to measure test precision and discrimination. CEC has, however, no power over the laboratories to declare a test stand unacceptable or unapproved and for most current European tests does not keep ongoing records of reference oil results to check for changes in test severity or precision.

There are great dangers in making assumptions about the correlation of field performance with engine (or rig-test) performance based on limited data and particularly on data using a single additive chemistry. Shear stability provides an illustrative example and Figure 76 shows plots of shear stability against polymer molecular weight. The first plot shows shear stability as seen in field performance and the second shear stability as measured in a laboratory test.

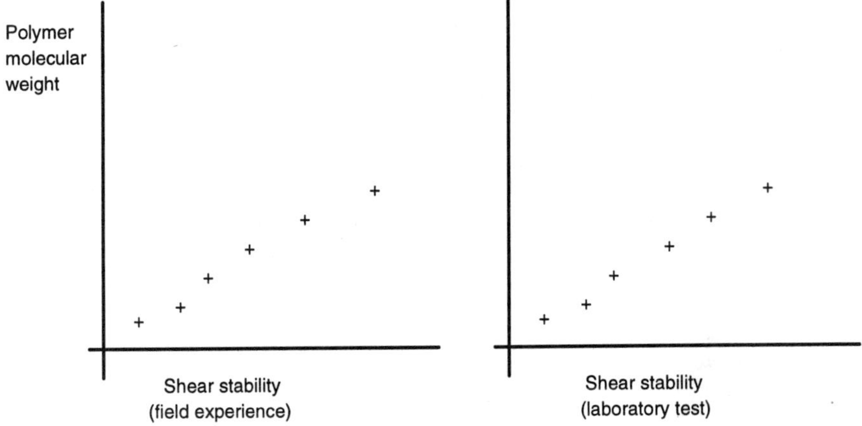

Fig. 76 Good lab/field correlation with a single polymer system.

For this product, which we will call chemistry A, the laboratory test discriminates well and can be used to predict field performance. For a different product, which we will call chemistry B, similar plots could be produced and the same conclusion drawn. Taken separately, both chemistry A and chemistry B can provide good correlations between polymer molecular weight and shear stability in both the field and in the laboratory test. It might be concluded therefore that the laboratory test may safely be used to predict field performance. However, when the two chemistries are compared directly it is found that they do not correlate with each other, and therefore the test is not of general validity (Figure 77).

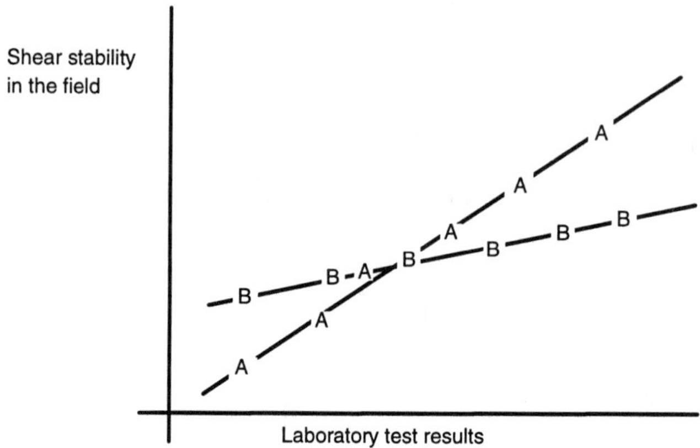

Fig. 77 Lab/field correlation for two different polymers.

Any emerging test for lubricant evaluation should therefore be checked at an early stage to ensure that it is not overly sensitive or insensitive to particular types of additive chemistry.

3.4.5 Reference Fuels

Reference engine fuels are required for a variety of reasons but primarily to test the performance of engines or evaluate the properties of other fuels. An important area also is the provision of reference fuels to evaluate the exhaust emissions and fuel economy of engines and vehicles for legislative purposes.

The quality of fuel used in engine evaluation of lubricants is an important variable, which must be eliminated as far as possible to permit evaluation of lubricant performance alone. Reference fuels to evaluate the performance of engine lubricants are a small but important section of the overall reference fuels business, and have special requirements. Because of the special nature of the market, very few companies are reference fuels suppliers, but examples include Phillips (U.S.) and Haltermann (Germany). Special requirements for reference fuels include:

- long-term stability
- geographical availability
- availability of appropriate quantities
- ability to be able to make equivalent future batches

For some other purposes, fuels sold to commercial specifications or national standards may be adequate for engine test purposes, but commercial variability is too great for this approach to be taken with lubricant testing. In almost all cases a batch is blended from carefully chosen components and stored until needed. Gasoline is particularly difficult to keep for a long time without deterioration. Often it will be stored under a nitrogen blanket. Typically, the supplier will work with the test sponsor (usually a technical society) to decide on the batch size and method of financing. Invariably, if future repeat batches have to be made, they will need to be engine tested to see if performance has been maintained. Sometimes it proves impossible to make an exact copy of an earlier batch and the test sponsor has to decide how to minimize the problem. Options include modifying the length of the test or changing the severity of the test targets.

Particular care is needed when choosing reference fuels to ensure that the qualities are chosen for the right reasons. As testing is expensive, any way which reduces test length is usually welcomed. Thus a high sulfur fuel will generate deposits and wear more quickly than a low sulfur fuel; an oxidatively unstable fuel will generate more sludge than a stable one. There is a temptation for test developers to choose reference fuels without consideration as to whether the mode of action is the same as in the commercial world. Often it can be found that the pattern of deposits found in a test engine is quite unlike that found in the field, and the validity of the test may then be challenged. As mentioned elsewhere, early consideration will have to be given to the validity of the older engine tests run on leaded gasoline or high sulfur diesel fuel when fuels in the field are changing rapidly.

3.5 Tests on Used Oils

Used oil testing serves different purposes to the testing of new oils, and therefore requires consideration in a separate section. Tests on new oils are generally to ensure conformity with a specification (e.g., viscosity, low-temperature properties, engine performance, etc.), while used oil testing provides information about the condition of the oil and its suitability for further use and/or the condition of the equipment in which it is being used. It is convenient to separate these two objectives, although one set of tests may well be sufficient to make a judgment on both aspects.(13)

3.5.1 Tests for Evaluation of Oil Condition

Oils have finite useful lives although these have been increasing considerably in recent years. A private motorist will usually safeguard both his warranty and his engine by adhering to the car manufacturer's recommendations, which are carefully judged to balance maintenance frequency against possible problems. The motor manufacturers tend to set conservative oil change periods because in the past unforeseen problems have arisen due to combinations of such factors as fuel quality changes, emission control additions, engine design modification, or changes in patterns of vehicle usage. The manufacturers tend to rely heavily on assessment of engine condition on strip-down after field trials, but also perform used oil analyses and rely on assistance from the oil and additive companies for oil analysis data if problems are found in the field.

In the commercial field, particularly for big fleet users, it is more common to change the oil only when it is unfit for further use, rather than relying on fixed oil change periods.(14) The economic incentives to do this increase as the size of the equipment increases as well as the size of the fleet, because of the cost of a new charge of oil. Users such as the railroads and power generators also find the practice attractive. Such users normally measure oil life by the numbers of hours run, and will lay down the hours at which oil should be sampled and examined, plotting oil condition against hours run and setting condemning limits at which an oil change will be ordered.

In service an oil becomes oxidized and contaminated with fuel, soot and other combustion residues, and possibly with coolant. Viscosity modifier, if present, may lose its thickening power, and other additives may be used up by their action in overcoming contamination or potential wear and corrosion. Used oil analysis can give an indication of the degree of deterioration of an oil, but it must be stressed it is not a precise science, and experience and judgment are required for the interpretation of test results.

Important tests are as follows:

Viscosity

Viscosity increases due to:

(i) Oxidation of the base oil
(ii) The build-up of suspended insoluble matter
(iii) Build-up of dissolved resinous material (from combustion residues)

Viscosity decreases due to:

(i) Fuel dilution
(ii) Shear breakdown of viscosity modifier

Flash Point

May decrease due to fuel contamination.
May increase due to gross coolant contamination.

Fuel Dilution

This is present to some degree in all used oils, and some depression of flash point will be found. Fuel dilution of the oil from a gasoline engine can be measured by a distillation test such as ASTM D 322 which is relatively simple and gives a direct answer. However, for diesel contamination, distillation would have to be done under a vacuum, making the test more complicated. There is also the possibility of some crossover between the heavy ends of the fuel and some of the lighter material in the lubricant or additives. Gas chromatographic methods such as ASTM D 3524 have therefore replaced vacuum distillation as a means of assessing diesel dilution. (A similar method also exists for gasoline dilution [ASTM D 3525].)

Insolubles Content

The standard test is ASTM D 893 in which a sample of the oil is mixed with either pentane or toluene and centrifuged. The solids are precipitated, dried and weighed. A coagulant may be added to assist if desired. The toluene insolubles contain solid carbon and soot, inorganic contaminants, and wear residues, and so represent essentially extraneous material which has entered the oil. The pentane insolubles also include resins and organic material coming from oil oxidation and polymerization of fuel residues. The difference between the two figures (pentane insolubles minus toluene insolubles) gives an indication of oil degradation and the level of deposit precursors.

The quantity of insoluble matter that an oil can carry depends on the detergency/dispersancy level and can be several percent for a highly formulated oil. Monitoring of the insolubles level will eventually show a drop in insolubles content, at which time an oil change is overdue, for the insolubles are forming a sludge in the engine and the oilways and not remaining suspended. Once this level has been determined for a given lubricant it can be used to set a lower figure as a condemning limit.

Acid Number (ASTM D 664)

For a non-detergent oil (e.g., transmission oils, hydraulic oils) the development of a significant acid number indicates oxidation of the base oil. For engine oils the strong acids arising from fuel combustion are neutralized by the alkaline detergents and are not measurable directly.

Base Number (ASTM D 4739 or D 2896)

A detergent oil when new has a set level of alkalinity or base number. Minimum base number requirements are often set against manufacturers' lubricant specifications, and oil companies will usually quote typical base number levels for a brand of lubricant.

In service the base number decreases as the additive is used up in neutralizing acids, principally those from the combustion process. Base number is therefore a measure of the amount of detergency remaining in an oil. However, it is not sure whether all the base number measured is still effective as detergent, or indeed if any of the neutralized material may still play a useful role. Different oils give different rates of decline of base number depending on which method is used. In setting condemning limits for base number it is therefore important that the method is specified.

For many years the simple titration method D 664 was preferred for used oils, giving lower values than the D 2896 method which measures both strong and weak bases. However, for used oils in particular the D 664 method was poorly repeatable, and base number determination by this method has now been withdrawn as an official method. The preferred method for used oils is now D 4739, a potentiometric method using slow titration which can provide estimates of both strong and weak bases present.

The D 2896 method may remain in use for used oils for some time, as it is a quicker method and is preferred for new oil assessment. With any method, condemning limits for used oils must be set in the light of experience.

Additive Element Determination

Attempts have been made from time to time to assess the residual effective additive content by measuring the additive elements present after an oil has been stripped of suspended solids. Ultra-fine filtration or precipitation of the suspended material followed by analysis of the purified oil have been popular approaches. However, so little is known about the relationship between residual dissolved additive and the effective detergency remaining that such methods must be viewed with caution. It would at least be necessary to develop historical results for a given oil in a particular service and relate these to known engine histories before it would be possible to set condemning limits in this way.

The determination of total additive content in a used oil is also difficult and unreliable. Sometimes there are indications that the total additive content may rise with service, possibly by concentration around the insolubles, while oil which is burned is relatively depleted. However, the determination of additive elements in used oils is subject to severe interferences whether this be by wet chemical or by spectrometric methods, and its main use is in confirming that the engine has been provided with the correct grade and type of lubricant.

Infrared Spectrophotometry

This technique is increasingly available in marketing service or plant laboratories, and can give useful information on the presence of undesirable contamination. It is normal to compare the used oil with a sample of the same oil in virgin condition, preferably in a double-beam instrument. With a suitable instrument the following can be detected:

- Oxidation products
- Water contamination
- Glycol contamination
- Fuel dilution

There is mutual interference between some of these contaminants and also with additives present in the oil, so that it is necessary to regard detection of these materials as possible indications of their presence rather than as a quantitative determination.

"Blotter Spots" and Paper Chromatography

The "blotter spot test" has been the subject of much controversy over the years, tending to be ridiculed by the laboratory purist and beloved by the experienced field service engineer. The test is extremely simple to perform, and undoubtedly can give useful information about oil condition if employed routinely, and if interpretation is based on prior experience of deposit patterns.

The blotter spot test consists of placing a drop of oil in the center of a laboratory filter paper or a piece of blotting paper. The drop spreads out and the pattern produced depends on the composition of the oil and its condition. A variation of the test is to use strips of paper, either immersing their ends in the oils to be tested or placing a drop of oil near one end. The oil will be drawn up by capillary attraction and again produces a characteristic pattern depending on its condition. The patterns can be developed (extended) by subsequently immersing the end of the strips in a solvent which spreads the pattern along the strip. This technique is usually given the more scientific title of paper chromatography, a technique which has been used extensively in the analysis of new lubricants and additives. The interpretation of the strips is essentially the same as for the blotter spots with spread out bands rather than the concentric rings of deposit in the latter case.

As for all used-oil analysis it is most desirable to build up a history of results for a known oil in a particular service, and to use past examples related to the oil/engine history to interpret the patterns. The following guidelines can, however, be given as a starting point:

(a) A new or hardly used oil will give a round brown spot with no or little banding.

(b) A moderately used oil will give a fairly uniform dark spot, the dispersant/detergent carrying soot and other insolubles close to the edge of the spot (an almost colorless ring outside the main spot may be due to fuel dilution).

(c) A exhausted oil will show a dark central spot with a large paler brown area, the insolubles in this case not being carried outward by the migration of the oil.

(d) An oil containing free water will show clumped dark central spots and an outer ring of wet and weakened paper.

Most oils will lie between conditions (b) and (c), and the judgment has to be made as to how close to (c) the actual sample is.

An interesting variant of this test was developed by Ford of Europe in connection with "black sludge" build-up in the rocker covers of many European vehicles in the late 1980s. It was postulated that blowby condensate in the rocker cover reacted with partially depleted oil to cause insolubles dump-out. In the modified blotter spot test the used oil is treated with collected condensate before dropping it on the filter paper. The test appeared to increase in severity (earlier dump-out indication) with the added condensate, and to correlate quite well with field test data on black sludge generation.[15]

3.5.2 Testing for Equipment Condition[16,17]

Some of the above tests may give results suggesting equipment malfunction. For example, presence of water and glycol may indicate coolant leakage into an engine, and excessive fuel dilution may indicate misfire or diesel injector dribbling. Unusually rapid degradation of new oil may indicate a mistuned engine or excessive blowby from worn piston rings or liners.

Useful information about the mechanical state of an engine can also be obtained by emission spectroscopy of the used oil. This can produce analyses of trace elements in the oil down to the parts per million level. A knowledge of the engine metallurgy and, if possible, of the chemistry of the coolant is desirable. As before, results should be compared to past patterns rather than interpreted on an isolated basis. A sample of the unused oil should always be analyzed to set a reference level from which the used oil results diverge. This is because some of the elements (e.g., copper, silicon, boron) may be present in significant amounts in the new oil. The method is usually applied to reasonably large diesel engines with typically low rates of wear, and railroad companies have been the chief users. For smaller engines and particularly for small gasoline engines the levels of wear metals have been found too high and too variable to be useful, even if the cost of such analyses could be justified.

The technique has also been applied to aviation (gas turbine) oil analysis,[18,19] particularly by the U.S. Air Force.

The following suggestion of typical levels (above a new oil datum) and their interpretation is given for guidance:

Trace Metal Analysis of Used Oils for Medium/Large Diesel Engines (Figure are parts per million [ppm] in the oil)

Element	Typical Levels	Warning Level	Possible Significance
Iron (Fe)	30 - 70	100	General wear, liner scuffing
Aluminum (Al)	10 - 20	25	Piston scuffing, ingested dust
Copper (Cu)	15 - 25	50	Cu/Pb bearings, bronze bushings
Lead (Pb)	20 - 40	50	Cu/Pb bearings
Tin (Sn)	2 - 10	20	Al/Sn bearings
Chromium (Cr)	2 - 10	20	Chromed rings; possibly coolant
Silicon (Si)	10 - 20	25	Ingested dust; piston wear
Boron (B)	5 - 10	25	Coolant leaks

Notes:
Large wear particles will not be suspended in the oil and will not appear in the analysis; however, there is a general correlation between the level of wear and the quantity of dissolved or suspended metal which appears in the oil.

Iron content is a general indicator of wear in an engine, but levels can be erratic and vary between different engines of the same model.

Aluminum is the most useful indicator of piston scuffing, although aluminum will also appear in the analysis if the air filters have allowed substantial quantities of dust to enter the engine.

Silicon is a constituent of some pistons but is normally an indication of atmospheric dust ingestion. Beware, however, that a certain level of silicon appears in fresh oil from anti-foam additives. It can also leach out of some seals.

Copper/lead/tin are the bearing metal indicators, but the presence of copper in some new oils makes this determination more difficult.

Engines run on leaded gasoline will of course show high lead contents.

Chromium appears mainly from the plating of some piston rings, but chromate inhibited coolant is sometimes used and a coolant leak in this case would show chromium in the oil.

Boron is a common coolant inhibitor, but note that many new oils contain borated additives.

The warning level figures in the above table should be treated with particular caution. The important warning signal is when the wear metal levels increase to double or more of a typical stabilized level which has been found previously. It is important also to plot the history of individual engines. We have known cases where seemingly identical railcar engines ran at iron levels from two to five times, respectively, the typical level of the rest of the fleet with the same engines. On stripping the high iron engines for examination, no significant problems were found to explain the high levels. On the other hand, there is anecdotal evidence and reports in the literature that incipient problems have been spotted before they became serious, and much subsequent expense and downtime has been saved.

References

1. "Petroleum Products, Lubricants and Fossil Fuels," Annual Book of ASTM Standards, Section 5, ASTM, Philadelphia.
2. Sell, G. (Ed.), Quality Assessment of Petroleum Products, Institute of Petroleum, London, 1961.
3. Hsu, S.M., "Review of Laboratory Bench Tests in Assessing the Performance of Automotive Crankcase Oils," *Lubrication Engineer*, p 722, December 1981.
4. Caines, A.J., "Modern Methods of Testing Petroleum Products," *Institute of Petroleum Review*, pp 465-470, 1965.
5. Hodges, D.R., "Recent Analytical Developments in the Petroleum Industry," Applied Science Publishers for the Institute of Petroleum, 1975.
6. "Petroanalysis 81," Proceedings of the Institute of Petroleum Conference, John Wiley, 1982.
7. "Viscometry and Its Application to Automotive Products," SAE Symposium, SAE National Automobile Engineering Meeting, Detroit, 1973.
8. Klamann, D., Lubricants and Related Products, Chapter 10, pp218-247, Verlag Chemie.
9. Mortier, M., "Laboratory Shearing Tests for Viscosity Index Improvers," CEC Paper CEC/93/EL22, 4th International Symposium on

Performance Evaluation of Fuels and Lubricants, Birmingham, England, 1993.

10. Watkins, R.C., "An Improved Foam Test for Lubricating Oils," *Journal of the Institute of Petroleum*, Vol 59, No 567, pp106-113, 1973.
11. McCue, C.F., Cree J.C.G., Tourret, F. (Eds.), Performance Testing of Lubricants for Automotive Engines and Transmissions, Proceedings of the 2-6 April 1973 Symposium in Montreux, Switzerland, Applied Science Publishers for the Institute of Petroleum.
12. Watanabe, S., "Japanese Engine Tests for Specifying Gasoline Engine Oils," Japan Lubricating Oil Society, Savant Conference, 1991.
13. Asseff, P.A., "Used Oil Analysis Significance," API Farm and Construction Equipment Forum, Chicago, 1975.
14. Matthews, J.E.D., "Assessing Engine Oil Condition," *Motor Management*, pp 17-21, July/August 1981.
15. Allcock, D., "Lubricant Sludge Tests," (interview) *PARAMINS Post*, Issue 7-4, February 1990.
16. Frassa, K.A., Sarkis, A.B., "Diesel Engine Condition Through Oil Analysis," SAE Paper No. 680759, Society of Automotive Engineers, Warrendale, Pa., 1968.
17. Salvesen, C.G., "Engine Condition Defined By Oil Analysis," NBS Spec. Publ. No 584, pp313-328, 1980.
18. Clark, B.C., Cook, B.J., "Oil Wear Metal Analysis by X-Ray," Martin Marietta Aerospace, Denver, MRC-82-509, AD A118477, 1982.
19. Niu, W.H., O'Connor, J.J., "Development of a Portable Wear Metal Analyser for Field Use," USAF Wright Aeronautical Laboratory Technical Report, 83-2807, 1983.

Chapter 4

Crankcase Oil Quality Levels and Formulations

In Section 2.2 we set out the benefits which additives can provide in improving oil quality, and in Section 3.3.2 the measurement of oil quality by means of laboratory engine tests was discussed in a general way. In this chapter we intend to show how oil quality has improved in a continuous manner, how this has been more and more expressed in terms of meeting laid-down performance criteria in laboratory test engines,[1] and how new oils are formulated to meet the latest requirements.

4.1 Evolution of Quality Levels

The first "performance" additives, in the form of metal soap detergents, were developed and used because a field problem existed and the engine manufacturer (Caterpillar Tractor Company), with oil industry assistance, worked on the development of a laboratory engine test to evaluate improved oil formulations.[2] In the diesel segment of the market, this pattern has continued, with the sequence of events usually following a similar pattern.

(i) An engine problem is found in the field.
(ii) An oil quality improvement is postulated as a solution.
(iii) The oil and additive industries are requested to work on it.
(iv) A formulation is found which solves the problem.
(v) Laboratory engine tests to screen suitable oils are developed.
(vi) The engine manufacturer or an industry body issues a new oil specification.

In the early development of passenger-car gasoline engine oils there was less pressure for oil quality improvements; the average user was content with frequent oil changes and an engine lifetime which, if short by today's standards, was enough to see the first owner through the time he kept the vehicle. For subsequent owners a thriving "proprietary additives" industry purveyed products to quiet noisy engines and reduce their oil consumption. These consisted mainly of polymer thickeners (viscosity modifiers) to boost oil viscosity. Anti-wear additives were also promoted by some companies, based on E.P. performance demonstrations (unrelated to the real problem of corrosive wear).

With little external pressure to improve oil quality, the oil companies tended to concentrate on packaging and oil appearance. Dyes and products to improve the fluorescent sheen of the oil were sometimes the only additives used. By the 1950s, expectations for engine durability had increased along with the average miles driven. Engine manufacturers then started to demand higher and more consistent oil quality for passenger cars. Field problems were becoming apparent, as users' expectations of engine durability increased, and the manufacturers wanted to show improvements in product life and servicing requirements. Field problems increased, however, either stemming from substantial increases in engine power, specific modifications to reduce air pollution, or simply from increased demands on the automobile by its owners. As these problems arose, they were tackled by OEM development of approval tests and the setting up of approval and specification organizations.

Of considerable significance in the late 1960s was the desire by the U.S. military to avoid problems with their large numbers of automobiles (more than of tactical vehicles) which were subjected to unusually severe mixed driving regimes. This large customer of the automobile manufacturers was both a strong driving force and a participant in the definition and development of higher-quality oils.

These developments led to greater uniformity and a considerable overall increase in oil quality. However, in the passenger car sector there are more cases of quality changes being led by oil companies' promotional campaigns than is the case for the commercial sector, particularly in Europe.

Longer oil drain intervals and use of synthetic base stocks are cases where the oil industry has led change in advance of engine manufacturer demands: BP pioneered double the normal oil drain intervals (against opposition from OEMs) by use of high dispersancy oils, Esso in France marketed part-synthetic "Racing Oil" at service stations, and Agip and particularly Mobil promoted the advantages of fully synthetic oils.

In North America, quality improvements have been more associated with customer and OEM demands, field problems leading to better oils which then permitted improvements in areas such as frequency of oil drain intervals. Figure 78 shows how oil drain intervals have increased with the general improvements made in oil quality. At any time there is of course a range of recommended values depending on engine types, the service duty and manufacturers' individual attitudes, but the overall trend is clear. Drain intervals in Europe, while initially starting at a somewhat lower level than in the U.S., have tended to be higher than the North American values. Since the mid-1980s there has been a general flattening out (and in some cases reductions) to a band around 10-20,000 km (6-12,000 miles).

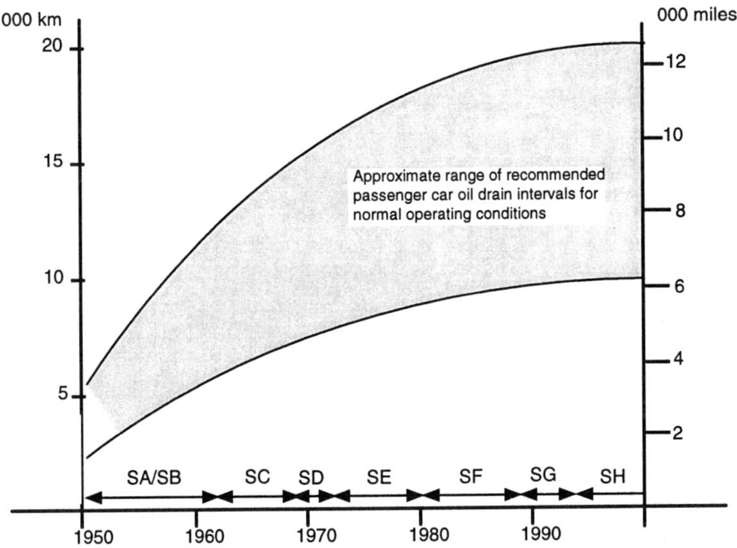

Fig. 78 Oil drain intervals versus API oil quality.

The extent to which engine oil quality changes over a forty-year period have made a dramatic impact on the vehicle, the consumer and the environment is illustrated more fully in the following table:

ENGINE OIL STRESS AND USAGE

Higher engine performance with reduced quantities of lubricant

Passenger car	A	B	C
Model year	1949	1972	1992
Power (kW)/rpm	25/4200	74/5000	96/5600
Power density (kW/L)	21	37	45
Oil capacity (L)	3.0	3.7	3.5
Oil consumption (L/1000 km)	0.5	0.25	0.10
Oil change interval (km)	1500	5000	15000
Oil flush at oil change?	yes	no	no
Total oil used after 15,000 km (L) (includes flush oil)	43.5	14.9	4.0
Typical fuel consumption (L/100km)	12	10	7

(Source: CEC Paper: "Lubricant Additives and the Environment" by W.G. Copan and R.F. Haycock)

These dramatic changes are illustrated graphically in Figure 79, in which the stress on the lubricant is defined as engine power divided by the total quantity of oil added to the engine (including top-up and oil changes) in unit time or unit distance.

In the passenger car example quoted, stress has increased about twentyfold in a forty-year period. The motor industry is getting three or four times as much power out of the same-sized engine and has reduced oil consumption to a fifth of what it was—indeed to the point at which top-up is seldom needed between oil change periods. Flushing oils are no longer used, and oil change periods have been extended tenfold. The motor industry has been able to take advantage of its engineering developments only because parallel developments in additive and lubricant formulation technology have enabled the lubricant to absorb the extra stress imposed on it.

In the following sections we will trace the development of passenger car and commercial (diesel) oils in terms of field performance requirements and the necessary additive treatments to meet them.

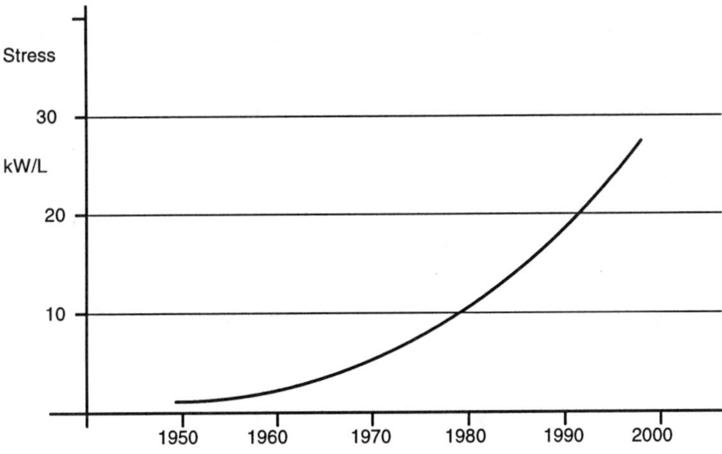

Fig. 79 Stress on the automobile lubricant.

4.1.1 Gasoline Engine Oils

Up to the middle of the 1950s the oils used in passenger cars usually contained no additives or at most a trace of anti-oxidant. The oils rapidly oxidized and became overloaded with contaminants such as soot and combustion acids and therefore required frequent oil changes, typical intervals being every 1000 miles/1500 km in Europe and somewhat longer in the U.S. The oil changes were necessary to remove contamination, but also permitted different viscosity grades to be used for summer and winter, therefore largely obviating the need to use multigrade oils (developed in the 1930s) for year-round use. Despite the frequent oil changes, engines suffered from rapid cylinder and piston wear largely as a result of acid corrosion. The wear then led to oil entering the combustion chamber and being burned with the production of carbonaceous deposits. The small high-revving engines in Europe had to be stripped and decarbonized with attention to the valve seats every 10-20,000 miles (16-32,000 km). An engine usually needed a complete reconditioning or replacement between 50,000 and 80,000 miles (80,000-130,000 km). The larger and slower-revving engines in North America tended to have both longer overhaul and total lifetimes.

Modern passenger car oils contain a complex mixture of additives, base stocks, and blending components which enable oil changes to be extended to around 12,500 miles/20,000 km, and engines generally outlast the life of the vehicle body shell. Average oil drain intervals in Europe in the early 1990s were 7500 miles/12,000 km while the U.S. figures were somewhat lower at 5000 miles/8000 km.

Engine design and metallurgy have undoubtedly contributed to both extending oil change periods and engine durability, but in each case the most significant development related to the use of oil additives. Alkaline oils reduced acid corrosion and dispersants held contaminants in relatively harmless suspension. Viscosity modifiers in multigrade oils helped reduce high-temperature wear and ZDDP anti-wear additives both reduced general wear levels and permitted the use of detergents in high-speed gasoline engines.

The improvement in oil quality has come about in a series of step changes. Many early quality levels can still be found in use in the developing countries or wherever the cost of the highest quality lubricants is considered unaffordable, whether this be for simple income-level or foreign exchange reasons. The concept of evolutionary rather than revolutionary changes being the route by which oil quality has developed is quite important, as suggested in the foreword to this book. Succeeding generations of oil quality have always built on the previous standard, increasing performance requirements in the same tests or substituting new tests when these were considered more relevant. Thus, in general, additive treat levels in oils have risen steadily at each quality level improvement, albeit with some small "learning curve" reductions during the lifetime of a given specification.

Some alternative approaches were indeed considered in the 1960s. Following from work on "detergent" gasolines (containing additives to keep carburetors and inlet systems clean), the concept of supplying most of the additives via the fuel was proposed. Small dispersant additions to gasoline produced major improvements in crankcase oil sludge-handling, and theoretically supplying additive from the top of the piston rather than from below should be advantageous. There was a need to develop certain new additives, but the idea foundered over the marketplace difficulty of tying together usage of special gasoline and special lubricant, with normal gasoline still being available from a multitude of suppliers, and possibly being used instead of the special grade. Other ideas proposed but not developed included supplying additives via a special oil filtration/oil doping canister, and precipitating rather than dispersing sludge in order to remove it by filtration. From time to time some of these ideas resurface and raise passing interest until the old objections re-assert themselves.

Filtration has been at the center of many ideas for lengthening oil life. In the 1950s metal edge-type filters were demonstrated which could yield a transparent filtrate from a thick and sludgy diesel oil. The problem with such devices was their limited capacity, being rapidly blocked and needing a messy and time-consuming cleaning operation at frequent intervals. As polymeric additives developed, it became clear that too fine a degree of filtration could remove valuable additives from the oil, and the present paper filter is a balanced compromise between transparency to additives and the ability to remove harmful impurities which could damage moving parts. (With modern oils, the filter is not designed to remove sludge, although it can become blocked with sludge if the oil is exhausted.)

Recently there has been some revived interest in fine filtration, but the original limitations and reservations still apply. All such ideas would require major changes to both hardware and maintenance practices. The convenience of national and international standardization of automotive products and practices carries as a penalty that revolutionary change is extremely difficult, if not impossible.

The development of additive treatment of oils took place mainly in the period 1940 to 1965, with developments since then being more concerned with formulating new oil qualities with increasing quantities of the types of additive already discovered, rather than invention of radically new types of additive. Pour depressant and viscosity modifier additives had been developed in the early 1930s. While pour depressants found some use in saving refinery dewaxing costs to provide motor oils of satisfactory low-temperature performance, the use of viscosity modifiers was thought of as a way to formulate "year-round" lubricants; however, with frequent oil changes it was more economical merely to use different viscosity oils for summer and winter. Therefore the viscosity modifiers developed at this early stage were not greatly exploited for year-round use although multigrades did offer easier starting for winter use and were therefore promoted by some oil companies.

Detergent additives had been developed in the 1930s for diesel engines (see Section 2.2) but their use was not extended to passenger cars until much later, although some of the so-called "Heavy-Duty" oils with detergency were used in U.S. commercial vehicles with large gasoline engines. Lower-quality oils such as "Regular" (containing no additives) or "Premium" (containing small amounts of such additives as anti-oxidants, bearing corrosion inhibitors, and boundary lubrication additives) were used to lubricate both passenger car engines and low-output diesel engines. The descriptive names used above ("Regular," "Premium," and "Heavy-Duty") are in fact the basis of the first American Petroleum Institute (API) Engine Oil Classification system, introduced in 1947.

Early trials had suggested that detergent additives could cause valve train wear problems in passenger cars, and there was also a fear that the introduction of a detergent oil into an already-used engine would lead to uncontrolled de-sludging, followed by blocking of the oilways. The wear problem was eventually overcome by the use of ZDDP as an anti-wear agent, at a higher treat level than its previous use as an anti-oxidant and bearing corrosion inhibitor. De-sludging proved not to be a major problem if a few extra early oil changes were made at the time of introduction of a detergent lubricant into an existing engine. In most cases, however, the use of new high-quality oils coincided with purchase of a new vehicle with an unused engine, and so such problems were not as frequent as originally anticipated. Use of conventional detergent additives remained very cautious, however, but in the 1950s the dispersant polyester viscosity modifiers appeared, which in dispersing the oil contamination allowed oil drain periods to be extended. This in turn permitted the year-round application of multigrade oils to be implemented. After the introduction of the potent ashless dispersant additives (succinimides, etc.) in the 1960s, it was possible to formulate oils of much higher dispersancy and/or to utilize the lower-cost polyisobutylene viscosity modifiers in multigrade oils. This eventually led in Europe, particularly in the U.K., to a rush to develop so-called "long-life" motor oils, and also to market wider multigrades such as SAE 20W-50s.

In 1952 the API defined new classifications for engine oils based on the service duty. The oils for passenger cars were described as being suitable for ML (light duty), MM (moderate duty), or MS (severe duty) service. The private passenger vehicle with a mixture of much low-temperature short-distance motoring and occasional long-distance high-temperature travel was considered to need MS quality oil. However, it was left to the oil marketers to decide which designation to apply to an oil, or what quality needed to be provided to match an MS designation. Regardless of their additive content many of the oils on the U.S. market were described as being of MS quality but in reality gave unsatisfactory performance. The U.S. motor manufacturers eventually decided something had to be done, and devised a series of performance tests in their own engines to define what they considered to be adequate MS quality. The proposals were first published in 1962 and were adopted by General Motors (GM) in that year and generally as mandatory performance standards for warranty purposes in 1964. This was a great step forward, and for the first time oils of consistent performance in relation to sludge and deposit control, oxidation, and wear performance were available with means of demonstrating their performance in a laboratory. The standards rapidly became accepted on a worldwide basis, although in areas such as Europe extra tests in local manufacturers' engines were often added to the basic requirements of the U.S. car makers.

ORIGINAL TEST SEQUENCES FOR MS QUALITY

Sequence	Purpose	Engine used
I	Low-temperature medium-speed scuffing	1960 Oldsmobile V-8
II	Rust, corrosion, deposits at low temperatures	1960 Oldsmobile V-8
III	Oxidation and deposits at high temperature	1960 Oldsmobile V-8
IV	Scuffing and wear	1962 Chrysler V-8
V	Sludge at low and medium temperatures	1957 Ford Lincoln

The Sequence I and Sequence IV tests were later dropped and wear measurements included in the Sequence III and Sequence V tests.

In the 1960s automobile-induced smog became a problem in U.S. cities, and as a means of alleviating this, positive crankcase ventilation (PCV) was introduced on new engines. Instead of crankcases having a breather tube open to the atmosphere, they were connected to the air induction system, and crankcase fumes were consumed in the engine instead of being vented to pollute the atmosphere. The recycling of active partially combusted products through the engine placed extra demands on the oil, and the quality level of the MS oils had to be increased. This was done for the 1968 model year. The oils were still called MS oils but were distinguished from the previous qualities by new manufacturers' designations such as Ford M2C 101-B and GM 6041-M. New Sequence tests II-B/III-B and V-B were developed, and Ford developed their own Falcon rust test.

These new oils were rapidly found inadequate for increasingly common problems in sustained high-power motoring, often involving boat and trailer towing at high speeds. A third higher level of MS quality was therefore proposed.

To relieve the ensuing confusion about what constituted MS quality, the API in 1970 introduced the new open-ended classification system which was jointly developed by API, ASTM and SAE, and which is still in use today. It is published as SAE J183, reproduced in Appendix 7. This gave gasoline oils the prefix S (for service station) followed by a series of letters from A onward indicating successive levels of increased quality. The 1968 MS level became SD and the new formulation chemistry to meet this level is in fact not vastly different from that in use today, although upgradings were made again in 1972 (SE), 1980 (SF), 1989 (SG), and 1993 (SH). These upgradings can be represented as more of a rebalancing and gradual improvement in certain areas to suit the demands of newly introduced engines rather than any fundamental changes in the type of formulation required. With the new API classifications an L-38 oxidation test was added to each quality requirement, and for some a Caterpillar piston deposit requirement.

The following table gives an indication of the type of treatment levels and components required to meet the various quality levels from SA to SH.

Typical additive treatment (mass %) to meet various performance levels for Gasoline Engine Oils

1947 API :	Regular Premium							
1952 API :	ML	MM	MS					
1972 API :	SA	SB	SC	SD	SE	SF	SG	SH
Additives								
Ashless Dispersant	-	-	1.8	4.0	5.0	5.0	5.5	6.0
Metal Sulfonate	-	-	0.6	1.0	1.8	1.2	0.8	1.7
Thiophosphonate	-	-	1.0	1.0	-	-	-	-
Phenate (Ca)	-	-	-	-	1.8	1.0	1.2	0.5
Other anti-oxidant	-	0.1	-	-	-	0.2	0.5	1.3
Anti-rust	-	-	-	0.2	0.1	-	-	-
ZDDP		0.2	0.6	0.8	1.0	1.3	1.3	1.4
Total Performance Additive	Nil	0.3	4.0	7.0	9.7	8.7	9.3	10.9

Notes.
The above treat levels are for illustration purposes only and refer to typical additives as sold and not to the active chemical ingredient content of such additives. Actual levels depend(ed) on the precise components chosen, the elapsed time since the classification had been introduced (the experience curve), and the skill of the formulator. In general, additive treating levels for a particular classification would reduce with time as formulation experience was gained, but other factors, such as changes in the severity of individual engine tests, also played a part. The dispersant is assumed to be of a succinimide type, the sulfonate of 300-400 base number, the thiophosphonate of 100-150 base number, and the phenate of 200-250 base number. The drop in typical treat levels between SE and SF is partly due to improved additive potencies but can also be ascribed to the introduction of unleaded gasoline as the test fuel for Sequence V-D and V-E tests, and use of a small, four-cylinder engine instead of a large V-8 for these tests, which resulted in a milder test requirement.

In addition to the above treatment the oil will require a pour depressant, some anti-foam additive and, if a multigrade, a selected type and level of a viscosity modifier. Additive performance packages may also sometimes contain extra diluent oil to assist pumpability of the package and its dissolution in the customer's base stock. The procedure for formulating most types of crankcase oil is given in Section 4.2.

The above table considered quality levels as expressed in terms of API performance levels, which are recognized worldwide and provide the framework for most oil quality systems. While there have always been broad similarities between engine oil formulations around the world, there have inevitably developed regional differences in detail requirements and formulation practice. These have generally arisen from engine manufacturers having different design philosophies, different histories of field problems, and different market conditions.

In Europe, high fuel taxes together with narrow roads in congested cities and higher permitted speeds on open roads (no restriction at all on German autobahns), led to development of small, fuel-efficient, highly rated engines long before such engines were seen in the U.S. Additional lubricant performance requirements were therefore demanded, including extra valve train wear tests and engine deposit tests. A requirement from 1984 has been for a level of high-temperature/high-shear viscosity which greatly exceeds that of a lubricant which would satisfy the U.S. passenger car market.

In Japan in recent years the OEMs have taken a position on the ash level of lubricants and require lower levels than are normally found in the U.S. or Europe at present. One of the driving forces, but not the only one, has been concerns about the effects of high ash oils on exhaust catalyst systems.

Regional as well as U.S./International specifications are discussed in Chapter 6.

4.1.2 Diesel Engine Oils

The development of standard quality levels for diesel engine oils occurred earlier and was more positive than was the case for the gasoline oils discussed above. The Caterpillar Tractor Company, having promoted the use of detergent additives, developed a single-cylinder test engine in 1940 (Caterpillar L-1) and this was adopted in the following year by the U.S. military for their specification MIL-2-104. The military commenced issuing approvals in 1944 against an updated version of this specification known as MIL-2-104B, establishing for diesel oils both a common level of quality and a means of demonstrating that quality level was actually met within an oil.

Caterpillar required higher standards of performance for their increasingly powerful engines, developed the supercharged 1-D test engine, and used this to define their requirements for their so-called "Series 2" quality. The military initially responded with a corresponding "Supplement 2" specification, but eventually decided that this was too extreme for general requirements and reverted to an intermediate "Supplement 1" level using the old L-1 engine but with higher sulfur fuel than previously. They renamed the lower level as MIL-L-2104A and the improved level as MIL-L-2104A (Supplement 1). With the Caterpillar Series 2, three levels of quality existed and these were reflected in the API 1952 service classification. This paralleled the "M" series for passenger car motor oils but used the letter D as prefix:

API Classification	Oil Implications
DG (good conditions)	MIL-L-2104A needed (possible to use straight oil changed frequently)
DM (moderate severity)	"Supplement l" oil needed
DS (severe service)	Caterpillar (Series 2) quality needed

In 1955 Caterpillar again increased their performance requirements with their quality level known as Series 3. This time the military followed Caterpillar with their specification known as MIL-L-45199 using both the 1-D and the new 1-G Caterpillar test engines. This was a very severe quality requirement, and this standard was to last for thirty years as the pinnacle of diesel performance. The Series 2 level became obsolete and has never been resurrected.

When the 1970/71 API performance classification was issued it coded the diesel oils as follows, using a prefix of C (for commercial):

API Performance Classification

CA	MIL-L-2104A
CB	"Supplement 1"
CD	Series 3/MIL-L-45199

(API CC was for a multi-purpose oil to be discussed later)

Formulations for these oils were originally based on mixtures of sulfonates of low basicity and phenates, at various treat levels similar to those shown below:

Oil Quality Level	Detergent Treat Level (mass %)
API CA	2%
API CB	2% - 3% possibly with 0.5% ZDDP
(Caterpillar Series 2)	(8%)
API CD	12% - 14%

During the 1960s various new additive types were developed which permitted these treat levels to be reduced, achieving the same standards of diesel performance but with an overall superior balance of properties within the oil. The new products were:

- Highly overbased sulfonates and phenates
- Thermally stable ZDDP for anti-oxidant and anti-wear performance
- Thermally stable ashless dispersant for improved sludge handling and deposit control

A 1969/70 API CD oil would have had a composition similar to that shown below :

Ashless dispersant	3.5 mass %
Overbased Sulfonate	2.0 mass %
Overbased Phenate	1.5 mass %
Aryl ZDDP	1.5 mass %
Total	8.5 mass %

It can be seen that this type of formulation is not dissimilar to that used for gasoline oils, and the target of formulating an oil to meet both the highest standards of gasoline and of diesel performance was soon in the minds of the oil formulators. The concept of such multi-purpose oils is discussed in Section 4.1.3.

Today's formulations for severe diesel lubricants use derivatives of the above technology, although most of the components will have been refined and improved in the intervening years. In particular the ZDDP is more likely to be a specially developed, thermally stable alkyl type. Overall treat rates have risen to meet higher severity standards. New requirements for diesel lubricants have mainly sprung from a requirement to reduce emissions. To reduce the "dead volume" between the piston top land and the cylinder wall, top land clearances have been reduced. This has led to carbon deposition giving wear and oil consumption problems and higher levels of detergency and dispersancy have been needed. Oil consumption problems in heavy-duty, two-stroke engines have also been amenable to formulation changes. At the present time there is a split between specifications for severe four-stroke engines (CG-4, etc.) and the two-stroke engines (CF-2 and CG-2, etc.).[(3)]

4.1.3 Multi-Purpose Gasoline/Diesel Oils

Evaluation schemes for gasoline oils (such as the API performance requirements) have sometimes included a diesel engine test as a measure of piston deposit formation. Initially this was the Caterpillar L-1 engine (later known as the Caterpillar 1-A) which is an unsupercharged engine of low severity. Testing in more severe engines with early gasoline oils produced unsatisfactory deposit levels. This was due to a lack of adequate thermal stability for many of the additives when subjected to the higher temperature regime of a supercharged diesel engine.

As mentioned earlier, the lower treat level "Heavy-Duty" oils were successfully used in the large commercial gasoline engines found in U.S. buses and trucks, but caused wear problems in high-speed automobile engines until these were overcome by use of increased quantities of ZDDP.

These wear problems were particularly severe in the smaller, higher-speed engines found in Europe, but were also seen in the U.S. The initial situation therefore was that gasoline and diesel engine oils were not interchangeable.

Operators of mixed fleets of gasoline and diesel vehicles had an obvious interest in a common oil for both types, which would permit them to carry lower stocks and avoid accidental misapplication. The U.S. military represented the extreme case of a mixed fleet user, having in fact more passenger cars than diesel-powered trucks, track-laying and fighting vehicles, although the latter could be said to be more important. Working with Caterpillar and the oil industry they developed a concept of an oil which would lubricate all except their most severe diesel engines, and a specification was developed which was finally published in 1964. This utilized the new Caterpillar 1-H engine, which was supercharged and intermediate in severity between the 1-A and 1-D engines. Gasoline performance was assessed in a sludge test in the Labeco engine, also used for a test measuring oil oxidation and bearing corrosion wear, while rust and general engine performance were measured in an automobile V-8 engine.

This specification was coded MIL-L-2104B which should not be confused with the earlier 2-104B. A noteworthy inclusion in the specification was the possibility of approval for a multigrade oil.

This oil quality immediately became extremely popular with civilian fleet users on a worldwide basis, and considerable quantities are still used today in locations where users or oil formulators are unable or unwilling to afford the use of higher additive contents. Despite later coding as API "CC," the oils are still frequently known as "MIL-B," although this specification has been obsolete since 1970.

Successful as MIL-L-2104B was, the U.S. military soon discovered that they had a need for a much higher-quality multi-purpose oil. They were making increasing use of large earthmoving and constructional equipment for which Caterpillar Series 3 (MIL-L-45199) oils were required. Also there were problems with passenger cars that spent large amounts of time puttering around a military base but which could suddenly be called upon to make a high-speed dash over some hundreds of miles to another location. Under such conditions engines frequently malfunctioned as sludge build-up was converted to carbon deposits under the higher-temperature operation. A program to develop a new oil was started in 1967 with the assistance of the engine manufacturers, the oil companies, and the additive industry, as well as API, SAE and ASTM.

It was found that by using additive treat levels approximately twice those of MIL-L-2104B and selecting thermally stable types of dispersant, ZDDP, and detergent additives, it was possible to formulate an oil that would meet both the 1968 "MS" gasoline performance and the Caterpillar Series 3 performance, and the military prepared to issue a new specification as MIL-L-2104C. Unfortunately while this program was in progress, the U.S. motor manufacturers had decided that an increase in oil quality from the 1968 "MS" level was needed in order to overcome some field problems. They were ready to announce the new gasoline quality level at the same time as the U.S. military program had finalized development of the MIL-L-2104C oil, in effect making its gasoline quality obsolete from the outset. After much discussion the U.S. military finally issued two new specifications, the MIL-L-2104C combining the severest diesel requirements with those of 1968 "MS" quality, and a new specification using the 1970 "MS" level (now known as API "SE") with the lower diesel performance level of MIL-L-2104B. The new specification was coded MIL-L-46152. The rationale was that the MIL-L-46152 would be used at military bases where passenger cars and light trucks predominated (non-tactical applications), and MIL-L-2104C would be used for tactical vehicles and in times of crisis would satisfactorily lubricate any passenger vehicles that had to use it.

As already discussed under Section 4.1.1, to sort out the confusion caused by three levels of "MS" quality, the API in cooperation with SAE and ASTM brought out their new oil classification system, which is the one in use today. By the time it was issued, oils of "CD/SE" quality, the so-called "universal" oils covering the requirements of both MIL-L-2104C and MIL-L-46152, were being formulated and marketed.

The basic technology developed at this time has remained valid up to the present time, the improvements in gasoline quality represented by successive specifications leading up to the current API SH quality, requiring some rebalancing of the basic formulation and addition of supplementary anti-oxidant. The diesel scene is more confusing, but again changes to meet the requirements of specifications up to and including API CF-2 and CG-4 require rebalancing and selection of optimum components rather than a fundamental change of technology.

The philosophy of multi-purpose oil merits some further discussion. The arguments in favor of the concept are powerful, namely the reduction of the number of grades of oil to be stocked by end users or oil distributors, and avoidance of errors or wrong application to critical equipment.

Against this, however, it must be said that formulation to meet the requirements of both types of engine is difficult if the higher levels of performance are sought in each case. This leads to higher treat levels of additive and therefore higher cost than would be the case for gasoline or diesel performance alone, and it may also lead to compromises in oil quality. Use of more rigorous testing standards may reduce the possibility of such compromises being made, and will inevitably lead to higher than ever additive treatment levels. There will probably continue to be an equilibrium between use of multi-purpose and single-purpose oils which will vary as new test requirements and new additive technology develop, and market forces re-value the cost/benefit profiles of the more sophisticated oils.

Despite misgivings in many quarters when first introduced, there is no doubt that the inclusion of gasoline levels of dispersancy into diesel oils has dramatically improved their sludge-handling properties and permitted much longer oil drain intervals. The advantage to gasoline oils (particularly those for use with unleaded gasoline) of the high levels of detergency introduced with diesel performance requirements is more difficult to see. Perhaps the original U.S. military concept of two compromise oils, represented at the time by MIL-L-46152 and MIL-L-2104C, is in fact a good one. Neither oil will cause a disaster in the alternative application, but will permit near-optimum formulation for the main area of intended service.

In Europe there has been a strong demand for multi-purpose oil to be sold for the passenger-car market due to the large number of diesel-engined passenger cars. These operate under quite severe conditions of high-speed driving and restricted cooling under streamlined body shells. API CD quality has been a requirement, although for turbocharged engines some manufacturers have an even higher requirement than the CD level. In 1984 CCMC (see Section 6.3.2) introduced a specific classification for passenger-car diesel oils. In 1993 and 1994 single-purpose passenger-car diesel oils became a significant element in the range of European service station oils.

Extreme European requirements tend to be higher than those of the U.S. for both gasoline- and diesel-powered vehicles, but multi-purpose oils can still be formulated using carefully balanced, selected components. Additive levels are steadily rising, however, and any future restrictions on additive content in terms of reduced ash, phosphorus, or metal levels will pose severe problems for formulators, and may render the production of multi-purpose oils covering the extremes of gasoline and diesel performance an impossibility.

Even within the heavy-duty diesel area, it may be difficult to meet the requirements of U.S. and European engine manufacturers' specifications at the same time.(4) European engines tend to have smaller piston to liner clearances than North American engines (typically 0.4-0.5 mm rather than 0.75-2.0 mm), the top rings tend to be closer to the crown, and the piston crown from the top to the first ring usually has no taper. These differences, translated into engine test requirements, tend (as a broad generality) to encourage European formulations to include higher quantities of metallic detergent-inhibitors than equivalent North American formulations. At the time of writing it is a major challenge to design a lubricant formulation to simultaneously meet the most severe U.S. and European specifications, especially if gasoline engine performance is also required.

The need for engines to be homologated to meet ever-tighter exhaust emissions targets will also have an impact on lubricant formulation, and in particular on the ability to formulate multi-purpose oils. The era of fully multi-purpose oils may therefore turn out to be 1970-1990, with divergence taking place thereafter.

The table below shows some typical performance additive treat levels in mass % for oil qualities ranging from API CC level to an oil of the API SH/CG-4 level.

API CLASSIFICATION	CC	SD/CD	SE/CD	SG/CE	SH/CF-4	SH/CG-4
Ashless Dispersant	1.5	4.0	5.5	6.0	6.0	7.5
Thiophosphonate	0.8	-	-	-	-	-
Overbased Sulfonate	0.5	3.0	3.0	2.0	2.0	2.0
Overbased Phenate	-	2.0	2.0	2.0	2.0	2.0
Anti-oxidant	-	-	-	0.3	0.6	0.6
ZDDP	0.7	0.7	2.0	1.0	1.0	1.3

4.1.4 Super Tractor Universal Oils (STUO)

Super Tractor Universal Oils (STUO) have been very popular in Europe and some other parts of the world in the 1980s but for various reasons have never been popular in the U.S., where separate engine and hydraulic oils are normally used. Recent specification updates with increased wear requirements have made formulation much more difficult.

However, these oils represented the pinnacle of multi-purpose or "all-can-do" technology, and merit discussion. They evolved out of the general-purpose Tractor Hydraulic Fluids (THF) which catered for all non-engine lubrication requirements of farm machinery (see Appendix 12). A STUO adds engine lubrication to the requirements and provides a general-purpose farm lubricant to use in all vehicle engines and the transmission and hydraulic systems of large tractors. (It does not provide full hypoid anti-wear properties for new automobile rear axles.) As in the case of the mixed fleet operator, this gives the farmer the chance to purchase oil in larger quantities than if several grades were needed, and avoids the possibility of erroneously selecting an unsuitable oil for an item of equipment.

The key areas which may require lubrication in a large tractor are:

The engine	(Diesel or gasoline)
The gearbox	(Manual or semi-automatic)
Power take-off (PTO) and clutch	(Small gearbox with oil immersed clutch)
Rear axle	(Spiral bevel gears)
Final reduction gears	(Epicyclic hub gears)
Wet brakes	(Oil-immersed brakes in rear axle)
Hydraulic system	(Pump and actuators)

The engine is always physically a separate system; the other areas may have separate systems or some may be grouped to share a common system (e.g., rear axle/wet brakes/final reduction gears).

An explanation of the gear oil requirements can be found in Section 6.1.

The basic lubricant quality required for each of these areas was originally as follows:

Engine	API CD or CE for diesels API SF or SG for gasoline
Gearbox	API GL-4 Allison C-4/Cat TO-2
PTO	API GL-4, and clutch to engage smoothly without slip
Rear axle	API GL-4 Good oxidation control Good load-carrying
Final reduction gears Wet brakes	API GL-4 Smooth operation and low wear Friction modifier to avoid judder, squawk, and grunt
Hydraulic system	High V.I. oil, good low-temperature performance Good oxidation and rust control Must be compatible with seal materials

It was possible to formulate an SAE 15W-30, API CE/SG oil with high oxidation resistance and meeting API GL-4 (from ZDDP and sulfurized additives present) which could be treated with a balanced amount of selected friction modifier to meet the somewhat conflicting requirements of PTO clutch and wet brakes (all additives in the formulation are selected for minimum adverse effects in the various lubrication areas). Formulations are similar to those for CE/SG type oils, with somewhat higher additive treat levels and some special additives to modify the frictional properties.

4.1.4.1 STUO Specifications

The key tractor manufacturers recognized the interest in this type of oil and issued specifications covering the concept. There is some conflict of recommended viscosity levels but their original performance requirements could all be met and an SAE 15W-30 viscosity did not pose any operational problems for any equipment. However, as stated earlier, increased wear requirements have made the original technology obsolete, and new compromises are being sought.

Key manufacturers' specifications:

Ford	M2C 159-C
John Deere	JD 27
Case/IH	JIC 187
Massey-Ferguson	M 1139

Some manufacturers such as Fiat have had unpublished internal specifications and approve a limited range of lubricants with their own testing. At the time of writing the Fiat and Ford New Holland tractor businesses have been rationalized under Fiat management, and it is not clear how future specifications will develop.

These various specifications include physical property requirements, engine performance, gear performance, various bench end rig-tests, and full-scale tests in tractors.

It remains to be seen if new technology for STUO will emerge, if the new wear requirements may be relaxed, or if the STUO era is over.

4.2 Formulating a Crankcase Oil

The previous paragraphs have discussed how various types of formulation arose, and indicated the type of additives used in them. Before proceeding to look at other types of oil, it is worthwhile to discuss the process by which a new oil is formulated from components. Crankcase oils are the most frequently reformulated type of oils and among the most complex, and so serve as a good example for the principles which can be applied in general to most types of oil.

The Specification

This is the starting point, and will cover both physical properties (viscosity range, etc.) and the performance requirements expressed loosely in general terms such as "turbocharged passenger car diesel engine oil" and particularly in terms of engine test passing limits. There may also be other physical or chemical limitations such as maximum zinc or phosphorus levels, and there may be appearance requirements (clear and transparent, dyed red, etc.).

Most specifications and oil formulations evolve out of earlier versions, and some ideas on a composition for the new oil usually exist based on prior technology. In the case of multigrade oils, the viscosity modifier may have a greater effect on the engine cleanliness than the detergent or "performance" additives have on the viscometrics, and therefore the physical properties of viscosity at high and low temperatures are normally considered first. A "dummy" or best guess performance additive treatment is used for this initial work.

4.2.1 Choice of Base Stocks

For many years conventional solvent-refined paraffinic petroleum stocks have been predominant in crankcase oils. However synthetic and unconventional petroleum base stocks are now being increasingly used and are essential components for meeting some specifications. Some marketers have found the ability to describe an oil as containing synthetic base stocks a valuable sales aid; examples of such base stocks include:

Esters

These have become very familiar base stocks and provide good high- and low-temperature viscosity performance, low volatility, high solvency, inherent anti-wear properties, and when suitably inhibited are very stable. They are normally completely miscible with petroleum base stocks, and are also relatively highly biodegradable, a property which is becoming more and more important. Use of large ester additions can create problems of seal swell, which is an increasingly difficult specification requirement.

Polyalphaolefins (PAO)

These are highly paraffinic, low-volatility base stocks with high viscosity indices. Their relatively low solvency power can cause some problems of additive solubility, but this property can be useful in boosting the potency of any viscosity modifier used. Suitably inhibited they are stable materials, giving low engine deposits. One disadvantage is their low biodegradability which is inferior to that of most mineral oils. Their seal compatibility characteristics tend to be opposite to those of esters, giving seal shrinkage, so a mixture may provide a solution to problems in this area. Formulations containing PAOs generally seem to be extremely good at holding large quantities of contaminants in suspension.

Other Synthetic Base Stocks

Polyglycols and polyethers are not at present widely used in crankcase oil formulations, but are important in formulating certain specialized oils such as hydraulic fluids. Other synthetic materials have insignificant usage.

Unconventional Petroleum Base Stocks

These include hydrocracked and hydrogen reformed stocks of a highly paraffinic nature. For a given viscosity they have lower volatilities than stocks containing significant naphthenic material, and represent a lower-cost alternative to PAO for those companies which are able to manufacture them. Catalytically dewaxed base stocks can provide very low pour points, but because of a consequent reduction in paraffinic material they have relatively poor V.I.s. Combinations of hydrocracking and catalytic dewaxing feature strongly in the design of new lube manufacturing plants. These types of base stocks were discussed in Section 2.1.2.

The reasons for including some of the above stocks in a formulation may range from a desire to promote a quality image and justify a higher price, to an inability to meet specification requirements without their use. At the present time such requirements include the restrictive volatility limits for SAE 5W and 10W multigrade oils contained in such specifications as those of the CCMC/ACEA in Europe and some factory-fill specifications in the U.S.

4.2.2 Choice of Viscosity Modifier

The most important criteria in the selection of viscosity modifiers are the level of shear stability and the level of diesel performance required. These are in conflict, as diesel performance is generally degraded by high polymer content, but improved shear stability is obtained by reducing the molecular weight of a polymer but adding more polymer to the finished oil to provide the desired viscosity lift. The thermal stability of polymers is also of importance and polyester-type materials may be less suitable for severe diesel performance. For the most severe requirements, it is necessary to choose long chain polymers with minimum side groups. They should be of the highest molecular weight which will permit the shear stability targets to be met. Ethylene-propylene and styrene isoprene copolymers are the molecules which have enjoyed considerable commercial success, although new polymer types are continually being developed. Polymethacrylates still have a significant market, particularly for specialized lubricants. It is of course important that the viscosity modifier does not cause excessive low-temperature thickening, thus giving problems in the mini-rotary viscometer or cold cranking simulator tests.

To formulate an oil against the physical requirements it will be necessary to know the viscosity and volatility of the various base stocks, the thickening potency and shear stability of the viscosity modifier, and the potency of the pour depressant. From this information trial blends can be constructed using a dummy detergent additive package and submitted for tests against the specification. A typical but not mandatory order of testing would be:

Property	If Failing, Modify:
Viscosity, 100°C	Base stocks, VM treat
Pour point	Pour depressant
CCS viscosity	Base stocks, VM
MRV viscosity	Base stocks, VM, pour depressant
Volatility	Base stocks, blending agents
Shear stability	VM molecular weight
High shear viscosity	VM type, molecular weight
	(VM = viscosity modifier)

4.2.3 Developing the Performance Package

Modifications to existing formulations may be required if there have been significant changes in the base stocks, the viscosity modifier, or in the performance specification requirements. The paragraphs below set out the main factors which have to be considered when selecting the types of components and their treat levels for use in a new formulation.

Dispersant

The treat level is normally set by the sludge performance specification which has to be met. At the time of writing this is usually defined by the Sequence V-E test, although in Europe the Mercedes-Benz M-102E/M111 tests are also important. For diesel lubricants the Mack T-8 soot formation test may also be a controlling factor.

A dispersancy credit may be available from the use of multi-functional dispersant viscosity modifiers to permit a reduction in overall additive treat level. Dispersants of adequate thermal stability benefit diesel performance, but the levels employed are normally set by the sludge requirement. For oils meeting the current severest level of gasoline performance, treat levels of 5-7% would be typical for modern dispersants.

For good diesel performance, dispersants of high thermal stability must be chosen, a property which depends not only on the chemical type but also on details of purity and manufacturing technique.

Metal Sulfonate

Originally basic metal sulfonates were the principal soot-suspending components in diesel formulations. However, with the development of highly overbased sulfonates and the incorporation of dispersants into diesel oils they are now seen primarily as a source of alkalinity. Total base number (TBN) can now be 400 or higher, while at the same time the soap content has tended to decrease. Magnesium sulfonates are sometimes now preferred to calcium, because they have lower ash levels for a given TBN and the rust performance as measured in the Sequence II-D test is superior. Sodium sulfonates have even better rust performance and lower ash, but can require extra anti-wear treatment due to their extreme affinity for metal surfaces. High treat levels of sodium additives have also been known to cause corrosion of aluminum pistons once these are de-oiled for servicing or test rating.

The sulfonate molecule tends to act as pro-oxidant and therefore formulations containing high levels of sulfonate soap require extra inhibition. The action of sulfonate is mainly in the lower piston areas, and treat levels of 1.0 to 2.5 mass % are typical.

Phenates and Salicylates

(Salicylates can be regarded as analogous to phenates, although with their extra carboxylic acid group they have two valencies available for bonding to metals, usually calcium.)

Both these additives are powerful inhibitors and contribute a great deal to deposit control in diesel engines, especially in the upper piston areas. They also assist in preventing oil oxidation and thickening, particularly in their sulfurized forms. Phenates are available at several levels of base number, from un-neutralized alkyl phenols (which are mildly acidic and react with some of the base number present from other additives) to 250 TBN versions. While magnesium phenates are known, the most popular additive is a 250 TBN sulfurized calcium phenate used in formulations at between 0.5 to 2.0 mass %.

Zinc Dialkyldithiophosphates (ZDDPs)

These compounds provide the principal anti-wear and oil anti-oxidant properties of the formulation. The chemical structure affects the anti-wear potency and the chemical stability of an individual ZDDP, and with four alkyl groups to every zinc atom many variations of structure are possible. In general, lower-molecular-weight and secondary alkyl types are less stable but have higher anti-wear activity, while high-molecular-weight primary types are more thermally stable but as a consequence have delayed or reduced anti-wear action. The former would tend to be used for specialty gasoline oils and the latter for diesel oils. Typical treat levels are 1.0 to 1.5 mass %. High treat levels can give tappet pitting with some metallurgies.

The older aryl (phenol based) ZDDPs, which for a time were used extensively in diesel oils, are not now generally employed, having been replaced by stable longer chain primary types made, for example, from C7 to C9 alcohols.

ZDDPs contribute to the ash level of a formulation, but a greater concern is their phosphorus contribution. Phosphorus is considered to be an exhaust emission catalyst deactivator, and many specifications now include maximum phosphorus limits.

Additional Anti-oxidants

The current limits for viscosity increase in the Sequence III-E test, coupled with phosphorus and ash limitations, mean that with most conventional base stocks an oil anti-oxidant additional to any ZDDP is needed. There are many supplemental anti-oxidants commercially available, with the traditional oil anti-oxidants such as hindered phenols and aromatic amines most commonly used. Mixtures of two or more different anti-oxidant chemistries are often particularly effective because different oxidation mechanisms can be inhibited by each type. Treat levels for these products in high-performance specifications would be about 0.5 mass %.

4.2.4 Evaluating and Finalizing a Formulation

The high cost of engine tests requires that care be taken in the order in which tests are run, so that late failures do not require rerunning too many tests with a revised formulation. Recent changes to pass/fail criteria ("statistical testing") and rules for "reading-across" of prior results after minor changes have made design of a test program both more important and more difficult. (See "The CMA Code of Practice" in Section 6.3.1.)

Normally the first properties to be considered are the physical properties, controlled by base stock selection, the use of synthetic or other special stocks, and the viscosity modifier and pour point depressant additives. At this stage a "performance package" or "detergent-inhibitor package" is chosen for the commencement of testing. The knowledge of the formulator with regard to the performance of existing packages and the individual and combined responses of the components available when used in the various testing environments is crucial here. The formulator ideally should have response curves obtained from statistical testing for the key additive components. The availability of this information and the formulator's ability to use it both to decide on the initial package and then modify it as necessary will very much determine the success and cost-effectiveness of the formulation program.

When the physical properties have been adequately met, testing proceeds to the performance targets which are normally associated with standard engine tests. For some tests there may be low-cost screening tests available, run either in simple laboratory equipment or in engines which are not qualified for approval testing. Tests are best run in order of increasing cost, so that if a test fails and the formulation has to be changed, it is the cheaper tests which have to be repeated. However, if new technology is being developed or a new quality level is being formulated, then those tests expected to cause problems would be run first in order to avoid late failures in a testing program. In such cases test order becomes a matter of personal judgment. A typical order of testing is given in the table below:

Typical Order of Testing a New Formulation

Property	**Test (example)**
Physical Tests:	
Viscosity	Kinematic at 100°C
Low-temp. flow	Pour point
Low-temp. cranking	CCS
Gelling tendency	MRV
Shear stability	Bosch pump
Volatility	Noack
High temp/shear vis.	TBS
Seal compatibility	(Various)
Engine Tests:	
Oxidation/corrosion	L-38, Petter W-1
Oxidation/deposits/wear	Seq. III-E
Rust	Seq. II-D
Sludge/deposits/wear	Seq. V-E, M-102E
Soot dispersancy	Mack T-8
Diesel deposits	Cat. 1N, DD6V-92T, Mack T-6 OM 364A, VW Int. T/D
Oil consumption	Cat.1K, NTC-400

If a formulation change has to be made, it may not always be necessary to rerun all the earlier engine tests. For example, if there was a late failure in a major diesel test, then depending on the specific changes made, an approval authority may allow read-across to a new formulation where the changes consist of small additions of already-present components or an uptreat in the total additive package. For example, it might be argued that a small dispersant addition would be generally neutral with oxidation and rust tests unaffected, while sludge control would be improved. On the other hand, an addition of extra sulfonate would improve the rust and the sludge performance but could possibly adversely affect the oxidation performance, and such tests would have to be rerun. In the case of a phenate addition, the oxidation performance would be improved, the sludge probably unaffected, but the rust performance might be harmed. A general increase in the amount of the total additive package used is normally considered beneficial, but care must be taken that restrictions such as maximum ash or phosphorus contents are not exceeded.

The CMA Code of Practice (see Section 6.3.1 for more details), introduced in the spring of 1993, sets out in detail the formulation changes which may be permitted for lubricants run in the tests contained in the latest API "S" category.

As soon as the broad structure of the new formulation is known, a representative blend should be made and subjected to tests such as appearance, stability, foam performance, etc. Problems in these areas can often be fixed by such methods as changes in the blending order or by addition of compatibility agents, but it must be determined that a formulation is blendable before spending large sums on the more costly engine tests. The task facing an additive supplier, who must preferably incorporate the additives into a single concentrated package, is considerably more difficult than that of a lubricant manufacturer who may be prepared to blend the oil from individual additive components. Most motor oil formulations are, however, blended from whole or partial packages provided by additive manufacturers.

The foregoing comments on formulation may suggest that a formulator needs no more than a recipe book, some response curves and a few designed experiments to carry out these tasks. If so, the challenge has been understated. The interactions between components are extremely complex. Although some generalization (such as effect of temperature or acid attack) will be equally relevant for different engine environments, nearly every new engine test brings unexpected challenges and is seldom introduced without lubricant reformulation.

It is not normally too difficult to meet the requirements of a single engine test in isolation. It is the combination of requirements—passenger car and heavy-duty; U.S. and Europe; engine and transmission—which introduce conflicts and compromises. The skilled formulator will know the ingredients and interactions but still occasionally fail to meet technical targets, irrespective of any cost constraint. The successes, like those of a master chef, sometimes seem to owe a little to art as well as a lot to science.

4.3 Specialized Crankcase Oils

In addition to the common types of crankcase oil discussed above, there are certain specialized grades which merit a brief mention.

Arctic Oils

These are low-viscosity, low pour point oils for use in extreme low-temperature conditions where flow properties of ordinary low-viscosity oils would be inadequate. Oils meeting U.S. Military Specification MIL-L-46167B fall into this category.

The base stocks can be synthetic, normally diesters, or a catalytically dewaxed stock. These latter stocks have only moderate V.I.s, but their low pour points make them very suitable for this type of oil. A standard performance package will probably be used. Most viscosity modifiers will be unsuitable for very-low-temperature use.

Performance requirements at low temperatures in the field are less severe than at normal temperatures, and can be "read-across" from tests on the additive treatment in oils of viscosity more suitable for the standard engine tests. Alternatively, as favored by the U.S. military, the arctic oils can be run at the normal test temperatures with some allowance being made for wear performance if necessary. The increasing availability of catalytically dewaxed base stocks now permits formulation of oils with excellent performance at both normal and low temperatures.

Storage Oils

If vehicles are used and then placed in storage for long periods, it is necessary to ensure that the engine does not suffer from corrosion while standing idle. All operating parts should be covered with a film of rust-protective oil, which should not drain away. It is also desirable that sufficient oil film remains to provide initial lubrication under boundary conditions when the engine is restarted. The most severe requirements relate to military vehicles which can be stored for periods of years without use, but other vehicles may be parked for quite long periods awaiting sale or may be shipped by sea, providing a somewhat hostile environment.

For long-time storage it was originally the practice to remove spark plugs (or injectors) and spray preservative oil into the cylinders while the engine was turned slowly by hand. Plugs were then loosely replaced and the engine given an overall spray of preservative oil. Such preservative oil was typically based on sodium sulfonates and lanolin (degras).

With the development of their MIL-L-21260 specification (currently MIL-L-21260D) the U.S. military specified an oil which could be used in the crankcase to drive the vehicle to the storage site, and then, provided a correct shut-down procedure was followed, would preserve the engine for an indefinite period. As well as a full set of performance tests the specification includes the following tests specifically concerned with corrosion.

- Sequence II-D
- Humidity Cabinet Panel Rust Test
- Salt Water Corrosion Test
- Acid Neutralization and Corrosion Test

These oils are difficult to formulate because many useful rust inhibitors used for normal storage oils adversely affect other areas of engine performance such as deposits and wear. Barium sulfonate was originally a valuable component but barium additives are not now used for toxicity reasons and sodium and/or magnesium sulfonates have to be substituted. Balanced quantities of other additives such as succinic acid derivatives and ethoxy condensates are used to maximize anti-rust performance without adversely affecting other parameters. All components have to be carefully selected. TBN must be adequate to rapidly neutralize combustion acids.

Engine manufacturers' requirements have generally been less severe, but initially had common features. They required initial fill oils which would act first as break-in oils, then preserve stored vehicles after they had been driven to a parking area. For measuring protection in humid climates or in sea shipment it was common to run test engines in the laboratory for a short period and then to place the whole engine in a humidity chamber for some weeks. Rusting of critical and internal parts was assessed at the end of this period.

Formulations for such oils required high sulfonate levels (usually barium) and auxiliary anti-rust agents. The break-in requirements were satisfied by use of selected ZDDPs (see below). A useful additive for cases where only limited running was contemplated before storage was high-molecular-weight polyisobutylene which helped to retain oil on cylinder walls and valve gear. Today the testing of complete engines in humidity chambers has largely been abandoned, along with most rust tests other than the Sequence II-D. Some manufacturers, such as Volkswagen, do retain extra corrosion tests in their initial fill specifications, but most appear to have found from experience that modern, highly formulated oils perform adequately without additional rust or corrosion requirements.

Run-in (Break-in) Oils

For many years it was traditional to run-in new engines on straight oils containing no anti-wear additive. This was to accelerate the removal of high spots (asperities) from rings and cylinder bores. For heavy-duty, large diesel engines, mild abrasives were even sometimes used to speed the process. These oils were changed to fully formulated oils after a fixed short period, regardless of the actual level of break-in achieved.

In the 1970s technology was developed which permitted the use of fully formulated balanced oils but with specially selected anti-wear additives (ZDDPs) which were active at the high temperatures generated around any asperities and helped to remove these by chemical wear without scuffing. After asperities were removed and mating surfaces were smooth, high temperatures were not created and the chemical wear ceased. This mechanism parallels that found in hypoid gear oils, where wear takes place and additive is depleted during break-in, but after this has taken place the action stops and in principle less active formulations would be adequate.

In the case of motor oils, the need for strong anti-wear additives is reduced as soon as break-in has been achieved. Formulations containing more stable ZDDPs which produce lower levels of piston deposits can therefore be used, in the form of conventional formulated oils.

More recently the finish of mass-produced engine parts has been improved to a level where little break-in is required. If local hot spots do arise then, if overall oil temperature is high, anti-wear action of conventional anti-wear additives will be triggered. This is why high-speed delivery runs of new vehicles seldom give any evidence of break-in problems. In fact, break-in may be more satisfactory if the engine is driven hard with high oil temperatures than if it is "nursed" with an oil based on normal anti-wear additives.

Break-in is now considered a non-problem, and instead of a first oil change being mandated at a low mileage it is common to run the initial fill lubricant until a standard oil change interval (e.g., 10,000 km) has been reached.

References

1. "Evolution Des Huiles Pour Moteurs et Leurs Methodes D'Evaluation," *Revue de L'Institut Français Du Pétrole*, Juillet - Aôut, 1970.
2. Raymond, L., "Today's Fuels and Lubricants and How They Got That Way," SAE Paper No. 801341, Society of Automotive Engineers, Warrendale, Pa., 1980.
3. Fetterman, G.P., Jr., and Shank, G., "Heavy Duty Diesel Engines and Lubricants—Evolving Towards the 21st Century," NPRA Meeting, Houston, Nov. 2-3, 1995.
4. Lavender, J., VME Construction Equipment GB Ltd., open letter to the oil and additive industries, reproduced in *PARAMINS Post*, Issue 4/94, p.13.

Chapter 5

Practical Experiences with Lubricant Problems

Based on the available literature, the history of automotive lubrication has been reasonably trouble-free in the period since 1960, if measured in terms of post-production hardware redesigns. Problems of all sorts have certainly occurred, most notably perhaps the severe sludge problems of the mid-1980s and improper formulation in the U.S. market, but probably thanks primarily to good relationships between individuals in the motor and oil industries, and to the technical societies, the majority have been solved prior to vehicle launch or lubricant commercialization. With some rare exceptions, the average vehicle user would have been very unlucky to encounter a lubricant-related problem.

Although industry learns by experience, old problems sometimes recur in a new guise, and we thought it worthwhile to record some typical examples of problems and solutions. Some of the problems were clearly related to the lubricant and were quite independent of the hardware. Some were hardware related, but were interpreted at the time as being lubricant related. Others were truly interactive and involved both. You may consider some of the examples to be too trivial, or the judgments at the time to be too obviously wrong for the issues to be worth repeating, but all of them took the time of experienced engineers and chemists, and cost money, and there seems little reason to suppose that similar mistakes could not occur again.

5.1 Problems of Use of Inappropriate Lubricant

Problems falling in this category include misblending, mislabeling and use of inappropriate quality.

Misblended Lubricant

Lubricants have certainly been misblended from time to time, but there are no examples in the literature where misblending has resulted in major problems to the automotive industry. A very few cases have been mentioned (mostly with hearsay evidence) and it is assumed that these will have been settled privately by the oil company and motor manufacturers concerned.

Mislabeling (Deliberate)

As lubricant quality is determined primarily by performance tests rather than by physical and chemical characteristics, it is very difficult to find out without great expense whether oils sold in the open market are actually what they claim to be in terms of quality level. In Europe there has never been a strong feeling within the industry that a significant part of the market has been deliberately mislabeled, although some very low-quality but correctly labeled oils have been offered to motorists for top-up purposes. In contrast, in the United States in the late 1970s and early 1980s, information reached the motor industry which suggested that there might be a significant problem of mislabeling—whether caused deliberately or inadvertently. In 1982 the U.S. army asked SAE to develop a monitoring program based on one designed in 1979 by API.

A Review Committee was formed by SAE to develop a detailed Oil Labeling Assessment Program (OLAP),[(1)] the Committee being made up of representatives of oil producers, engine manufacturers, additive suppliers, users and the U.S. army. The program was initiated in 1987 and, although funded initially wholly by the U.S. army, was subsequently funded by the industry.

The program's objectives were to:

- Obtain and analyze annually a representative sample of automotive engine oils sold in the North American marketplace.
- Identify those samples that did not have the required viscosity characteristics or that appeared to have a substantial deviation in additive type and/or level from that found in similarly labeled products.
- Attempt to resolve questionably labeled products through correspondence with the product's marketers.

Results of the surveys have been published regularly by SAE.(1) The surveys have differentiated between two types of questionably labeled oils: those where the viscosity was not as stated and those where the performance additive content seemed out of line with the Review Panel assessment of what would be necessary to support the advertised claim. Over a six-year period up to 1993, an average of 11% of all oils sampled were considered to be questionably labeled. A majority of the concerns were related to viscometrics. There is evidence that the program has had a beneficial effect on quality in the U.S. market, as there has been a general improvement over the six-year period, particularly regarding the most severe deviations, namely oils with insufficient additive content and oils with poor low-temperature viscometric performance that could lead to engine pumpability failures in very cold weather.

Inappropriate Quality of Lubricant

"Inappropriate" here means having the wrong viscosity, detergency, antiwear characteristics or ash content, etc., compared to the needs of the engine.

The motor industry has limited opportunities to persuade the user of the value of appropriate lubricants, the primary ones being comment in the owner's handbook and on the vehicle itself. The private owner may not read the handbook, and both private and commercial users may be influenced to buy lower-cost products. Passenger car manufacturers have a particular challenge to convey messages to owners of existing vehicles if they wish the owners to take advantage of latest lubricant technology. Articles in the popular motoring and trade press are probably the most effective weapons, together with news bulletins to the owners where these are known.

Problems in general are few, and they are mostly confined to the commercial market where owners and their oil company suppliers will try to minimize the number of lubricant grades in stock and occasionally misjudge. In principle the problems should be fewer in the U.S. (which nominally has a single quality passenger car lubricant market) than in Europe which has a wide variety of available qualities. Some results of a Ford of Europe survey of vehicle owners' understanding of lubricant quality were published in 1992.(2) An initial survey in the U.K. in 1988 showed that a very low percentage of vehicle owners were using quality as recommended in the vehicle handbook (SAE 10W-30, API SG). In 1990, following an aggressive program of dealer education, Ford conducted another survey and was pleased that over 90% of dealerships were themselves using and selling API SG/CD lubricants. However, they noted that viscometrics were still inappropriate (SAE 20W-50 instead of 10W-30) and initiated a dialog with the oil industry to encourage marketing of lower viscosity lubricants.

5.2 Lubricant/Design Interactions

This area seems to provide the greatest number of examples of practical problems. In many of the examples the lubricant was not at fault but was considered for some time to be so. The following represents a small but typical selection.

Bore Wear

A major interaction between a vehicle manufacturer and the European oil industry took place in the late 1960s which still has repercussions today and has greatly influenced both the quality specifications for European lubricants and how these are set.

A leading supplier of prestige vehicles in mainland Europe had become also the dominant supplier of taxicabs in Europe and was displacing the previous American domination of this market in the Middle East and developing countries. For reasons of economy, easy maintenance, and longevity, the vehicles in Europe were mainly diesel-powered. Performance of these vehicles was limited in terms of acceleration, but engines' lives were on the order of 200,000 to 400,000 km. The normal oil recommendation was for a MIL-L-2104B type oil or DEF 2101D, Supplement 1.

In the late 1960s the power output of the engine used was increased somewhat by modifying the engine, and complaints rapidly arose of high oil consumption and engines being worn out at less than 100,000 km and in some cases at less than 50,000 km.

The initial reaction of the oil industry and other motor manufacturers was that this was simply an engineering matter. The problem was variously ascribed to:

- Reduced block rigidity
- Increase of top bore cooling
- Insufficient cooling for increased power output
- Block crystallinity
- (and probably others)

The engine/vehicle manufacturer was naturally anxious to find a quick solution to the problem and mounted a major investigative program. There was a great spread in the engine lifetimes, and evidence of a lubricant effect could be surmised, with one particular oil appearing to prevent the problem, while other brands reduced it. At this stage the oil and additive industries became involved, albeit with some continuing reluctance. A taxi field test was developed by the manufacturer, later to be succeeded by a bench test with both hot and cold running periods.

Use of "Series 3" type oils were beneficial to bore wear but were not generally available at service stations and were considerably more expensive. They also tended to give dirtier engines, because at this time the MIL-L-2104C type oils were still under development, and a "Series 3" type oil typically contained a 12% mixture of sulfonate and phenates with little or no dispersant and no anti-wear additives. The specified MIL-L-2104B oils were typically of 4% to 6% total additive treat, and included both dispersant and anti-wear additives. After much testing it was confirmed that certain oils could prevent the problem, and the manufacturer implemented an approval program for oils which were shown to be satisfactory in the engine test. The higher treat, higher base number oils of MIL-L-2104B type were satisfactory, and so were the majority of the new MIL-L-2104C oils. The secret seemed to be a balanced oil of reasonably high base number which also contained a potent anti-wear additive. We are unaware if any precise mechanical reason for the sudden decrease in engine longevity was ever highlighted, but improved oil quality provided a satisfactory solution.

The engine test underwent two subsequent updates and became enshrined in European specification requirements for motor oils. The test measured both bore and valve train wear and engine cleanliness and later became the significant test for valve train wear in European specifications, including those for gasoline oils! The test is well organized, documented, and is very familiar to all European engine testing laboratories, but its relevance to modern oil specifications has been questioned.

As a result of this initiative, European oil quality was undoubtedly raised, particularly for the passenger-car diesel market. A major oil approval system was developed which dominated European oil formulation for twenty years, and is only now declining in favor of regional and international specification and approval procedures.

Black Sludge

This particular problem which occurred in gasoline engines in the mid-1980s deserves special mention for a variety of reasons.(3) It is an excellent example of the complex nature of an engine/lubricant interaction problem, and it occurred almost simultaneously in many regions of the world.

The problem spawned major warranty claims, articles in national newspapers, new engine tests, the withdrawal of certain fuels and lubricants from the market and the opportunist marketing of commercial "fixes."

Even with the benefit of hindsight, the problem is not fully understood and probably had somewhat different origins in different places although certain features were common.

Black sludge is a dark heavy deposit, sometimes several millimetres thick, which can adhere to metal surfaces in the engine. It adheres strongly enough to the metal surface that it will not be removed by a gentle wiping action. In consistency it can vary from a grease through rubber to carbon. In most examples it built up first in the rocker cover and other cold parts of the engine and then spread into the sump. Eventually an oil way would be blocked and the engine would seize.

The phenomenon suddenly became a major concern in Germany and the U.K., with significant numbers of engine failures also being reported in the United States and in other areas of the world.

The following factors seemed to be associated with the problem:

1. A foul-air-vented engine design, in which blowby gases were vented from the crankcase into the rocker cover
2. Fuel quality, at least in Germany
3. Driving pattern
4. Oil change interval
5. Engine lubricant quality

The association with particular engine models and the foul air venting was clearly demonstrated by evidence of field failures produced by the motor industry. Some foul-air-vented models gave numerous recorded engine failures and other fresh-air-vented ones gave none. A V-8 engine gave sludge in the foul-air-vented bank, but not in the fresh-air-vented bank.

A common feature to all the field problems where some history could be established was a large percentage of low-speed, low-temperature driving, which would encourage water and acids to collect in the rocker cover. In many cases there was also an identifiable regime of high-speed driving which could cause deposits, formed under low-temperature conditions, to be baked hard.

In addition, at least in Germany, a significant percentage of the problems were associated with a specific commercial gasoline containing an additive which had an adverse reaction in the above-mentioned operating conditions. When the fuel was withdrawn from the market, the problems reduced sharply.

Oil change interval was not directly implicated but the problems occurred at a time when oil drain intervals in Europe were broadly being increased from a typical 10,000 km (6000 miles) to a typical 15,000 km (9000 miles) and thus putting more stress on the lubricant.

Lubricant quality was involved in several ways. One particular marketed lubricant was shown to be associated with poor sludge performance and was in due course used as a poor reference in the CEC M102E black sludge test. In general it was clear that even if poor lubricant quality was not causing the problem then good lubricant quality could help to reduce it in the short term—and far more rapidly than any change in engine design. Lubricant dispersancy was a key parameter, and it became evident that it was the ability to cope with large quantities of water, and polar organic contaminants derived from the fuel and lubricant rather than just carbonaceous materials, which separated one oil from another in terms of black sludge performance.

Ford of Europe demonstrated this point elegantly(4) using a modified version of the blotter spot test to study lubricant performance in police fleet tests.

The Mercedes-Benz M102E engine test was developed rapidly in order to allow sludge-preventing lubricants to be identified, approved and marketed. The test was initially developed and sponsored by Mercedes-Benz and was used as part of their approval process for passenger car lubricants. The development was continued under the auspices of CEC and the test became part of the CCMC G4 and G5 lubricant classifications.

Debate still continues regarding whether the M102E test evaluates oils significantly differently from the (American) Sequence V-E test and regarding whether either of them really correlates well with black sludge generation in the field. Interestingly, although field evidence was widespread, it has proved extremely difficult to generate similar sludge under accelerated laboratory conditions without using unstable and unrepresentative test fuels.

Mayonnaise

A cosmetic concern which appears fairly regularly in passenger car operation is the detection of a light-colored creamy emulsion in the rocker cover. The sight of the "mayonnaise" emulsion usually concerns the private motorist, but it is not known to have caused any engine failures, although breather blocking can cause startability and driveability problems. The emulsion is usually formed by driving under low-temperature conditions and disappears with a brief period of high-speed driving. Mayonnaise can block PCV systems (see below) leading to oil loss.

The tendency to develop mayonnaise emulsions grew as lubricants with greater dispersancy were marketed in the 1970s. The incidence of mayonnaise production may be controlled by careful lubricant formulation including the use of supplementary demulsifiers.

Crankcase Ventilation

Exhaust gas ventilation systems which take a tube directly from the crankcase to the intake manifold need to be designed to avoid interaction with cold air which may freeze the water in the pipe, pressurize the system, and lead to oil being blown out via seals or the dipstick tube. Emulsion will do the same for most PCV systems. The fuel pump breather hole is a common oil exit.

Muddled Thinking

A manufacturer of passenger cars had a valve train wear problem in a new engine design which generated significant warranty claims. The problem could be alleviated by careful choice of lubricant. It was felt within the engine development department of the company that a well-known and well-used lubricant approval test used by the company should have protected against such a problem occurring. We consider this to be muddled thinking. The existing engine test would adequately rank oils under the conditions of the test. But it does not follow that such a test will give protection to an engine of different design or that oils will be ranked in the same order if design or test conditions are changed. A motor manufacturer cannot possibly test all marketed lubricants, so the most viable approach is to use a standard engine test for approval of suitable lubricants and then use a borderline oil from this test to approve final engine design.

Catalytic Converter Blocking

In the 1980s at least one European car manufacturer recorded problems of catalyst front face blocking in vehicles exported to California. Automotive three-way catalyst converters are flow-through devices typically having honeycomb structures. Unburned lubricant leaving the combustion chamber will be trapped in the early part of the converter and although most will eventually be volatilized, the metallic elements will remain as an ash. The blocking was manifested by poor engine performance and the problem was solved by reducing oil consumption. It had not been anticipated because initial test work had used the higher-viscosity grade SAE 20W-50 oils common in the European market rather than the SAE 10W-40 oils which were more representative of the Californian market.

Catalyst Converter and Oxygen Sensor Effectiveness

Apart from the catalyst converter front face blocking mentioned previously, the rare metals used in exhaust catalyst manufacture may be partially poisoned by phosphorus and other elements present in crankcase lubricants. Oxygen sensors placed in the exhaust gas to assist electronic engine management may also be similarly affected.

There has been pressure on the oil industry to reduce the level of phosphorus in crankcase lubricants but this has been difficult to sustain when there are no recorded field problems directly associated with lubricant quality affecting catalyst durability. There will be an ongoing challenge for the oil and additive industries to meet the continuing demands for increased catalyst durability without sacrificing engine life and in a cost-effective manner.

Pressures will continue to grow as greater demands on catalyst converter durability are required by the legislators. In 1993 the U.S. Environmental Protection Agency required catalytic converters on new cars sold in the U.S. market to have a demonstrated life of 160,000 km, increased from 80,000 km.

At the present time improvements in oil consumption control and developments in catalyst tolerance to phosphorus poisoning have allowed phosphorus levels in crankcase lubricants to stay relatively constant at around 0.1% mass, but there will be continuing pressure on the oil and additive industries to develop low-phosphorus technology.

Oil Consumption Effects—1

Engine problems are often associated with extremes of oil consumption within production variation. Engines with relatively high oil consumption can be run continuously without oil or filter change if topped up regularly, and engineers seeking to test engine sensitivity to lubricant qualities such as piston deposit and ring sticking tendencies (or to determine recommended oil drain periods) should ensure that they select engines with oil consumption at the low end of the expected range.

Oil Consumption Effects—2

In about 1983 a European passenger car manufacturer advised the industry of a field problem of engine knock when accelerating after a sharp deceleration. It was eventually traced to a design fault related to too much oil entering the combustion chamber down the valve guides. Crankcase oil has an octane number of zero. The solution was to control oil consumption and fit a valve to control the effect of deceleration. Modern carburetor and fuel injector system designs help to prevent this type of problem.

Oil Consumption—3

Lubricant quality was blamed by a European manufacturer who suffered high oil consumption in overhead cam design engines. The problem was actually caused by pumping too much oil to the top of the engine and was solved by use of a small restrictor to the oil feed to the top of the engine. The problems were indirectly caused by the manufacturer operating a policy of sizing the oil pump according to the size of the sump without paying enough attention to details.

Pre-Delivery Problems

Several vehicle exporters from Europe have reported high exhaust emissions and wear problems with new vehicles. These problems were originally incorrectly related to oil consumption but later shown to be associated with fuel dilution giving up to 20% increase in sump oil volume. (A typical vehicle exported by sea may expect 50 cold starts before it reaches the customer.) The Californian authorities recognized the problem and allowed vehicles to be driven before running emissions tests. Multipoint fuel injection greatly reduces the problem.

Valve Deposits

There are several reported cases of lubricant-related deposits being formed on intake valve stems when conditions are such that the base stock can volatilize leaving behind the viscosity modifier polymer. The problem may be reduced by limiting the flow of oil down the valve guides but valve sticking and deposit formation will occur if the flow is reduced too much. Unusually, the problem is associated with careful drivers as engine temperature is critical and the problem does not occur with hot or cold running. The simplest cure for the problem is to use detergent-containing gasoline. As the deposits will shrink on standing it is most important that they are examined as soon as the engine test is finished.

Lip Seal Compatibility

This is a classic example of an interaction issue and it has generated very strong emotions in the oil industry and particularly in the petroleum additive industry. The main geographical area of concern has been Europe. For many years it has been recognized that petroleum products have an effect on elastomers and may cause them to change their physical properties. It has been equally recognized that elastomeric seals used in engines should be compatible with the lubricants used in those same engines. Feelings of discontent began to arise in the early 1980s when there was a proliferation of seal compatibility tests issued by the motor industry and a proliferation of types of elastomers used. Particular attention focused on fluoro elastomers which were gaining favor for use as engine seals. It became clear that these elastomers reacted adversely to some nitrogen-containing additives used almost universally in automotive lubricants.

The main concerns voiced by the oil and additive industries were:

1. That it was unfair that lubricants should be forced to meet elastomer compatibility tests, rather than elastomers meet lubricant compatibility tests;
2. That laboratory glassware tests promoted by the motor industry did not correlate with field performance, and that, in particular, tests on fresh oils (as required) gave much more severe results than tests on used oils (which were more relevant to the real world);
3. That no evidence of problems in the field had been provided by the motor industry to justify use of the seal compatibility tests in specifications;
4. That the glassware tests were imprecise.

The additive industry was particularly concerned that it was forced to adopt non-optimum formulations and to use expensive "fixes" in order to pass glassware tests that it believed were irrelevant to marketplace needs.

The motor industry of course took a different view of things. While generally sympathetic to oil industry concerns, it has considered it prudent to take a fail-safe attitude and has given priority to other issues. The competitive nature of the oil market has meant that as long as one company can meet the technical targets (whether relevant or not) then others are obliged to meet them. In 1995 the problems still remained largely unresolved.

Bore Polishing

This wear phenomenon emerged during the 1970s as a major concern in highly rated diesel engines used under overload conditions such as in logging operations in Finland. Although cylinder liner surface finish is important and ceramic impregnation can help, lubricant detergency can play a key role. Several standard engine tests have been developed in Europe to evaluate lubricants for bore polishing tendencies, and in the U.S., the Cummins NTC 400 test measures oil consumption control, a closely related phenomenon.

The first recognition of bore polishing as a problem gives a good example of how an individual may have a major influence on lubricant quality and test procedures.

SISU is a truck builder in Finland. They purchased engines from an external supplier and found many bore polishing problems associated with the very severe service. SISU is small compared to most vehicle manufacturers and has few means of exerting pressure on the outside world. However, thanks largely to the personal efforts of their then service manager, the company managed to galvanize the European oil industry to address their problem. New engine tests, notably the Volvo TD 120 and Ford Tornado, were developed to measure lubricant effects on bore polishing.

Other vehicle manufacturers acknowledged the problem and a new generation of high-performance lubricants entered the market. Several engine manufacturers (including the one which supplied the problem engines) introduced bore polishing requirements into their lubricant specifications and in due course the CCMC diesel classification D3 was introduced with bore polishing performance as its primary requirement.

Production Problems

A European car manufacturer had a significant number of warranty claims related to bearing failure in early vehicle life. The problem was traced to a fault in the crankshaft hardening process which caused the crankshaft to bend after it had been fitted to the engine. The manufacturer was concerned to find a short-term solution to protect vehicles in production and waiting to be sold and called one of us to inquire if a lubricant might provide assistance.

Test work indicated that the addition of an extra anti-wear additive to the crankcase oil might prevent at least some of the failures. As a consequence an urgent program was developed to manufacture the additive, package it into 20 mL plastic sachets, and distribute the sachets to the production plant and all distributors of the vehicle. The contents of one sachet were put into each new engine at point of sale and on the production line until the fundamental problems were overcome. The action appeared to be highly beneficial but the benefit could never be quantified.

Scuffing

In the 1970s a European manufacturer produced a novel four-valve engine with direct-acting valves. Prototype endurance tests were satisfactory but later experience gave finger-follower scuffing at stresses below design limits. The problem was solved by making a hole in the finger-follower pad, providing more lubricant and thus lowering the operating temperature. The potential problem had not been identified in prototype testing because the oil had been controlled at too low a temperature. A contributing factor was that there was a wear protection gap in the temperature range 150-170°C. The particular ZDDP used ceased to be effective above about 150°C and the sulfur compounds present in the formulation became effective only above about 170°C.

Tappet Pitting

While scuffing is caused by lack of anti-wear protection, tappet pitting (a fatigue phenomenon) has been demonstrated in some pushrod engines using certain metallurgies lubricated with oils containing too much ZDDP of an inappropriate type—and particularly oils containing aryl ZDDP (usually satisfactory in diesel engines) in gasoline engines. Twin concerns to protect against scuffing and pitting encouraged some specification setters to quote maxima and minima for zinc and/or phosphorus concentration in lubricants. Some manufacturers check the robustness of new engine designs by testing with oils of high and low concentration of ZDDP.

Process Changes

A problem of bearing failures during the first few hours of running of engines of a particular type was attributed to the lubricant as factory-fill quality had coincidentally changed from API SE to SF. The manufacturer switched back to the previous oil formulation without success.

The problem was eventually traced to a change in the rate of grinding the crankshaft which led to a "fish-scale" surface finish. Application of more rigorous quality control procedures would probably have prevented the problem.

Limitations of Rig-Testing

A vehicle manufacturer noticed abnormally high wear during cam and tappet rig-testing. Moisture is needed to activate the ZDDP and in dry conditions oils might need to be pre-conditioned.

Most lubricant test rigs are designed to run on fresh lubricants. In some cases the lack of contaminants such as combustion soot, wear or oxidation products in the oil can lead to rig-tests being unreliable and usually mild relative to field experience. Diesel soot in particular can have a dramatic effect on wear, and it is usually rash to extrapolate from fresh oil wear tests for diesel lubricants.

Lubricant chemistry may respond in different ways in rig-tests than in field experience. As an example, different viscosity modifier chemistries may each show field correlation with shear stability measured in rig-tests as polymer molecular weight is changed within a chemical family, but the rig-test will not necessarily correlate with field experience when different chemistries are mixed.

Pre-ignition

A European luxury car manufacturer recorded pre-ignition problems which mostly occurred in chauffeur-driven cars operating a mixed 70% town/30% motorway driving regime. The problem was traced to use of a lubricant with a high ash content which failed the Fiat pre-ignition test. The lubricant had been recommended primarily because of its suitable viscometrics but the high ash content had been overlooked. The lubricant had been designed primarily as a mixed fleet oil for the diesel market.

5.3 Inadequate Test Procedures

While lubricant test procedures are never perfect, once introduced they are usually maintained until a new measuring technique or a change in hardware design renders them obsolete. Rarely is a test procedure found to be fundamentally flawed and requiring urgent modification.

Such a situation did occur after the mini-rotary viscometer (MRV) test procedure to measure lubricant pumpability had been introduced into the SAE viscosity classification. The procedure was developed from practical experience with a variety of differently formulated lubricants and involved subjecting the lubricant to a complex heating and cooling cycle prior to measurement of viscosity. A safety factor had been built into the viscosity classification in that an oil should not be able to pass the cranking requirement (cold cranking simulator) if it could not also pass the pumpability test. It should ensure that a too-viscous oil would fail to allow an engine to start, rather than allow it to start but then cause bearing failure by inadequate pumpability.

However, in the winter of 1980-81, field pumpability problems were reported in the U.S., Canada and Sweden. Subsequent investigation showed that in most cases the oils had unusual compositions and had slipped through the test which should have failed them. A crash program of work to develop a new cooling cycle culminated in the issuance of a modification called the TP-1 procedure, which has successfully protected against pumpability problems in the field ever since. However, as modern automobiles crank more easily than many of their predecessors, the U.S. OEMs are concerned that the safety margin provided by the MRV - TP-1 may be eroded.

5.4 New Marketing Initiatives

Oil companies who have something genuinely new to bring to the market have a dilemma. They need the good will of the motor industry, but by prior discussion risk leakage of their proprietary information and possibly the loss of competitive advantage.

On balance we believe that the most successful route is for the oil company to take the motor industry into its confidence. Two examples illustrate the point:

(i) A major international oil company proposed to launch a new very-high-quality synthetic passenger car crankcase lubricant. They took the motor industry into their confidence and ran many cooperative field tests. The motor industry, seeing the benefits of the product, were anxious to support it and gave help when minor problems occurred. One vehicle manufacturer was so enthusiastic for the product quality that a lubricant specification was written specifically to encourage greater use of similar quality oils.

(ii) Another international oil company developed a low-viscosity fuel economy lubricant. They decided not to work in advance with the motor industry but to take the market by surprise. Although the product had many valuable features it did give some bearing failures and in more cases caused oil pressure warning lights to be activated and gave oil consumption problems. The motor industry felt disinclined to be cooperative and the product was withdrawn. An indirect benefit was the heightened interest in high-temperature/high-shear viscosity which subsequently became a classification parameter within CCMC and SAE.

Another example of the difficulties of introducing new concepts is given by the experience of introducing copper anti-oxidant technology (see Section 2.2). Products containing copper were first offered to the market in 1980. Comprehensive field testing had shown no negative effects from this new concept, but because copper is commonly measured as a wear metal in used lubricants (particularly diesel oils) it was considered wise to alert the motor industry in advance of any oils being placed on the market. Forewarned, the OEMs could be prepared for any unusual complaints from vehicle operators about sudden high copper contents in oil, and could advise on the importance of using appropriate reference oils. In practice the motor industry was pleased to have been advised and the introduction caused relatively few problems, although the interference with used oil analysis was and continues to be a minor inconvenience.

Some engine manufacturers have special tests which they require to be run before "new technology" is accepted. Caterpillar would require a multi-cylinder OL-1 engine test for new heavy-duty diesel technology, and Mercedes-Benz used to require field testing for all new lubricant technology and an M102E test for all new gasoline engine lubricant technology. Clearly the judgment on what constitutes new technology will rest primarily with the oil or additive company.

References

1. McMillan, M.L., Stewart, R.M., "The SAE Oil Labeling Assessment Program - Six Year Cumulative Report," SAE Paper No. 922296, Society of Automotive Engineers, Warrendale, Pa., 1992.
2. "Crankcase OM Quality Change in UK," Ford Dealer Survey, *PARAMINS Post*, Issue 9-4, p.13, February 1992.
3. Haycock, R.F., *et al.*, "Understanding Black Sludge Formation," *Petroleum Review*, December 1988.
4. Allock, D., "Lubricant Sludge Tests" (interview), *PARAMINS Post*, Issue 7-4, February 1990.

Chapter 6

Performance Levels, Classification, Specification and Approval of Engine Lubricants

This section aims to give a general introduction to how and why engine lubricants are classified around the world, and how they are approved for commercial use. It gives some history in order to show how we have reached today's situation, discusses the most influential classifications and specifications used today, and speculates on future developments. As technical societies play a vital part in the development of classifications, their role is explained in some detail. Tabulations of Classifications and Specifications are given in Appendix 7.

6.1 Definitions

Classification of lubricants is very closely interlinked with specification and approval and it may be helpful to give the definitions we use and which we believe represent the common understanding of these expressions.

a) Classification—The division of lubricant qualities into broad types to meet separate but broad user needs. A classification will usually include a description of the type of engine in which the oil is designed to be used, sometimes the type of vehicle, and sometimes (e.g., in the API Engine Service Classifications) may include reference to the warranty years to which the classification applies. Any special features, such as lower than typical viscosity, energy-saving characteristics, etc., will be highlighted. Classification is usually carried out by technical societies.

b) Specification—This is a much more precise definition of an oil than its classification. It will usually apply to an oil for use in a limited range of equipment or a limited end-use market. Test methods will be defined in detail, including physical properties such as viscosity and volatility, and also performance tests in rigs or laboratory engines. Chemical tests should preferably be limited to a few minima or maxima, such as TBN or phosphorus content. Against each test will be the limits to which the oil must conform. Engine lubricant specifications are most often written by users, vehicle manufacturers, or other original equipment manufacturers (OEMs).

c) Approval—This is the process by which an engine or vehicle manufacturer or a lubricant user formally confers acceptance of an oil for use in his equipment. User groups would include military authorities, utility companies, railroads, postal services, or other organizations controlling large vehicle fleets. An approval involves two parties and the presentation of evidence of performance from one to the other, with exchange of paperwork and an ongoing commitment by both parties. Usually the receipt of an approval will be advertised in some way by the oil company.

All the above are intended to fulfill two major aims: helping the lubricant purchaser to select the correct grade of oil for his application, and to protect both the equipment supplier and the lubricant manufacturer against wrong oil application and consequent claims from the end user. These aims can be met by relating the perceived performance level of an oil to the applications to which it is suited, and therefore the ability to measure performance is key throughout.

Another word which deserves definition is "sequence." This is used in at least two ways. In the U.S. it is conventionally used to describe the operating conditions of an individual engine test such as the Sequence II-D, III-E, V-E, etc. (see Section 4.1.1). Several sequences together form a classification.

In Europe the term has been used by CCMC and ACEA to describe a combination of individual tests, and in this sense the meaning is very close to that of "classification" (see Section 6.3.2).

6.2 Performance Measurement

6.2.1 Performance Parameters

In Section 1.7.2 the general properties required in a crankcase lubricant were discussed, with an indication of how various additives could be used to enhance the performance of straight petroleum base stocks. The role of additives was amplified further in Section 2.2. We will now deal with the complex subject of how the performance of crankcase oils is demonstrated in the laboratory and how the various tests are used to specify a desired level of performance for an oil.

Oil performance is always relative—almost any liquid will lubricate an engine for a time, but this time needs to be in terms of months rather than minutes! The expected lifetime of an oil (between oil drain intervals) has continuously increased and, concurrently, engine longevity has also been increased. Both changes have largely been the result of oil quality (performance) improvements demanded by the engine builders. With certain arguable exceptions each new oil specification calls for increased performance in one or more areas over its predecessor. Performance improvement can imply either improved engine life between major overhauls, or extended oil life between oil change intervals.

Such improvements can consist of maintaining existing drain intervals or engine lifetimes for new engine designs (or duties which are more severe) and where without improved lubricant quality the times would have to be reduced. More rarely, oil quality changes can result directly in enhancement of engine performance. This tends to be in areas such as fuel economy, oil consumption, exhaust emissions, etc., where the oil characteristics come directly into play. While friction reduction can release small extra amounts of power measurable in terms of fuel consumption, there is no way a lubricant can influence factors such as peak power output in a major way, except by continuing to satisfactorily lubricate an engine which has been uprated.

The performance of crankcase oils is almost entirely measured by testing in engines, usually on test beds but sometimes over the road. Exceptions include viscosity retention under shear, where diesel injector rigs are principally used, and various compatibility tests such as seal swelling tendency. Such tests are performed on new oils and do not reflect the chemical and physical changes which an oil undergoes in service; they represent test expediency rather than test realism. They also are more relevant to specifications than classifications. In contrast, dynamometer engine tests attempt to simulate service conditions and protect against service problems.

It is important to distinguish between the performance of a sample of oil and its detailed specification. While the specification may well include performance requirements, it will also include physical and chemical requirements which are not directly performance related. Many of the physical requirements such as flash point or haze will be included to guard against misblending or contamination. Other requirements such as high-temperature viscosity may represent a compromise between the needs of different types of equipment.

Chemical requirements must also be regarded as secondary. While the alkalinity of a high-performance diesel oil may provide a clue as to its performance level, the total balance of additives is much more important than the base number alone. Consideration of individual elemental contents is even less rewarding.

In the early days of engine testing, it was envisaged that various tests would be developed, each concentrating on measuring one aspect of performance with a high degree of discrimination. Thus early tests evaluated diesel piston deposits, oxidative oil thickening, or valve train wear on an individual basis. With the introduction of the "MS" Sequence tests in 1962, however, testing in production multi-cylinder engines became common, particularly at that time for gasoline testing, but starting a trend which has continued and which recently has resulted in many multi-cylinder diesel engines being used for performance tests. Such tests on commercial engines are in effect sponsored by their manufacturers, who undertake to maintain supplies of engines and spares for many years. These manufacturers naturally wish to see oils perform well in all areas in their engines, and the inclination has been therefore to specify minimum performance criteria for several parameters, and not just for a single aspect of performance which the test may be designed to highlight.

Although the oil companies (and the additive companies who frequently formulate for them) ultimately perform and fund much test development and therefore act as a moderating influence by arguing against undue test proliferation, the situation described above has led to many performance areas having to be demonstrated in several tests, although there is usually one test which has been designed to focus particularly on each aspect of performance, and one leading performance area for assessment in each test. A growing tendency is for the engine manufacturer providing the test engine to demand in his own specification a wider assessment of performance in his engine than is required for the national or international specification for which the test was developed.

The four key performance areas used for assessment are shown below, with the typical parameters measured to indicate the performance level achieved:

Oil Related:
- Oxidation
- Thickening
- Bearing corrosion (from oxidation)
- Viscosity
- Seal degradation

Engine Wear:
- Valve train wear and pitting
- Bearing wear
- Ring wear
- Liner wear
- Bore polishing

Engine Deposits:
- Carbon and Varnish:
  - Piston grooves and lands
  - Ring sides and rear
- Sludge deposits:
  - Rocker cover
  - Sump
  - Timing case
  - Oil ring clogging
  - Oil pump screen clogging

Rust and Corrosion: Valve lifter rusting/corrosion
Bearing corrosion

There are many different tests for measuring each performance parameter, and details of the performance factors covered by each of the key tests are included in Appendix 7. Measurements can be physical measurements (such as for wear) or carefully controlled visual ratings (of deposits, etc.). This was discussed in Section 3.3.2 on engine testing.

We will now look at the performance specifications for several oil types, and see how performance is specified in terms of tests.

6.2.2 Performance Requirements for Gasoline Engine Oils

The most widely used criteria for gasoline engine oils are contained in the API "S" series of oil classifications and their associated SAE J183 performance criteria. Other national, international, and industry organizations frequently use these as a basis for their own specifications.

Taking the API SH Classification as a reference point, the table below shows the tests used for measuring the various performance areas:

PERFORMANCE AREA	TESTS	ENGINE TYPE
Bearing corrosion/ oil oxidation	CRC L-38 Sequence III-E	Labeco 1-cyl. 1986 Buick V-6
Valve train wear	Sequence III-E Sequence V-E	1986 Buick V-6 Ford 4-cyl. E.F.I.
Piston deposits	Sequence III-E Sequence V-E	1986 Buick V-6 Ford 4-cyl. E.F.I.
Sludge	Sequence III-E Sequence V-E	1986 Buick V-6 Ford 4-cyl. E.F.I.
Rust	Sequence II-D	1977 Oldsmobile V-8

There are several things worth noting in the above table. One is how old test hardware tends to be retained, often after normal production has ceased. This is understandable in view of the high cost and time necessary to develop a new test. In some cases such as the Labeco laboratory test engine, the design is very different from that of modern commercial engines and some have argued that it has lost commercial relevance.

Another interesting fact is that the pattern of a GM engine for the Seq. II low-temperature rust test and the Seq. III high-temperature test, coupled with a Ford engine for the cycled high/medium/low-temperature Seq. V test, has been continuous. (Chrysler participation with a Seq. IV test and Ford's demand for their own rust test ceased after API SD.) This is partly a matter of the availability of engines already found suitable for a given type of test, and partly inertia in the test development system.

Generally, performance requirements have become more severe as the API classifications progressed from the 1964 SC to the 1993 SH level, with gradual changes to more modern engines. There was, however, a discontinuity and probable drop in sludge performance requirement when the Sequence V-D test replaced the Sequence V-C for API SF. Yet at the same time this change added a low-temperature wear requirement in an overhead cam engine. Not only was there a change from a traditional large V-8 engine to a small, four-cylinder overhead camshaft engine of European type, but there was a change from leaded to unleaded gasoline in this test. The Seq. V test with its cyclic procedure and rating for wear, piston deposits and sludge has always been a key test in measuring gasoline performance of oils, and the precise effect on oil quality of this double change is difficult to assess. It can be argued that it satisfies better the current U.S. market needs, which is the primary purpose of the API classifications and the corresponding specifications.

The above tests apply just to the definitions of gasoline oil quality in the U.S. as defined in the API classifications. For multi-purpose oils with diesel performance other tests may need to be added.

In the case of European specifications, tests have very often been added to basic U.S. Military Specifications or API classifications to ensure that performance in what was initially a very different type of engine was satisfactory.[1] These transatlantic differences are less now than previously, but the differences in engine sizes and designs have been succeeded by differences in operating practices (particularly related to long-distance, high-speed driving in Europe and longer oil drain intervals) so that specifically "European" tests are still called for. These tests generally, although not exclusively, call for a comprehensive rating of most engine parts.

Originally this was done by individual engine and vehicle manufacturers, notably Ford UK and later Ford Europe and Mercedes-Benz. With the setting up of the CCMC to coordinate the efforts of the purely European manufacturers it was originally intended to set up a purely European set of engine tests and not rely on the API Classification tests at all, but this was not successful and most U.S. tests remain in the European Sequences plus key European tests.[1]

The sequences set by the CCMC therefore followed the API pattern:

CCMC-G1	Effectively API SE plus:	Ford Cortina Fiat 132 for pre-ignition OM616 for valve train wear (a Mercedes diesel engine)
CCMC-G2	Approximates API SF plus above European tests	
CCMC-G3	As G2 but for a low-viscosity oil	
CCMC-G4	Effectively API SG plus:	Mercedes M102E for sludge deposits Peugeot TU-3 valve train wear
CCMC G5	A low-viscosity version of G4	

With the issue of the ACEA sequences in 1996 (see Appendix 10(c)), the CCMC specifications will all be effectively obsolete.

In the European context, the M102E requirements are considered to be more severe than those of the Sequence V-E in API SH. In G-5 the Seq. III-E viscosity requirements (measuring oxidation) are also more severe than in API SH.

ACEA as the successor to CCMC published their proposed new sequences at the end of 1994. The passenger car gasoline oil sequences are intended to be effective from mid-1996 onward and comprise the following:

ACEA A1-96	Low-viscosity oil	Peugeot TUM3 valve train scuffing and wear Peugeot TUM3 ring stick, piston deposits and viscosity increase Mercedes M111 black sludge and wear Seq. III-E , Seq. V-E
ACEA A2-96	Multigrade oils	Tests as for A1-96
ACEA A3-96	Superior quality	Tests as for A1-96, tighter Seq. III-E limits

Appendix 10(c) contains details of these sequences.

6.2.3 Diesel (Commercial Oil) Tests

The tendency to have several engine tests measuring the same properties is even more marked in the case of diesel lubricants, and has accelerated in the last decade. The argument is that diesel engines vary more widely in design and power ratings, and tests in different types of engines are necessary to ensure that the spectrum of engines in the field is adequately covered. This begs the question whether the engines should be designed to suit the oil, or the oil to suit the engine (see also Section 6.4.2). Some have argued that the proliferation of multi-cylinder diesel test requirements has come about because engine manufacturers, on both sides of the Atlantic, have had engine durability problems and have demanded new oil qualities which are able to minimize these. The countervailing argument is that without increases in oil quality, engines would not have reached the high levels of reliability at high specific power outputs which are seen in the field today.

The primary lubricant requirement for a diesel engine to operate satisfactorily is that the piston rings must continue to provide a good gas seal. They must be free to move radially, without excessive vertical play, and should not wear rapidly. To achieve this, carbon build-up in the ring grooves should be minimized. (Refer to Figure 19, Chapter 1.)

One source of carbon formation is progressive build-up of lacquer deposits which are baked to solid carbon. To reduce testing time the early Caterpillar diesel tests (L1, 1D, 1H, 1G) set severe standards for varnish formation on piston lands and skirts as well as requirements for actual groove-filling by carbon. As indicated earlier, a Caterpillar piston from a failing test is actually in a far better condition than one operating satisfactorily in the field. This may be regarded as unrealistic, but the alternatives are even longer tests (the Caterpillar tests take 480 hours), or more severe (and therefore probably unrealistic) operating conditions. One way this has been done in the past was to increase the fuel sulfur content in order to increase the rate of acid generation. With fuel sulfur levels in the field declining markedly this approach will become far removed from most field situations. This then is the climate which tends to promote dissatisfaction with diesel testing. However, the alternative of abandoning laboratory engine tests and reverting to the even more difficult field assessment can be regarded as unrealistic owing to cost and time considerations.

As new diesel engine designs have developed, needs for new types of lubricant tests have evolved. Highly rated engines with reduced top land clearances have given problems with bore polishing (a form of wear), which can lead to high oil consumption before general deterioration of performance. Use of small diesel engines of relatively high speed in passenger cars has also led to development of tests in this type of engine for bore wear and valve train wear. CCMC PD 2 is an example of a specification employing such tests.

Important diesel engine tests in key specifications are given below, with an indication of the parameters assessed:

API CA/CB	Caterpillar L1 for top groove fill and deposits groove 2 and below
API CC	Caterpillar 1H/1H2 for top groove fill, deposits groove 2 and below, ring side play
API CD	Caterpillar 1D for top groove fill, deposits groove 2 and below
	Caterpillar 1G/1G2 for top groove fill, deposits below and ring side play
API CD-II	Caterpillar 1G/1G2 for top groove fill, deposits below, and ring side play.
	Detroit Diesel 6V-53T for deposits and ring, liner and valve condition

API CE	Caterpillar 1G2 for top groove fill, deposits, and ring side play Mack T-6 for deposits Mack T-7 for oil viscosity increase Cummins NTC-400 for oil consumption, piston deposits, cam follower spindle wear
API CF-4	Caterpillar 1K for deposits, oil consumption, ring stick, and scuffing Cummins NTC-400 for oil consumption, piston deposits, cam follower spindles wear Mack T-6 for deposits Mack T-7 for oil viscosity increase
API CG-4	Caterpillar 1N for piston deposits, ring sticking, scuffing, oil consumption Mack T-8 for soot-related viscosity increase, filter plugging, and oil consumption GM 6.2L for soot-related valve train wear
CCMC D-1	Caterpillar 1H2 for top groove fill and deposits below Mercedes OM 616 for cam and cylinder wear
CCMC D-2	Caterpillar 1G2 for top groove fill and deposits below Mercedes OM 616 for cam and cylinder wear
CCMC D-3	Mercedes OM 352A for bore polish and top groove deposits Mercedes OM 616 for cam and cylinder wear
CCMC D-4	Mercedes OM364A for bore polishing and piston deposits Mercedes OM616 for valve cam wear and cylinder wear Mack T-7 for viscosity increase
CCMC D-5	As for D4 but more severe OM364A limits
CCMC PD2	VW 1.6 TC Diesel for ring sticking and piston deposits (to be replaced) Mercedes OM602A. Overall cleanliness and wear (under development) OM 616 for valve cam wear and cylinder wear

New ACEA diesel specifications called E1 to E3 for HD diesels and B1, B2, B3 for passenger car diesels introduce new tests from Mercedes, Volkswagen and Peugeot (see Appendix 10(c)).

Many diesel specifications also include gasoline engine tests for oxidation, rust, and sludge. This arises partly from the fact that many diesel oils are considered to be multi-purpose oils, but also because these properties are important for diesel engines but suitable tests in diesel engines have not been developed.

A problem for regions outside North America which also make use of API quality levels is that, as soon as new quality levels are established, the old ones are declared obsolete and tests to confirm these superseded levels are not supported in terms of stands, fuel, parts and rating validation.

Prior to the 1980s there existed more generally a situation where a range of quality levels existed, and to an extent a consumer could choose the oil quality level he could afford, and balance frequency of oil changes against oil costs. Today so many engines have critical requirements that in general the best possible oil is needed to provide satisfactory engine life. As oil and additive technology is developed, new oil qualities are demanded, and if the oil technology is not immediately available then engine manufacturer demands plus internal oil industry competitive pressure lead to a race to develop it. The formulation complexity and cost of new oil qualities is thus rising gradually; but there are exceptions. While most engine manufacturers recommend only one oil quality for a particular engine service, some, for example Mercedes-Benz, Volvo and MAN in Europe, will in effect give an incentive to oil companies to develop higher-quality oils by granting approvals for oils for extended drain use or for use in very severe service conditions, while accepting older quality oils as suitable for normal use.

The situation in the Asia Pacific region is particularly interesting. Taking the region as a whole, the total lubricant consumption has exceeded that of N. America since the beginning of the 1990s. As well as Japan and Australia the area includes China and India, areas of vast population where rapid growth of lubricant sales is anticipated. Throughout the region, but particularly in these two countries, there are large volumes of lower-quality lubricants sold, about half of the gasoline engine oils in China and India being below API SE quality. These are still expected to account for one-quarter of sales by the year 2000. The use of old-technology engines and an inability to afford expensive lubricants makes it difficult to boost quality levels rapidly.

The heavy-duty diesel market throughout the area is split between API CD and API CC qualities. Even in Japan, penetration of post API CD qualities is expected to be slight up to the year 2000.

Japanese tests to validate oils of approximate API SE quality are already in place, but action may also need to be taken over API CD and even API CC validation by retention of Caterpillar 1G and 1H parts and rating expertise.

6.2.4 Problems with Engine Test Procedures

There are many complaints today about the individual test procedures used in company or industry specifications. They include:

- too expensive
- insufficiently precise
- parts or fuel unobtainable
- inconsistent parts quality
- irrelevant to today's market
- incompatible with another requirement
- unnecessary duplication of another test

Let us consider how some of these issues arise, and what may be done to avoid them.

a) Too expensive—The cost for a single engine test running up to 500 hours may cost up to about $110,000 at 1994 prices for a large diesel engine test (see Appendix 8 for list prices). Much of the cost is for the engine itself including replacement parts and most of the remainder is for the fuel consumed. Once a test is accepted as being necessary there is limited opportunity to have more than a marginal influence on the cost.

b) Insufficiently precise—This is a general criticism of most engine tests, but particularly of worldwide diesel engine tests and of

European engine tests of all sorts. Precision issues were discussed in more detail in Section 3.4.

Because engine tests are relatively so expensive compared to other forms of research and development testing and because results are so often tied to commercial approvals, some of the rechecking which would normally apply in R&D is not always followed. Sometimes a "pass" may achieve a commercial value which is unrelated to its technical merit if it happens to be based on a "lucky bounce." Conversely, good technology may sometimes be scrapped because the engines were unknowingly "running severe" at the time of decision-making.

Use of back-to-back comparisons with industry reference oils can go some way to making the best of imprecise tests—but at a significant cost penalty compared to running reference tests more rarely, for example one in ten tests.

There is often little common ground between those who argue that industry cannot afford the time and money to develop precise tests and those who argue that industry cannot afford not to develop precise tests.

Some of the problems of precision arise because of the different needs of the test users. There has always been a conflict between acceptable speed of development and acceptable test precision. OEMs with engine problems have understandably wanted to settle for an imperfect test giving a quick and practical answer. Oil formulators have felt the need for good precision far more strongly. Without good test precision, the oil formulator may be frustrated by inability to distinguish real trends from statistical "noise" in his experiments and will lack the time or money to conduct designed experiments. There is thus a continuing search for low-cost screening tests to predict the results of full-scale tests.

c) Parts or fuel unobtainable—Providing components for test engines is not usually the highest-priority job for an engine manufacturer.

Very few companies are in the business of providing test fuels. In spite of today's sophisticated analytical techniques it is very hard to duplicate the qualities of a provisional batch of fuel and batches are often engine tested and rejected—at great cost and inconvenience to everybody involved. Large volume production and long-term storage is possible for diesel fuels where fuel specifications remain relatively constant.

For gasoline there are more difficulties. It is unstable in storage (although nitrogen blanketing or use of floating roof tanks helps) and it therefore is impractical to have large production runs. There have been regular problems to produce repeatable fuel batches for the Sequence V and M102E tests.

d) Inconsistent parts quality—This single problem has caused enormous waste of resources. The root cause is usually a lack of recognition by the sponsoring organization or parts supplier of the importance of quality control.

The OEM fuels and lubricants specialist may find it hard to understand why special tolerances may be requested, when stocks are available which meet production tolerances.

Sometimes components such as piston rings have been sourced from two different suppliers without telling test laboratories. This may be perfectly acceptable for routine engine production but gives significantly different results under lubricant test conditions. Every effort must be made to minimize component variation.

e) Irrelevant to today's market—Some tests seem to continue even when nobody can be sure of their usefulness. This may be due to inertia, to the size of the hurdle to develop a replacement or because the specification owner does not have to pay the test costs. It may also be due to a wish to continue to avoid a problem seen in the past and "cured" by use of the test. The best chance of eliminating irrelevant tests comes from regular and rigorous reviews within the technical societies and sponsoring organizations.

f) Incompatible with other needs—This may be a legitimate complaint or more often an expression of frustration at a difficult technical challenge. As we discussed in the previous section, formulation invariably involves compromises and some compromises are more difficult to achieve than others. If two important specifications contain apparently incompatible requirements the oil company has the option to try to meet both or to offer separate products. This luxury does not exist when the challenge is in the same specification. Examples of such a challenge include meeting friction/fuel economy and wear needs together, or solving seal compatibility problems without compromising something else.

g) Unnecessary duplication—The cause of this is often political rather than technical. Perhaps two companies or industry groups start work on a common problem at the same time. Both developments reach fruition at the same time and neither group is prepared to "lose face" or write off experience by backing down—so industry gets both tests. Sometimes the tests arise from different regions in the world and there is strong pressure to use the local test. Sometimes there may be a genuine wish to consolidate, but nobody is quite sure whether the tests are equivalent or not, and nobody has enough of a vested interest to carry out the work (such as a field test) to find out.

Justification for New Test Development

In 6.2.2 we discussed the understandable desire to stay with known established tests; the above paragraphs indicated how dissatisfaction with tests can arise and pressures mount to develop new tests or improve old ones to make them more relevant to today's needs. Test development is a long and expensive process, and what amounts to a cost/benefit analysis needs to be done before rushing into major programs. The paragraphs below give some thoughts on this process.

a) Justification

Are different metallurgies or other new materials such as ceramics involved? Are there novel design features which have impact on lubricant formulation? Are there higher temperatures or pressures than previously encountered? Are legislators calling for some new parameter to be measured? Fuel economy is a relatively recent example; exhaust emissions performance or catalyst compatibility will probably follow.

Driving habits of the public or vehicle users can also change (e.g., more long-distance driving), and to have this reflected in test conditions, changes will eventually be necessary.

b) Type of Test
Which is most appropriate, a rig-test, a bench engine test, or a field test? The arguments for each are discussed in Chapter 3 and have also been given by their advocates elsewhere.(2)

c) Complexity and Cost
Should a mono-cylinder or a multi-cylinder engine be used? Modern practice strongly favors multi-cylinder engines and first CCMC and then ACEA have formally stated their preference for multi-cylinder tests in most situations, arguing particularly the importance of using modern design, "real world" engines. But multi-cylinder engine tests and especially those carried out using large diesel engines are extremely expensive and some testing experts argue that not much more value is obtained from six cylinders than one. If engine design is a key factor in the test, then there may be no alternative to the use of multi-cylinder engines. If a more fundamental factor such as top ring temperature is being studied, then a single-cylinder test may in principle be more cost effective.

Few mono-cylinder engine tests have been introduced since the 1960s, but exceptions include the Caterpillar 1K and 1N tests. The Austrian consultant engineers, AVL, market a mono-cylinder diesel engine which offers the flexibility to be used with the piston configuration of several standard multi-cylinder engines, and tests have been developed to simulate various U.S. and European expensive multi-cylinder tests. In 1993 the concept received recognition when MAN announced that they would accept test results from the AVL engine with MAN cylinder head components for

MAN QC approval in lieu of the very expensive multi-cylinder MAN engine test.[3] In the mid-1980s work by the oil industry and the British Ministry of Defence on a replacement for the obsolete Cortina high-temperature deposit and ring sticking test used in the CCMC G1, G2 and G3 specifications, led to the development of the single-cylinder W1 Cortina test. For the reasons given previously the test was not favored by CCMC, but it continues to be a Ministry of Defence requirement.

6.3 The Organizations Involved and Their Roles

Which organizations are involved? A newcomer trying to understand the world of lubricant classification is liable to get swamped in the complexity, and in a jungle of initials and acronyms for organizations which superficially all seem to overlap or do the same job. The titles of organizations are given in full when they are first mentioned and acronyms are used subsequently. A glossary of acronyms is provided in Appendix 2.

The interrelationships between the organizations are also complex and are discussed looking in turn at the three main regions of influence—the U.S., Europe and Japan.

6.3.1 The United States

6.3.1.1 The Society of Automotive Engineers (SAE) Viscosity Classification

The first property of engine lubricants to be classified was viscosity and the SAE J300 viscosity classification[4] has been the basis since its inception in the early 1900s. It has not remained unchanged, however. It has been adopted throughout the world and is being raised to an ISO standard. The latest version is reproduced in Appendix 9.

The essence of the classification is to indicate viscosity at both low (engine winter starting) temperatures and high (operating) temperatures. The viscosities are indicated by two numbers, with higher values showing greater viscosity. The low-temperature viscosity is indicated with a W (for winter) and is given first. The correct way to indicate SAE viscosity is SAE xxW-yy, e.g., SAE 10W-30.

Viscosity at low temperatures was originally estimated by extrapolation from measurements at higher temperatures but for a long time low-temperature cranking viscosity and pumpability have been measured using the Cold Cranking Simulator and Mini-Rotary Viscometer, respectively. Kinematic viscosity at 100°C continues to be measured by a U-tube method.

Because of the general recognition that high-temperature/high-shear (HTHS) viscosity measured at 10^6 reciprocal seconds and 150°C is more representative of conditions in operating bearings than kinematic viscosity at 100°C, HTHS viscosity has recently been included as a part of SAE J300.

6.3.1.2 Performance Classifications and the "Tripartite"

By far the most influential of all organizations to define and develop automotive crankcase lubricant qualities in the world has been the informal tripartite of three large organizations—SAE, ASTM (now known just by the acronym but formerly called the American Society for Testing & Materials), and The American Petroleum Institute (API).

These organizations (each having fields of interest well beyond the confines of automotive lubricants) have for many years interlinked, each having a defined role but with individuals and companies often being represented on two or even all three of the organizations.

In the early 1950s, the American Petroleum Institute (API) introduced a system for classifying the various types of service conditions under which engines operate. These were designated as ML, MM, MS (for spark-ignition engines) and DG, DM, DS (for diesel engines), but no performance standards were initially specified. These classifications were covered in more detail in Sections 4.1.1 and 4.1.2. Performance standards came with the introduction of the Sequence tests in the mid-1960s, and then, in order to provide more precise definitions of oil performance and engine service, in 1969-70 API, in cooperation with ASTM and SAE, established a new engine service classification for engine oils. ASTM defined the test methods and performance targets. API developed the service letter designations and "user" language. SAE defined the need and combined the information into an SAE Recommended Practice in the SAE Handbook for consumer use. The current revision of that document is called "Engine Oil Performance and Engine Service Classification" (other than "Energy Conserving"), SAE J183 JUN91,[5] reproduced in Appendix 7.

The current API Engine Service Classification is divided into an "S" series, covering engine oils for use in passenger cars and light trucks (mainly gasoline engines), and a "C" series, for oils for use in commercial, farm, construction and off-highway vehicles (mainly diesel engines). An oil can meet more than one classification, for example, API SG/CD or CE/SG. The API SG category was formally adopted in March 1988 and met the requirements of 1989 model year passenger cars. Similarly, the API CE was adopted in 1988 and recommended by all American heavy-duty engine manufacturers. In 1990 it was replaced by API CF-4 for four-stroke engines. API SH was issued in 1993, differing principally from API SG in the procedures for running tests.

The complete API system is described in API Bulletin 1509, "Engine Service Classification and Guide to Crankcase Oil Selection."

API does not grant approvals but will certify lubricant quality. In 1983 it developed its Service Symbol (shown in Figure 80) and issued licenses to oil companies who wished to use the symbol to promote their products. The central part is reserved for the SAE viscosity grade, the top half of the outer ring shows the API performance classification and the lower half of the ring is reserved to indicate whether the oil meets an "energy conserving" or fuel economy standard. Only "currently supported" performance qualities are allowed to be shown in the top part of the ring. Performance levels such as API SD which are considered by API to be obsolete may not be shown. Today a further major change has taken place in API's role and its Engine Oil Licensing and Certification System (EOLCS)[6] represents a step change in oil quality management, which will be discussed later.

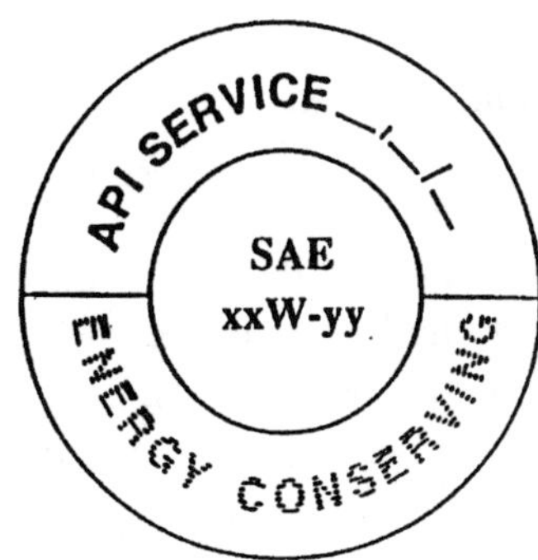

Fig. 80 The API Service Symbol.

6.3.1.3 The U.S. Military and the SAE Lubricants Review Institute (LRI)

The role of the U.S. military has been mentioned already in Section 4.1 as the introduction of new combinations of requirements forced major changes in the way diesel engine oils were formulated. The MIL-L-2104 and MIL-L- 46152 series of specifications were major influences on other specifications all over the world. Oil companies worldwide required their products to be approved to U.S. Military Specifications and other users of all sorts such as bus and truck companies insisted on seeing evidence of U.S. military approval from potential suppliers. Very few of the approvals granted were ever used for the intended primary purpose of tendering for army business.

A special feature of the military approval process which gave it value to third parties was the reviewing procedure. A review board chaired by the U.S. army and including lubricant and engine testing experts from the principal U.S. engine manufacturers met five times a year to review submissions from oil companies or (more usually) additive companies acting on behalf of oil company customers. All engine parts had to be available for inspection and rating by the experts and their professional advisers. The parts, as well as mountains of accompanying paperwork, were examined with great rigor and it was the quality of the review process which gave value to the eventual approvals. Tests considered unsatisfactory would be required to be rerun.

In 1977 the "MIL-Board" was placed under the auspices of the Society of Automotive Engineers and renamed as the Lubricants Review Committee of the Lubricants Review Institute,[7] although most of the functions have remained unchanged. In our view, since the late 1980s, a combination of factors has caused the influence of the LRI to diminish. Changes in engine testing policies and the strengthening of the API classification (see later) have given oil companies and end users less cause to feel the need for the security of a U.S. military approval. An alternative view is that the LRI has merely rebalanced its efforts, to place more emphasis on combat vehicles.

In Europe, the development of the CCMC sequences (see later) gave oil companies the opportunity to use marks of quality which many saw as being more relevant to the European market. In the U.S. the Department of the Army was increasingly questioning whether the cost and bureaucracy associated with the approval process was needed and whether it would be better use of the taxpayers' money to tender for oil to the appropriate and latest API quality, called a "commercial item description" or CID in army terminology. First MIL-L-46152 and then MIL-L-2104 series specifications were largely abandoned.

In mid-1995 the LRI still reviews MIL-L-2104F approvals, gear oil approvals, and SAE 10W oils for tracked vehicles used in arctic conditions.

6.3.1.4 Conflict and Change

For many years the United States, primarily through the "tripartite," has been the acknowledged world leader in setting automotive crankcase lubricant quality standards, but in the second half of the 1980s concerns were expressed which ultimately led to conflict and change.

A virtue and a weakness of the "tripartite" has been inherent in its constitutional framework. The relevant SAE and ASTM committees particularly include representatives from the motor, oil and petroleum additive industries who may sometimes have conflicting objectives. Meetings are usually open to anyone to attend and speak, and voting memberships on key committees are carefully balanced between "producers" and "users."

Balance was not always achieved in the past, and this was one of the causes of changes made in the late 1980s.

Negative votes are addressed carefully to see if they are "persuasive" and are resolved wherever possible. In most respects these committees are an excellent example of democracy in action. But democracy has its draw-backs. The resolution of negative votes in a hierarchy of committees takes much time. A pressure group or even an individual company has great ability to disrupt or delay. A commercial interest may be hidden very effectively in a fog of technical arguments and often a neutral listener may be quite unaware that a debate which seems to be about test methods may actually be a battle for markets.

The Motor Vehicle Manufacturers Association (MVMA) felt the most frustration and in a preface to a document proposing a new way of doing things[(8)] spelled out many of their concerns in detail. They pointed out that new classifications were taking something like six to eight years from conception to reality, and in many cases the final performance require-ments were lower than requested at the start of the protracted program. This response time was unacceptable in a rapidly changing world when the OEMs in MVMA were trying to shorten the development time for new vehicle models from about seven years down to three or four. A particular concern was that OEMs were required to produce "a basket of failed parts" before the oil industry would take them seriously. The OEMs argued for prevention rather than cure. Some wanted lubricants to help meet future standards with future engines where by definition there would be no field experience.

Eventually in 1989/1990 the motor industry threatened to take unilateral action and develop independent specifications and approvals. The API argued in return that the existing system had served the consumer very well in the past and that there were no significant field problems. Moreover, they argued that as the oil industry rather than the motor industry sold oils to the consumer and took legal responsibility, they had to have a major say in defining the quality of products sold under their brand names. Signifi-cant concessions were eventually made by both industries and at the time of writing peace is at least partially restored, although the underlying differences of opinion still remain.

6.3.1.5 The CMA Code of Practice

We have already mentioned how concerns have been expressed, particularly by MVMA, about the quality of the tripartite process and in particular about its slow speed of reaction to motor industry need. Concurrently, questions have also been raised about the definition of "passing" quality when applied to lubricants tested in engines. Should it mean "always able to pass," should it mean "capability to pass has been demonstrated," or something in between? The answer is not self-evident when the poor precision of engine tests compared to laboratory glassware tests is considered. With rare exceptions (e.g., the British military) there has not been a requirement to pass each test every time or even on average.

The British military authorities when granting lubricating oil approvals have always required the sponsor or test laboratory to state their intentions and formulations in advance of running an engine test for approval purposes and to provide all results (passing or failing) to the authorities. There are strict limits on the number of attempts (usually two) which may be made to obtain a pass on a particular formulation. While several other European military authorities followed the British lead, other specification and approval authorities, including the much larger and more influential U.S. military as well as individual OEMs, have traditionally not required such information. Failing attempts did not need to be reported and only passing results (usually accompanied by relevant engine components) needed to be presented.

This has meant that in some circumstances formulators could make many attempts at each of the engine tests in a specification. The incentive to run multiple tests was great if (a) the test was very imprecise, or "bouncy" (see Section 3.3.3), or (b) the volume of business at stake was great and a price competitive formulation was needed, or (c) a lot of money had already been spent on a particular formulation, the particular test was near the end of a long sequence and reformulation would be too time-consuming or expensive to contemplate. In practice time and money rather than anything else limited retesting. The result of the general practice is that many people on all sides of industry have felt uncomfortable when an oil which is "approved" might be quite unlikely to pass all tests first time if tested at random against the specification.

Another gray area concerns the formulation itself. Many oil specifications, especially those for high-quality branded lubricants for an international market, may well require as many as twenty engine tests, plus bench tests and chemical and physical limitations, all with conflicting formulation needs. It is usually impossible for a formulator, even without financial constraints, to get the formulation "right the first time." How then should formulation changes be handled? Is it reasonable to start all over again because of a small change to the concentration of one component in a formulation? In practice some "case law" has built up in both the U.S. and Europe to provide guidance. For example in the U.S., formulators would explain their predicaments to the Lubricants Review Institute, acting on behalf of the U.S. military. Depending on circumstances, a running change in a formulation might be accepted without any retesting or could require several tests to be repeated. Thus an increase in ashless dispersant level would usually be considered to have a positive effect on sludge performance and be neutral in other tests, while an increase in mono-functional viscosity modifier would be considered detrimental to diesel engine piston deposit control.

Changes in base stock composition cannot usually be made without invalidating approvals. Some international oil companies have gone to great lengths to demonstrate interchangeability of base stock sources to the authorities. Other companies may have considered this unnecessary and made their own judgments on validity.

Different formulating companies will almost certainly have made different judgments about the need for disclosure of formulation changes, and within one company the judgment may have been different depending on the nature and geographical location of the approval authority, and indeed upon the experience and character of the individual making the decision.

Apart from the concerns for change from MVMA previously mentioned, the growth of interest in "quality practices" and ISO 9000 series accreditation raised fresh queries in people's minds about industry practices. These were strongly articulated at a European industry conference in 1989,[(9)] from which serious European interest grew.

In due course such views received support from most members of the additives industry and a panel was set up by the petroleum additives industry within the U.S. Chemical Manufacturers Association to develop a Code of Practice for engine testing. The panel commenced operation in 1990 and the Code was implemented on March 30, 1992.[10]

The Code of Practice originally covered the engine tests used in the latest API S category (currently the Sequences II-D, III-E, V-E, V-I, CRC L-38) and the Caterpillar 1G2 when used in combination with an "S" category designation. In early 1994 the diesel tests Caterpillar 1N, Caterpillar 1M-PC, Mack T-8, GM 6.2L and Detroit Diesel 6V92TA were added. In principle any other tests showing adequate precision and control may be added.

There are five main features to the Code:

1. Engine tests are registered with an independent monitoring agency prior to running the tests. They may be run only in referenced stands which have met statistical acceptance criteria.

2. The test sponsor may select the laboratory in which the test is to be run, but may not select the individual test stand which must usually be the "next available" stand.

3. A concept of Multiple Test Acceptance Criteria (MTAC) is introduced. For an oil to be considered a pass on its first test, all controlled parameters must pass individually. If a second test is run, the mean value of each parameter must be a pass. If three or more tests are run, one complete test may be discarded and the mean value of each parameter from the remaining tests must pass.

   Individual test results are adjusted for severity changes in the test procedure as identified by the Lubricant Test Monitoring System using control charting maintained by the ASTM Test Monitoring Center.

4. The concepts of "minor" formulation modifications has been introduced and details of how and in what circumstances changes

to an original formulation may be made have been documented in great detail. All subscribers to the Code will now have a consistent set of rules from which to work.

5. A "candidate data package" is provided to the lubricant marketer by the sponsor of the test program if this is not the marketer. This provides not only all the test results but an audit trail which can demonstrate the nature and validity of any formulation changes and details of all scheduled tests and severity changes.

Use of The Code of Practice is voluntary, but all major petroleum additive companies have formally committed that all tests currently covered by the Code and initiated by them will be run in accordance with the Code, even if required for example for a European application to meet a CCMC classification.

6.3.1.6 The API Engine Oil Licensing and Certification System (EOLCS)

The API Engine Oil Licensing and Certification System was borne out of the frustrations and conflicts mentioned in 6.2.4. There are five key elements:

1. The definition of performance standards. These will be defined in a manner similar to the old "tripartite" system, with SAE, API and ASTM carrying out their traditional roles. There is provision for ILSAC (see 6.3.5.2) to take over from ASTM if ASTM is unable to achieve consensus.

   The system defines physical, chemical and performance characteristics of engine lubricants. In 1995 it covered categories API SG/SH/CD/CD-II/CE/CF/CF-2/CF-4 and CG-4.

2. The protocol for engine testing. This essentially says that each engine test must have been run according to the CMA Code of Practice. Note that whereas use of the Code is, in general, voluntary, in the context of EOLCS it is mandated if the particular test has been covered by the Code. In mid-1995 the tests fully covered

by the Code were II-D, III-E, V-E, V-I, L-38, GM 6.2L, T-8, 1MPC, 1N.

3. A licensing procedure. The lubricant marketer is responsible for product performance and must certify that each viscosity grade of each brand meets requirements. There are detailed rules for allowing interchange of base stocks and for allowing engine test read-across in certain circumstances from one viscosity grade to another. There is a sliding scale of fees with a minimum of $500 per grade per company with an extra $1000 for each million U.S. gallons sold above the first million gallons.

4. Certification marks. Two marks are available to be licensed, the API symbol shown earlier as Figure 80 and the API Certification Mark (Figure 81), formerly called the ILSAC Certification Mark.

   The API Symbol may be placed anywhere on the product packaging and the details contained therein may change according to the nature of the product and as performance levels change. The Certification Mark must be on the front of the container. This mark will not change, but must be used only if the product meets the current ILSAC performance level (see Section 6.3.5.2). If the product is suitably qualified, the marketer may use either or both symbols.

Fig. 81 The API Certification Mark.

5. Conformance audits. All licensed products will be subject to review for conformance. Companies whose products fail may suffer enforcement penalties ranging from suspension of the product license to a requirement to remove the product from the market.

6.3.2 Europe

6.3.2.1 Individual Manufacturer Specifications and Approvals

European OEMs have differed in attitude towards lubricant classification and approval. Fiat Auto has recommended only oils marketed by their associated company Fiat Lubrificante. The Peugeot Group and Renault have had special contracts with individual oil companies. Some companies (e.g., Saab-Scania) have had special technical requirements for service fill oils but have not formally approved products or published lists of them. Ford of Europe was one of the originators of lubricant approval systems but later decided that they were not cost effective. In 1995 Ford operates global standards for factory and service fill based primarily but not exclusively on SAE viscosity and API classification. Mercedes-Benz, Volkswagen and Volvo Truck have been among the companies who have recently had the greatest influence on the lubricant marketers because they have established specifications, granted approvals and published lists of approved lubricants. The Mercedes-Benz Betriebsstoff - Vorschriften/ Specifications for Service Products (Fuels, Lubricants, etc.) updated regularly by Mercedes gives a comprehensive guide to lubrication requirements for their different vehicles and components under a variety of service conditions. Approved oils are listed, and such listings seem to have considerable marketing value to the oil companies concerned.

In the last few years the need for individual OEM specifications and approvals has been reduced because of the existence of CCMC classifications (see below).

Further information on European OEM specifications is given in SAE J2227[11] reproduced in Appendix 7. At the time of writing Volkswagen has taken an initiative related to their own approvals which may have profound impact on testing procedures elsewhere. They have required that all approval tests be conducted within a limited range of independent laboratories which have been required to produce several reference oil test results and to demonstrate adequate discrimination between the reference oils. Candidate test oil results will be judged primarily against the reference oil results obtained on the same test stand.

Volkswagen has argued that testing quality will be enhanced by taking these steps. The lead take by some engine manufacturers certainly benefits engine development and all engine owners.

6.3.2.2 CCMC

The Comité des Constructeurs du Marché Commun or Committee of Common Market Automobile Constructors (CCMC) was a trade association founded in 1972 and lasting until late 1990.[(12)] In 1991 it was replaced by a differently structured organization, ACEA (Association des Constructeurs Européens d'Automobiles).

The development of the CCMC crankcase lubricant sequences has had little in common with activities of the "tripartite" in the U.S. The motor industry (CCMC) developed its plans in private and was not restricted by the need for consensus with the oil industry. However, in 1974, partially in anticipation of the formation of CCMC and the need for a dialog, the Technical Committee of Petroleum Additive Manufacturers (ATC) was formed. Similarly, in 1976, a lubricant marketers organization (Association Technique de l'Industrie Européene des Lubrifiants or ATIEL) was formed specifically bearing in mind the need for a discussion with CCMC.

The Fuels and Lubricants Working Group within CCMC first published their sequences in 1975.[(13)] These covered additional requirements over and above the API designations and were thought necessary to bridge shortfalls in operating small engines in a European environment. Most of these additional test requirements have been developed by the Coordinating European Council for the Development of Performance Tests for Transportation Fuels, Lubricants and Other Fluids (CEC),[14] based on prototype tests recommended by engine manufacturers.

The sequences were extensively revised in 1984 and 1991. The final sequences include:

G-4	minimum quality for passenger car gasoline engines
G-5	low-viscosity fuel economy oil for passenger car gasoline engines
PD-2	for passenger car diesel engines
D-4	minimum quality for commercial vehicle diesel engines
D-5	as D4 but for more severe service or extended oil drain

More details of the CCMC sequences are given in Appendix 10(b).

Performance parameters such as high-temperature/high-shear viscosity and volatility were recognized widely for the first time and had a significant and positive effect on quality in the European marketplace.

There were several important differences between API categories and CCMC sequences. The U.S. classifications were backed up directly by working groups monitoring test quality; the CCMC sequences were based (mostly) on CEC test procedures but the links between the organizations were much less formal. In Europe a classification acknowledged the existence of a passenger car diesel market, in the U.S. one did not exist. In Europe CCMC recognized the possibility to provide differently classified lubricants for the same equipment depending on type of service or oil change interval; in the U.S. there is only one recommended minimum quality for a specific application at any one time.

CCMC has never given approvals, although often encouraged to do so. The organization was reluctant to incur the financial costs of developing and maintaining the necessary infrastructure, and was not a legally established body with the power to enforce any such system.

6.3.2.3 ACEA

ACEA has objectives which are broadly similar to those of CCMC, but a different construction. Unlike CCMC, it includes U.S.-owned companies which manufacture in Europe (G.M., Ford) and also European companies (Volvo, Saab-Scania) whose headquarters were outside the original Common Market.

ACEA seems likely to continue the CCMC policy of issuing lubricant "sequences" but not granting approvals. Like CCMC it favors using multi-cylinder European engine tests developed by CEC wherever possible. There are sequences for gasoline engines, light-duty diesel engines and heavy-duty diesel engines. Each of the three sequences contains three variants, making nine in all. Details of the sequences are given in Appendix 10(c).

6.3.2.4 CEC

In Europe CEC has a parallel but much more limited role than ASTM in the U.S. First its scope is limited to performance tests (essentially engine and rig-tests), and generally bench tests are left to other organizations such as the Institute of Petroleum (IP) or the German standardization organization Deutscher Normenausschuss, which awards DIN (Deutsche Industrie-Normen) classifications. Second it develops tests but does not set limits on those tests, and leaves it to commercial organizations to set limits. The senior body (the Council) is comprised of members elected from its ten constituent national European Organizations. Reporting to it are Committees including an Engine Lubricants Technical Committee (ELTC) and a Transmission Lubricants Technical Committee (TLTC), each with voting membership from national organizations. The working groups reporting to the committees have membership from individual companies and it is really only at this working level that the organization has much in common with its trans-Atlantic cousin.

CEC has done and continues to do much useful work, but as in the U.S., the existing processes are undergoing review. Council meetings and Technical Committee meetings are held semi-annually, and decision-making can be slow. The organization and voting by national bodies has some strengths but also weaknesses. Decisions have been requested of people who lacked the detailed experience to make effective judgments. Many of the CEC lubricant engine test procedures are less rigorously defined and maintained than their U.S. counterparts.

The organization recognizes these concerns and has done much in the early/mid-1990s to improve its practices. It is cooperating closely with ACEA, ATIEL and ATC.

The use of an Executive Committee has assisted technical decision-making and a full-time Technical Director is now employed. New protocols for test procedures have encouraged, and produced, some new tests of high quality.

6.3.2.5 Europe: The Future for Classifications

Change is inevitable in Europe in the next few years but it is still too early to say exactly how the changes will evolve. Most parties see that quality classifications cannot be built on poor test procedures, and initially the focus is likely to be to identify the most important bench, rig and engine test parameters and develop new tests or improve existing CEC tests to far more exacting quality standards.

At the same time as individual test quality is being improved, motor, oil and additive industries have together developed a process for managing the development of new performance specifications. This is called the European Engine Lubricants Quality Management System (EELQMS). Where practicable, industry has drawn from U.S. experience. There is an ATC Code of Practice which currently contains European Tests and which is aligned closely with the CMA Code described in 6.3.1.5. Likewise ATIEL has developed a lubricant Code of Practice which is broadly based on API EOLCS described in 6.3.1.6. Copies of the ATC and ATIEL Codes of Practice are available free of charge on request.[15]

There is one significant difference between practices in the U.S. and Europe. The importance of auditing and quality monitoring is equally recognized on both sides of the Atlantic, but Europe with its larger number of smaller but more diverse laboratories has favored placing relatively more emphasis on "the laboratory" via ISO 9000 quality systems and EN 45001 laboratory accreditation rather than on "the test stand."

ACEA has issued replacements for the CCMC sequences for implementation from January 1, 1996, onward. ACEA may possibly issue licenses to use a logo in the manner of API but will be unlikely to accept the financial burden of administering an approval system without cooperation from the oil and additive industries. The ACEA sequences are shown in Appendix 10. In mid-1995 it seemed likely that ACEA would encourage the use of high-quality testing protocols but essentially rely on oil company integrity when making ACEA quality claims.

6.3.3 Japan

Japanese vehicle manufacturers in general have relied on the API classification system to recommend engine oils for service fill applications in the passenger car area.

There is a parallel with Europe (CEC) in that the Japanese Automobile Standards Organization (JASO) has developed engine test procedures without in the past having given pass/fail limits. The parallel is not complete because the Japanese vehicle manufacturers often do not require these tests for their own in-house specifications. The engines tests have an unofficial parallel with the tests used in the API "S" sequences and are used to evaluate oils of different quality levels. In 1993 JASO modified its previous policy and quoted limits for a Japanese industry standard (JIS K2215) which includes a specification for the minimum quality engine oil for Japanese automotive gasoline engines.[16]

Within Japan the OEMs have generally developed proprietary factory fill specifications and then selected oil companies to develop the required engine oils using proprietary engine tests. These oils are also marketed by the OEMs as so-called "genuine oils." Such oils, predominantly of SAE 10W-30 viscosity, have about one-third of the local Japanese market. There is reluctance to reformulate these oils until there is an engine change, and virtually no formulation modifications are accepted without extensive testing.

Since 1990, the Japan Automobile Manufacturers Association (JAMA) has worked with the Motor Vehicle Manufacturers Association of the United States (MVMA) to develop a new performance standard called ILSAC (see Section 6.3.5.2). It is intended that the ILSAC specifications will in due course contain at least one Japanese engine test. As a result the Nissan KA24E low-temperature valve train wear test has been developed and round-robin testing has been completed by JASO, but so far the results have not been published.

There have been no Japanese industry diesel specifications developed to date, although there are preliminary proposals for the development of a light diesel engine oil specification.

The successful introduction of the JASO 2T Engine Oil Standards has given the Japanese industry a prototype which may well be followed in due course on the crankcase oil side, with the main application being the rapidly growing Asian automotive markets where Japanese OEMs are expected to dominate.

More information on Japanese tests is given in SAE J2227[11] (Appendix 7).

6.3.4 Other Countries

In the former Soviet Union and Eastern European areas, and in the Middle East and North Africa, standard test procedures were developed based on locally manufactured engines. These procedures were intended to yield performance levels parallel to the API and British Military designations by which quality in these countries was usually described. The principal test method developers were the National Oil Companies, and the driving force was often the lack of hard currency to purchase hardware and fuel from the U.S. and Europe.

In fact, the correlation between API performance levels and these national test procedures was very weak or non-existent, and discussion of oil quality between such organizations and oil or additive companies used to the API system was fraught with problems. In many cases there was also insufficient understanding of the degree of control needed on test hardware and fuels, with commercial spare parts and pump fuel being commonly used.

The cost and practical difficulties of setting up U.S. tests locally, and the time delays and cost of running tests in the U.S., will lead to continued use of parallel test procedures for a long time to come. The situation is not unlike that in Europe in the 1950s when the Petter W-1 and Petter AV-1 tests were developed in local low-cost engines as stand-ins for the Labeco L-38 and Caterpillar 1-A tests. These tests and some derivatives were widely used for a quarter of a century, and can still be run in some laboratories today. As happened in Europe, the situation is expected to change gradually in favor of international standardization, but local tests and standards will continue for some time.

6.3.5 The International Scene

6.3.5.1 International Organization for Standardization (ISO)

For many years the ISO classification for viscosity of hydraulic oils has been as well recognized as the SAE viscosity classifications for crankcase and gear lubricants, but at the time of writing ISO has not ventured into the more complex world of performance standards for automotive lubricants. However, in the mid-1980s an ISO working group was set up with the objective of developing worldwide performance standards.

Some initial proposals based around the API "S" and "C" series classifications relate to equipment and use, and (unlike fundamental properties of matter) these will change with geographical area and with time. ISO of necessity works very slowly and, bearing in mind the implied need to follow rather than lead regional (e.g., U.S.) developments, will almost inevitably be out of date most of the time. At a time when the motor industry is looking for faster response times from technical societies to changing needs, we doubt whether the concept of using ISO for crankcase lubricant performance standards is a sound one.

6.3.5.2 International Lubricant Standardization and Approval Committee (ILSAC)

Some of the concerns for change in the way classifications are developed have already been mentioned. One expression of those concerns has been the development of the International Lubricant Standardization and Approval Committee (ILSAC), led originally by MVMA and now by the American Automobile Manufacturers Association (AAMA) in conjunction with JAMA.

The MVMA's opinion was that as vehicle design becomes increasingly international and as trading barriers break down, it makes sense to have worldwide lubricant specifications. With this thought in mind, they made overtures to the other major passenger car manufacturer federations JAMA and CCMC. Details of their discussions have not been made public but their suggestions were broadly welcomed by JAMA but rejected by CCMC. (It is understood that there may have been a majority within CCMC in favor of cooperation, but the CCMC constitution required unanimity.)

At the time of writing ILSAC has issued and revised one passenger car engine oil performance standard called GF-1[(17)] (Appendix 10), has published the draft for a second to issue in 1996, and issued the principle for a third to issue in about the year 2000. The GF-1 specification is based on API SG but in addition includes an energy-conserving requirement that limits the specification exclusively to low-viscosity SAE OW, 5W and 10W multigrade oils. For this reason GF-1 standard lubricants are not widely marketed in Europe as the viscometric requirements tend to conflict with those of CCMC G-4 and G-5.

The API (formerly ILSAC) Certification Mark is shown in Figure 81.

6.4 General Comments

So far in this chapter we have considered the primary crankcase lubricant classifications that exist in the world but not looked in any depth at the underlying needs or philosophies which may cause organizations to initiate change or to deviate from each other in their approaches. These areas are now considered in more detail.

6.4.1 Advantages and Disadvantages of Establishing a New Quality Level

Let us consider a hypothetical case of an engine design which is close to production. The engine will run adequately on conventional lubricants in the marketplace, but seems to stay cleaner and show less wear when using Brand X from oil company Y. An experimental formulation from an additive supplier has given the best performance in terms of fuel economy and engine cleanliness but the additive company cannot guarantee that the product will be placed on the market. The formulation is known to contain some components which will increase cost and for which toxicological data are still pending. The engine manufacturer's lubricant expert is being asked for a recommendation to put in the vehicle handbook.

The engine manufacturer's existing policy is to recommend oils of the latest API and ACEA qualities and in addition to suggest that certain name brands have been found suitable.

It will be necessary to consider:

- How will the consumer ultimately benefit?
- How much effort would be needed internally to change existing policy? Does the company have the resources to administrate a specification and approval procedure?
- Can the company just recommend Brand X? Does Brand X have a large enough market? Is there a legal infringement?
- Do the advantages shown by the experimental formulation justify establishing a new specification? Can these advantages be demonstrated in a formal engine test sequence? If so, is there a standard test available or does someone have to develop one?

- Does the company have the power and influence to persuade oil companies to offer products to meet its needs either by offering something special or by reformulating their standard products?

- How will competitors in the industry react? Will they be antagonistic or supportive, seeing some benefit for themselves?

- Whatever is done, will the buying habits of the ultimate consumer actually be influenced?

- Should consideration be given to having different oil change period recommendations, depending on lubricant quality, to encourage other oil companies to upgrade lubricant standards to those of Brand X?

- Should the additive supplier be encouraged to offer the experimental formulation for sale to oil companies?

- Can a specification be written around the important new qualities of Brand X or the experimental formulation without having to name the product? (In this connection the Petroleum Additive Industry in Europe has a clear policy. ATC strongly deprecates the use of chemical limits such as quoting minima or maxima for certain elements, in lieu of defining performance tests. Such action may be a convenient short-term expedient but will probably not have technical validity and in the longer term might inhibit innovation and product development.)

Every case will be different, but many of the above questions will arise quite frequently. Users will have dilemmas similar to those of OEMs. There may be a fine balance between ordering something "off the peg" compared to having something "tailor-made" to meet individual needs. One of the biggest users of all, the U.S. military, is moving away from writing its own specifications toward buying commercial products. (See Section 6.3.1.3, U.S. Military and the LRI.)

Every OEM has to consider whether or not use of his own hardware should be encouraged in a test to be incorporated in an industry specification. On the one hand, incorporation should give the best possible guarantee that oils in the marketplace will meet the needs of his own equipment. On the other hand, use of hardware will probably imply a commitment to helping with test development and making arrangements to produce specially well-controlled and segregated batches of components such as piston rings and bearings.

There will almost certainly be a requirement to maintain spare parts for a guaranteed period—usually ten years from the initiation of the test procedure. This may be very burdensome if the engine ceases to be produced commercially. An example of engine test life outlasting commercial life is given by the Ford Cortina test used in the CCMC G series classifications and in British Military Specifications. Operators have been reduced to scouring scrapyards for certain critical components.

There are diverse views within the motor industry on these topics. In the U.S., most of the major OEMs have developed engine tests for use either within an industry specification framework (GM, Ford, Caterpillar) or for their own use (Mack). In Europe some OEMs, notably Mercedes-Benz, Volkswagen and Peugeot, have been active in engine and rig-test development and have encouraged use of these tests for company and industry specifications. They have made commitments to supply spare parts.

Others have taken a different approach for practical or philosophical reasons. One very senior manager in Europe took the view that use of his company's hardware in an industry test would be seen as providing evidence of a design weakness or a field problem, and therefore refused to allow potential tests to be offered to industry.

An engine or vehicle manufacturer wishing to set up an approval system should consider how significant the commitment may be in terms of money and manpower. Even the simplest system will involve recordkeeping and correspondence with oil companies from all parts of the world. If tests are to be specially run, then somebody must view the used components and check that the recorded values of deposits, wear, etc. (the "ratings") are consistent. If field tests are to be required then people must be available to help set the test up and examine disassembled engines after test—often at awkward times and in inconvenient locations. Parts have to be stored for future reference. Somebody has to be available to listen to stories of hardship and make judgments where something unusual has occurred. Judgment seems to be needed most often when tests are expensive!

For a variety of reasons including cost, most manufacturers have not committed the resources to operate comprehensive lubricant approval programs. Some companies used to run them for many years but then stopped the practice. Mercedes-Benz, who for many years has run probably the most comprehensive OEM lubricant approval system in the world, stopped giving passenger car lubricant approvals although approvals were maintained for commercial vehicle lubricants. In 1995 Mercedes-Benz announced that passenger car lubricant approvals would be re-introduced in 1996.

Some manufacturers give approvals but use an honor system; that is, they rely on the integrity of the presenter of the data (usually an oil marketer or additive company) and do not require that they examine used test engine parts themselves. Mack is an example of such an organization.

6.4.2 Motor Industry and User Quality Level Philosophies

There is no simple way to define the quality of a lubricant. Quality as defined by an engine manufacturer, a user and an oil company will reflect their different viewpoints, and each of these may change depending on location and nature of the end-use market.

Let us first consider the OEM. His primary objective will be to recommend a lubricant which will enable the hardware to run to its design limits without mechanical failure. He will also be concerned about his competitiveness in the market, reflected for example by convenience to the user and cost of ownership of the engine or vehicle. He wants a long oil change interval but he would like the lubricant to be inexpensive and, more particularly, widely available. He may need to be concerned about meeting legislation (see Section 10.2). He may want to use the qualities of the lubricant to exploit the built-in design features of the equipment, whether it be power, fuel economy or engine life.

There is a complete spectrum of possible philosophies ranging from "my equipment must be capable of working satisfactorily with any oil that it can encounter in the marketplace" to "the oil is a design component like any other, and should be specified to give optimum performance." An example of a company who for good reason would be identified closely with the former quotation is Perkins Engines. Perkins provides engines for an extraordinary range of applications in a wide range of markets including underdeveloped countries. They often have little or no control over the use of the engine, and there may be no choice of fuel or lubricant. In those circumstances it makes sense to have a robust design which is insensitive to lubricant quality. An example where the other quotation may be more applicable is given by a Grand Prix racing engine. The type of use is extremely well-defined and the engine needs to provide maximum performance for a very limited time. The oil may be considered completely as a design component with no regard to cost, or to marketing and distribution issues.

OEMs who have the market share or image, and who take the trouble to operate lubricant approval schemes, have the opportunity to move closer to the second philosophy; those who do not must rely on broad industry classifications and the desire (brought about by competition) of oil companies to meet expressed needs. Most U.S. and European OEMs have traditionally tried to exercise control over lubricant quality while Japanese OEMs have been closer to the "robust" philosophy as exemplified by Perkins. Some OEMs have been able to overcome service problems by specifying new improved oil qualities (see Section 5.2).

How about the user? Major purchasers of lubricants such as the armed forces, local authorities, utilities and transit companies may set their own standards. In many cases Military Specifications have been cited by the other organizations.

The armed forces may have a closely defined range of equipment to deal with, but will be concerned about coping with extremes of climate and long storage life. To simplify logistics they will want the minimum possible number of products. They cannot risk having quality problems.

The fleet user, even more than the OEM, will be influenced by cost and convenience. He does not want problems with interpretation of used oil analysis because of unusual lubricant composition. He does not want disposal problems because of toxicity. Operators are becoming concerned about environmental factors such as biodegradability. Some users such as car rental companies may have a wide range of vehicles to maintain, while a bus company may well have a complete fleet from one manufacturer. Their approaches to defining lubricant quality may thus be quite different.

The oil company too will have to decide on the markets in which it wishes to compete, and define the qualities of its branded products. Its primary concern will be to maintain overall profitability, whether sales be to the retail market, to fleet users or factory fill. The oil company cannot readily offer for sale hundreds of products to meet the whims of each OEM or user who may have established a specification. Economies of scale will force him to offer a limited range of products for each broad market and he will consider all possible combinations to find the best market fit. Should lubricants for light vans be a separate market? If not, should they fit with passenger cars or heavy-duty vehicles? Can the same branded product be used for a light commercial market in the U.S. where the gasoline engine predominates, as in Europe which is predominantly diesel? Most international oil companies have found the need to compromise between a wish for worldwide rationalization and the need to meet too many specific requirements, and will market a mixture of worldwide, regional and national formulations.

The oil company will want to distinguish its products from those of its competitors by promoting features that make it different. It may not want to promote "approved by OEM X, Y or Z" unless the OEM is very prestigious, as such claims may enhance the image of the vehicle manufacturer rather than that of the oil company.

Such natural self-interest has sometimes generated conflict between the motor and oil industries (see Section 6.2.4), but in general there is enough common ground between the needs of OEM, oil company and user to suggest that they need to work in reasonable harmony.

References

1. Cahill, G.F., "Evolution of the CCMC Engine Lubricant Sequences," CEC Symposium Paper, Ref. FL/31/89, 1989.
2. Bouvier, J., Gairing, M., Roberts, D.C., "Approaches to Lubricants Testing" (interviews), *PARAMINS Post*, 8-4, February 1991.
3. Gross, T., *et al.*, CEC/93/EL01, CEC Symposium, Birmingham, May 1993.
4. "Engine Oil Viscosity Classification," SAE J300 DEC94, Society of Automotive Engineers, Warrendale, Pa.
5. "Engine Oil Performance and Engine Service Classification (Other Than 'Energy Conserving')," SAE J183 JUN91, Society of Automotive Engineers, Warrendale, Pa.
6. *Engine Oil Licensing and Certification System*, API Publication 1509, Twelfth Edition, January 1993.
7. Outten, E.F., "Lubricants Review Institute (LRI)," *PARAMINS Post*, 1-4, p.10, February 1984.
8. Preamble to draft North American Lubricants Standardization and Approval System, issued by MVMA, Oct. 22, 1990.
9. Tebbe, S.G., "Value Added and Ethics—Prerequisites for Quality," CEC Symposium, Paris, May 1989.
10. CMA Product Approval Code of Practice.
11. "International Tests and Specifications for Automotive Engine Oils," SAE J2227 AUG95, Society of Automotive Engineers, Warrendale, Pa.
12. Cahill, G.F. and Cucchi, C., "CCMC Comment" (interview), *PARAMINS Post*, 6-2, p.4, August 1988.
13. CCMC Ref: FL/29,30,31,/89.
14. "The C.E.C.A Brief Description of Its Organization, Objectives and Activities," CEC, March 1989.
15. Copies of ATC and ATIEL Codes of Practice are available from: Print Controller, RAM Technical Print, Kingsway Park Close, Kingsway Industrial Park, Derby DE22 3FT, United Kingdom, Ph: + 44 1332 345 950, Fax: + 44 1332 202 466.
16. Watanabe, S., "Japanese Engine Tests for Specifying Gasoline Engine Oils," preprint of 8th International Colloquium "Tribology 2000," Technische Akademie Esslingen, Vol. 3, 23.6-1, 1992.
17. *The ILSAC Minimum Performance Standard for Passenger Car Engine Oils*, October 12, 1992.

Chapter 7

Other Lubricants For Road Vehicles

7.1 Gear Oils

7.1.1 Introduction

Gears exist in many forms, one simple and obvious division being between cases where the shafts are parallel (spur and helical types) and those where the shafts are at right angles to each other (bevel, hypoid, and worm gears). Examples are illustrated in Figures 82 to 86. Other angles between the shafts are possible, but seldom found in automotive equipment.

An important feature of gear lubrication is that the contact between the lubricated surfaces is intermittent, permitting them to be flooded with fresh lubricant between contacts. There is, however, a high load placed on small

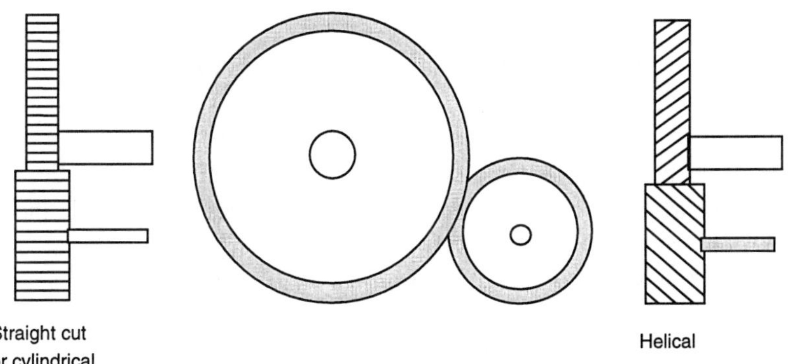

Fig. 82 Spur gears (diagrammatic).

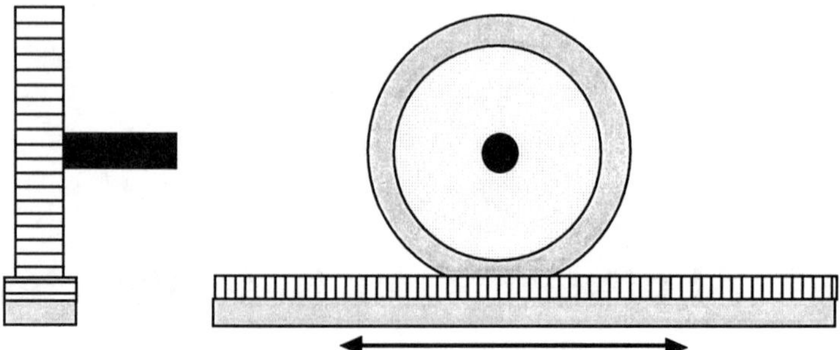

Fig. 83 Rack-and-pinion mechanism.

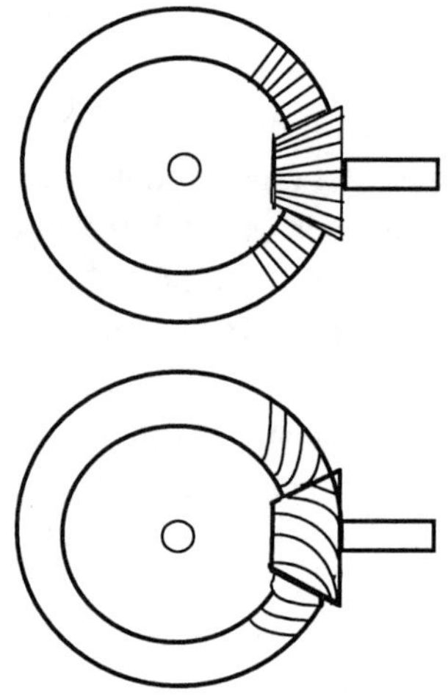

Fig. 84 Bevel and spiral bevel gears.

surface areas of the teeth when power is being transmitted across the contacting surfaces. The teeth must be sufficiently large to handle the power and torque levels, must be made of sufficiently hard material, and the lubricant must be designed to minimize wear at the contact areas.(1) Some improvement to the load-carrying ability of uncompounded mineral oils is normally required if a significant amount of power is being transmitted through the gears.(2)

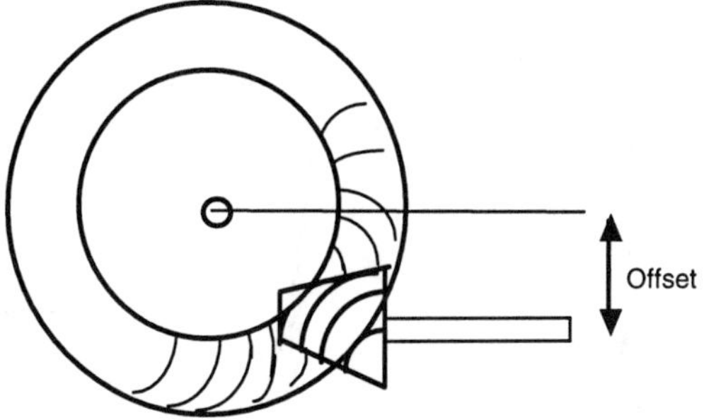

Fig. 85 Hypoid gear.

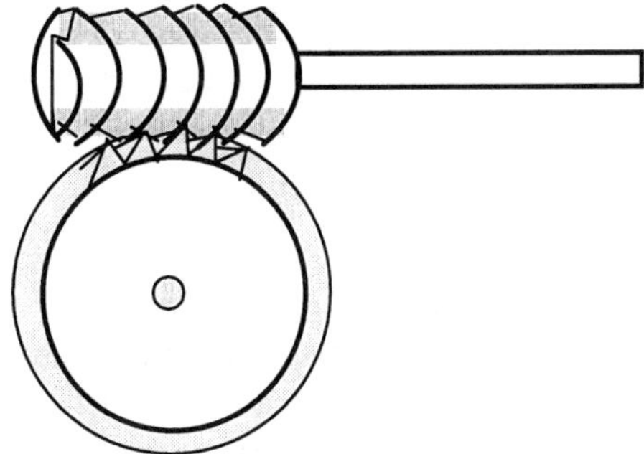

Fig. 86 Worm wheel and pinion.

From the lubrication point of view a more significant division of gear types is by the amount and the type of sliding that occurs as the teeth come into and out of mesh. There is always some sliding action at the contact surfaces as the teeth come into and out of mesh.. For spur, helical, bevel and spiral bevel gears this is at right angles to the lines of contact across the gear teeth. As the centers of each pair of engaged teeth pass each other (as the so-called "pitch circles" intersect) the contact becomes of a purely rolling type, and as the teeth disengage the direction of sliding reverses (Figure 87).

The area of tooth contact where sliding takes places is determined by the number and the geometry of the teeth, the deformability of the metal, and the precise set-up of the gears relative to each other.

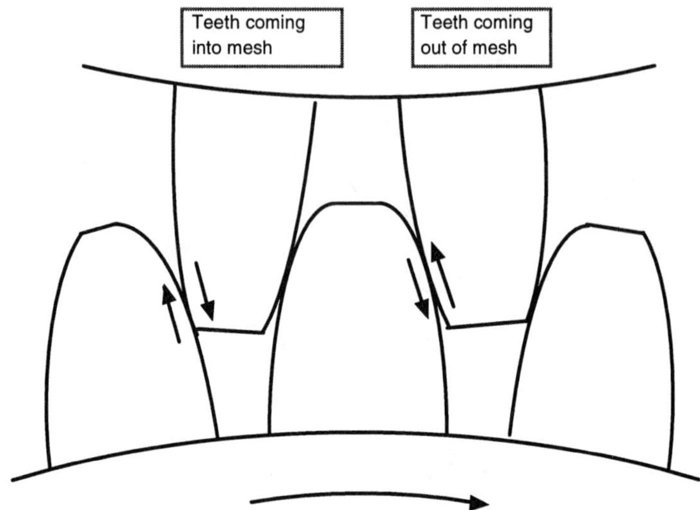

Fig. 87 Sliding between gear teeth as they mesh.

Gears with parallel shafts, or shafts whose lines intersect when these are projected, have this relatively small amount of sliding at right angles to the line of tooth contact. If the shaft lines do not intersect, then considerable sliding also occurs laterally along the line of tooth contact. This happens in both hypoid and worm gears, the former being used in the final drive of many vehicles. This lateral sliding imposes more severe lubrication requirements, and hypoid gears in particular have a very high requirement for EP properties. Worm gears have an even greater sliding action, but the high temperature this generates causes normal EP agents to be too active and therefore unsuitable, particularly if the worm is made of bronze, a metal attacked by EP agents. High-V.I. compounded oils, containing fatty materials and possibly mild EP agents, are normally used for worm gearing, with polyglycol-based lubricants being used in some industrial applications. The use of worm gears in rear axles of automotive equipment is now rare. They are widely used in lightly loaded auxiliary applications such as wiper and window mechanisms. They are also used in small, self-propelled cultivators.

7.1.2 Additives

Gear oil additives of many types have been used for the different types of gears. Load-carrying is of obvious importance, but for some metallurgies the need to avoid chemical attack or corrosion of the metal is a constraint. Gearboxes may run hot, and the action of the gears may tend to aerate the oil, so strong oxidation inhibition is a normal requirement. Seals and any internal finishes of a gearbox must not be seriously affected by solvent or chemical action of the gear lubricant, and its viscometric performance must also be adequate for the operational temperatures expected.

Natural fats and fatty acid esters provide good boundary lubrication for relatively lightly loaded gears, and lead soaps such as lead naphthenate were initially found to be good for somewhat heavier duties, but are now obsolete. Sulfurized fats and esters provide extreme pressure properties, but the presence of "active" or simply dissolved sulfur will cause staining or corrosion on bronze gears or bushings and brass bearing cages.

The wide selection of possible chemical types which have been historically used to impart EP properties to gear oils can be categorized by reference to their active elements. Some common types are listed below:

Sulfur Compounds(3)

The simple solution of elemental sulfur in mineral oil provides good EP properties, but is corrosive to yellow metal.

Sulfurized fats and fatty esters provide good lubricity with EP properties.

Sulfurized olefins and polyolefins are low-cost, general-purpose EP agents, and are now very widely used in automotive and industrial applications.

Disulfides (especially dialkyl disulfides) are potent EP agents in laboratory tests and have been proposed for use in industrial applications.

Sulfochlorinated olefins are very powerful EP agents, and are less corrosive than many purely sulfurized EP agents. They are now considered obsolete.

Chlorine Compounds

As well as the sulfochlorinated materials mentioned above, chlorinated hydrocarbons from kerosene to waxes provide excellent EP properties, but can give rise to corrosion problems if combinations of moisture and high temperatures are encountered. There are also now general environmental concerns about the presence of chlorine in lubricants (see Chapter 10).

Phosphorus Compounds[(4)]

These have been widely used as anti-wear and EP agents. Tributylphosphate and tricresylphosphate have been widely used in various industrial oils, as have phosphites, phosphonates and phosphoric esters. Tricresylphosphate has given rise to health concerns because of its benzenoid structure. Phosphorized fats (or phosphorized synthetic fats such as glyceryl oleate) are also very useful in providing good lubricity along with EP properties.

Phosphorized and phospho-sulfurized olefins and esters are a valuable source of EP performance and have found use in automotive gear oils.

The most widely used phosphorus compounds with EP properties must surely be the ubiquitous zinc dialkyl dithiophosphates, which combine both phosphorus and sulfur in the molecule. These bring anti-oxidancy as well as lubricity and EP properties to a formulation, but may be too active (i.e., they decompose too readily) for the most severe applications. They were used in some API GL-4 types of gear oil at one time, but are unsuitable for the more severe GL-5 and above. In automatic transmissions and manual gearboxes lubricated with engine oil ZDDP is usually the main load-carrying agent. Amine dithiophosphates, which have the advantage of being ashless, can be used in more severe applications (GL-5 and above).

Other Additives

As well as load-carrying ability provided by one or more EP agents, a gear oil may contain the following:

- *Metal deactivators* to prevent corrosion of brass, bronze, or other non-ferrous metal parts.

- *Anti-rust agents* to prevent corrosion of gears and housing under moist conditions.

- *Dispersants* to keep parts clean by suspending sludge and oil degradation residues.

- *Friction modifiers* to control clutch and/or braking operations by modifying the ratio of static to dynamic friction in such devices.

7.1.3 Automotive Gear Oil Formulation

The basic functions of an automotive gear oil are as follows:

- Friction reduction
- Wear reduction (including shock loading and reverse loading)
- Heat removal
- Bearing lubrication
- Corrosion protection, if moisture present

7.1.3.1 Manual Gearboxes

These are normally based on straight spur or helical gears, and, in the case of front-wheel-drive vehicles with in-line engines, bevel and spiral bevel gears may also be present within the gearbox for the differential and final drive. With a transverse engine, the arrangement is normally called a transaxle and spiral bevel gears are not needed. Lubricant for such gears requires only moderate EP properties, and this can be provided by a percent or so of ZDDP. Motor oil is a frequent recommendation for transaxle units, combining a suitable level of ZDDP with excellent cleanliness and anti-corrosion characteristics. The EP requirement is about that of API GL-4 (see Section 7.1.5), and an alternative approach is to use an oil with about half of the additive treatment of a hypoid oil, which provides greater gear protection. In a synchronized gearbox the lubrication of the synchromesh cones may be as critical as gear protection, and the use of a powerful additive may not suit some types of geometry and/or metallurgy. The mating surfaces of synchromesh cones are normally bronze/steel or molybdenum/steel, and the lubricant must not cause excessive slippage or glazing of the cones. New materials are under development which may result in additional or alternative restrictions on the type of additives that are acceptable.

Automatic transmission fluid (ATF) is increasingly being specified for the lubrication of precision manual gearboxes, giving better shifting at low temperatures. This also provides controlled friction characteristics and good anti-wear oxidation and corrosion performance, but the viscosity may not be adequate for older types of boxes. The friction of the fluid needs to match the needs of the specific gearbox. Some ATFs have friction that is too low. ATF is discussed as a separate topic in Section 7.2.

Use of motor oil or ATF in a manual gearbox may mean that viscosity-modifying polymers will be present. These will initially provide useful protection against high-temperature thinning of the oil, but the shearing action of the gearbox will eventually cause permanent viscosity loss in the oil. However, under the high shear regime in the area of tooth contacts the oil suffers from temporary viscosity loss anyway, and the effect by permanent shear may not be as great as simple viscosity tests might indicate. It can be argued that the base oil viscosity is more relevant for gear lubrication than for motor oil, and therefore wide multigrade oils may not be entirely suitable for gearbox use if formulated with viscosity modifiers designed for engine lubrication. Part and wholly synthetic SAE 75W-90 and 75W-140 multigrades are coming into favor despite their relatively high cost, and may contain very shear stable viscosity modifiers.

7.1.3.2 Final Drives (Hypoid)

Since its development in the 1920s by Packard, the hypoid rear axle has been the major concern of automotive gear oil developments. It was favored for the low driveline resulting from the offset pinion and a greater tolerance to shock and reverse loading than the alternative worm gear. Due to the high degree of sliding between the gear teeth, for long service life it requires a lubricant of strong EP capability, plus good lubricity, especially during the running-in period.

Early lubricants were based on lead soaps and dissolved sulfur, but these were of restricted load-carrying ability and were aggressive to bushings and bearing cages. Under high-torque conditions the active sulfur could actually promote chemical wear, and the oils were not particularly stable.

New compounds and mixtures were developed, and blends of the sulfurized fats, lead soaps, and chlorinated hydrocarbons were popular for many years. After the development of ZDDP, combinations of this additive with chlorine/phosphorus/sulfur-containing hydrocarbons found use in "universal" gear oils, which were designed to be suitable for both the hypoid axle of passenger cars and the spiral bevel rear drives in many trucks.

Increased performance was demanded for universal gear oils, particularly by the U.S. military, and these formulations were succeeded in their turn by sulfur/phosphorus additives based on sulfurized olefins and phosphorus esters. These were supplemented in many cases by nitrogenous materials such as triazoles and thiadiazoles which act as corrosion inhibitors and also provide inhibition and some additional anti-wear performance. This is a common type of formulation in use today.

Gear oil formulation is a difficult balancing act between the various requirements in the standard specifications. Precise formulations are proprietary in nature, and the manufacturing processes for the individual additives and the purity of these is also at least as important as the nominal mix of ingredients developed in a formulation program.

In addition to EP, anti-oxidant, and anti-corrosion additives, a modern gear oil formulation may contain cleanliness (dispersant), seal compatibility, and anti-foam compounds. If a limited-slip differential is to be lubricated, additional friction control additives such as sulfurized fatty material may be required to prevent chatter or noise. The requirements can differ between different manufacturers' units.

Synthetic base stocks are appearing in some very heavy-duty axles, following their adoption for some industrial gear oils. While specifications written around synthetic base stocks have appeared for certain gearboxes, we are not aware of any major demand for synthetic rear axle oils that has arisen in the passenger car market up to the present time. As OEMs move to extended drain or fill-for-life for gearboxes, use of synthetics for combined rear axle and gearbox oils could grow. Fuel economy may provide an additional incentive.

7.1.4 Gear Oil Testing

The routine bench tests such as viscosity, pour point, foaming, corrosion, etc., are essentially the same as for motor oils, which have been covered in Section 3.2.

Two other important types of tests are used for gear oil evaluation: EP bench testers and axle rigs. The latter correspond to the engine tests for crankcase oils.

7.1.4.1 EP Testers

This is a group of relatively compact machines for giving rapid and inexpensive assessment of the wear-prevention characteristics or extreme pressure ratings of lubricants. They usually consist of a rotating spindle against which a stationary member is pressed, the wear on one or both parts being measured after running for a given time and at a given load. Alternatively, the load required to produce a given level of wear or actual seizure of the components may be determined.

The Timken Tester (D 2782)

Developed by the roller- and ball-bearing manufacturer, the tests consist of a rotating outer ball-bearing race, with a steel block pressed against it by a weighted beam. The lubricant is circulated over the contact area, and the load is increased stepwise until a wear scar appears on the block. The load value before scoring commenced is known as the "Timken OK Load" and is a measure of the EP properties of the lubricant.

The Falex Tester (D 2670 and D 3233)

This is an EP tester of similar function to the Timken machine, but utilizes a rotating spindle against which two opposed V-blocks are pressed. The load on the blocks is increased by a ratchet mechanism until seizure occurs and the spindle can no longer be rotated, by which time the blocks will be appreciably scarred. A target value for the load before this occurs may be set.

The Four-Ball Machine (D 4172 and D 2783)

Developed originally by Shell, the apparatus consists essentially of three standard ball-bearings clamped together on a base plate, and a fourth ball-bearing clamped to a spindle and resting on the other three. The top ball is rotated via the spindle and pressure applied. Two models of the equipment exist, of which one is a lighter, more-sensitive model for use at lower loads and fixed speeds, temperature and times. The average diameter of the wear scars on the lower balls is measured to evaluate anti-wear properties under boundary lubricant conditions. A more robust version of the machine is used with higher and increasing loads to act as an EP tester. Wear scar diameters are plotted until seizure occurs.

The three EP test machines described above are operated under specified conditions with standardized and uniform test pieces and provide a reasonably reproducible measure of the EP performance of oils. However, because the geometry and nature of the test pieces is different in each case, they may not always rank a series of oils in the same order of EP performance.

The FZG Machine (D 5182)

This machine is somewhat intermediate between the small testers above and those in the next section. Considerably more expensive to operate because of the test piece costs, it is the main method of evaluating gear lubrication characteristics in the laboratory. The particular apparatus was developed by a German gear research institute, but uses the same principle as the earlier I.A.E. and Ryder gear machines in that it is what is known as a four-square gear tester. Two sets of gears are connected together by two parallel shafts, one of which is a torsion bar and can be used to increase the load on the gear teeth. The apparatus is run with two test gears dipping into the lubricant to be evaluated, and after running for a fixed period the apparatus is stopped, the test gears examined, and the load on the gears increased by further twisting the torsion bar. The load is increased continuously after each period of running until the gear teeth show significant damage. Either the damage load stage or the load for damage-free operation may be reported.[(5)]

7.1.4.2 Axle Tests for Hypoid Gear Wear[(6)]

In these tests the axle is driven by a powerful V-8 engine and loading is applied by a dynamometer. There are two tests in common use at the present time, but others are under development.

CRC L-37 High Torque Axle Test

The axle is first run under high-speed, low-torque conditions for 100 minutes. The gears are examined via an inspection hole and if satisfactory are run for 24 hours under low-speed, high-torque conditions. At the end of the test the gears are dismantled and rated for discoloration, tooth rippling, ridging, or pitting and wear.

CRC L-42 High-Speed Shock Loading Test

The axle is driven at high speed and periodically subjected to shock loading. At the end of test the scoring from the coast sides of gear and pinion is evaluated against results from reference oils.

7.1.4.3 Other Tests

In addition to the EP tests in axles listed above, there are other gear oil tests for other properties which also use axle rigs. The most important are:

CRC L-60	A 50-hour oxidation test in which the gears are motored while air is blown through the oil. The increases in oil viscosity and insolubles content are measured.
CRC L-33	An axle rust test in which gears are first motored in the presence of oil and water to thoroughly coat all surfaces, after which the axle is placed in a humidity cabinet for seven days and rusting assessed at the end of this period.

Other approval authorities or axle manufacturers may have individual axle tests in relation to their own equipment, but the above are those recognized by the U.S. military, ASTM, CRC and worldwide.

7.1.5 Gear Oil Specifications and Quality Levels[7]

Early quality requirements were generally set by manufacturers' field tests, the U.S. military being the first to set a performance specification based on standard tests. Their original specification, MIL-2-105 (later MIL-L-2105), was for a universal gear oil for hypoid and other final drives and for manual gearboxes, and was issued in 1943. In 1962 a major increase in performance level was demanded and MIL-L-2105B required approximately twice the additive content of the earlier specification. The MIL-L-2105C and MIL-L-2105D specifications represented more modest improvements, providing for multigrade gear oils among other changes. Plans are for a new MIL-L-2105E specification to be issued shortly, and this is expected to be of a significantly higher performance standard.

In 1969 the API produced a gear oil classification which is still in use (see SAE J308 in Appendix 11). This defined six different qualities for automotive gear oils:

API Designation	Lubricant Type/application
GL-1*	Straight mineral oil for some manual transmissions
GL-2*	Mild EP for worm gears
GL-3*	Mild EP for spur and spiral bevel gears
GL-4	Medium EP (equivalent MIL-L-2105) moderately severe hypoid for older vehicles
GL-5	High EP (equivalent MIL-L-2105B) for all hypoid axles and some manual transmissions
GL-6*	Extra high EP (GL-5 + higher anti-scoring) for high offset axles[8]

* Effectively obsolete/inactive

Of the original six designations which contained performance tests, only those for GL-5 can now be run in their entirety due to non-availability of test parts for other tests. The test requirements for GL-5 are as follows:

Test	Type	Parameter
CRC L-37	24-hr dynamometer (high-torque/low-speed)	Gear scoring/wear
CRC L-42	2-hr dynamometer (high-speed with occasional shock loading)	Gear scoring
CRC L-60	48-hr oxidation test with motored gears	Viscosity inc./ insolubles
CRC L-33	7 days in Humidity Cabinet after initial motoring phase	Rusting (rating)
ASTM D 130	Copper corrosion	Corrosion of Cu strip
ASTM D 892	Foaming tendency	Foam level on test

Viscosity grade is determined by viscosity at 100°C and the temperature at which the low-temperature Brookfield viscosity reads 150,000 cP (between

–12°C (SAE 85W) and –55°C (SAE 70W)). The SAE viscosity classification J306 is reproduced in Appendix 11.

The industry has for some years been working on new classifications. One is API MT-1 where MT stands for manual transmission. The market for this new category is anticipated to be primarily manual transmissions in heavy-duty trucks. Better quality was needed because of a requirement for better synchronizer performance and to minimize seal compatibility problems caused by deposit formation.

The API MT-1 requirements exclude the L-37 and L-42 tests but include the additional L-60 ratings and seal compatibility tests, plus new tests for synchronizer and anti-wear performance. The copper corrosion requirement is also more severe. There is also an FZG anti-wear test.

A second proposal, provisionally called ASTM PG-2 (and which will probably be given a number in the API GL series), is designed as a high-performance hypoid lubricant category to have higher performance than API GL-5, especially in the areas of seal compatibility and improved anti-spalling performance.

The category will probably include all current API GL-5 tests, a Mack spalling test (not yet developed), thermal stability, and cleanliness requirements from the L-60 test and seals tests.

Detailed specifications and limits are included in Appendix 11.

The SAE Viscosity Classification for Gear Oils now includes seven grades as follows:

	Winter Grades				**Summer Grades**		
	70W	75W	80W	85W	90	140	240
100°C Viscosity Min. (cSt)	4.1	4.1	7.0	11.0	13.5	24.0	41.0
100°C Viscosity Max. (cSt)	-	-	-	-	<24.0	<41.0	-
Max. Temp for Viscosity = 150,000 cP (Brookfield) (°C)	–55	–40	–26	–12	-	-	-

Multigrade gear oils are possible, and the U.S. military approves SAE 80W-90 and 85W-140 grades. They require duplicate gear performance to be run on every submission based on low base oil viscosity (SAE 75W and the multigrades) whereas "additive approval" exists for more usual grades. SAE 75W-90 part-synthetic gear oils are beginning to be seen, and are a Volkswagen requirement.

Base stocks for most hypoid gear oils consist of mixtures of 500 or 600 neutral stocks mixed with bright stock, and containing pour point depressant as necessary. As discussed above, synthetic stocks are now being used to formulate wide multigrade gear oils.

OEM requirements are normally based on the API classification and/or the U.S. Military Specifications. Some manufacturers may have additional tests, and in Europe the inclusion of an FZG gear test and a four-ball EP test is quite common. The general European level is at present similar to API GL-5 but is expected to increase with the introduction of new oxidation tests and more severe requirements from Peugeot, Renault, ZF, Mercedes-Benz and Volkswagen. In the U.S. the requirements of Eaton Axle and of Mack Truck specifications are more severe than API GL-5 and the Mack GO-H/S specification calls for the use of synthetic base stocks. All OEMs and the U.S. military require extensive field testing of any new additive technology, whether developed for an existing or new quality level.

Other national (especially military) gear oil specifications exist, but do not have significant impact outside national boundaries.

7.1.6 Limited-Slip Differentials

These are particular types of rear axle designs which eliminate one-sided wheelspin associated with application of power through a conventional differential in situations of poor tire adhesion. They usually contain friction clutches which come into action when a substantial speed difference arises between the two sides of the differential.

Each design differs in detail and in lubrication requirements, which are basically for a friction-modified oil formulated to eliminate shudder and noise when the clutch mechanism engages. Normal boundary lubricant additives are typically used, but the preferred type and concentration varies between different axles. Some manufacturers have performance tests to rate candidate lubricants, and some leave lubricant specification to the vehicle assembler. To satisfy limited-slip differentials, two approaches have been used. The first is to formulate a complete rear axle oil with built-in limited-slip additives, and the second is to provide a supplementary additive to be added to the rear axle after charging with normal hypoid oil. The first approach is possible when the additional requirements are not excessive, but for difficult axles (which can be individual units) doping with supplementary additive to achieve acceptable noise levels may still be required.

Electronic traction-control devices utilizing automated braking systems are expected to supersede the older mechanical limited-slip differentials.

7.2 Automatic Transmission Fluids (ATF)

The percentage of vehicles with automatic transmission has always been much higher in North America than elsewhere in the world. In the U.S. it reached over 90% and has been quoted at 87% for 1993 despite the increasing percentage of small sub-compact vehicles. In Europe, penetration has always been low, although after the concept of the small four-speed automatic transmission was pioneered it began to approach 10%, but has been set back by considerations of fuel consumption and the environment. Conventional automatic transmissions were 5-10% less efficient in fuel consumption than a manual gearbox on the open road, although in city traffic they can approach the fuel efficiency reached by the average driver with a manual gearbox. Improvements such as lock-up devices have also reduced the penalty in recent years. In Europe the market share for automatics varies greatly with engine size. In 1993 it was only about 2% for cars below 1.5-litre engine capacity, 10% in the 1.5- to 2.5-litre range, rising rapidly thereafter to over 80% for cars with engine capacity over 3.0 litres. Including CVTs (see below), the penetration of automatics in Europe is expected to reach about 20% in the year 2000. In Japan, in 1993, about 77% of new vehicles had automatic transmissions.

In other parts of the world the penetration of the automatic transmission varies significantly, ranging from virtually zero in some countries, to a surprisingly high figure in those countries where the majority of automobiles are imported luxury cars for top officials or businessmen.

7.2.1 Development of Automatic Transmissions[9]

After some earlier attempts with friction drives on both sides of the Atlantic, the first transmission comparable to the present dominant automatic type was the Oldsmobile Safety Transmission of 1937 which used a simple fluid coupling and semi-automatic gear changing. This was superseded in 1939 by the first fully automatic transmission, the Oldsmobile Hydromatic, using planetary gear trains and manual and hydraulic controls for the clutches and bands of the units. Following problems with the earlier unit after a recommendation for the use of engine oil as the working fluid, a new type of fluid with better low-temperature characteristics was specifically designed for the new transmission by GM Research, who worked closely with Mobil. Buick pioneered the replacement of the simple fluid coupling with a torque converter type in its 1948 Dynaflow, which resulted in a transmission very similar to those of today. Developments at GM were closely tracked by those at Ford and Chrysler who developed similar transmissions.

With the introduction of the torque converter more heat was generated in the transmissions, and an improved type of fluid was required. In 1949 General Motors issued their "Type A" specification for a fluid, which in turn was succeeded by "Type A Suffix A" in 1957. These were initially the standard specifications for transmission fluids, but Ford in 1960 and Chrysler in 1964 subsequently issued their own specifications, while GM again introduced an improved fluid known as DEXRON™ in 1967. Ford's and General Motors' fluids have been upgraded several times in the intervening period, mainly in terms of improved oxidation and stability requirements, with the Ford fluid initially having significantly different frictional characteristics from the GM fluid (see below). More recent changes to Ford's requirements have brought their fluids more in line with those of GM.

Transmissions in general have increased in complexity in the last decade. As well as the mechanical lock-up clutches referred to earlier to improve fuel economy, electronic control and controlled-slip clutches are being introduced. These will both improve shifting performance and give some further efficiency improvements.

In Europe, automatic transmissions were manufactured by vehicle makers such as Citroën and British Leyland, and Warner Gear (as Borg-Warner) enjoyed a significant share of the market with transmissions similar to those of GM and Ford which were imported for limited used by their affiliates. These earlier European manufacturers have now ceased production, and the significant players are ZF, Mercedes-Benz, Renault and local production by Ford and GM.

Japanese manufacturers have tended to manufacture their own transmissions. Toyota has had a major share of the car market and their transmissions are of the GM type.

A low-cost type of automatic transmission was developed by Van Doorne in Holland, which used an old principle of belt drive between variable cones. This provided continuously variable gear ratios, but was limited to relatively small power throughputs and even then suffered from breakage problems with the rubber belts. The car manufacturer using the system (DAF) was taken over by Volvo, who persevered with the concept for a time, but it has now been dropped in its original form. The concept has been taken up in Europe by Ford, Fiat and Rover, but using metal belts instead of those of rubber/fabric composition. Using metal belts it is possible for the belt to be driven in compression rather than in tension, and this aids belt life. Advantages claimed for this type of transmission include the continuously variable ratios, a high power transfer efficiency, and relatively low complexity compared to a conventional automatic transmission. Vehicles with this transmission are on sale, but uptake does not appear to be very large. There is a general prejudice against automatic transmissions in Europe arising partly from the traditional higher cost of vehicles with automatic transmission, but also from the desire of youthful (and would-be youthful) drivers to obtain the maximum performance from a vehicle in terms of acceleration, under "hands-on" conditions.

7.2.2 Characteristics of a Conventional Automatic Transmission[10]

The paragraphs below describe how a simple conventional automatic transmission operates. Modern transmissions will incorporate fuel economy devices such as lock-up roller clutches, and control may be partially or wholly electronic, but the principles of the basic mechanism remain the same.

The most important feature of such transmissions is the use of constant mesh epicyclic or planetary gears, controlled by a series of clutches and restraining bands. Epicyclic gearing had been used in the Ford Model T and some later vehicles and is known for its quiet running and the ability to obtain various speed ratios without taking gears in and out of mesh. A planetary gear set is illustrated in Figure 88.

Various speeds can be obtained from this single gear set by feeding power through a series of input and output clutches, and by immobilizing one or more members of the set by means of brake bands which envelop cylindrical extensions of the various members. The geometry of clutches and brake bands is shown diagrammatically in Fig. 89. The following options are possible:

<u>Neutral</u>
With no bands applied and the output braked. The planetary gears will idle around when power is fed in.

<u>Low Gear</u>
The central sun gear is driven and the ring gear is held stationary by a band. The planet gear carrier rotates slowly and can be connected to the final drive by a clutch.

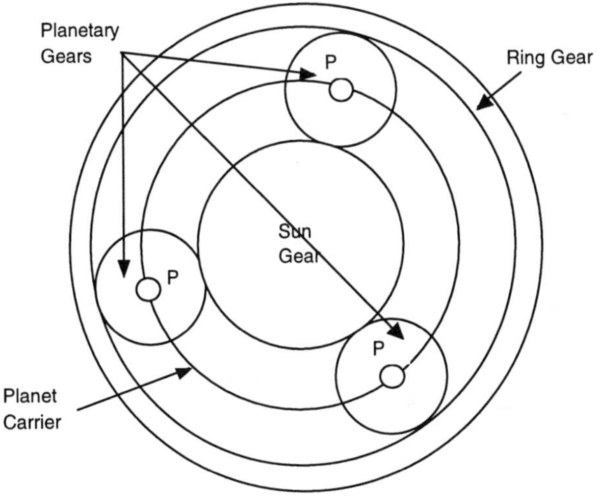

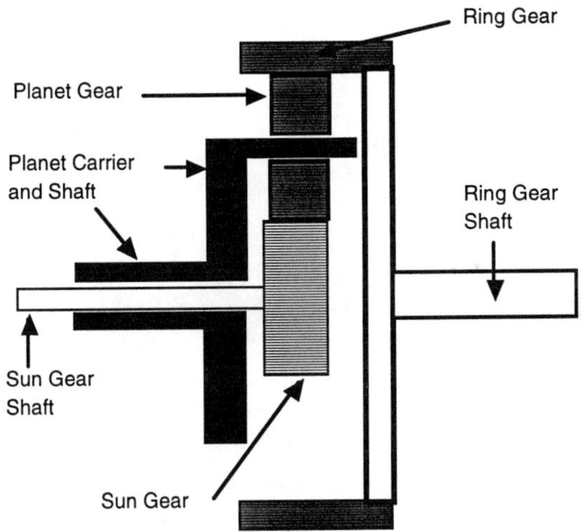

Fig. 88 Epicyclic (sun and planet) gears.

Intermediate

If the ring gear is driven, and the sun gear is held stationary, then the planet gears and their carrier revolve at an intermediate speed below that of the ring gear.

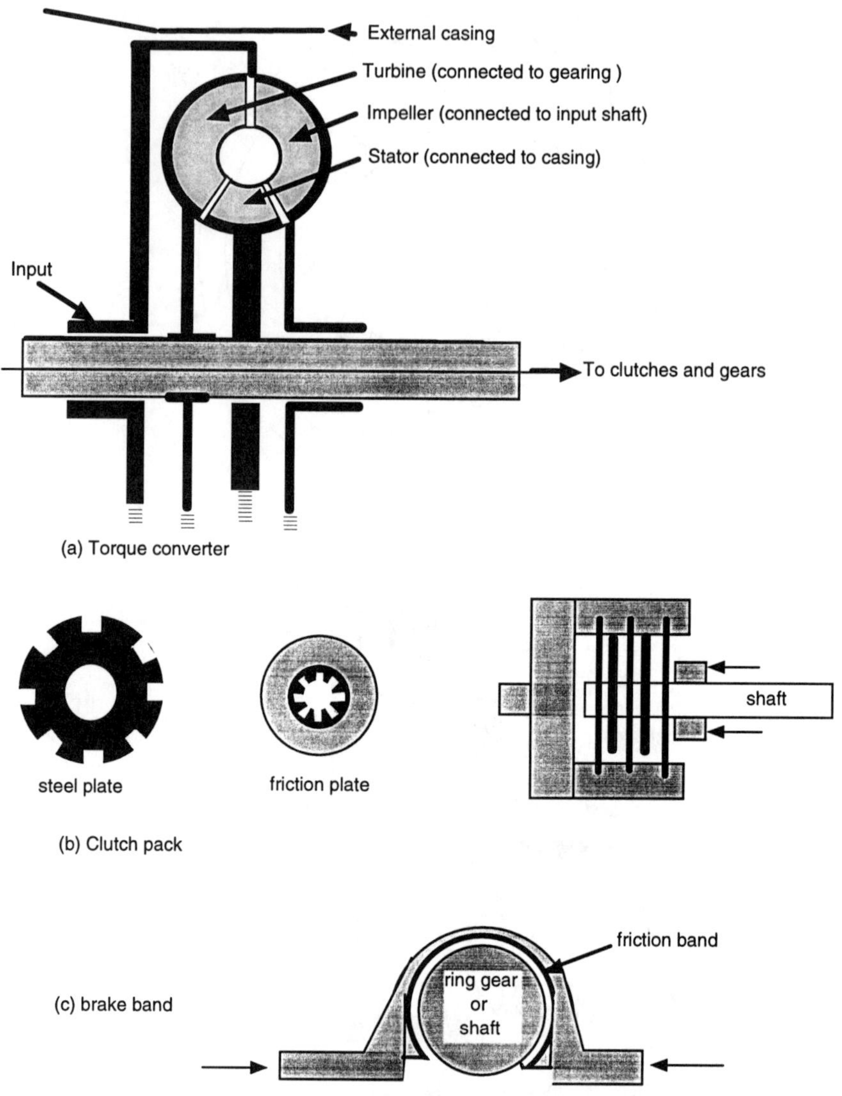

Fig. 89 Diagrammatic illustrations of torque converter, clutches, and brake bands.

<u>Top Gear</u>

If any two of the members are locked together, then the whole unit is locked and rotates at the input speed in direct drive. The input drive and the output can be taken from the most convenient points in this case.

<u>Reverse Gear</u>
If the sun gear is driven and the planetary gear carrier is held stationary then the ring gear rotates slowly in the reverse direction to the sun gear.

It can be seen from the above that all the necessary outputs for a three-speed transmission are available from a single planetary gear set. It is common, however, to use two sets of planetary gears with a common sun gear shaft. This permits the possibility of more speed ratios, and gives greater flexibility of design for positioning bands and clutches.

Clutches are often annular disc clutches, consisting of stacked metal and composition plates. For some functions one-way sprag clutches can be used on an output shaft. The brake bands are placed circumferentially around drum extensions of the gear set members and, like the clutches, are operated by hydraulic pressure. The clutches and bands in a simple transmission are operated by a control unit usually known as the "valve body" which directs pressure from an oil pump to the various hydraulic cylinders, to actuate the clutches and bands and to obtain the desired gear ratios. The valves in the valve body are controlled by:

- The selector lever
- A speed governor
- A "kickdown" mechanism linked to the accelerator
- A vacuum diaphragm load sensor

Power is fed into the gear train by the torque converter. This smoothes the input power and provides torque multiplication as the gears are accelerated following an upward change. The torque converter therefore assists in keeping engine revolutions and power output within a desirable range.[11]

A simple two-element hydraulic coupling, known for a long time, provides a smooth transfer of power but operates at relatively low efficiency. The torque converter with a third stator element provides torque multiplication for acceleration. A cutaway view of a torque converter is shown in Figure 89.

Externally, the device resembles a large donut with an input shaft at one side and an output at the other.

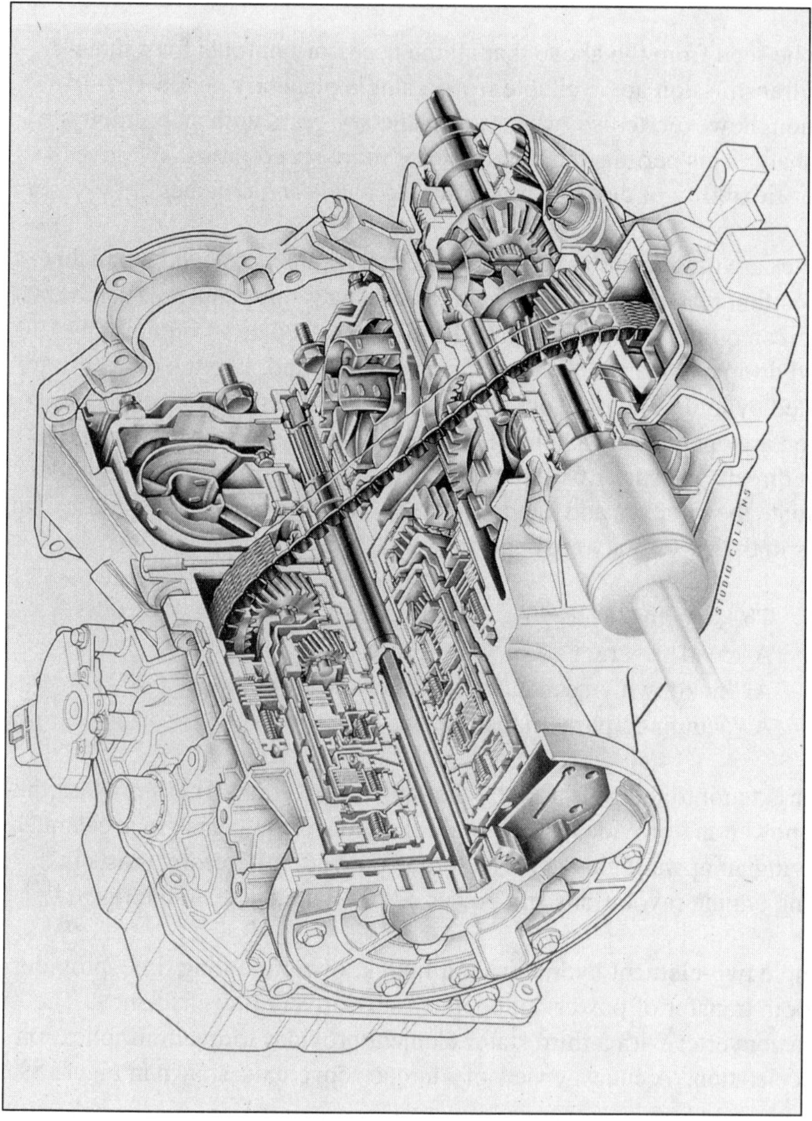

Plate 8. Modern automatic transmission (courtesy of Ford Motor Co.)

The vaned discs on the pump (or input) side are connected directly to the torque converter casing and revolve with it, being directly connected to the engine. The rotation of the pump sends a current of oil toward the periphery where it is deflected by the vanes toward the turbine sector, which rotates more or less strongly depending on the oil pressure developed by the spinning pump. The direction of the oil flow is reversed by the vanes of the turbine, and on returning to the pump would tend to slow it down. However, the blades in the stator reverse the flow again, and thereby the oil assists the pump rotation and the torque being transmitted is effectively increased.

Other important units inside the transmission casing include a powerful oil pump and a pressure regulator, a simple oil filter, a speed governor, and the valve body. Externally, some form of oil cooler is usually added to take away the heat generated in the torque converter and around the bands and clutches.

7.2.3 Requirements of an Automatic Transmission Fluid[(12)]

To overcome the heating and churning action within the transmission casing, an oxidation-inhibited and non-foaming oil is obviously essential. To maintain cleanliness within the transmission some dispersant is added, and some viscosity modifier prevents excessive thinning of the fluid when it becomes hot. (Often these additives are combined in the form of a dispersant polyester type of viscosity modifier.) The oil must also be compatible with the elastomeric seals of the transmission and must not corrode any of the metallic components. It should also inhibit the transmission against internal corrosion from atmospheric moisture when it is left standing.

Of vital importance to the smooth operation of the transmission is the effect of the oil on the friction surfaces of the clutches and bands. Not only must the fluid not attack the friction surfaces, but it is required to provide the correct frictional characteristics as the clutches and bands come into engagement. The required frictional characteristics are specified by the transmission manufacturers and must not change significantly over the life of the fluid. They are evaluated in bench clutch pack friction tests, and in "shift feel" testing on the road. As indicated earlier there was for a long time a major difference between the frictional requirements of Ford and those of General Motors and most other manufacturers. To produce a compact transmission with a possibility for eventual "fill-for-life" capability, Ford,[13] until the mid-1970s, used relatively small frictional surfaces and required these to have a high holding capacity with a fast and positive lock-up on the clutches and bands to prevent overheating from slippage. This required a static coefficient of friction which was higher than that of the dynamic coefficient.

Other manufacturers, with the notable exception of Borg-Warner and Toyota (up to 1984), preferred to have smoother and more prolonged changes produced by a fluid with a static coefficient of friction lower than that of the dynamic coefficient. This property is obtained by the use of additives known as friction modifiers to the fluid. Typical friction curves are shown in Figures 90 and 91.

Following complaints of "clunk" on low-speed changes in new cars, Ford modified their policy in 1978 and introduced a friction-modified fluid for factory-fill purposes (M2C138-CJ). For top-up purposes their existing fluid was considered acceptable. In 1981 a modified specification M2C166-H was issued for both factory- and service-fill, to be succeeded by their MERCON™ M2C-185A in 1987. There are now only minor differences between the frictional requirements of General Motors' DEXRON™ II and DEXRON™ III specifications and those of the Ford MERCON™ fluids.[14]

For the larger commercial vehicles there are automatic and semi-automatic transmissions for which the fluid requirements are set by makers such as Allison with their C-4 requirements, and Caterpillar with their earlier TO-2 and new TO-4 specifications. The frictional requirements of the C-4 and TO-2 types of fluid are similar to those for passenger car applications, and both these specifications can usually be met either by ATF, Tractor Hydraulic Fluids (see later), or by motor oils of suitable viscosity ranges.[15, 16]

The recent Caterpillar TO-4 specification was designed to provide improved performance in Caterpillar equipment by significantly reducing the

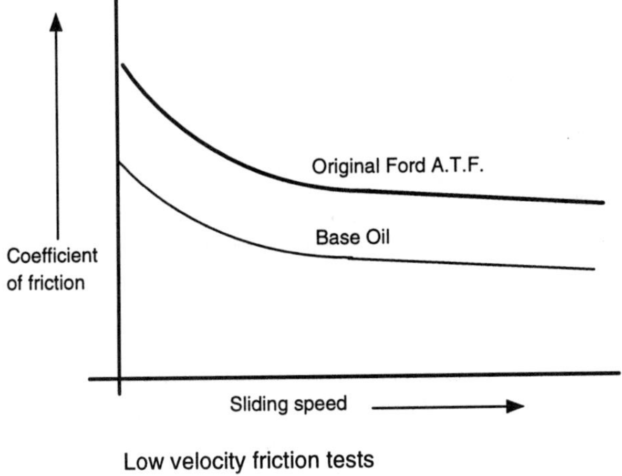

Low velocity friction tests

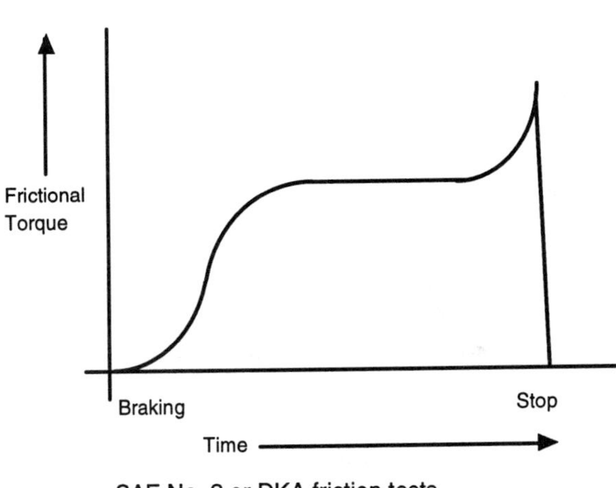

SAE No. 2 or DKA friction tests

Fig. 90 Ford's original frictional requirements (high static friction).

acceptable range of frictional characteristics. It therefore is now essentially a specialty lubricant. SAE J1285, "Powershift Transmission Fluid Classification," is included in Appendix 12, together with the basic requirements of the principal European heavy-duty transmission manufacturers.

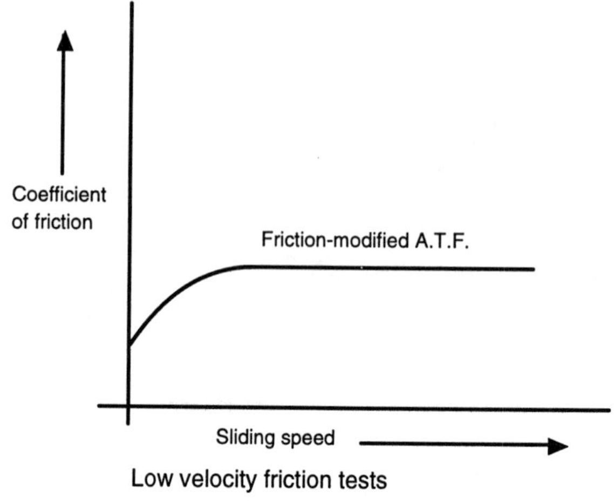

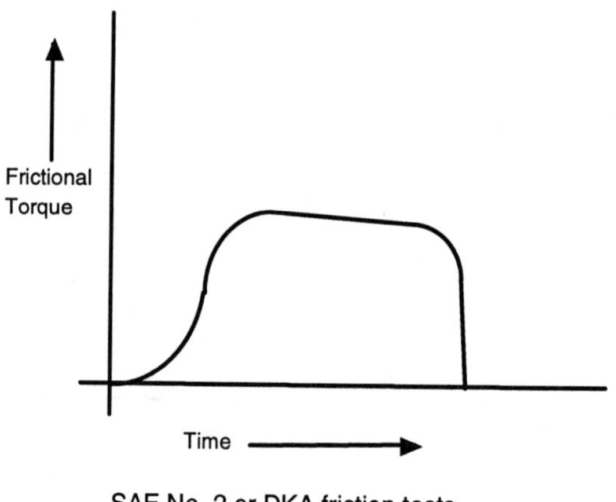

Fig. 91 GM-type frictional requirements (low static friction).

7.2.4 ATF Testing

Tests required on a new ATF formulation include physical properties, bench tests in transmissions and test rigs, and a performance evaluation on the road. A typical specification would include the following:

- Viscosities at 40°C and 100°C
- Brookfield viscosities at various temperatures to –40°C
- Flash point
- Copper corrosion
- Rust protection
- Foam tests
- Elastomer seal compatibility tests
- Wear test (vane pump)
- Friction tests on clutch and band materials
- Transmission oxidation test
- Transmission cycling test
- Shift performance on the road

The physical tests for these oils are the same as those used for motor oils or gear oils. Various tests in full-sized transmissions are used to evaluate ATF. Most manufacturers have proprietary tests, run on both laboratory test beds and on the road. Examples of tests not covered elsewhere include the following:

Bench transmission tests:

4L60 oxidation test (Formerly the THOT test)—In this GM test the transmission is held at an elevated temperature and air is passed through while it is motored electrically. At the end of the test the conditions of both the oil and transmission are rated. Ford and other key manufacturers have similar tests.

4L60 cycling test—This test uses a full drive train of V-8 engine and transmission which is repeatedly accelerated and decelerated for 20,000 cycles. At the end of the test the conditions of both the transmission fluid and of the transmission itself are evaluated.

Lubricity testing:

The Vickers Vane Pump Test (ASTM D 2882)—This is a widely used test for lubricity, used for testing certain transmission oils as well as for its original purpose for hydraulic oil evaluation. A 12-vane rotary pump is run for a fixed period, and at the end of the test the weight loss of the vanes plus the contact ring is measured.

Tests for frictional characteristics:

The best known are the SAE No.2 Machine in the U.S. and the DKA machine in Europe. These evaluate the frictional characteristics of transmission fluids under both static and dynamic conditions. These devices measure the torque required to overcome the friction of clamped plates of friction materials at high and low speeds of slip, with repeated unclamping and reclamping to lock-up. The change of friction characteristics after many thousands of lock-up cycles is evaluated as a measure of fluid durability.[17]

Road testing:

A skilled tester evaluates the "shift feel" of the test oil versus reference oils in a specific transmission and vehicle combination.

Field tests and strip-downs may also be performed if considered desirable.

7.2.5 ATF Formulation

At the present time automatic transmission fluids are normally petroleum based although some synthetic varieties exist. A complex blend of additives is required to provide the mixture of properties required to meet the various specifications. The formulation must be carefully balanced, and additives must be selected with care so that the properties of the fluid do not materially change during its lifetime. The principal types of additives used are as follows:

Oxidation inhibitors

These prevent thickening and deterioration of the oil and the production of sludge and varnish. Typical types are ZDDPs, hindered phenols, amines and sulfurized compounds.

Metal deactivators

These assist the anti-oxidant additives by passivating the wide variety of metal surfaces in the transmission. Some ZDDPs confer passivity, and some sulfides and certain nitrogen compounds are also widely used in this role.

Corrosion inhibitors
These are to prevent attack on bearings and other transmission components which may come into contact with oxidized oil. ZDDPs, phenates and sulfonates are useful.

Rust inhibitors
These prevent the rusting of the ferrous parts when the transmission is lying idle in moist atmospheric conditions or with condensed water inside the casing. Organic acids, amines, and sulfonates can be employed in this role but can also influence the frictional characteristics.

Anti-wear agents
Apart from the ZDDPs, organic phosphates, sulfurized and chlorinated compounds and some amines have been used. All are likely to have considerable effect on the frictional characteristics of the fluid.

Viscosity modifiers
These reduce the lowering of viscosity with increase in temperature. Many types of polymers may be used, but these will often be of lower molecular weight and more shear stable than are usually employed in motor oils. The multi-functional dispersant polymers have been widely used, alongside other polyester types.

Dispersants
These keep the friction surfaces free of deposits and prevent build-up of sludge in the valve body, which would impair operation of the transmission. Some dispersancy may be provided by the viscosity modifiers, but succinimide dispersants are also frequently added. In the past, thiophosphonate detergents have also been used, but are not utilized in modern formulations.

Friction modifiers
These are adsorbed onto the surface of the friction materials and reduce the coefficients of friction, particularly that of static friction. Without friction modifiers, gear changing tends to be harsh, and fluids of this type, formerly favored by Ford, are known as *hard fluids*. Use of friction modifier produces a more slippery fluid, which is known as a *soft fluid*, and gear chang-

ing is smoother. Typical additives used include the fatty oils, fatty acids and amides, and organic phosphorus acids or esters.

Seal swell agents
A basic transmission fluid may cause seals to either swell or shrink slightly, depending on their nature and the constituents of the fluid. It is normal to provide a small amount of swell for the elastomers specified by the transmission manufacturer. If necessary aromatic base oils, organic phosphates, or other compounds may be added to promote a slight seal swell.

It can be seen from the above that the formulation of an ATF is a more complex procedure than that of a motor oil, with antagonistic factors having to be balanced against beneficial ones to a greater degree. Formulations use many components similar to those in motor oil formulations, although the balance is different, additional components will be present, and the total treat rate will be somewhat less. Fortunately the bench and transmission tests required to be performed on an ATF are in general less expensive than the full-scale engine tests required for the development of a motor oil, and approval programs are therefore less costly.

7.2.6 ATF Approvals and Specifications

Approval procedures are more rigorously defined in the U.S. than in other areas of the world. In general, oils are submitted under code to the approving OEM, having been tested at an independent laboratory. Repeat tests are not permitted without prior approval. If bench tests are satisfactory the OEM will conduct "shift feel" tests against a reference fluid at his own proving ground.

Details of important ATF specifications are included in Appendix 12. The current international specifications in use include the following:

- GM6297-M (DEXRON™ III). This has recently superseded GM6137-M (DEXRON™ IIE). Current DEXRON™ quality is taken as a reference quality or model fluid by many other manufacturers. Tests for approval include physical tests plus friction tests on both band and plate clutch materials. Wear tests are carried out using the Vickers Vane pump, while oxidation and cycling tests are carried out in the 4L60 transmission.

- GM Allison C-4. This is for heavy duty, including military transmissions. It includes specific friction tests and a wear test, as well as standard bench tests (corrosion, oxidation, seal compatibility, physical properties).

- Chrysler MS 7176. This specification is very similar to DEXRON™ II which is specified as an alternative if the Chrysler fluid is not available.

- Ford ESP-M2C33F and G. These are the old Ford fluids without friction modifier, and may still be specified for aftermarket or specific auxiliary units such as power steering pumps.

- Ford ESP-M2C138-CJ. Ford's first friction-modified fluid, it is still specified for the AOD transmission, although DEXRON™ II has been used in this transmission.

- Ford ESP-M2C166-H. An improved version of the CJ fluid with better friction characteristics and used for factory-fill worldwide by Ford.

- Ford ESP-M2C185A (service fill) and WSP-M2C185A (factory fill). These were early MERCON™-quality fluids which have frictional characteristics close to the GM DEXRON™ fluids.

- Ford MERCON™. Ford's new automatic transmission fluid specification for service-fill from Sept. 1, 1992.

- Ford SQM-2C9010A. This was Ford of Europe's specification for the specific transmissions manufactured by Ford in Europe. It has a degree of friction modification to give approximately equal coefficients of static and dynamic friction. It is being replaced by M2C 166-H.

- Other European Manufacturers. The most significant manufacturer is ZF, a specialist gear and transmission supplier. Mercedes-Benz and Renault also manufacture their own transmissions. European specifications are not well formalized, and approval is given by OEM and/or transmission manufacturers based on their own in-house and field testing programs. Basic quality is usually DEXRON™ II or DEXRON™ III plus wear tests in the FZG gear machine and friction tests in the DKA clutch pack rig. (Renault, however, uses the SAE No. 2 machine.) Details of European requirements are tabulated in Appendix 12.

- Japanese Manufacturers. While Toyota in the past required a hard Ford-type fluid, Japanese requirements are now for an essentially DEXRON™ type fluid. New slip control lock-up systems are under development, and these are expected to require improved performance in the areas of anti-shudder, torque-carrying capacity, shear stability, and friction durability. JAMA are working with AAMA with the intention of developing an ILSAC ATF standard.

7.2.7 Tractor Hydraulic Fluids

These are fluids for the specific application of lubrication for gearing and hydraulic systems in agricultural tractors. They are related to and precursors of the Super Tractor Universal Oils discussed in Section 4.1.4, but have no requirement for engine lubrication. Units to be lubricated therefore include:

- Gearbox (automatic or manual)
- Power take-off (PTO) and clutch
- Rear axle
- Reduction gears
- Wet brakes
- Hydraulics

The performance requirements for these are:

- API GL-4 or above for manual gearboxes, PTO, rear axle, hub reduction gears, etc.
- Allison C-4/Cat.TO-2 for automatic gearboxes
- High V.I. and good low-temperature performance
- Friction properties to suit gearbox, PTO, and wet brakes
- Good oxidation resistance
- Compatible with all seal materials

Suitable formulations can be constructed using good quality base stocks with inhibitors and detergents to provide stability, anti-corrosion, and cleanliness, and an adequate level of ZDDP or other suitable EP agent to provide the load-carrying ability. A balance of friction modifiers is required to meet the various friction requirements, particularly the wet-brake performance.

Test requirements include both standard bench test requirements for gear oils and hydraulic oils, and tests in tractor units for factors such as PTO, gearbox, and brake noise performance. Key specifications are included in Appendix 12.

7.3 Greases

7.3.1 Introduction

A general definition of a grease is that it is a semi-fluid lubricant obtained by combining a solid thickening agent with a liquid lubricant. The advantages of such a lubricant over the purely liquid form are mainly as a consequence of its ability to remain in place while the mechanism is in operation. These advantages can be listed as follows:

- Reduced need for re-lubrication.
- Reduced drip and splatter or splash.
- Simplified equipment design.
- No need for oil sump or oil pump.

- Easily held by simple seals.
- Prevents dirt ingress to the mechanism.
- Reduces mechanical noise by muffling and by filling clearances in worn parts.
- Can be combined with both liquid additives dissolved in the oil and solid lubricants dispersed in the grease structure.
- Hence can provide excellent anti-wear and EP properties.

The main disadvantages of grease are its inability to remove heat rapidly, and the lack of a flushing action to remove wear debris and contamination. Simple conventional greases have a relatively limited operating temperature range; specialized greases can satisfy extreme operating conditions but at rather high cost.

Greases as we know them today are not much more than one hundred years old. As animal and vegetable oils provided liquid lubricants before the petroleum era, so animal fats such as tallow (beef fat) and lard (pork fat) provided semi-solid lubricants. There is also some evidence from prehistoric chariots that a combination of lime and animal fat was sometimes used as hub lubricant. The idea was probably conceived as a means of holding a reasonably effective solid lubricant (lime or powdered chalk) in position with fat, but reaction between the alkaline solid and rancid fat would have produced a type of calcium soap grease in situ.

The first commercially made greases were also based on lime. This was combined with wood rosin, producing a calcium soap. The soap was dispersed in oil by vigorous agitation, and most of the excess water of reaction driven off by heating. It was found that excessive dehydration prevented the formation of a satisfactory grease, and it is now known that simple calcium greases require stabilization by 1-2% of water or some other stabilizing material.

Hydrated calcium greases were dominant for many years, but the need to retain water as a stabilizer set a limit on their high-temperature operability which became more and more of a problem. They have now largely disappeared from the market. Many different soaps produced from different metals and alternative combinations of organic acids were evaluated, and many new types of grease were developed. The variations can be described in the following terms:

Alternative Metals

Apart from calcium, greases are usually based on sodium, aluminum, or lithium. Barium greases are known, but are not now popular. "Mixed soap" greases are possible, combining different alkaline metals with a common fatty acid. The choice of metallic constituent strongly influences the properties of the grease,(18) but choice of the acid constituent and particularly the use of mixed acids to form complex greases is also very important (see below).

Alternative Fats and Fatty Acids, for Soap-based Greases

Natural fats and oils are glycerides, that is, a combination of a fatty acid with the tri-hydric alcohol glycerol. When an animal fat or vegetable oil is boiled with an acid solution it is hydrolyzed to yield a variety of natural fatty acids which depend on the starting materials. Typical acids are the C15 palmitic acid and the C18 stearic and oleic acids. These acids are not generally water soluble, and can be filtered off from the hydrolysis medium. The softer unsaturated acids such as oleic acid can be hydrogenated to yield higher melting and more stable materials. The fatty acids can be reacted with alkali to yield the soap plus water of reaction. Broad spectrum soaps can be produced by direct reaction of alkali upon the fat or oil, when saponification takes place and glycerol is liberated. These reactions are shown below:

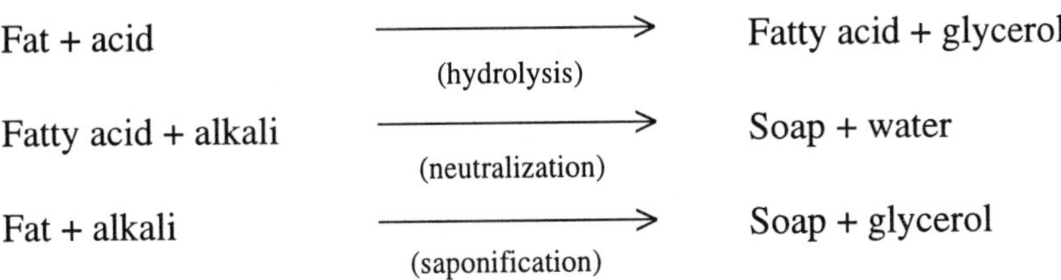

Palmitic, stearic, and tall oil acids have been widely used in the past for grease manufacture but a particularly valuable acid is 12-hydroxy stearic acid, which is produced from hydrogenated castor oil. This is now very widely employed in greases, particularly in lithium greases.

Complex Soaps

Mixtures of different soaps with a common metal may result from the saponification route or may be produced deliberately, but of greater significance are the so-called complex greases. These are made by co-crystallizing conventional soaps with a soap or salt formed from the same metal with either a low-carbon-number carboxylic acid (acetic, lactic, benzoic, etc.) or inorganic salts such as carbonate, chloride, phosphates, borate, etc. Complex formation dramatically improves the properties of a grease, particularly the drop point which is an indication of the maximum working temperature.[19]

Alternative Fluids

Apart from petroleum oils, greases can also be made with esters, polyolefins, alkylates, polyglycols, polyphenyl ethers, polyhalocarbons and silicone fluids (see Section 2.2). These can provide improved lubricity, oxidation stability, and wider working temperature ranges. Some of these fluids can be made into grease with conventional soap thickeners, but many of them require special thickening agents, often the solid type discussed below.

Non-Soap Thickeners

After metal soaps, solid inorganic materials are one of the most common thickeners, and they can be used to produce greases from most lubricating liquids. The common solids include treated clay (bentonite, montmorillonite), silica aerosol, and carbon black. Solid thickeners must be very finely divided, and must have a crystal form which will not promote wear. Suitable particle shapes would include spherulites and platelets. Clay materials usually have to be treated with materials such as quaternary amines in order to render the surfaces oleophilic.

Substituted ureas and polyureas have recently become the most common non-soap thickening agents and have high oxidation resistance. In North America and Japan they are commonly used in constant-velocity joints of front-wheel-drive vehicles. (In Europe, conventional greases are still used.)

Pigments such as phthalocyanines have been used as thickeners for some of the more inert lubricating fluids.

Additives

Many greases contain additives to improve their performance considerably over that of the simple combination of fluid plus thickener. Liquid or solid additives can be added to the grease during manufacture, usually following the "cook-out" or heat-soak stage (see Section 7.3.3). Possible additives include:

- Oxidation inhibitors
- Metal deactivators
- Metal corrosion inhibitors
- Rust inhibitors
- Film strength/lubricity additives
- Anti-wear additives
- Additives
- Viscosity modifier polymers
- Tackiness polymers (high mol. wt.)
- Water repellants
- Dyes
- Odorants

7.3.2 Characteristics of Common Greases

Calcium Greases

Hydrated calcium greases are low-cost products of a smooth consistency, typically described as "buttery." Their main limitation is an effective maximum working temperature of about 80°C, due to the drop point of about 80-90°C. Calcium greases would normally require addition of anti-oxidant and rust inhibitor additives. Very little grease of this type is sold today.

Anhydrous calcium greases have a stabilizer other than water, and are usually based on 12-hydroxystearic acid. They have drop points of up to 150°C and tend to be expensive.

Better are the calcium complex greases, first developed in 1940, in which the calcium atoms are combined both with a fatty acid and (typically) with acetic acid, which gives both a more open and a more stable structure to the grease. Such greases have much improved high-temperature performance with drop points up to 300°C, and they also can have built-in E.P. properties, good corrosion resistance, good shear stability, and good water resistance. Calcium complex greases are of moderate cost, less than that of the anhydrous type of calcium grease.

Sodium Greases

Developed before the turn of the century, these were first seen as the solution to the operating temperature problem found with hydrated calcium greases. The drop point of sodium greases is between 150°C and 200°C. They are low-cost products, but unfortunately suffer from water sensitivity.

Aluminum Greases

These have a drop point between 100°C and 120°C and often a useful "stringy" consistency. They have good oxidation, water sensitivity, and rust performance. They suffer from poor shear stability and poor pumpability, and are relatively high-cost products, being normally made from aluminum alcoholate rather than the hydroxide. A trimer of aluminum isopropoxide is now commercially available which eases the preparation of aluminum soaps.

Lithium Greases

These were first developed at the U.S. Naval Testing Station in 1940 and are now the most popular type, outnumbering all other types of grease in terms of usage. The majority of lithium greases are based on 12-hydroxystearic acid, the fatty acid made from hydrogenated castor oil. Ordinary lithium greases have drop points in the range of 180°C to 190°C, while complex lithium greases have drop points between 260°C and 300°C. They have good shear stability and water sensitivity, and are of moderate cost. The complexing agents are diabasic acids such as azeleic or sebacic acids.

Polyurea Greases

These have high drop points above 250°C, but not very good shear stability. The water sensitivity is satisfactory but the greases normally require rust inhibition. Their outstanding property is their oxidation resistance.

Solid-Thickened Greases
The organo-clay greases based on treated aluminum silicates are the most common types. They can be used to form greases with most fluids which will wet the surface of the particles. Drop points are high (260-300°C+) so such greases are useful for high-temperature applications. Normally used for the more exotic lubricating fluids, these greases tend to be expensive.

Carbon Black Greases
These greases have a high solids content and are of themselves effectively a combination of both solid and liquid lubricants. Their physical stability is not generally good, but they have found uses in extreme environments. One application has been in the nuclear industry where combinations of polyphenylethers and carbon black have been used as radiation-resistant greases. They are no longer commercially important, having been superseded by other solid-thickened greases.

7.3.3 Grease Manufacture

For a long time grease manufacture was more of an art than a science, with materials being mixed together in a molten state and then often being poured out into open trays to solidify. The fatty acid raw materials tended to be of variable quality, and laboratory batches were made when new raw materials arrived, in order to adjust the recipe. Past experience was used as a guide for the mixing and the setting of the grease. Today, raw materials are more tightly specified and the manufacturing process is carefully controlled.

A key relationship in grease manufacture is the harmony between the solvency of the lubricating fluid and the solubility of the thickener. This influences the mixing time and in particular the rate of cooling which is necessary to achieve the best structure in the grease. As the solution of soap in oil is cooled, the soap molecules come out of solution and crystallize into chains. The rate of cooling and the amount of agitation determines the length of the chains and the way they intermesh to form the grease structure.(20)

While continuous grease-making processes have been devised, normally manufacture takes place in a grease kettle, which is a heated vat with hoppers for addition of solid materials and overhead pipe work for the removal of water and other vapors as needed. The kettles are provided with means of cooling at a predetermined rate. The soap is formed either by neutralization or by saponification, and part of the fluid lubricant is mixed in. The whole is heated until the reaction is complete and the physical structure of the soap has formed in the "cook-out" stage. It is then cooled. The reaction temperature, the amount of mixing, and the cooling rate are all important in obtaining the correct fiber structure. Initial cooling is achieved by addition of the balance of the oil, but cold water circulation through cooling pipes is usually needed. The solubility of the soap in the fluid determines the holding temperature at which much of the soap is allowed to come out of solution and form the grease. Milling of the grease breaks down excessively long soap fibers and is used to meet the targets for consistency. Excessive milling can, however, destroy the grease structure.

Aluminum greases are often still poured into trays to solidify, as very rapid cooling is necessary.

For clay and other solid thickeners the solids and fluid are intimately mixed together, possibly by milling, until the suspension is uniform and the fluid has thoroughly wetted the solid surface. This can take place at ambient temperature.

Polyurea greases are made by in situ reaction and polymerization of amines and isocyanates. (This is a dangerous process and has to be carefully controlled.)

7.3.4 Grease Testing

Tests on greases are different from oil tests because of the normally solid or semi-solid nature of greases. The following notes cover the key tests; ASTM test method designations are given where applicable.

Consistency

The firmness of a grease is measured in terms of penetration number. A weighted cone is dropped onto a flat smooth surface of the grease in a container, and the extent to which the cone sinks into the grease is known as the penetration number. A typical method is ASTM D 217.

Many greases change consistency when manipulated. "Worked penetration" means that the grease is repeatedly forced in a standardized manner through holes in a "grease worker" to soften it before the penetration test is performed.

Greases are usually classified in terms of consistency by their National Lubricating Grease Institute (NLGI) number. These are as follows:

NLGI No.	**Worked Penetration ***
000	445-475
00	400-430
0	355-385
1	310-340
2	265-295
3	220-250
4	175-205
5	130-160
6	85-115

*After 60 strokes. Figures are tenths of a millimetre

Roll Stability (D 1831)

The stability of grease in service is important, and many tests have been devised to measure it. In this test instead of using the standard "grease worker," the grease is worked in a drum mechanism before determining the penetration. Water can also be added in the test to determine its effect on grease stability.

Apparent Viscosity (D 1092)

A measure of the grease's flow properties is obtained by forcing the grease through a capillary by means of a piston. As grease is non-Newtonian, the temperature and shear rate are measured for different flow volumes.

High-Temperature Flow (D 3232)
A "trident" probe in a Brookfield viscometer measures the flow resistance at high temperature.

Low-Temperature Torque (D 1478)
A ball-bearing is packed with grease and cooled to 54°C. The torque required to turn the bearing is measured on start-up and after 1 hour of running.

Drop Point (D 2265)
This test is in effect the opposite of a pour point of a liquid lubricant, giving an indication of the high-temperature operability limits of a grease. The grease is placed in a small cup adjacent to a thermometer and heated until the first drop runs out of a hole in the bottom of the cup. (Greases must be used at temperatures below the drop point, or they will tend to run out of the bearings and lubrication will be lost.)

Oil Separation (D 1742)
A test in which grease is placed on a wire screen and oil which separates can be measured.

Evaporation (D 972/D 2595)
Tests in which the tendency of the fluid constituents to evaporate at elevated temperatures is measured.

Oxidation Stability (D 942)
This is measured by sealing grease in a "bomb" containing oxygen and measuring the oxygen absorption.

Water Wash-Out Test (D 1264)
In this test, not now considered to correlate well with service conditions, a bearing packed with grease is sprayed with water and the amount of grease lost at the end of the test is a measure of its resistance to water wash-out.

Water Spray-Off Test (D 4049)
In this test grease is coated on a metal plate and its ability to resist removal by a water jet is measured.

Corrosion Tests
Corrosion of ferrous metals is assessed in D 1743, and copper corrosion by method D 4048.

Wheel Bearing Tests
There are many proprietary tests in which ball or roller bearings are subjected to long-term evaluation under load. The key ASTM standardized tests, in which wheel bearings are packed with grease and run under specified conditions by means of an electric motor drive, are as follows:

D 1263 Leakage Tendency	In this test the loss of grease after running under specified conditions for a given time is evaluated.
D 3527 Life Test	In this test the time to failure of lubrication is determined by measuring the torque resistance of the bearing, and determining when this starts to increase.
D 4290	This is an accelerated version of the above test, giving a shorter time to failure, and hence a more rapid test.

High-Speed Wheel Bearing Tests (D 1741 and D 3336)
These assess grease performance under high shear conditions in a rapidly rotating bearing.

Chemical Tests on Grease
These are used more for identification of unknown greases or for checking for cross contamination. Quality control of grease manufacture relies more on the physical tests such as drop point and penetration, and performance tests such as the Timken test. Methods such as flame photometry, atomic absorption, and X-ray fluorescence may be used when appropriate for determining the type and quantity of soaps present, such as sodium, lithium, aluminum, etc.

EP Tests
The tests for extreme pressure performance performed on liquid lubricants are equally applicable to greases. The most common are the Timken test

(D 2509) and the four-ball test, which may be used for lubricity under ASTM D 2266 or for determination of EP failure load under method D 2596.

Fretting Wear (D 4170)
This assesses ability to prevent wear prevention in conditions where the motion is oscillatory in nature.

7.3.4.1 Grease Specification and Approvals

The major roller bearing manufacturers (SKF, Hoffmann, etc.) have specification and test rigs and offer guidance on grease requirements rather than giving approval of specific brands.

Equipment manufacturers will recommend greases in terms of metal type, drop point, etc., and general properties such as EP or graphited. Some may approve specific greases for use in their equipment, and in some cases make a solus recommendation. For some sophisticated equipment, Military Specifications may be quoted as quality guidelines.

The NLGI classifies greases as either for chassis and general use or for wheel bearings:

For chassis use	For wheel bearings
LA	GA
LB	GB
	GC

Severity of duty increases going from A grades to C. See Appendix 13 for performance descriptions of these grades.

The U.S. military has many grease specifications and issues approvals. The more general ones are as follows:

MIL-G-23549	Grease, general-purpose
MIL-G-244139	Grease, multi-purpose, water-resistant

MIL-L-15719	Lubricating grease, high-temperature
MIL-G-819371	Grease, ultra-clean

7.3.5 Use of Grease in Motor Vehicles

Grease is used for chassis lubrication, in the power train, and for wheel bearing applications. Very small quantities are used in instruments. "Sealed-for-life" joints are now generally used for suspensions and steering gear, and the repeated use of liberal quantities of cheap greases in these applications has been replaced by one-time additions of superior products.

The requirements of general-purpose automotive greases include high drop point, good mechanical shear stability, good resistance to water wash-out or spray-off, and wear and corrosion protection. Good low-temperature properties are also needed for vehicles standing outdoors in low winter temperatures.

Constant-velocity driveshaft joints require special greases with load-carrying ability. Inboard joints, subject to heating from engine and gearbox, require good high-temperature performance. Outboard joints can have extreme high-temperature requirements if close to disc brakes, or relatively mild if well separated and in a good air stream.

Extremely high temperatures can be experienced in mechanical clutch mechanisms and front hub bearings adjacent to disc brakes.

Requirements for automotive greases are summarized in SAE J310 in Appendix 13.

New designs of outboard constant-velocity joints which are integral with the hubs will pose even more severe high-temperature requirements combined with load-carrying ability.

In the interest of both good high- and low-temperature performance, it is likely that more synthetic-based greases will be used in future vehicles, with petroleum-based greases remaining for general-purpose use.[(21)]

References

1. Bartz, W.T., Lubrication of Gearing, Expert Verlag (1988), translated Moore, A.J., Mechanical Engineering Publications, London, 1993, ISBN 0 85298 831 1.
2. Pilon, E.L., "Gear Lubrication—1," *Lubrication*, Vol. 66, No. 1, 1980.
3. Elliot, J.S., Hitchcock, N.E.F., Edwards, E.D., *Journal of the Institute of Petroleum*, Vol. 45 (428), p219, 1959.
4. Sanin, P.S., *et al.*, *Wear*, 3, p200, 1960.
5. Michaelis, K., "Testing Procedures for Gear Lubricants with the FZG Gear Rig," *Industrial Lubrication and Tribology*, Vol. 26, No. 3, pp91-93, 1974.
6. Tourret, R., Wright, E.P. (Eds), Performance and Testing of Gear Oils and Transmission Fluids, Proceedings of the International Symposium of the Institute of Petroleum, 1980.
7. Alliston-Greiner, A.F., Plint, M.A., "Gearbox Oil Tests and Procedures: A Critical Survey," CEC/93/TL02, CEC 4th International Symposium, 1993.
8. Korosec, P.S., Norman, S., Kuhlman, R.E., "Gear Oil Developments Beyond GL-5," SAE Paper No. 821184, Society of Automotive Engineers, Warrendale, Pa., 1982.
9. Lacoste, R.G., "Automatic Transmission Fluids," *Lubrication*, Vol. 54, No. 1, 1968.
10. Crouse, W.H., Automotive Transmissions and Power Trains, 4th Edition, McGraw-Hill, 1971.
11. THM-200: Principles of Operation, 1st Edition, Hydramatic Service Dept. (GM), 1975.
12. Deen, H.E., Ryer, J., "Automatic Transmission Fluids—Properties and Performance," SAE Paper No. 841214, Society of Automotive Engineers, Warrendale, Pa., 1984.
13. Ross, W.D., Pearson, B.A., "Automatic Transmission Fluid Type F Keeps Pace with Fill-for-Life Requirements of Modern Automatic Transmissions," SAE Paper No. 680057, Society of Automotive Engineers, Warrendale, Pa., 1968.
14. Deen, H.E., O'Halloran, R., Outten, E.F., Szykowski, J.P., "Bridging the Gap Between DEXRON™ II and Type F ATF," SAE Paper No. 790019, Society of Automotive Engineers, Warrendale, Pa., 1979.

15. Potter, C.R., Schaefer, R.H., "Development of Type C-3 Torque Fluid for Heavy-Duty Power Shift Transmissions," SAE Paper No. 770513, Society of Automotive Engineers, Warrendale, Pa., 1977.
16. McLain, J.A., "Oil Friction Retention Measured by Caterpillar Oil Test No. TO-2," SAE Paper No. 770512, Society of Automotive Engineers, Warrendale, Pa., 1977.
17. Friihauf, E.J., "Automatic Transmission Fluids—Some Aspects on Friction," SAE Paper No. 740051, Society of Automotive Engineers, Warrendale, Pa., 1974.
18. Greases, NLGI Handbook.
19. Boner, C.J., Modern Lubricating Greases, Scientific Publications Ltd, 1976.
20. Boner, C.J., Manufacture and Application of Lubricating Greases, Reinhold, 1954.
21. Sorli, G.E. "Beyond the looking glass, grease in the 90s," *NLGI Spokesman*, XLIV(2), 64-65, 1980.

Chapter 8

Other Specialized Oils of Interest

8.1 Two-Stroke Oils

The different types of two-stroke oils were reviewed briefly in Section 1.7.2. The requirements of the two-stroke engine used in heavy-duty trucks are included in the requirements of diesel oils which have already been discussed in Section 4.1.2. The requirements of those railroad engines operating under two-stroke conditions will be covered in Section 8.3, and the detailed requirements of large marine two-stroke engines are considered to be outside the scope of this book. This section will therefore deal with the smaller types of two-stroke engines, although these range from single-cylinder engines of 50 cc or less to large multi-cylinder outboard engines of several litres capacity. Because the lubrication requirements vary considerably between the different applications, it will be necessary to consider these separately.

8.1.1 Automobile Engines

While today many two-stroke Trabant and Wartburg vehicles still exist in eastern Europe, and three-wheel motorized rickshaws can be found in many developing countries, the early attempts to use two-stroke engines in passenger-carrying vehicles must be considered to have been failures, the main problems being smoke emission and unreliability. To a large extent both of these can be attributed to the use of unsuitable lubricants in the past; the users were unwilling to pay anything more than a low price for lubricants and most oil companies were initially unwilling to devote major development programs to a minor sector of the lubricant market.

Potentially, two-stroke engines should be ideally suited for smaller passenger-carrying vehicles because of their relative simplicity and their high power-to-weight ratio. They produce relatively low amounts of nitrogen oxide emissions and therefore have the capability of providing very low exhaust pollution if combined with a catalytic converter. For these reasons there is now a considerable interest in the development of a new generation of two-stroke engines for automobiles. However, there are many problems to be solved in overcoming the very high hydrocarbon emissions of the conventional small two-stroke engine and solution of these problems could result in weight and size penalties which would nullify the key advantages of the two-stroke in these areas. A useful summary of the situation was given by Subaru at the Japanese SAE meeting in spring 1994.[1]

To overcome the problem of leakage into the exhaust system of incoming air/fuel mixture, either direct fuel injection or an external scavenging system or both will probably be employed. Exhaust or scavenge valves will probably be of the poppet type rather than ports opened by piston movement. In France the Institut Français du Pétrole (IFP) has developed its IAPAC engine with conventional crankcase compression and an exhaust poppet valve to the scavenging system.[2,3] while the Subaru and Toyota designs utilize supercharging with piston/port valves in the case of Subaru and a totally poppet-valved system for Toyota. The Orbital Engine Company has developed a design of fuel-injected engine based on the design of Dr. Sarich which operates in a similar manner to the I.F.P. design.[4] It has been licensed to several OEMs including Ford and Fiat. It is a dry sump design using a total loss lubrication system and we understand the results are very promising. (It is not an orbital engine in the mechanical sense.)

The Toyota and Subaru systems use a wet sump design, and we also understand work at G.M. and Chrysler is focusing on four-stroke type lubrication systems for their two-stroke developments. The U.S. manufacturers are basically looking for low-cost, low-polluting engines, whereas the Japanese appear also to be attempting to exploit the basically smoother characteristics of the two-stroke engine versus a four-stroke, and developing it for the luxury and high-performance market.

In all recent developments emissions limitation was an early goal, but practical difficulties may result in this being little better than for the best four-stroke designs. Engine and exhaust temperatures tend to be low at idle and low throttle openings, and two-way catalysts to reduce hydrocarbon and CO emissions will take a long time to reach operating temperature. On the other hand, temperatures are very high at full throttle, and catalysts can be damaged through overheating. This is compounded if carbon deposits laid down on the catalyst during low-temperature running are simultaneously burned off. Catalyst performance and life are therefore problems. Engine designs which improve scavenging would make the control of hydrocarbons and CO easier, but increase the level of NO_X emissions, one of the reasons for choosing a two-stroke design.

No precise details of the lubrication requirements of these various new engines are known at the present time, but it would be expected that the total loss lubrication systems would require lubricants similar to high-quality 2-T motorcycle or outboard oils, while those employing a wet sump design would be expected to use a fluid similar to high-quality passenger car motor oil. In each case it may be necessary to use special base stocks in order to keep the engine deposits at a minimum and to reduce the level of oil pollution in the exhaust stream.

We will now go on to consider the more traditional two-stroke applications.

8.1.2 Mopeds, Motor Scooters and Lawnmowers

These traditionally employ single-cylinder, air-cooled engines of 50 to 200 cc capacity, operating engine speeds of up to 6000 rpm, and normally use a fuel/oil mixture of about 25:1 ratio, although 12:1 or even 6:1 was not unknown in the past. For silencing, and to act as a partial throttle against over-speeding, the mopeds and motor scooters have a restrictive and complex exhaust system, while lawnmowers tend to have much simpler or even vestigial systems. Lawnmowers run at high powers for longer continuous periods, and do not have the same degree of air flow over the cooling fins and therefore tend to run hotter than the road machines.

This class of engine traditionally has been of relatively low compression ratio, and scavenging of exhaust from both engine and silencer are rather poor, particularly at low or idle speeds. To prevent deposit build-up, strong detergency based on metallic types is considered necessary by most formulators, although these can be combined with some ashless detergents in addition. High detergency reduces misfiring from carbonaceous spark plug fouling, but high levels of ash also contribute to spark plug fouling and can give rise to pre-ignition from combustion chamber deposits. The desirable level of metal-containing detergents therefore depends on the type of engine and the operating conditions. Any future requirements for catalyst systems on this type of engine would restrict or prohibit the use of such detergents.

Piston/cylinder and big-end bearing lubrication are very important, the latter particularly as there is no oil sump and plain bearings are often used. The traditional aid to lubrication is a proportion of bright stock in the base oil, the bright stock being relatively involatile and tending to collect around the lower piston and run down the connecting rod. Unfortunately, its involatility means that it enters the exhaust system where it tends to form deposits, and also leads to an oily and smoky exhaust. Better carrier lubricants would be esters or alkyl benzenes, which burn cleanly and provide a clean exhaust. Esters in particular provide particularly good lubricity. Unfortunately in the past both of these synthetic lubricants were considered too expensive for the cost-conscious market represented by these applications. Polybutene thickener in a light base stock may be an intermediate cost solution which is increasingly adopted, as polybutene of moderate molecular weight tends to decompose without leaving heavy deposits. As in other areas, introduction of legislation such as mandatory air pollution requirements will change the perception of the quality level of this market. Already emissions legislation in Austria, Switzerland and Taiwan requires that mopeds and chainsaws are fitted with catalysts. In Thailand, 2T oils have to pass a smoke emission test (along with performance tests) in order to obtain a license for marketing.

8.1.3 Chainsaws

This application uses very similar engines to those of the above, although they tend to be of higher compression ratio and have much less exhaust throttling. They are, however, very liable to over-speeding, and need a clean exhaust due to the proximity of the operator. Polybutene and ester lubricants provide clean exhaust emissions, with esters also providing a high degree of lubricity which can be used to reduce the oil-to-fuel ratio, thereby improving the emissions further. Oils containing synthetic thickeners instead of bright stock also do not need such a high degree of detergency for deposit control. Use in environmentally sensitive locations is creating a demand for biodegradable formulations for the motor oil as well as the chain oil. Exhaust catalysts are required in some countries and this requirement will spread. Chainsaw manufacturers are pressing for higher-quality and special-purpose oils for their equipment, and this will result in further diversity in types of 2T oil.

8.1.4 Motorcycles

In the last twenty years two-stroke motorcycles have seen considerable development, leading to high-speed, highly tuned, multi-cylinder designs with high specific power outputs. Lubrication is still on a total loss basis, but is either at much leaner oil-to-fuel ratios or is by positive oil injection from a metering pump linked to the throttle. This latter system requires that an oil remains sufficiently fluid down to low temperatures to avoid starvation on start-up in cold weather. Due to the high levels of power output, high levels of detergency and lubricity are required for this application, with metal-based detergents again showing superiority over ashless formulations. Polybutene has been found successful to provide a good level of lubricity and an exhaust with a low level of visible smoke. Ester-based oils are again very satisfactory, but considerably more expensive than polybutene-thickened petroleum oils. Biodegradability is becoming a requirement for motorcycle oils in Europe, due to the possibility of oil from leakage, spillage, or oily exhaust emissions being washed from road surfaces by rain into ground waters. The subject of biodegradability will be discussed more thoroughly under the outboard motor section.

Japan has taken a lead position in the development of new motorcycle oils and specifications and polybutene-based formulations were pioneered there.[5]

Newer moped and scooter designs tend to follow motorcycle developments with engine speeds up to 9000 rpm and variable oil injection.

Racing motorcycles impose even more severe demands than normal road machines, except that long-term deposit build-up is not so critical, as these engines are frequently stripped and rebuilt and exhaust systems are simpler. Castor oil-based fluids traditionally used at one time have been largely replaced by synthetic products, combining selected synthetic esters with balanced detergent and anti-wear packages. Ingredient cost and the availability of special formulations are not problems in this area.

8.1.5 Outboard Motors

Some early outboard motors consisted of little more than air-cooled, moped-type engines driving propellers through simple driveshafts. The market is now dominated by sophisticated multi-cylinder water-cooled engines initially developed mainly in the U.S., with both specific and severe lubrication requirements. Outboards are required to run for long periods at maximum speed and power output, are frequently over-revved if the propeller leaves the water, but can also spend long periods at idle or puttering around harbors. Increasingly, outboards have oil injection systems and the use of a fuel/oil mix is declining. Due to the large fuel (and therefore oil) throughput when the engines are run for long periods at high power outputs, the oil-to-fuel ratio is kept low. Ashless formulations are specified in order to avoid pre-ignition problems, but detergency is very important and special amide ashless detergents have been developed specifically for outboard use. Salt water corrosion protection at low oil-to-fuel ratios is specified to protect the system components in marine environments.

Smoke emission is still a major concern, as is the case for all two-stroke applications, but the greatest problem in the environmental area is biodegradability. Not called for at present in salt water environments or anywhere in the U.S., it is mandatory in many countries for the use of outboards on inland lakes and rivers. Biodegradability testing is still in its infancy and subject to wide variations due to standardization difficulties. Doubts have been cast on the validity and utility of some test methods. It is clear, however, that mineral oils are usually not sufficiently biodegradable for inland water applications, and that synthetic base stocks may be required. Polyolefins are even less biodegradable than mineral oil so that base stocks need to be chosen with care. Esters appear to be the most biodegradable type of base stock and with their other advantages will probably dominate the outboard lubricant scene in the future, with the possible exception of the U.S. where biodegradability concerns have been slow to surface.

The testing of outboard motor oils is quite specific and complex, and will be dealt with in more detail later.

8.1.6 Two-Stroke Oil Tests and Specifications

Europe could be said to be the original home of the moped and motor scooter applications for two-strokes, while the U.S. was historically the dominant supplier of sophisticated outboards. More recently Japan has been participating significantly in supplying both types of market. Asia in general is now the largest consumer market for two-stroke oils, estimated at about a quarter of a million tons per year. Of this, India, with the largest 2T bike/moped population, consumes about 60,000 tons/year.

Initially, manufacturers such as Motorbecane in France, Piaggio in Italy, and the Outboard Marine Corporation in the U.S. developed their own tests in order to recommend or approve suitable two-stroke oils for their engines. The European tests were coordinated and refined by the CEC, and the BIA (Boating Industries Association, representing the majority of the U.S. outboard manufacturers) became responsible for outboard oil testing. In the mid-1980s, the API and CEC set up a provisional two-stroke lubricant performance and service classification list, covering the whole range of two-stroke applications. This classification is described in SAE J2116 in Appendix 14, and is summarized in the table below.

Original/API designations		Application and Problem Areas
TSC-1	TA	Mopeds, lawnmowers, small generators and similar engines prone to deposit-induced pre-ignition and to exhaust system blockage
TSC-2	TB	Motorscooters, motorcycles, and some chainsaws prone to scuffing, ring-sticking, pre-ignition, and power loss
TSC-3	TC	Chainsaws, motorcycles, snowmobiles, etc., including some water-cooled non-outboard engines. Prone to ring-sticking and deposit-induced pre-ignition damage
TSC-4	TC-W	Water-cooled outboard engines very prone to pre-ignition damage

Tests for TC-W were well-established BIA tests, but for TA, TB, and TC the test engines included French Motorbecane, Italian Piaggio, and Japanese Yamaha engines. Engine and parts supply were a major problem and the classification never worked satisfactorily as a means of standardizing quality because of test problems. At the present time the TA and TB qualities are effectively obsolete and unsupported, and the status of the other two is as follows:

API TC (Revised)
Detergency test in Yamaha Y350M2
Lubricity test in Yamaha CE 50
Pre-ignition test in Yamaha CE 50

API TC-W
Succeeded by TC-WII
To be replaced by TC-W3 (see below)

The API TC classification is expected to have limited life and is currently regarded as the lowest acceptable level of 2T oil quality.

In Europe, the OEMs conduct their own in-house tests and approve oils which they find satisfactory. These tests, comprising both laboratory engine tests and field tests, are not generally available to the market and development of new formulations is difficult.

In 1989 the Japanese Standards Organization (JASO) set up a working party of OEMs, lubricant manufacturers, and additive companies to produce a new standard for the industry. This was presented in 1990 and consists of three levels of quality, FA, FB, and FC. These represent typical global 2T oil quality levels, with lubricity and detergency quality increasing from FA to FC while exhaust blocking and smoke emission improves. Japanese manufacturers quote FC as their minimum requirement, but in practice their own "Genuine" or approved oils are above this level. The engine test requirements are summarized in Appendix 14.

The JASO specifications were originally intended just for use in the Asia-Pacific area, but worldwide interest has been such that a new global series of specifications has been proposed, to be developed jointly by JASO, ASTM and the CEC, which may be issued under the ISO banner. This will eliminate the lowest JASO quality FA but adds a new higher level with greater detergency requirements. The tests would be those in the JASO specifications and the same reference oils could be used.

European OEMs, coordinated by CEC, have examined the JASO qualities and found the FC level to be inadequate, with oils which fail in European field tests passing the JASO detergency test and the reference oil JATRE-1 giving poor performance in European in-house testing. A CEC task force is therefore working to develop more severe forms of the Japanese tests which will correlate with European requirements. This could lead to the proposed new global GD level with the relationships to the JASO specification as follows:

Global Specification Proposal	-	GB	GC	GD
JASO Specification	FA	FB	FC	-

Their proposals are to be balloted in London in May 1996. A new severe test in a Piaggio engine is also under development for the GD or a higher category.

The above specifications do not include outboard oils. These have been administered by the National Marine Manufacturers Association (NMMA) of the U.S., as successors to the BIA. TC-WII succeeded TC-W and is now being superseded by a new specification TC-W3. The following tests are included in these two specifications:

Engine Tests	TC-WII	TC-W3
General Performance	OMC 40HP	OMC 40HP
Lubricity	Yamaha CE50S	Yamaha CE50S
Pre-ignition	Yamaha CE50S	Yamaha CE50S
Detergency		OMC70HP
Detergency		Mercury 15HP
Other Tests		
	Miscibility	Miscibility
	Rust	Rust
	Filtration	Filtration
	Fluidity	Fluidity
		Compatibility

It is worth noting that in the outboard oil tests it is normal to run general performance tests at the standard recommendation of 50/1 fuel/oil ratio, while lubricity tests are run at the more extreme condition of 150/1 and pre-ignition tests are run at 24/1, again to increase the severity. The testing regime can therefore be considered extremely severe in relation to normal practice.

The tests and performance requirements for the various categories of two-stroke oils are included in Appendix 14.

8.1.7 General-Purpose Two-Stroke Oils

From the earlier discussion of the lubrication requirements of the various two-stroke applications, it is obvious that a single formulation giving good performance in all applications is very difficult to arrive at. Most engines use oil which is mixed with the fuel, although ratios differ widely between applications. Some engines use oil injection, but here again some may recommend the straight lubricant and others may recommend some fuel or solvent pre-dilution. The smaller and less-efficient engines are less satisfied with ashless detergents than the prestigious high-performance types, but perhaps the greatest difference is the traditional price differentiation in the market with the less-sophisticated equipment not considered as justifying high-priced lubricants. The general consensus is that a low ash formulation containing some metal detergent is desirable for most land-based applications, and this is desirably based on synthetic base stocks such as esters to improve performance and to permit use at low oil-to-fuel ratios. For the price-sensitive sector a formulation may have to be based on mineral oil plus polybutene in order to achieve low smoke levels in cruder equipment. For the outboard sector ashless oils are imperative to avoid destructive pre-ignition and the main development is in the area of synthetic base stocks, particularly esters for biodegradability. Outboard oils will normally successfully lubricate motorcycles and snowmobiles, although the oil-to-fuel ratio may have to be increased to above 1/50. For higher ratios exhaust smoke may then be excessive, since esters are not particularly low in smoke emission. The detergency may be adequate with ester-based products in the lower categories using high oil-to-fuel ratios, but as already mentioned the biggest problem in this sector is cost. Overall, most oil marketers will find it desirable to market at least two qualities of two-stroke oil, probably an outboard type and a low ash type for other applications.(6) If emission requirements eventually lead to general use of catalytic converters on 2T exhausts, then ashless formulations will become universal, resulting in new compromises between engine performance and oil quality which are difficult to predict at this time. Many existing engine designs would not respond well to use of completely ashless oils, but emissions legislation could enforce their use.

8.2 Gas Turbine Oils

Automotive application of the gas turbine is now confined to military use or experiments with hybrid-powered trucks, where its high power/weight and high power/volume ratios are very valuable and the negative factors of low fuel efficiency and high cost can be tolerated. Derivatives of the aviation type of gas turbine, such as those used in small propeller jet aircraft, are utilized.

The lubrication requirements of a gas turbine are in a sense extremely simple. There are no reciprocating parts to give rise to difficult wear control problems, and the oil is not contaminated by combustion products. The main roller-type bearings are fed with oil under pressure, and a scavenge loop returns it to a remote sump for recirculation after filtration and cooling. This cooling can be via an external oil cooler or by heat exchange with the incoming fuel. If integral gearing is to be lubricated, this will be fed with its own supply of lubricant from a tapping off the main gallery, coarse as well as the normal fine filtration being used to trap any gearbox debris (Figure 92).

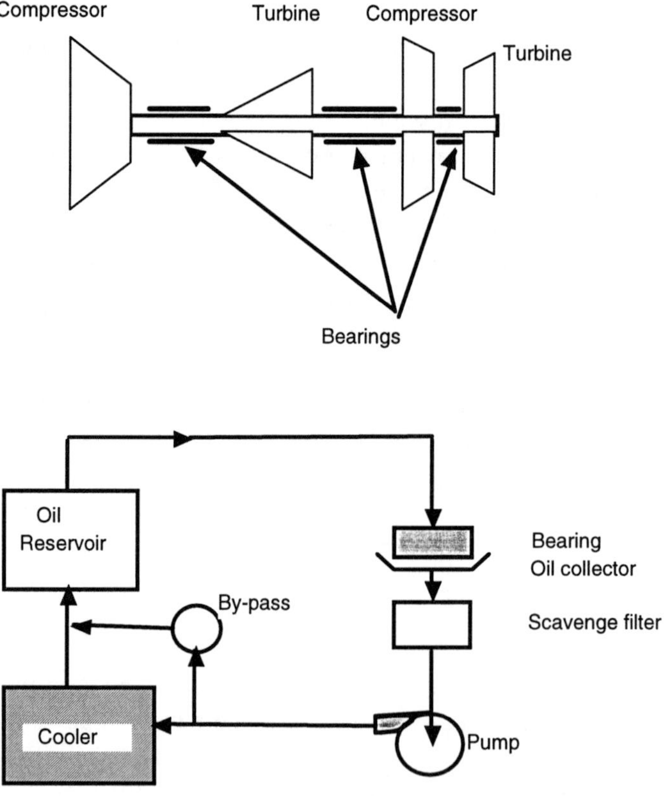

Fig. 92 Gas turbine lubrication.

Heated by the main turbine disks, the center and rear bearings run very hot, and oil may be heated in the bearings to over 250°C. Heat soak-back on shut-down is a particular problem. The lubricant therefore needs to be of good oxidation stability, low volatility, high viscosity index and, with a need for good oil flow at low starting temperatures, it must also have good low-temperature viscosity and pour point properties. Inhibited ester-based lubricants are exclusively used for the aviation type of gas turbine, combining the above properties with a high degree of natural lubricity. The early type of lubricant used diesters with a viscosity of about 3 cSt for pure jet uses, with a different version thickened with complex esters to a viscosity of about 7.5 cSt being used for propeller jets with integral gearboxes. Anti-oxidants such as phenothiazine or phenylamines and their derivatives were used to give high thermal stability, and metal deactivators and anti-foam agents were also used. These early lubricants have now generally been superseded by newer types, of superior performance, based on hindered-ester fluids. These are made from polyhydric alcohols and mono-acids rather than the mono-alcohols and diacids of the type 1 fluids. The natural viscosities of these so-called type II or type II1/2 fluids is around 5.0 cSt, and they can be used for gearbox lubrication. Oil consumption of a gas turbine is minimal, but it is essential that the same lubricant is used for top-up as is used for the initial fill of the equipment. In particular, accidental topping-up with a mineral oil-based fluid would be potentially disastrous. Recommendations as to the type of fluid and approval of specific brands is given by the equipment manufacturers. The U.S. military also has specifications covering this type of lubricant.

8.3 Railroad Oils

For many years the expression “railroad oil” has implied internationally a lubricant that can satisfy the particular requirements of General Motors EMD railroad engines. These are supercharged, high-power, two-stroke diesel engines of unique design found on railroads around the world, being by far the most commercially successful type of railroad engine. A particular feature of this engine design is that for many years certain bearing parts, particularly the wrist-pins (the junction between the piston and the connecting rod) and some turbocharger parts were silver plated. Although alternative designs and metallurgies are now being introduced, the silver-plated bearing is still found in the field. It was a most successful design, providing minimum friction but requiring that compatible lubricants were used. The most significant aspects of this were that the lubricants had to contain no significant zinc content, and only certain detergent additives could be used. There is little problem in formulating lubricants to satisfy only EMD engines, but other designs often require anti-wear additives (such as those normally based on zinc compounds) and also a high level of detergency. As other engine designs have increased in power output and fuel quality levels tended to reduce, it became more and more difficult to formulate economic multi-purpose lubricants which could satisfy both EMD and other types of engines used alongside one another on some railroads.

The detergent quality of oil required by EMD depends on the sulfur content of the fuel in use:

Fuel sulfur %	Oil TBN (D 2896)
Less than 0.5	7 - 13
0.5 - 1.0	10 - 13
over 1.0	20

This type of recommendation relating fuel sulfur content with the lubricant alkalinity is quite common among manufacturers of medium and large diesel engines with worldwide operations. Fuel sulfur contents vary widely even within close geographical areas, but there is a tendency for them to be low in highly industrialized countries (less than 0.5%, and likely to decline due to particulate emission considerations) and high in the developing countries. In particular South America, North Africa, and the Middle East have typical diesel fuel contents of above 1.0% sulfur. Because the most highly overbased detergents (calcium and magnesium carbonate overbased) are unsuitable for EMD oils, a lower base number type of product has to be used and the treat levels even for the common 13 TBN oils tend to be high. Oxidation stability is provided by using large amounts of phenate-based detergents, and anti-wear performance is provided by the use of stable non-zinc materials such as chlorinated hydrocarbons.

EMD lubricants are normally SAE 40 in viscosity, and have to pass silver and copper corrosion tests and a laboratory engine test before they can be considered for field trials. Approval is given to a lubricant only after a minimum of one year of successful operation in at least three modern locomotives.

The greatest population of EMD locomotives is in North America, but substantial numbers also exist in South America, Africa and the Middle East. There are very few in Europe and these can be readily provided with their own special lubricant. There is, in fact, less reluctance to use dedicated lubricants for different types of engines in Europe than appears to be the case for the U.S., where a single "all-can-do" oil is often preferred at a maintenance site.

The main competition for EMD in North America comes from General Electric (GE), who demands high TBN oils and has no zinc limitation. Although high TBN conventional diesel oils can meet the requirements of GE, well-formulated EMD oils can also meet the requirement, although not at the lowest possible cost. The dangers of using the incorrect lubricant in an EMD engine usually dictate that a general-purpose railroad oil satisfactory for these engines is used wherever there are mixed fleets.

European railroad engines tend to be medium-speed designs of conventional four-stroke type, although the Napier "Deltic" engines used for a time by British Railways were two-strokes of a unique configuration, consisting of a bank of triple cylinder units arranged in a triangle with a crankshaft at each corner and containing opposed pistons. This engine produced a tremendous amount of power from a very compact unit, but its extreme mechanical complexity made servicing difficult. It also contained various sensitive bearing metals, and the oil was required to be of high thermal stability, high detergency, and to be uncorrosive to the various metals used in its construction.

Lubricants for European railroad engines today consist in the main of conventional super high-performance diesel oils similar to API CF-4 and

CCMC D5 categories, with TBN being increased if necessary for reasons of high fuel sulfur content. Sulfur content in many areas is now expected to fall, but in others it will remain high for the foreseeable future. Many engine manufacturers, for example Poyaud whose engines were used by French railways among others, had individual severe requirements for lubricating oils and an approval system of their own, which again tended to lead to provision of separate oils for individual engine types. Engine manufacturers also often have different viscosity and other requirements for their engines, which makes it difficult to design general-purpose oils for a railroad system. Also with the current rapid development of trans-European railroad systems based on electric power, it is unlikely that pressures for multi-purpose railroad oils will grow in Europe.

In addition to railroad use, it is worth recording here that many railroad engine designs, particularly those of EMD, find application in the area of oil drilling rigs.

8.4 Hydraulic Oils

These are used in a very wide variety of applications in automotive equipment, including brakes, power-steering mechanisms, transmissions, and suspensions, as well as more obvious uses in construction equipment. Off-road uses range from mining and tunnelling machines to bridge and barrier actuators. The variety of lubricant types is correspondingly complex ranging from mineral oils through complex mixtures of synthetic chemicals to water-based emulsions.

Automobile hydraulic fluids fall into two classes: either synthetic chemical fluids of the polyalkylene glycol or borate ester types for simple piston-operated systems (e.g., braking systems), or inhibited mineral oils for continuously pressurized systems, such as those associated with hydraulic suspensions. The latter are becoming increasingly popular in Europe, having been pioneered by Citroën over 40 years and now being employed more generally in both luxury cars and off-road vehicles to provide ride and/or height control.

The main requirements for any hydraulic fluid are good oxidation stability, suitable viscosity characteristics covering low ambient to high operating temperatures, and compatibility with the chosen seal system. The relation between the hydraulic fluid and the material of the seals in the system is of fundamental importance in all hydraulic applications, and neither should be changed without giving careful consideration to the other.

Where hydraulic pumps are used to generate the operating pressure in the system, consideration must be given to their requirements. Anti-wear agents and rust and other corrosion inhibitors may be needed to ensure trouble-free operation.

Shock Absorber Oils (damper oils) have extreme requirements. They must be fluid down to all conceivable ambient temperatures to avoid damper or vehicle damage, but must retain sufficient viscosity at operating temperatures of 100°C plus (internally). Synthetic fluids such as polyethers or esters would be useful, but most manufacturers want a low-cost fluid. In the past naphthenic base stocks with low-molecular-weight polymers (as shear-stable viscosity modifiers) have been used. With declining availability of naphthenic stocks and more realistic performance tests, paraffinic stocks are now favored. Catalytically dewaxed stocks could be very useful in this area. Oils need to be inhibited against thermal breakdown and oxidation, and to have anti-foam and boundary lubrication additives.

Automatic transmissions, discussed more fully in Section 7.2, contain hydraulic converters and actuators and are therefore hydraulic fluids although they have additional special requirements. They normally consist of a series of epicyclic gears controlled by brakes and clutches immersed in the lubricant, and apart from the usual requirements of oxidation stability and suitable viscosity, fluids used in automatic transmissions must provide the necessary frictional characteristics for smooth and successful operations of the brakes and clutches. Because of their ready availability, ATFs are frequently used in a purely hydraulic role, for example as power steering fluids and in suspensions.

Tractor hydraulic fluids are multi-purpose products for all tractor applications except engine lubrication. They were discussed under ATF in Chapter 7.

For larger hydraulic equipment, such as in earthmovers and vehicle hoists, mineral oils are normally used, sometimes the same oils as are used in the engine. The same basic requirements apply, namely suitable viscosity characteristics, seal compatibility, and good oxidation stability, with good anti-rust performance also being expected. Oils for hydrostatic drives require similar characteristics to those used in piston rams and actuators. Most equipment uses mineral oil-based products with good viscosity/temperature characteristics inhibited against rust and oxidation.

In cases where fire is considered a hazard, hydraulic fluids may be required to be either non-flammable or to resist combustion. Oil/water emulsions are growing in popularity, while various synthetics have been used for critical applications. In the past a particularly popular type has been the phosphate esters, although for reasons of cost and some technical limitations these are now declining in popularity as water-based types are becoming more successful.

8.5 Air-Conditioner (Refrigerator) Lubricants

It is well known that past refrigeration systems contained CFCs (chlorofluorocarbons), and that escape of these materials into the atmosphere was having a damaging effect on the ozone layer which protects us from ultraviolet radiation. An automobile air-conditioner contains a refrigeration unit which basically consists of a compressor that liquefies the refrigerant and forces it through an expansion valve. Here the refrigerant is atomized into a lower pressure area and vaporizes, extracting heat from the surroundings. It is then re-liquefied in the condensing section and returned to the compressor. The compressor requires lubricating, and for domestic refrigeration systems and for car air-conditioners this is done by mixing a proportion of lubricant into the refrigerant. While complete solubility of the oil in the refrigerant is not essential, it must be sufficiently soluble to provide adequate lubrication of the compressor when the mixture of the two is being compressed.

It has been estimated in the U.S. that nearly half of the CFCs escaping into the atmosphere come from automobile air-conditioning systems. There are three main reasons for this:

i) Compressor (on engine) connected to unit by flexible pipes—leakage possibility.
ii) Damage to units or piping in crashes.
iii) Lack of care or knowledge in servicing or scrapping.

A typical refrigerant for an automobile air-conditioner has been the material known as R12 which is dichlorodifluoromethane and for which naphthenic mineral oils provide a suitable lubricant. Unfortunately, R12 and similar compounds are highly reactive with ozone.

To overcome the attacks on the ozone layer the refrigerant will be changed to either R22 (a so-called CHFC, difluorochloromethane), CHFC mixtures, or better, R134A, which is a chlorine-free fluorocarbon, namely tetrafluoroethane. None of these refrigerants is sufficiently miscible with mineral oil for this to be a suitable lubricant, and synthetic refrigeration oils are required. Alkyl benzenes are suitable for R22 but these are inadequate for R134A, for which polyalkylene glycols and polyol esters are suitable. Probably these are the most likely types of lubricant which will be employed in future air-conditioning units.

In the U.S., where the majority of automobile air-conditioners are produced, the change has already taken place to utilize R134A and polyalkylene glycols.

8.6 Industrial Lubricants in Automobile Plants

It is proposed to give here a brief review of the key types of industrial lubricants likely to be found in plants manufacturing automotive equipment. While these are not in themselves vehicle lubricants, they may be of interest to many of those who are engaged in vehicle manufacturing but have no direct knowledge of factory lubrication.

There are many more different types of industrial lubricants than there are types of vehicle lubricant, but they are individually usually less complex than crankcase oils or automatic transmission fluids. The industrial lubricants change much less rapidly in terms of both quality levels and individual formulations than is the case for automotive oils, and the oils which are used in circulating systems have longer service lifetimes than crankcase oils.

General-Purpose Lubricating Oils for Machinery

These are based on paraffinic petroleum base stocks, containing anti-oxidant and rust inhibitor additives. The anti-oxidant has traditionally been a phenolic type, although use of ZDDP, while a more expensive alternative, confers some anti-wear properties in addition. Rust inhibitors are long-chain organic acids and their salts, or certain amines, and function by coating the metal surfaces and repelling water.

This type of oil is normally used inside a controlled environment such as a factory building, and not subjected either to the heating and cooling nor to the contamination which occurs in an engine. Because of the relatively constant temperature operation, high- and low-temperature viscosities are not as critical as in the case of motor oils, although if the same oil doubles as a general-purpose hydraulic oil then high viscosity index and low pour point can be provided.

Industrial oils are classified according to viscosity under the ISO classification, in which viscosities are measured at 40°C. Typical grades are ISO 46 which corresponds approximately to the SAE 20 grade for motor oils, ISO 68 which is between SAE 20 and SAE 30, and ISO 100 which is between SAE 30 and SAE 40.

Industrial Hydraulic Oils

As indicated above the same oil can often be used for general-purpose hydraulic use as is used for general lubrication. Hydraulic systems require a pump to maintain pressure in the system, and it is the needs of the pump which normally dictate the quality of hydraulic oil required. With vane-type pumps in particular, a moderate level of EP performance is necessary in order to minimize wear. Tricresylphosphate (TCP) has been traditionally used as an anti-wear agent, but for vane pumps working at high speeds and pressures this is barely adequate, and formulations containing ZDDP are now preferred. However, some high-performance pumps, usually of the multi-piston "swash-plate" type, contain silver surfaces and ZDDP is unsuitable. For this type of pump, sulfur/phosphorus EP agents at low levels of treatment are frequently employed.

As well as anti-wear performance, hydraulic oils need to have good thermal stability and oxidation resistance, also anti-corrosion and yellow-metal compatibility properties, good foam control but low air-entrainment properties, good demulsification, and need to be compatible with typical seal elastomers. A choice of good quality base stocks and selected additives allow this to be achieved at relatively low cost.

Water-based hydraulic fluids (emulsions) have been promoted as safer materials for use in factories, as well as in the obviously dangerous environments such as mining. While such products are obviously safer in the case of fire, there are reservations about anti-wear performance, leakage problems, and their stability in the case of accidental contamination. However, the use of such fluids is expected to increase markedly.

Other types of fire-resistant hydraulic fluids include water/glycol/polyglycol solutions and phosphate esters. These are expensive and in the case of phosphate esters give rise to environmental health concerns.

Slideway lubricants (for machine tools) were mentioned in the discussion on basic lubrication (Section 1.4.4). A good quality hydraulic oil is needed, and unless hydrostatic lubrication is used, suitable boundary lubrication additives to prevent "stick-slip" motion must be incorporated.

Turbine Oil

Lubricants for steam turbines and large industrial-type gas turbines are made from special base stocks selected for their stability and high V.I., treated with mixtures of phenolic and amine anti-oxidants and with anti-rust and anti-foam additives. The oils are very stable, are in a large circulating system, and not exposed to contamination or extremes of temperatures, and therefore have very long lifetimes often measured in decades.

Industrial Gear Lubricants

These vary widely depending on the application. For straight, helical, and bevel gears, which have little sliding between the teeth, "straight" oils can be used which consist of good quality base stocks containing just anti-oxidant and anti-rust additives. For highly loaded applications, such gear oils can be compounded with sulfur or sulfurized fats to improve their load-carrying ability. Lead naphthenate, once widely used, is not now considered a desirable constituent.

Hypoid gears, such as are found in the rear axles of vehicles, are not normally encountered in industrial use, although if gearing has offset shafts which do not intersect, then a degree of tooth sliding will be present. In such cases automotive-type gear oil additives containing sulfur/phosphorus EP agents will be used.

Worm gears are quite common in industrial applications, and with their high level of sliding between the components can generate large amounts of heat if lubrication is inadequate. To overcome local heating, worm gear lubricants have a high level of anti-oxidant treatment, and are compounded with fatty oils or sulfurized fats to provide lubricity and reduce the temperature generated at the teeth surfaces. Use of more powerful EP agents is not normally possible, because these would break down under the influence of the local heating and would cause heavy chemical wear of the teeth, and corrosion of the worm gear if this is made of bronze as is commonly the case.

For difficult cases of gear lubrication, especially for small gearboxes, synthetic oils can be used and polyglycol ethers are a common choice.

Greases are also widely used in small gearboxes. They stay in position and minimize leakage so that continuous lubrication is not required. They can have EP properties or contain solid lubricants such as molybdenum disulfide, and can be based on synthetic as well as mineral lubricants. The gear or equipment manufacturer therefore has a wide choice of products to choose from in specifying a suitable lubricant for his equipment.

Metal-Working Oils

These are used for processes such as drilling, turning, milling, and grinding. A lubricant is required both to lubricate the sliding of the tool against the workpiece and the removed chip, and also for the extraction of the large amounts of heat generated. In many operations quite intense frictional heat is generated, and to avoid tool damage and workpiece distortion it is necessary to remove this heat as efficiently as possible. Water is much more effective at removing heat than oil, although it does not have the same lubricating ability. The most common metal-working oils are therefore emulsions of oil in water, the water serving to remove the heat, and the oil providing lubricity between tool and workpiece. Under the intense heating at the cutting edge the emulsion tends to break down leaving the oil behind, particularly if the water is flashed-off by the heat.

Emulsions come in several types, of which the white milky emulsions are the most common. These have relatively large oil droplet sizes, or have relatively high oil contents. The so-called transparent or more correctly translucent fluids contain very finely divided oil droplets, and are more stable and consistent in performance than the milky emulsions.

These emulsions are produced by mixing so-called soluble oils with water. The water:soluble oil ratio for the milky type may range from approximately 10:1 to 50:1. The translucent type of cutting oil contains a much higher level of emulsifier, and the dilution ratio can be as much as 70:1 for operations such as grinding, where heat removal is the most important consideration.

The soluble oils, more correctly known as emulsifiable oils, can be quite complex packages, and contain much more than just base oil.

Depending on the sophistication of the formulation, the actual base oil content can range from as little as 20% of the emulsifiable oil to around

90%. Originally naphthenic base oils were frequently employed because of their easier emulsification, but increasingly the more widely available paraffinic stocks are taking their place. For difficult applications synthetic stocks can also be used, but their high cost makes them suitable only for specific operations.

Within the emulsifiable oil is an emulsifier package chosen from among the following types, depending on the base oil to be emulsified and the other additives which will be present in the finished oil:

Emulsifier Type	Example
Anionic (typically sodium salts)	Sulfonates Carboxylates Sarcosines (acylated amido-carboxylates) Phosphates
Cationic	Amine Salts Imidazoline Salts
Emulsifier Type	**Example**
Non-Ionic	Polyglycol Ethers Esters of Polyhydric Alcohols Alkyl Phenols

The sodium sulfonates and the fatty acid sodium salts are both cheap and make excellent general-purpose emulsifiers. By modifying their chain lengths a balance can be struck between emulsifier characteristics and anti-rust performance. Other emulsifiers are chosen if considered more suitable for the oils to be emulsified, or if the highly alkaline nature of the cheaper products would cause problems of workpiece staining or corrosion. An important factor is also the so-called hydrophilic/lipophilic balance of an emulsifier in relation to the mixture to be emulsified.

While many emulsifiers have anti-corrosion properties, in many cases it is necessary to boost these to meet the highest performance standards. Typical additives used to improve anti-rust and other anti-corrosion properties are as follows:

Sodium nitrite (now declining due to toxicity fears)
Borax
Boric amides and esters
Dicarboxylic acids and esters
Alkanolamines and alkanolamides

Another requirement for a cutting oil is the control of bacterial growth. Because bacteria grow well in uninhibited oil and water emulsions, a bactericide is normally added to formulations to prevent this from occurring and causing smells and deterioration of the emulsion. Some formulations, particularly those containing boron additives, do not encourage bacterial growth and a specific bactericide may not be needed.

EP agents can be added to the emulsifiable oil to provide improved lubricity and load carrying of the oil in the package, which normally results in improved surface finish of the workpiece and longer tool life. Typical EP agents are sulfurized hydrocarbons or sulfurized fats and fatty oils.

Many of the additives required to be emulsified are in themselves water rather than oil soluble. To combine them in the emulsifiable oil package a proportion of water is required which may be as high as 50% of the total. To combine this with the base oil and the emulsifier system and other additives a coupling agent such as alcohol or glycol may be needed adding further to the non-oil content of the package. Thus if a so-called emulsifiable oil is mixed with water in a ratio of 50:1, then the actual amount of base oil dispersed in the resulting emulsion may be at a ratio as low as perhaps 200:1. The actual ratio is unimportant as long as the mixture of oil, water, and additives performs well.

Low dilution (high oil content) and EP emulsions have extended the range of operations for which emulsified oils are suitable. However, for the most severe operations, more lubricity is needed and “straight” cutting oils, often of the EP type, have to be employed. The EP agents range from dissolved sulfur for ferrous metal operations, through sulfurized fats and fatty oils, to chlorinated oils and waxes. The following is a list of various types of metal-working operations in the order of increasing severity:

Grinding
Sawing
Simple turning
Planing
Drilling
Milling
Tapping
Broaching

As well as the type of operation, the metal being machined is also of critical importance. Good machinability of a material results from a combination of several factors including the hardness, the work-hardening tendency, the reactivity with cutting fluid and its additives, and the thermal conductivity of the metal. Some metal alloys are specifically formulated to improve their machinability. For the most difficult to machine materials, tool life and service finish is very dependent on the correct choice of cutting fluid. A balance has to be struck between the need for heat removal, which favors water-based fluids, and lubricity, which favors the straight oils with EP additives. For some really difficult cases special water-based solutions may be employed for maximum heat removal, and these may contain active chemical agents to provide EP properties and also necessary anti-corrosion inhibitors. The use of neat chlorinated solvents has been seen in the past, but these are now not considered for open factory operations because of their health hazards.

An abbreviated list of metals which are frequently machined is given below starting with the most difficult and proceeding in order of improving machinability:

Titanium alloys
Nickel alloys
Stainless steel
Alloy steels
Carbon steel
Copper
Bronze

Cast Iron
Aluminum alloys
Brass
Zinc alloys
Magnesium alloys

The above is only an approximate guide, as different alloys of the same metal can have different machining characteristics.

Pressing, Drawing and Stamping Lubricants

This refers to the production of shaped articles by cold forming of sheet materials. For pressing and shallow forming, metal deformation is relatively small and movement of the metal against the punch and die is limited. In such cases a light coating of oil (often the preservative oil already on the sheet) is sufficient.

As the depth of draw increases, the amount of metal deformation increases, as does the heat generated in the process. A lubricant must aid the movement of metal against the punch and die surfaces, preventing any tendency to weld or stick to either. It must withstand the high temperatures generated and the pressures applied, and not react excessively with the metal of the press parts or workpiece. Depending on the severity of the operation, the following may be used as lubricants:

Emulsions
Mineral oils (incl. EP)
Fatty oils
Waxes
Polymer solutions
Liquid polymers
Sheet polymers

For oil-based products, the viscosity is important and use of additives can improve drawing characteristics and surface finish. As for most industrial oils, considerable technology exists which would be out of place to discuss here in detail.

References

1. Kataoka, R., "Prospects for Two-Stroke Engine Development for Cars," JSAE No. 9431049, 1994.
2. Duret, P., *et al.*, "A New Two-Stroke Engine with Compressed-Air Assisted Fuel Injection for High Efficiency Low Emissions Applications," SAE Paper No. 880176, Society of Automotive Engineers, Warrendale, Pa., 1988.
3. Duret, P., Moreau, J-F., "Reduction of Pollutant Emissions of the IAPAC Two-Stroke Engine with Compressed Air Assisted Fuel Injection," SAE Paper No. 900801, Society of Automotive Engineers, Warrendale, Pa., 1990.
4. Plohberger, D., *et al.*, "Development of Fuel Injected Two-Stroke Gasoline Engine," SAE Paper No. 880170, Society of Automotive Engineers, Warrendale, Pa., 1988.
5. Hosonuma, K., *et al.*, "Effect of Base Oil on Two-Cycle Engine Oil Performance," SAE Paper No. 911275, Society of Automotive Engineers, Warrendale, Pa., 1991.
6. Bristow, J., "Modern Two-Stroke Lubricants," presented by Goebels, K., at the Esslingen Technical Academy meeting, Stuttgart, 1985.

SAE Special Publications SP-849, SP-883, SP-901,and SP-942 on two-stroke developments will also be of interest.

Chapter 9

Blending, Storage, Purchase and Use

In the earlier chapters we have concentrated on the constituents of various types of lubricant, how these relate to the quality of the oil, and how this quality is measured and controlled. We will now look at how lubricants are produced and important aspects of purchasing and usage. Each topic has its own complexities, but we will try to give sufficient information to aid understanding, without getting lost in too much excessive detail.

Before an oil can be blended for sale, the composition must be fixed, and this is discussed first. For production of oils of consistent good quality a supply of consistent raw materials must be arranged, and this is dealt with before actual blending methods are reviewed.

Several parties or organizations will be involved in the production and marketing of an oil, and we will be referring to the following:

<u>The oil company</u>: The organization that will decide on the oil's composition, offer the product for sale, and whose brand name or logo normally will appear on the packaging.

<u>The oil marketer</u>: May be the same as above, or a re-seller such as a parts supplier, supermarket or store, or an OEM. Re-branding of purchased finished oil imposes greater responsibilities on a retailer, but seldom implies major control over compositional decisions.

<u>The lubricant blender</u>: The organization in whose plant the base oils and additives are mixed to produce the finished oil ready for sale. The blender will test the oil to check that the mixture is correct, but would not normally

do performance tests. He may or may not package the lubricant for the marketer.

The base stock supplier(s): The provider of the untreated base grades of lubricating oil including unconventional materials.

The additive supplier(s): The petrochemical company or companies supplying the additives which provide the required performance characteristics of the oil.

The oil formulator: The organization that decides the exact composition of the oil in terms of base stocks and additives to be used and their proportions in order to provide the required performance. Normally a great deal of testing will have been done before the composition is finalized. The formulator will normally be either an oil company or an additive supplier.

Several or even all of the above may be parts of a single company or corporation, but in practice they will often operate on an "arm's length" basis. Often they will also perform the same functions for other non-corporate outside organizations.

9.1 Deciding on Oil Composition

It is most important that the oil marketer decides at the outset exactly the type and quality of oil that it wishes to market. Assuming that this is a retail crankcase oil the following quality, specification and performance requirements need to be decided:

- Type of oil—gasoline, diesel, or multi-purpose?
- Market sector targeted—passenger cars, commercial vehicles, or both?
- If passenger cars, is there a significant diesel segment?
- If commercial, does this include large trucks and off-road equipment?
- Quality levels (e.g., API SH, API CF-II, CCMC PD2)
- Are specific OEM approvals or API licensing required?
- What viscosity grades?

- Any interest in factory-fill supply? (will need a specific approval)
- Any interest in supply to military authorities? Will approvals be needed?
- Will base stocks be purchased on a long-term or "spot" basis? Are they "in-house"?
- Any particular cost restraints?

Having decided on the initial answers to the above questions, the next step is for the oil company (responsible for the oil composition) to discuss the proposed oil with one or more additive suppliers. An additive supplier will probably already have additive packages suitable for formulating the oil required and can advise on the type of approvals or qualifications instantly available in relation to the use of specified base stocks. It is important to realize that if a specific approval is required whether it be for an OEM, for API licensing, or for a military authority, then either base stocks which already have approval with the same additive package must be used or an expensive new oil approval program will have to be run on the new base stock and additive combination. Approvals can therefore depend on relationships between oil company and the supplier of the base stocks. If the lubricant company blending the oil manufactures its own base stocks, and require these to be used, then a new approval program is very likely to be needed, although in some cases a "package approval" may be granted based on the approval authority's available evidence of the performance of the additive in other base stocks.

Some OEMs approve additive packages and are familiar with a wide range of commercial base stocks, and will approve what they consider to be a reasonable combination of the two without further testing. Military authorities generally approve specific additive plus base stock combinations, but a base stock purchaser can often obtain a "reblend" approval if the original approved oil manufacturer agrees (often the base stock supplier).

If no specific approval is to be obtained for the new oil then the oil marketer (oil company or re-branding retailer) must assure himself that the oil which is blended is of the quality claimed in the marketplace. Again, the additive supplier will be able to provide data validating the quality of his additive package, and be able to suggest the quality claims which could be made for an oil formulated with a chosen combination of base stock and additive package. The desired quality claims on packaging and advertising copy must be considered carefully in relation to the approved or licensed status of the oil. In their Engine Oil Licensing and Certification System (EOLCS), the API suggests some forms of words which may be used to describe an SG oil which is not licensed or approved, provided the manufacturer is confident it is of SG quality. A common form of words in Europe is "This oil meets or exceeds the requirements of API . . . ."

It is important to note first that there is no mention of approval or licensing in such claims for the oil, but the claims of quality level can be quite specific. The oil marketing company must at all times be in a position to defend itself against misrepresentation suits by any local standards or other authorities, and should therefore seek key demonstration tests (in the base oils to be used) from the additive supplier. The latter should present a case for the quality of the oil to the lubricant marketer in a similar manner to the case which would be presented to an OEM for an official approval. If standard U.S. passenger car engine tests are involved, then the requirements of the API Engine Oil Licensing and Certification System and the CMA Code of Practice (see Section 6.3.1.5) will need to be followed.

This type of claim language would apply to other quality levels, and would be generally accepted in most marketing areas with the onus on the oil marketer to substantiate the quality if called upon to do so.

It may be desired to market different viscosity grades of the same brand, in which case they should be treated as different oils, although approval requirements for supplementary viscosity grades may be simplified. Formal licensing procedures must be followed if the API Service Symbol or Certification Marks are to be displayed on the product packaging.

In suitable circumstances, the oil company may conduct its own formulation or approvals using purchased additives (possibly from an in-house affiliate).

When the composition of each of the viscosity grades of oil to be produced to the chosen quality level has been agreed on, then the oil company should prepare manufacturing specifications for the oils so that these can be passed on to the blending plant. The additive company can probably again be of considerable assistance in this.

To ensure consistent oil quality, both the manufacturing specification for the finished oil and the purchasing specifications for the components must be adequate.

9.2 Purchasing the Components

9.2.1 Component Specifications

As well as financial and stock drawing arrangements, the purchasing agreements should include specifications for all the components to be used in the blends, with provision of samples of each material for reference purposes. This applies to all base oil and additive suppliers. It may be that viscosity modifier, pour depressant and performance package additives are supplied by different additive companies and if the oil has a part synthetic content then this will probably be supplied by a different company from the main base stock provider. The reference samples are very important and should be stored in a cool dark place. They are used for monitoring future deliveries of the materials, and as reference materials in any dispute with the supplier. If a laboratory infrared spectrophotometer is available then infrared traces should be obtained from each sample. Infrared spectra are commonly used as "fingerprints" by which commercial deliveries of both additives and finished lubricants can be compared with reference samples. (In large organizations with sophisticated laboratories, the use of Fourier Transform infrared spectrophotometers with built-in computers permits automated comparison of spectra.)

Proper purchase/supply specifications must be drawn up with all component suppliers. For base oils the following is a representative list of tests which might be found in a typical purchase specification:

- Appearance (clear and bright)
- Density
- Color
- Viscosity

- Viscosity index
- Flash point
- Pour point
- Ash content

Some description of base stocks is also advisable, for example "high-V.I. solvent-refined base stocks from Middle-East crude." A change of crude source or type of refining is not normally acceptable, as approvals will be invalidated. Apart from the viscosity clause, most of the requirements in a base oil specification are simple maxima or minima.

Specifications for additives can pose greater difficulties, especially for a performance (detergent inhibitor) package. The specification typically would include the following requirements:

- Appearance
- Viscosity (typical at least, a range desirable)
- Elemental Contents:
  e.g.: calcium
  magnesium
  phosphorus
  sulfur
  zinc
- Plus typical properties (density, flash point, pour point, etc.)

The declaration of performance quality is normally associated with the additive code or name, and is contained within the purchase contract rather than the purchase specification.

For viscosity modifiers and pour depressants minimum performance targets for a dilute blend in a specified test oil should be specified. There should be a general description of the purpose and quality level provided by the additive package, and guidelines on the correct handling of the materials should always be sought. Many additive packages require heating to reduce viscosity before they can be pumped into a blending vessel, and overheating may easily decompose the additive, possibly with the production of toxic fumes.

It is the additive supplier's responsibility to provide handling and storage guidance and information on hazards and toxicity of his products in some form of safety data package.

The agreement of specification limits for the elemental contents between the additive supplier and the lubricant blender may take a little time. Reference should be made to Section 3.4.2.1 in which there is some discussion of the statistics of specifications between supplier and purchaser. The supplier has limitations on the control of manufacturing which he can achieve, and his own testing will add additional variability to the measured quality of the material supplied. The lubricant blender also has testing variability, but also needs additional flexibility to allow for the inevitable quality variations of his own manufacturing. The finished oil must, however, be manufactured to a specification that ensures the finished oil is of the quality claimed, and hence the purchase specification between lubricant blender and additive supplier is squeezed by the output specifications of the two parties. A satisfactory agreement must be made if business is to proceed smoothly, and, if the two parties have not previously worked together, it may be necessary to have a period of cooperative sample exchange and analysis in order to arrive at common ground on the analysis of the elements of the package. Experience has indicated that this can be done, but considerable time may elapse before such discussions are successfully concluded and a satisfactory specification is developed.

When developing the specification for the oil to be marketed, which is an essential early step, consideration should be given to the type of limits and specification ranges which the approving authorities require for their approved oils. As an example, for the elemental content of lubricants the U.S. military requires limits of –10% to +20% of the declared value for the approved oil, while the API oil licensing scheme sets limits of –10% to +15% for elemental contents over 0.01%. (These limits may initially appear generous, but analytical precision is not good at lower elemental levels, and when blending tolerances are considered, these ranges will severely restrict additive supplier limits.)

Finally it is worth mentioning that if several additive packages are being obtained from one supplier for the formulation of several types of oil, it is sometimes possible to arrange for a large percentage of the additives to be supplied in the form of a single part-package. This may permit delivery of a large proportion of the additives supply in bulk with consequent cost savings, the remainder being provided in drums to boost the base package to the necessary quality levels for each oil. The viability of this approach depends not only on the storage facilities at the oil blending plant, but also on the specific composition of the various additive packages being supplied. This system would not be applicable if there were different additive suppliers involved, and it is not always possible even if the different formulations are provided by the same supplier.

9.2.2 Checking Incoming Materials

The amount of inspection depends on the source of the raw material and how it is delivered. For example, base oil delivered by pipeline from an adjacent base oil manufacturing plant and owned by the same organization may require very little checking, but base oils supplied by sea will require careful inspection to ensure that they are the correct materials and that they have arrived uncontaminated. Additives pose particular problems because they are relatively difficult to test and analyze, yet the supplier may be remote and not in a close relationship with the blender. Supply may be via an agent, and the product could have spent some time in storage, possibly not under ideal conditions.

The quality control process starts with the agreement of specifications for the purchased products as discussed in Section 9.1. When these were agreed, lines of communication for rapid discussion of quality problems concerning incoming material should have been set up, and samples of representative on-grade material obtained for reference purposes. More than a few millilitres of each product is required, for these samples will be used to visually check incoming material and also as references when there is some doubt about quality. Suppliers should also be asked for their own certificates of analysis to be supplied with the material when it is delivered, so that their inspections can be compared to the agreed specifications and to test results as they are obtained in the blending plant laboratory.

Tests on base oil samples should follow those given in specifications (see Section 9.2.1) but not every test may need to be done. The following may be considered:

- Appearance
- Viscosity
- Flash Point
- Pour Point
- Viscosity Index
- Specific Gravity/Density
- Ash Content

The above tests are ranked in order of urgency and in terms of the information that they may give. Appearance may immediately indicate if the oil being received is very different from the sample which was provided by the supplier, which should be viewed alongside the new material. If the oil is noticeably hazy then it is possible that it is contaminated, possibly by water. If water is suspected a "crackle" test can be performed, but this is not an entirely reliable indication of water content. The viscosity will indicate if the correct grade has been supplied, and the flash point will tell if the material has been contaminated by light products during shipping or pipelining operations. The pour point may vary somewhat between shipments but should always be in the same area; it can be useful for distinguishing between paraffinic and naphthenic stocks. The viscosity index will likewise confirm whether or not a specified quality of high-V.I. base stock has been supplied, and the density can indicate if the base stock has been made from a different crude than was originally the case. Finally ash content is another measure of possible contamination, possibly by additive materials. In most cases contamination by additives or fuel products would be detected by the appearance test. It is suggested that the first three tests be performed on material during delivery, and the full spectrum of tests is performed on a full tank of oil after delivery has taken place. Of course if any adverse indications are seen, pumping should stop immediately until the quality has been investigated further and the cause of the discrepancy ascertained. Initial delivery into a holding tank is desirable, with product being transferred to the main bulk tank after testing.

Additives may be delivered in bulk or in drums, and if in bulk preliminary tests should be performed on the delivering tanker (sea, rail or road) before any material is off-loaded. With the cargo should travel samples from the additive plant as the material was loaded, and these and material from the delivery should be compared with the stock reference samples. A wide variety of additives can be distinguished visually by color and apparent visual viscosity when viewed in a glass container. The key tests on additives are as follows:

- Appearance
- Viscosity
- Elemental Contents/Sulfated Ash
- H_2S

Infrared absorption is a more sophisticated test than visual appearance, and can give much information about consistency of additive package composition. However, an agreement between supplier and customer on how results will be interpreted is most desirable.

As well as the possible supply of an incorrect grade, appearance will also indicate whether the additive has been contaminated by water, in which case it may appear milky and form a noticeable sediment on standing. Viscosity will vary somewhat from batch to batch of the same material, but should lie within a reasonable band for correctly manufactured material. The key parameters of an additive are usually the elemental contents such as calcium, magnesium, phosphorus, zinc, sulfur, etc. These are best measured individually by such methods as atomic absorption, ICP spectroscopy, or X-ray fluorescence, but if such equipment is not available then a carefully performed sulfated ash content will normally provide sufficient reassurance.

Additives such as viscosity modifiers and pour point depressants will have to be checked by dissolving a small proportion in a base oil and measuring the viscosity and pour point of the blends. In relation to supply specification and discussion of potency of such products with the supplier, it will be necessary to use a standardized test oil for making the blends and not stock base oil from tankage.

Additives will in most cases be supplied in drums, and it is essential that at all times these are stored correctly. Additive manufacturers will give guidance on drum storage and also on the storage of additives in bulk. Drums should not be stored on their ends because of the possibility of water collecting in the drum head around the bung, and detergent additives in particular must be stored so that they do not become overheated. Performance package additives stored in dark drums in tropical sunshine are very likely to decompose or at least lose their efficiency.

Drums, like other packages, are normally sampled by the statistical rule which says that the number of packages examined must be equal to the cube root of the total. In this connection it should be noted that every batch of material contained within one delivery should be treated as different material. When drums are opened for sampling, note any pungent odors from the barrels particularly if these are unusual. DO NOT SNIFF AT THE BUNGHOLE AFTER REMOVING THE BUNG. If the additive has partially decomposed due to overheating it may be evolving hydrogen sulfide (H_2S) as well as malodorous compounds, and should be treated as hazardous. On all deliveries of detergent inhibitors or performance packages it is worthwhile to run a simple test for H_2S liberation in the laboratory and compare this with results obtained on the standard samples. To do this a small amount of additive is placed in a flask and put in a boiling water bath, and a filter paper moistened with lead acetate solution is placed on top. If the additive is decomposing, the paper will rapidly blacken. Some additives may give slight coloration after five minutes and this can be ignored.

Checks on Synthetic Base Stocks

These can be very similar to the checks on petroleum base oils, but the specific gravity is probably considerably more useful in the case of synthetics as a means of identifying that the correct product has been delivered. For some products a saponification number is a measure of quality, and the suppliers will be able to suggest other characterizing tests.

9.3 Oil Blending

The process of producing finished lubricants from base oils and additives is invariably described as oil blending rather than oil manufacture because there is no significant chemical reaction which takes place and it is a simple mixing operation. There are two significantly different types of process which can be used for this mixing operation: batch or continuous in-line blending, with considerable variations between plants of each type. Small plants invariably use batch blending, but for high-volume production either automated batch blending or continuous blending can be used.

9.3.1 Batch Blending

This is the most common type of process in which ingredients are placed in a large vessel or blending tank where they are mixed until homogeneous, then are transferred to intermediate storage and a new batch is commenced. Figure 93 is a simplified flow diagram of a batch blending operation.

A common design of batch blending plant in the past was the so-called "vertical" design in which the blending vessel is located on a floor above the main holding tankage, into which the product can flow by gravity. Above the blending vessel is another floor from which raw materials are loaded into the blending vessel. This type of plant requires a strong and relatively expensive building to contain the blending vessel, and there is often little space on the upper floor to contain materials such as additives, all of which have to be raised to this loading floor. With improvements to the capacity and reliability of pumps, the horizontal type of plant is now favored, provided that there is sufficient space to accommodate it economically. In this type of plant materials are pumped laterally from one place to another, and working space is less restricted. In all plants where lines are used for more than one product, provision for line flushing or pigging must be provided.

In either case base oil supply is required into the top of the blending vessel, usually by a dedicated pipeline from a bulk storage tank. Blending agents and additives are normally provided from bulk supply only in large and particularly in continuous plants, and normally will be taken from drums.

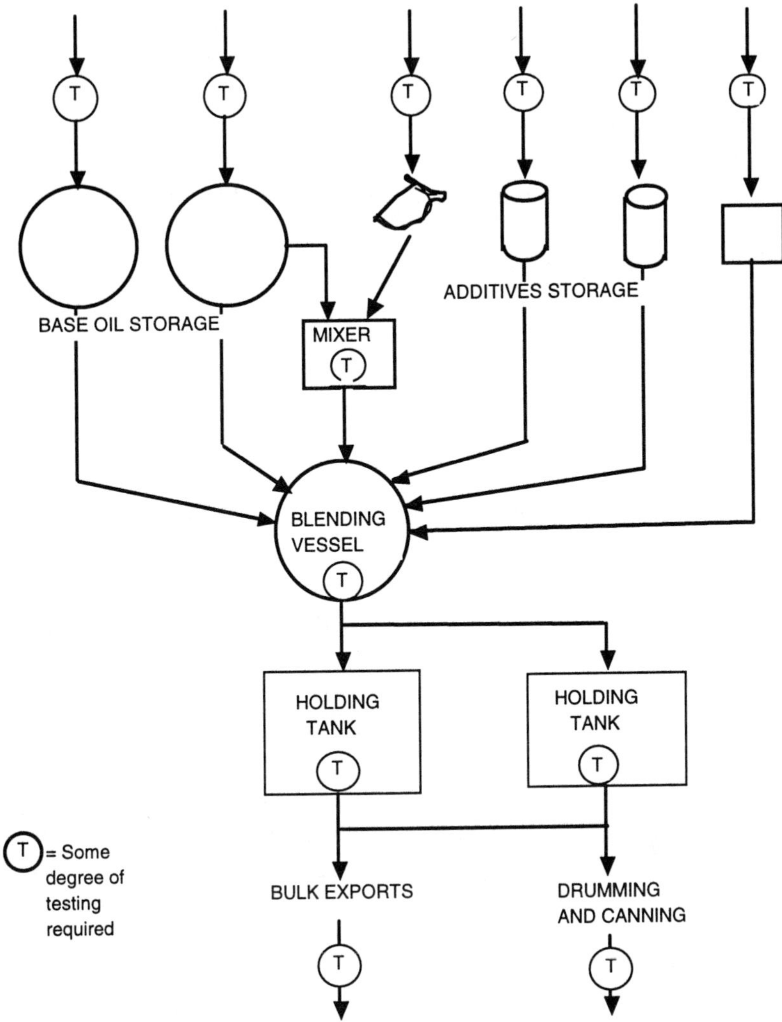

Fig. 93 Batch blending of lubricating oil.

This can be by a small transfer pump, or the barrels can be emptied into a floor tank or pit and then pumped into the blending vessel. For some types of oil, solid additives are used and these will be unloaded from bags into hoppers above the blending vessel. Some viscosity modifiers are available as rubber crumb in which case this would be treated in the same manner, but care must be taken to avoid agglomeration. For piped products in cooler climates outflow or suction heaters on the bulk storage tanks are desirable, and the tanks themselves may be required to be heated to keep the contents at a reasonable temperature. In the case of additives it is important that only low-pressure stream, hot water or hot oil is used for heating, as excessive temperatures can lead to decomposition of many additives. Heating coils must always be sound and inspected from time to time because water can rapidly accelerate additive decomposition, and if carried from a base oil tank into the blending vessel can ruin a blend.

Hot oil is increasingly being used in heating coils to eliminate internal corrosion and to minimize product loss if leaks do occur. In the case of drummed additives, it will be necessary to provide either a drum oven or electrically heated jackets which can be used to heat the contents of the drums to a temperature at which they will flow easily. Again care must be taken particularly with heating jackets not to cause local overheating of the additives.

To make a blend, in the case of simple plants, about 80% of the base oils are added to the blending vessel which must be clean or contain only a heel of a similar oil from the previous blend. The base oils are heated to around 60°C and then the additives are added. If there are no very viscous additives the temperature can be lower and these may be added in the order of increasing viscosity, the blend being agitated throughout the addition process. In the case of some viscous viscosity modifiers, it may be desirable to heat the base oil to approximately 80°C, add the viscosity modifier and mix well, and then cool the batch before adding the more temperature-sensitive detergents. For some sensitive blends (such as automatic transmission fluids) there may be a preferred order of component addition at specified temperature maxima, and in these cases the advice of the additive supplier should be followed. In all cases great care must be taken not to overheat additive components.

The measurement of the components can be either by volume or by mass/weight. Weight is now generally used, with the blending vessel on load cells or alternatively the base oils may be measured by hydrostatic gauge and the additives weighed in by difference on a drum scale. The proportions of additives and base oils are of course different depending on whether the measurements are to be by volume or by weight, and it is important that the correct figures are used.

Mass flow meters (Coriolis meters) are becoming available which can measure mass flow through a pipe, in which case proportions are as for weight.

It is important that the ingredients are thoroughly mixed together and that the blend is uniform. The simplest method, applicable to medium-sized blends, is to agitate the blend by blowing in compressed air through a perforated pipe, sparge ring, or jet nozzle at the bottom of the tank. A more sophisticated version of this is the "Pulsair" system in which a programmed series of air bubbles provides efficient agitation. The most common method is to have some form of mechanical agitator which is usually some type of propeller mixer. For making small blends of solid additives in oil, a high-speed turbine mixer is often employed. Mixing can be assisted by a pump-around system which means that product is withdrawn from the side or bottom of the tank and pumped back into the top, providing a constant circulation of product. A variation of this, which is an efficient blending system in its own right when set up correctly, is to pump the material back through a jet nozzle placed in the bottom of the tank which agitates the product vigorously when the pump-around system is switched on.

The length of time required for mixing will depend on the agitation system and the viscosity of the blend, and will usually be judged by previous experience. For new and possibly difficult blends it is advisable to turn off the agitation and, after the product has settled, to compare the appearance and possibly the viscosity of top and bottom samples from the blending vessel. At this stage the blend is not completed because not all of the base oil may have been added at the beginning of the process. Assuming that the full amount of each additive required to produce the full volume of oil was programmed, it will be necessary to add the remaining base oil. The reason for holding this back is that the viscosity of the finished blend must lie within certain limits, and there will always be differences in individual batches on both base oil and additives in terms of their viscosities. It is therefore necessary to measure the viscosity of the semi-finished blend and to calculate the adjustments needed in terms of heavy and light base oils to meet the target viscosities. Calculated correction amounts of base oil can be mixed in fairly rapidly, after which the blend viscosity should be checked again and possibly a last adjustment made. The amount of adjustment necessary, and therefore the quantity of oil held back, can be minimized by experience and by a knowledge of the actual viscosities of the components being used in the blend and how these differ from nominal values.

When it is sure that the blend is homogeneous and on-grade for viscosity, if it cannot be held in the blending vessel it should be transferred to a holding tank from which samples are sent to the laboratory for full release testing. If this testing indicates that the blend is either too rich or too lean in additive content, then at some later stage it will have to be pumped back into the blending vessel and the blend adjusted once again. If there is a large holding tank and it is required to consolidate several blend batches into this tank, then they should all be blended to meet the specification, but space left in the tank for a final adjustment batch. The holding tank is mixed before testing, and if this is found to be off-grade then the final batch should be blended to correct the contents of the holding tank and bring the whole tank on specification.

When the contents of the holding tank are released as being on specification, then the lubricant can be passed to the filling line where after passing through a filter, drums or cans are filled with the product. Of course it is important that the product does not get contaminated in the filling system, and therefore this should be thoroughly drained between products and preferably pigged or flushed with a small quantity of base oil. Drainings from washings of lines and the blending vessel and other tankage are often used as fuel, but it is possible with care to segregate them and incorporate them into future blends of the same type. Most motor oils can be considered the same type of oil, and these must be separated from gear oils and from industrial oils. In many blending plants separate blending vessels will be used for exclusively blending certain types of oil in order to avoid cross-contamination.

Quality control of samples from the filling line is usually at the discretion of the chief chemist or quality controller, and with a correctly operated system a visual check may be all that is necessary to ensure that the correct grade of material is being filled into the drums or cans and that these are correctly labeled for the grade involved. With a single filling line and manufacture of different types of oil it may be necessary to check the initial material from the filling head for contamination before the product is filled into packages for sale. Quality control is discussed more fully in Section 9.4.

9.3.2 Automated and In-line Blending

The above described the operation of a typical manually operated blending plant. For production of large quantities of a relatively few grades of oil, it is possible to automate this process by having all the mixing vehicles on load cells and providing all the products to be used as blend components in the form of fluids which can be charged by pipeline. This may mean that the more viscous additives need cutting back with suitable base oil either by the additive supplier or on site.

Because the number of incoming pipelines is limited, in the case of complex blends, such as when oils are being blended from component additives rather than packages, it may be necessary to combine some additive streams by pre-mixing. The addition of both base oils and additives to the mixing vessels can be computer-controlled and a continuous stream of consistently on-grade product produced.

A more elaborate scheme has used small pre-blend vessels like robots, which move around tracks to different filling heads, their contents being mixed together either by tank mixing or by in-line homogenization. The whole operation is run by a computer.

Another type of automated plant is the proportional or in-line blender. This is used by many oil companies for producing the very large volume of motor oils and certain industrial oils which constitute a major part of their business. As for the automated batch blending process, all components have to be available as non-viscous liquids, which frequently means cutting back on site with a base oil which is a component of the blend. This may be done in small batch-type blenders. A limited number of components are metered into a mixing line with their rates of flow adjusted so that all products are continually metered in quantities designed to finish up with the correct blend mixture. This of course is achieved by computer control. In the mixing line a vortex is created and the various input streams are mixed so that they emerge at the end of the line as a virtually homogeneous blend. This is pumped to a holding tank where usually a final mix is performed by a pump-around system, and the product is sent for quality control testing. For a complex blend, if there is not a sufficient number of entry ports into the mixing line for all the components of a blend, then some of these will be combined in the pre-blend tankage used for cutting back the viscous additives, so that all the components can be continuously and simultaneously charged.

The system is ideal for a relatively small number of very large-volume blends, and once set up can produce continuously on-grade product. As relatively high mixing energy is applied to blends, checks should be made that any polymers used do not suffer unacceptable levels of shear breakdown.

9.4 Quality Control

If on-grade raw materials are utilized, then production control consists merely of seeing that the correct amounts of the raw materials have been correctly mixed together. The base oils have to be in the correct proportions, the correct percentages of viscosity modifier and pour depressant additives have to be incorporated, and the right amount of detergent or other additives have to be included so that the oil meets its performance targets.

Each batch or consolidation of batches should be fully tested in the case of manually operated plants, while for automated blending where accuracy has been fully demonstrated, statistical sampling may be employed. For any type of blending, particular vigilance must be observed after a change of oil grade or type.

For single- and multigrade crankcase oils the following are recommended:

Single-Grade Oil	Multigrade Oil
Viscosity (100°C)	Viscosity (100°C)
Pour point	Pour point
Elements or ash	Elements or ash
	Cold cranking simulator or MRV

The viscosity is obviously a very important property, but it must be remembered that for the multigrade oil a low viscosity can either be due to a misblend of the base oils or to an insufficiency of viscosity modifier. The pour point can normally be adjusted by extra additions of pour point depressant if it is initially deficient. To determine if the correct amount of detergent additive has been added, a carefully performed sulfated ash is acceptable although analytical determination of at least one of the elements (calcium, magnesium, etc.) is desirable. For the multigrade oil it is desirable that some form of low-temperature test other than the pour point is performed. Low-temperature specification testing is very time consuming but consideration should be given to setting up a Cold Cranking Simulator or MRV apparatus. (Strictly, the CCS is a requirement for certifying SAE W grade conformity.) Another possibility is to have a non-specification test such as kinematic viscosity at –20°C, compared to a reference blend which has been made in the laboratory and thoroughly checked against the final oil specification. This latter reference blend is desirable anyway to use as a standard for the tests, and possibly for providing to a bulk purchaser of the oil. For additive content checks, side by side determinations of the sulfated ash on a reference oil as well as the oil being blended would be one way of increasing the value of this test.

For non-crankcase oils elements other than those identifying detergents will have to be checked. These would include zinc, phosphorus, sulfur, and possibly chlorine or other elements. Purchasers of finished lubricants may require the infrared spectrum to be consistent batch-to-batch. Checks on an infrared spectrophotometer may therefore be needed to ensure this is so, before oil is released for sale.

Due to the difficulties of test repeatability and standardization it is not recommended that laboratory performance tests (such as the four-ball EP tester) be used for production control purposes.

It is not proposed here to go into detail on control chart procedures for the production of lubricants, but this is always a worthwhile procedure. They should include features such as the results of first-time blends, and the numbers of reblends required to bring material into grade. Consistent errors such as undercharging of viscous additives can be identified and corrected for, and a more consistent product will be produced more efficiently.

Comprehensive literature exists on statistical quality control techniques.

9.5 External Monitoring Schemes

Oil companies and their blending plants should be aware that, as well as customers who purchase lubricants in bulk, other bodies may test lubricants which have been provided by the blending plant. In the future the most noticeable may be the Engine Oil Licensing and Certification Scheme (EOLCS) which has been set up to administer the award of the API license. Oils bearing the API symbol will be taken from the field in a random manner and analyzed for conformity with the original qualified composition. Both physical properties and elemental content will be examined, and guidelines are laid down as to the permitted variation before the supplier is called in to account for any differences. The SAE Oil Labeling and Assessment Program in which U.S.-marketed oils have been checked for apparent conformity to declared quality is due to finish but may continue in some form.

It has been suggested in some countries that trading standards officers may examine oils on the market and request justification of their quality claims from the oil blender and/or the additives supplier. Performance tests would not be envisaged, but they could check for physical properties and for the correct amount of the declared additive being present in the oils on the market. Such action has already taken place in cases where members of the public have complained about apparent inadequacies in the quality of lubricants which they have purchased.

9.6 Storage of Lubricants

Lubricants can be stored in tanks, barrels, small containers or cans. The requirements in each case are that the material should be protected from contamination, should not deteriorate in storage, and that safety, health and environmental matters are taken care of. For barrels and small containers it is important that the contents are easily identifiable from the markings, and that all grades are easily accessible for shipping purposes.

9.6.1 Bulk Storage in Tanks

Fixed-roof tanks of various sizes may be used to store lubricants. Before use they should be thoroughly cleaned, be scale free and may be coated internally with a proprietary protective coating that is of a type compatible with the lubricant to be stored. The tank should initially be dry, and filling arrangements must be such that ingress of water to the product is eliminated. Because base oils may contain a little water, or condensation may take place in moist climates with fluctuating temperatures, and in case of accidents, the bottom of the tank should be sloping and provided with a drain from which water and any bottom residues can be removed. For viscous oils and in cooler climates it will be necessary to provide some heating. For lighter oils an outflow (suction) heater is sufficient, but if bulk heating is necessary the heating coils must be sound and should be fed with hot oil or water preferably, or fed with only low-pressure or exhaust steam to prevent local overheating.

The tanks should stand on an impervious concrete platform and be bunded (provided with a catchment area for the product, should the tank leak). Watch must be kept for leaks at valves and flanges, and these should be eradicated as soon as possible. Lagging (insulation) of transfer lines is common, but the tanks themselves are not normally lagged for finished oils, but regular use of viscous base oils may justify lagging to save energy.

If horizontal-type tanks on piers are used, the same general remarks apply, with the tanks being slightly sloping toward a draw-off point for water and bottom samples. Where draw-off lines are common to different grades, a pigging system should be installed.

9.6.2 Barrel Storage

Very large volumes of product are stored in barrels, and this is the most difficult and potentially the most hazardous form of storage. Ideally, barrels should be stored on their sides and with the bungs below the liquid level. This is to prevent water from collecting in the tops of the rims and being drawn in as the barrels cool, and keeping the bung seals moistened with product guards against leakage. However, storing large numbers of barrels horizontally is hazardous unless a large amount of special-purpose racking is used, which can be extremely costly. Barrels stored one on top of the other should never be more than two barrels high, and the ends must be securely chocked. Products must be stacked so that access to the lower layer is not required until the first layer has been removed. In no cases should barrels be directly laid on the ground, but should be on battens or an impervious base. Provision should be made for stock rotation on as close to a "first-in/first-out" basis as can be achieved.

The safe storage period or "shelf-life" of products must be considered, and product should not be left for excessive periods at the bottom of a stack.

Barrels are frequently stored in fours on pallets. Full drums should normally be stored no more than two pallets high, although empty drums can be stored in higher piles if adequate care is taken in placing and removing the top layers. In particular, forklift trucks should have strong safety cabins and not be of the open type. Vertically stored drums must be protected from rain, and if they cannot be stored in a warehouse they should be sheeted over with tarpaulins or plastic wrapped. In hot climates the drums should be protected from direct sunlight by light-colored screens or roofing.

Leakage must be prevented as much as possible. If taps are used on barrels then drip trays must be provided and care taken that the taps are functioning properly and are shut-off after use. Small hand pumps are sometimes used to withdraw material from a barrel placed on end, but again care must be taken that this does not leave a trail of oil when removed from the barrel. To mop up accidental oil spills proprietary crystalline materials are available which are much less hazardous than the traditional sawdust. Care must be taken when removing barrels with forklift trucks that the barrels are not accidentally punctured, thereby producing both an environmental problem and a health risk. Barrels are often returnable or reusable and therefore care should be taken at all times not to damage them unduly in handling operations.

9.6.3 Cans and Small Packages

These should always be kept under cover in a warehouse or other suitable building. Small containers will normally be packed in multiples in cardboard containers. Where the stock is for in-house use, provision must be made for unpacking and disposing of cartons, and not allowing these to become oil soaked where they will present a hazard.

9.7 Purchasing Lubricants

In this section we do not propose to go into the mechanics of placing an order, handling the paperwork, or questions of payment and credit. We hope rather to provide useful comment on questions of the quality and the specifications for the goods being ordered, in the hope that these will provide useful guidelines for the technical staff to pass on to the purchasing function. There are many parallels between purchasing additives and purchasing finished oils, and reference will be made to Sections 9.1 and 9.2.

A purchaser is in a chain of organizations, starting with the providers of basic raw materials and proceeding through a series of operating companies which add value to the product until it is finally sold to the end user. Sometimes there is a degree of integration backwards from the point of sale, through various manufacturing operations, even as far as the raw materials. In such cases it may be that several steps in the production/

purchasing chain may share common technical advisors and laboratories, although an arm's-length approach is frequently adopted between oil retailing, oil blending, base oil manufacture, and additive manufacture, even if these are all contained within the organization of one large company. In many cases, however, no ties exist at all, and it is imperative that the purchasing and providing organizations have a common understanding of what is being requested in a technical sense so that acceptable material is delivered without misunderstandings or acrimony.

9.7.1 Quality Considerations

These differ depending on whether the purchaser is an actual user, or if the purchase is for resale. In this case, the considerations are those given in Section 9.1 for lube blenders, and the blending operation is in a sense a contracted operation.

For small users, particularly if equipment is under warranty, the recommendations of the equipment supplier should be followed. These may be for specific oil brands or for quality levels (API CD/SH, etc.) and viscosity grades. For equipment out of warranty, equivalent quality grades and viscosities can be obtained from the most advantageous supplier, provided that his quality claims for the lubricant are supported by suitable evidence or his reputation as a supplier.

The smallest users, with a few vehicles or other items of equipment to lubricate, will purchase oil in packages with a drum as the largest likely delivery. Testing in such cases is impractical, but even for the smallest user like the private car owner it is worthwhile to examine supplies for visual appearance and the way the oil pours from the container. By comparing with previous supplies, the rare possibility of an error of filling or labeling or of contamination can largely be eliminated.

For purchasers of oil in bulk, whether for resale or personal use, some level of quality testing is most desirable. Discussion of the quality to be supplied has taken place with the oil/blender/marketer. Evidence of the quality is not readily deduced from analysis of deliveries, so justification of quality claims and a specification of the oil to be supplied should be obtained at the time the contract is finalized, along with a reference sample of the oil. Deliveries should be checked against the supply specification either in full or by "short testing" for such properties as viscosity, flash point, and sulfated ash. Comparison should be made visually, and if available by infrared spectrophotometer, against the reference sample supplied at the time of signing the contract. More complete testing is justified for high-volume critical supplies, and where a sophisticated testing laboratory is available to the purchaser. The considerations are very similar to those given in Section 9.4 on quality control of blends at the blending plants. A balance has to be struck between insufficient checks for quality assurance and spending excessive time and money on tests which contribute little to quality assurance.

Needless to say, recourse should be made to the oil supplier immediately if significant deviations from the supply specification or the reference sample are noted. The supply specification must not be wider than that implicit for any oil approvals which are held by the oil, and preferably should be narrower to allow for testing variations between the blender and either the purchaser or a third party who may check the oil (see Section 9.2.1 for the analogous case of additive purchasing, and Section 9.5 on external monitoring).

For large-scale purchasers of oils in cans or drums, it will be impossible to check every package. Lot or batch numbers should be noted for each delivery and each batch sampled statistically (the cube root of the number of packages is a common procedure). If the tested material is satisfactory, a watch should still be kept for any visual anomalies as each package is opened.

9.8 Oil Use for Small Users

The lubricant used should follow the recommendations of the equipment supplier. In the case of older, worn equipment, it may be desirable to increase the higher-temperature (summer grade) viscosity to reduce oil consumption, but the winter grade recommendation should be adhered to, to avoid starting difficulties. Arguments to use lower-quality oil or extend drain periods should be resisted, but there may be no reason to use oil of higher than the minimum acceptable quality.

Oils of the same quality level from different suppliers may be used in an engine if this is unavoidable. At one time there were real fears of incompatibility between oils from different suppliers, but for many years oils approved by the U.S. military have had to be mutually compatible and have been formulated with broadly similar although not identical additives. Mixing of similar volumes of two or more different oils in bulk is still not to be recommended, but mixture in an engine should give no problems. Of course, the precise quality level cannot be guaranteed for a mixture, but should not differ substantially from its constituents.

Oil should be dispensed directly from small packages, or via clean cans or dispensers if purchased in drums. Care should be exercised to prevent cross contamination between oil for different purposes, and dispensing equipment for crankcase oils, transmissions oils, and hydraulic oils should be segregated.

Oil spillage and drips must be minimized, and if cotton waste or sawdust is used to mop up a spillage, this must be disposed of safely and rapidly, and not allowed to accumulate to cause a safety hazard.

Used oil, and particularly crankcase oil drainings, must be considered hazardous and disposed of to a used oil collection point. In no case should it be dumped on waste ground or into public sewers.

9.9 Use of Lubricants in Large Plants

In many factories such as automobile plants there are often hundreds of lubricants recommended by individual machine manufacturers for the lubrication of their equipment. Following these recommendations blindly can lead to both needless expense and to such complex lubrication procedures that the quality of maintenance actually suffers. It is almost always preferable to purchase a few different types of lubricant in large quantities, enjoying bulk delivery where possible, or volume discounts, and eliminating some of the risks of using the wrong lubricant in a machine. Obviously for new machinery the situation with regard to warranties needs to be considered, but in many cases the type of oil is specified rather than a particular brand, and much can always be done by negotiation.

Before lubricant rationalization can take place, a plant survey of the lubricating oils in use and the recommendations for each piece of equipment must be made by competent technical staff. These can either be from the user's own technical organization, or quite commonly from a lubricant supplier who wishes to obtain a major portion of the plant's supply. Requirements must be grouped by broad categories of lubricant, by industry standard categories or quality levels, and by viscosity grade. Any special restrictions or limitations for individual pieces of equipment must be carefully noted. With the availability of so many multi-purpose formulations both for crankcase and industrial oils, the division of the requirements into a few types of oil should be relatively easy, but there will be problems with regard to viscosity recommendations. If the viscosity recommendations are made in terms of a range of operating temperatures, then these should be related to the actual operating temperature of the equipment. Very often quite a wide tolerance is permitted. In cases where slow-speed, heavily loaded conditions apply it is better to err on the side of higher viscosity rather than lower, whereas if the main function of the lubricant is to remove heat, then lighter oils are preferable to heavier ones. In cases of doubt the equipment manufacturer should be consulted.

When the lubrication survey has been successfully completed, there should be relatively few major grades required, and a small selection of special products for requirements which cannot be rationalized. The major grades will normally be supplied in bulk by road tanker, and stored on site in tanks within the factory building. These tanks are much smaller than the large storage tanks at a lubricating blending plant, and are typically rectangular in design, mounted on piers, and slightly sloping. Internal heaters, usually of the electric immersion type, may be installed if the location is not normally held at the factory working temperature. Heating will also be required if any viscous oils are stocked so that these can be handled or pumped easily. Bulk oil temperatures should never exceed 70°C, and the surface temperature of the heaters should not be allowed to exceed 120°C. Some of the lubricants may be piped directly to the point of use or to an intermediate dispensing point. In many cases, however, it will be necessary to take quantities of lubricant from the main storage area to the machinery to be lubricated, and this is often done in the form of small hand trolleys containing tanks of oil with dispensing pumps which can transfer the oil into the machinery and collect used or waste oil if necessary. Equipment should be labeled with the maintenance schedules and the grades of lubricant to be used, and responsibilities carefully designated for the supply of lubricant and for the application to the machine. The total lubrication scheduling system should be controlled from manual display boards or a computerized system, ensuring that machinery receives the correct lubrication at the correct service intervals and that quantities used and stock levels are continuously monitored.

Handling procedures must at all times be designed to prevent contamination of the lubricants either with extraneous dirt or with other incompatible grades. In particular, contamination of turbine or hydraulic oils with traces of crankcase oil cannot be permitted, as this will adversely affect the water-shedding properties of these lubricants.

9.10 Complaints and Trouble-Shooting

Complaints can arise internally within a company, for example that a machine is not behaving well with the lubricant that has been supplied, or they can be external complaints by a customer to the supplier of the lubricant. While attitudes and approaches may differ somewhat, the basic rules for dealing with complaints is essentially the same in both cases. For maximum coverage we will treat the subject as though it is the latter form of complaint and look at some ground rules as to how the lubricant supplier can handle it.

9.10.1 Complaint Procedure

A policy for handling complaints should be developed and communicated to all personnel involved. Rapid action, replacement of material, and compensation for damage can often prevent the loss of a customer who has complained. However, too-ready acceptance of responsibility for a problem can become expensive, particularly if compensation for mechanical damage and loss of business is concerned. A compromise is to immediately replace any material which is suspected of being either contaminated or off-grade, but to resist suggestions that mechanical damage is directly associated with lubricant quality until an exhaustive investigation has taken place. Field workers investigating complaints must know precisely what offers they can make to an aggrieved customer.

It is most important that complaints be properly documented, and it is well worthwhile to develop special forms for the purpose. Details of oil grade, delivery date, batch number and customer storage conditions need to be ascertained, plus information on the engine or equipment type, duty, and age or condition. A sample of the oil which is the subject of a complaint needs to be obtained, and a sample of used oil if the complaint concerns used oil condition or equipment malfunction. The taking of samples requires particular care and attention. A representative average sample must be obtained, or specific top, middle and bottom samples in the case of non-homogeneous material. A copy of the equipment manufacturer's recommendations should also be obtained, or his contact address and telephone number.

9.10.2 Laboratory Examination of Samples

It is usually clear by this time whether the complaint is relatively trivial or if it has potentially serious implications. In the case of contaminated unused oil, the only testing necessary may be a visual examination without specific identification of any contamination or sediments. If the sample appears normal it should be compared with the retained sample which was taken at the blending plant before delivery. If they appear identical, correctness of grade can be further confirmed by tests such as TBN, sulfated ash, or metals analysis. If desired, contamination can be checked for by flash point, water content, and filtration of the sample. If the unused oil appears to be in order, and the complaint concerns its performance in an engine, the first action it to check whether the grade is one recommended for that engine and the service conditions. The laboratory should examine used oil samples for contamination, fuel diluent, insolubles, and TBN. From these results it will be possible to assess the overall condition of the oil, the mechanical condition of the engine and/or the length of time the oil has been in service. A judgment will then have to be made based on experience as to whether the problem is one of oil quality, engine condition, or the service conditions.

9.10.3 The Usual Causes of Complaints

1) Wrong grade supplied or used—easily detected in the laboratory, if not from drum labels.

2) Contamination—usually detectable by appearance or smell; can take place at the supplier's site, in transit, or at the user's premises.

3) Wrong recommendation—to be assessed by a sales engineer after considering engine or equipment type and service.

4) Poor engine or equipment condition—can be assessed by examination of the used oil for fuel dilution, sludge, wear metals, etc.

5) Extreme service conditions—detectable from insoluble contents, viscosity, wear metals, etc., provided that maintenance procedures are known. The user may also provide information on the type and severity of service.

6) Poor maintenance—can produce symptoms similar to either or both of the above cases, and has to be assessed more from a knowledge of the customer's operations.

7) Mechanical failure—often the basic origin of the complaint, and does not need specific confirmation. However, the parts should be examined as carefully as possible to try to determine a sequence of events (e.g., deposits leading to seizure or mechanical breakage).

Complaints of mechanical failure, blaming the detergency of the lubricant as the primary cause, are frequently unjustified. An engine maintained in good condition and operated within its normal duties should operate successfully without problems on the specified quality of lubricant. International standards of lubricant quality contain sufficient safety margins that normal quality variations will produce no detrimental effects. (An exception was the specification of low-temperature flow qualities after standing at very low temperatures. New tests on cold cranking and low-temperature flow had to be developed after a series of engine failures due to low-temperature operability problems.)

A common cause of problems is poor maintenance which can lead to either filter blocking, or in other cases excessive fuel dilution and consequent loss of oil viscosity. Diesel injector maintenance is particularly important for a heavy-duty diesel engine, and in the past many mechanical problems blamed on the lubricant have turned out to be caused by dribbling injectors producing excessive fuel dilution.

Complaints of short overhaul lifetimes due to poor lubricant performance are harder to deal with. Assuming that a maker's recommended quality of lubricant is in use, either the equipment is old and/or poorly maintained or the service conditions are more severe than the user realizes. (It is a common misapprehension that engines spending much time idling stress lubricant less than when working hard. The opposite may often be the case.) One solution here may be to provide a higher-quality lubricant for a trial period at no increased cost, and to monitor the effect with the user's cooperation.

Chapter 10

Safety, Health, and the Environment

10.1 Introduction

As is the case for most aspects of modern life, and particularly for the manufacturing industry, matters of safety, health and environmental impact cannot be ignored when considering automotive lubricants.

In this chapter we attempt to cover the key issues which should be considered by an oil company before placing a product on the market, or even before embarking on an expensive research and development program. Some of these issues are also relevant to the automotive industry who may use these products themselves and, more importantly, require others to use them in their equipment.

We are not attempting to replace the need to discuss these topics with experts in toxicology, industrial hygiene, legal aspects, or other related disciplines, but rather aim to highlight the areas where such expert advice needs to be sought.

Most oil companies and petroleum additive companies will have checklists to assist in the management of new product commercialization. In the areas of safety, health, and the environment (commonly abbreviated to SHE), questions on the checklist could include the following:

- Are there any special concerns about the chemical components that are intended to be used or may ultimately have to be used? Are these chemicals "new materials" or are they already commercial products?

- Will there be safety, health, or environmental risks in the manufacture, transportation or use of either the components or the finished product?
- Will the product be required to carry a warning label that might discourage potential purchasers or users?
- Will there be any hazards associated with the products of combustion, either in an engine or in waste disposal?
- Will there be any other waste disposal issues?
- Are there any other features of the product which could have (or be considered to have) a negative environmental impact?
- Is there some feature of the product that might be affected by forthcoming legislation?
- Will the product provide any advertisable environmental benefits?

The following sections provide some information on how all of these questions may be addressed.

10.2 Notification Laws for New Substances

In most developed countries it is no longer permitted for a new chemical to be produced and placed on the market, either alone or in a formulation, without prior notification to and approval from national authorities. It is therefore possible, particularly if an organization's links between marketers and research groups are weak, to spend considerable time and money working with new components that would require notification under the law when the cost of notification procedures in relation to the potential market and lifetime of the product would prevent an adequate return on the investment.

In the U.S., Canada, Australia, Korea, The Philippines and Japan, the essential step is to notify the authorities <u>before the chemical is manufac-</u>

tured—a *pre-manufacturing notification* (PMN). In the European Union, the same initials are used (PMN) for their requirement of a *pre-marketing notification.*

In many cases, and particularly if the product is polymeric, the decision as to whether notification is required may not be clearcut, and expert advice may be needed. Minor but otherwise unnecessary changes to a chemical product may be justified to ensure that a PMN is not required.

The costs of developing and presenting a PMN could well reach $1 million. The time required to perform the required testing for a new product could be up to 18 months.

Table 1 provides a comparison of the key points in some key regional notification laws.

10.3 Classification and Labeling

Most industrialized countries have requirements for classification and labeling of products which are placed on the market. The most widely recognized requirements are those of the United States and of the European Union. In Europe there is separate legislation relating to Dangerous Substances[1] (usually pure chemicals) and to Preparations (blends or mixtures). Such legislation is not usually of direct concern to the motor industry but is of importance to the oil industry.

In the European Union Dangerous Substances Directive there are fourteen categories of "dangerous," divided into five physico-chemical properties and nine categories based on toxicological properties. These are as follows:

- Classification based on physico-chemical properties:
  - Explosive
  - Oxidizing
  - Extremely flammable
  - Highly flammable
  - Flammable

Table 1. New Chemicals - Notification Laws - Comparison of Key Points

	EU Dangerous Substances Directive	US TSCA	Australian National Industrial Chemicals Notification & Assessment Scheme (NICNAS)	Japanese Chemical Substances Control Law (MITI)
Basic requirement	• 60-day pre-marketing (or pre-import) notification	• 90-day pre-manufacture (or pre-import) notification	• 90-day pre-manufacture (or pre-import) notification	• 3-month pre-manufacture (or pre-import) notification
Trigger	• >1 t/yr	• Any quantity not intended for R&D	• >1 t/yr	• >1 t/yr
Testing requirements	• Base set • Reduced test sets required for <1 t/yr and >10 kg/yr	• Not specified (but powers to request extensive testing prior to manfacture or import)	• Base set • Reduced test requirements for quantities between 50 kg and 1 t/yr • Testing requirements: - OCED MPD (minimum pre-marketing data set) - Reduced test requirements for quantities below 1 t/y	• Biodegradation (if substance not biodegradable then OECD base set type package required) • Ames test required under Industrial Safety Hygiene Law (ISHL)

Table 1. New Chemicals - Notification Laws - Comparison of Key Points (Cont.)

	EU Dangerous Substances Directive	US TSCA	Australian National Industrial Chemicals Notification & Assessment Scheme (NICNAS)	Japanese Chemical Substances Control Law (MITI)
Testing requirements (cont.)			• Testing requirements: - Commercial evaluation category available (max. 2 t/y for 2 y)	
Inventory	• EINECS • CAS numbers • EINECS registration numbers • 100116 chemical substances • Static (i.e., PMN substances are not added)	• TSCA • CAS numbers • Dynamic (i.e., PMN substances are added)	• AICS • CAS numbers • Inventory: - AICS - CAS numbers - Has confidential and non-confidential sections - Dynamic (i.e., notified substances are added)	• Existing Chemical Substance List • MITI numbers • About 20,000 chemical substances/ groupings • Static (separate list of new substances) • Over 5000 new substances

(continued next page)

Table 1. New Chemicals - Notification Laws - Comparison of Key Points (Cont.)

	EU Dangerous Substances Directive	US TSCA	Australian National Industrial Chemicals Notification & Assessment Scheme (NICNAS)	Japanese Chemical Substances Control Law (MITI)
Triggers for further testing	• >10 t/yr (50 t total) "may require" Level 1 • >100 t/yr (500 t total) "shall require" Level 1 • >1000 t/yr (5000 t total) "shall require" Level 2	• Not specified	• 10, 100, 1000, 10,000 t annually	• Not specified
Polymers	• Polymers exempt providing monomers and other reacting chemicals are listed and polymer definition is met	• Partial exemption	• Polymers: - Biopolymers and new synthetic polymers must be notified but not existing synthetic polymers	• Partial exemption

Table 1. New Chemicals - Notification Laws - Comparison of Key Points (Cont.)

	EU Dangerous Substances Directive	US TSCA	Australian National Industrial Chemicals Notification & Assessment Scheme (NICNAS)	Japanese Chemical Substances Control Law (MITI)
Confidentiality for chemical substance	• Partial	• Yes	• Confidentiality for chemical substances: - Notified new substances will be added to AICS after 5 years but can be kept in confidential section for 6 years thereafter	• No

- Classification based on toxicological properties:
  - Very toxic
  - Toxic
  - Harmful
  - Corrosive
  - Irritant
  - Carcinogenic
  - Mutagenic
  - Toxic to reproduction
  - Dangerous for the Environment

As automotive lubricants generally have a low level of toxicity, concerns are generally minor, but there will be marketing pressure not to have to label a product as hazardous if it can be avoided. If a choice is available, the marketer will choose the product which needs the lowest level of labeling. An example of such a situation was the requirement (initially in France) to label lubricants containing barium with a skull and crossbones. There was no evidence of any significant risk to the population at large or the users of lubricants, but marketing concerns forced a rapid replacement of barium-based additives by those based on other metals.

In some cases it may be important for the lubricant formulator to consider the exact composition of the formulation and not just the nature of each component. If, for example, a component is an irritant, then the need to label the finished lubricant (a "preparation") as irritant will depend on the percentage of the component in the formulation.

10.4 Toxicology of Lubricants

10.4.1 Base Stocks

The most useful general reference on this subject known to the authors is the CONCAWE report "Health Aspects of Lubricants."[(2)] Readers are encouraged to consult this document for more detailed information and further references.

Conventional mineral base stocks have low toxicity by inhalation and by skin absorption. Eye irritation is not usually a problem, but skin irritation leading to dermatitis may occasionally occur from prolonged exposure to automotive lubricants, although these oils do not represent nearly as great a hazard as soluble cutting oils. Inhalation of vapor or oil mist for short periods may cause mild irritation of the mucous membranes of the upper respiratory tract.

There have been numerous studies to investigate the carcinogenicity of mineral oils.[3] Estimation of the nature of the hazard varies with method of refining and is summarized by the International Agency for Research on Cancer (IARC). However, there are no clear boundaries between the classifications and the evidence is sometimes ambiguous, although it seems generally accepted that carcinogenicity is related to polycyclic aromatic (PCA) content. Most synthetic base stocks also show low levels of acute toxicity and irritation.

10.4.2 Additives

Individual additive manufacturers are responsible for providing full and accurate toxicological data on the products they market, but to avoid confusion in the marketplace the petroleum additives industry has tabulated existing data to provide standard information on acute toxicity for generic classes of products. Most of this work has been carried out and published by the Technical Committee of Petroleum Additive Manufacturers in Europe (ATC).

In general, engine oil additives present few if any physical hazards and, based on their low mammalian toxicological properties, the majority of lubricant additives are not classified as dangerous. To facilitate risk assessment and labeling, ATC has developed a generic classification of lubricant additives,[4] identifying 35 classes of product divided into six broad classifications. Each of these classes has risk and safety phrases associated with it.[5] With some exceptions[6,7] caused by minor differences in chemistry, all similar additives from different manufacturers will be classified in the same way.

Irritancy and sensitization potential are topics of importance to the additive industry. Of the major components used in automotive lubricants only zinc dialkyl dithiophosphate and some long chain calcium alkaryl sulfonates will be classed as irritant under European legislation and must be classified as "Dangerous." Some further details are shown in Table 2. Some in the industry consider the calculation method for irritative properties to be unreasonably strict, particularly because the calculation method does not distinguish between "weak" and "strong" irritants. The lack of a scientific basis for the regulatory thresholds for irritancy may contribute to this conservatism.

Some but not all sulfonates have been shown to be skin sensitizers in laboratory tests.

Recent legislation requires evaluation of the ecotoxicity of products on the market. The Petroleum Additives Panel of the (U.S.) Chemical Manufacturers Association (CMA) and the ATC have joined forces on a program of ecotoxicity testing looking at 13 fundamental classes of lubricant additives to establish whether products should be classified as "Dangerous to the Environment." While, in general, the majority of additives show a low level of toxicity, early results indicate that ZDDPs are harmful to aquatic life and may possibly be classified as "toxic." Calcium long chain alkaryl sulfonates may be regarded as "harmful."(8)

10.4.3 Unused Lubricants

Not surprisingly, experience has shown that the toxicological properties of fully formulated lubricants are closely related to those of the base stock and additive components. The human health and environmental hazards are also closely related.

Measured toxicity of mixtures is generally found to be close to the arithmetic sum of component toxicities, i.e., $t_{total} = (\%_a \times t_a) + (\%_b \times t_b) + ...\ \%_n \times t_n$.

Table 2. Acute Toxicity of Petroleum Additives (Source: CONCAWE Report 5/87)

Additives[a]	Typical LD50		Typical Irritancy Classification [b]	
	Oral (rat) mg/kg	Dermal (rabbit) mg/kg	eye	skin
Zinc alkyl dithiosphosphate	≥2000	>2000	irritant	irritant
Zinc alkaryl dithiophosphate	≥5000	>2000	not irritant	not irritant
Calcium long chain alkaryl sulfonate	>5000	>3000	not irritant[c]	not irritant[c]
Calcium long chain alkyl phenate	>10000	>2000	not irritant	not irritant
Calcium long chain alkyl phenate sulfide	>5000	>2000 (rat)	not irritant	not irritant
Polyolefin amide alkeneamine	>10,000	>2000	not irritant	not irritant
Polyolefin amide alkeneamine borate	>2000	>3000	not irritant	not irritant
Olefin/alkyl ester co-polymer	>10,000	>3000	not irritant	not irritant
Poly alkyl methacrylate	>15,000	>3000	not irritant	not irritant
Polyolefin	>2000	>3000	not irritant	not irritant

Notes:

(a) The additive names are based on the nomenclature system developed by the Technical Committee of Petroleum Additive Manufacturers in Europe.[4]

(b) Eye and skin irritancy potential is classified according to the EU criteria.

(c) This comment is anticipated to be updated.

Experience from our company has been that the measured toxicity of a formulation is likely to be lower than that calculated from component toxicities, suggesting that additives may cancel each other out in some way. Evidence is so far insufficient to make safe predictions about individual formulations.

For automotive lubricants, the concentration of ZDDP is usually a key determinant of toxicity, and this concentration is usually sufficiently low that automotive lubricants need not be classified as dangerous.

At the time of writing, in the European Community, for health hazard evaluations the legislative authorities allow toxicity to be measured on the product as marketed or calculated based on the toxicity of the component.[9] However, for environmental toxicity the authorities are still reviewing whether this approach is appropriate.

10.4.4 Used Lubricants

The toxicity of used engine oil is significantly different from that of unused oil, owing to contamination by fuel and combustion products.[10]

The major concern has been carcinogenicity caused by the polycyclic aromatic (PCA) hydrocarbons contained in fuel and generated by fuel combustion. The literature also refers to these products as polynuclear aromatics (PNAs) and polyaromatic hydrocarbons (PAHs).

Animal testing studies[11,12,13] have indicated that used oil from spark-ignition engines is clearly carcinogenic, whereas that from diesel engines is probably no more carcinogenic than the unused lubricant. Because of the carcinogenicity concern, there was concerted action in many countries[14] in the 1980s to require lubricant containers to give warnings about the hazards of used oil. Concurrently campaigns were launched by the motor industry to warn those populations most at risk (e.g., garage mechanics) to take appropriate hygiene precautions.[15] It is generally considered that use of gloves and barrier creams together with regular washing with soap and water should provide adequate protection. We are not aware of any epidemiological evidence to indicate that skin cancer is actually higher than average among populations considered at particular risk.

10.5 Biodegradability

A few paragraphs are included here on biodegradability because of its close relationship to environmental impact of lubricants.

Biodegradation, the destruction of chemical compounds by the biological action of microorganisms, is described by several definitions:

- *Primary* biodegradation: Any biologically induced structural transformation in the parent compound that changes its molecular integrity.

- *Ultimate* biodegradation: Biologically mediated conversion of an organic compound to inorganic compounds and products associated with normal metabolic processes (primarily carbon dioxide and water).

- *Acceptable* biodegradation: Biological degradation to the extent that toxicity or other undesirable characteristics of a compound are removed.

As a broad generation, primary biodegradation tests are quicker, cheaper and less meaningful than ultimate biodegradation tests.

There has been considerable controversy within the industry in the early 1990s about the applicability of specific test methods for lubricants. Much of this seems to relate to the lack of a detailed definition. Methods that measure only primary biodegradation should not be "passed off" as if they measured ultimate biodegradation and, in general, great care should be taken to avoid confusion of test methods. One of our anonymous reviewers has commented that, in the U.S., select cases for misleading claims based on biodegradation have been pursued by the Federal Trade Commission since 1990.

There are also some strong differences of opinion between those who dislike the use of primary biodegradation tests at all for evaluating lubricants, and those of a more pragmatic nature who believe that well-chosen primary degradation tests can be helpful and cost-effective in appropriate cases. The U.S. EPA and most U.S. experts seem to be in the former camp whereas there is more support for the latter approach in Europe. There is some evidence to support both points of view.[16,17,18]

There are many ultimate biodegradation tests including the OECD Modified Sturm Test and the EPA (Gledhill) Shake Flask Test. Different methods may be appropriate for different materials depending on whether the likely form of environmental contamination is to water or soil and depending on material solubility in water and other solvents.

Making another generalization, vegetable oils and ester-based lubricants will usually show significantly more biodegradation than conventionally refined lubricants derived from crude oil, but the exact relationship will depend on the test method used. The extent of hydrofining will affect mineral oil biodegradability.

The best known primary biodegradation test is CEC-L-33-A-94 (formerly CEC-L-33-T-82). This test measures loss of oil in an aqueous medium by an infrared method. It was originally developed to measure the biodegradability of two-stroke lubricants for outboard motors operating on large European lakes, but work[16] from 1991 to 1994 has suggested a further potential application as a screening test for biodegradability of high-risk and total-loss lubricants in soil. There has, however, been considerable concern that this test has been used and quoted for lubricants and applications for which there may be no field correlation data or evidence of relevance (see also Section 10.9).

10.6 Lubricant Effects on Automotive Emissions

For many years there have been growing concerns about the environmental effects and health hazards of automotive exhaust emissions. Controls on emissions of gaseous pollutants and particulate matter from new vehicles have been in place in most industrialized countries for many years and have been regularly strengthened.[19] With the exception of the contribution to diesel particulates, the lubricant makes a very small contribution to these emissions on a percentage basis relative to fuel,[20] and the percentage is likely to get even less as engine designs are improved further to reduce lubricant consumption.

As air quality becomes a growing concern worldwide, legislation on allowable vehicle exhaust emissions becomes ever tighter with carbon monoxide (CO), hydrocarbons (HC), mixed nitrogen oxides (NO_X) and diesel particulates (PM) being controlled on new vehicles in most of the world market. Engine design and the use of after-treatment devices are the primary routes to improvement of regulated emissions, and fuel quality is also a very important parameter.

As improvements continue to be made in the above areas, lubricant quality has gradually become more significant, especially in the diesel market, where under some conditions the lubricant may provide over half the total contribution to diesel particulates.[21,22]

There are two distinct areas of development activity: (a) lubricants for low-emission engines, and (b) low-emission lubricants.

The first area generally involves the need for lubricants to be designed to operate in very low oil consumption engines. These engines are also designed to minimize unburned and partially burned fuel and typically have a very high top ring to minimize "dead space" where unburned gases can accumulate. The developments place challenging and conflicting requirements on the lubricant. Lower oil consumption means less top-up, and the lubricant will need more alkalinity to cope with acids formed during combustion. However, increasing metallic components in the oil will lead directly to increased diesel particulates. The need for increasing alkalinity will be very significantly offset by the use of low-sulfur diesel fuel (0.05% maximum) which has been mandated to be used in the U.S. from 1993, in the European Community from 1996, and Japan from 1977. Lubricant formulators may not be able to take general advantage of this fuel change because some standard lubricant evaluation engine tests still use diesel fuel with 0.3% sulfur fuel, and because very high sulfur fuel is still sold commercially in many parts of the world where air quality is not yet a high priority.

The second area is currently explored very little although some consortia have studied the subject(23) and some papers have been published, that by M. Dowling being particularly comprehensive.(24) It is demonstrated that physical and chemical properties of lubricants can affect regulated emissions measured in legislated test cycles. Viscosity, volatility and additive chemistry are among the important variables. However, no two engines respond in exactly the same way, and different test cycles give sometimes conflicting results, and hence it is difficult to provide general guidelines on lubricant formulation to reduce exhaust emissions.

10.7 Disposal of Used Lubricants

The literature on disposal of used automotive lubricating oil is not vast,(20,25,26,27) but there are several alternative routes available, including re-refining and reuse, and burning as fuel. It seems generally agreed that a significant percentage—perhaps 30%—of all used crankcase oil is unfortunately unaccounted for. It is presumed that most of this lost oil is dumped illegally on land or into drainage systems. Such dumping is a major source of land and water pollution, and is receiving increasing attention from legislative authorities.

Many countries have introduced incentives to encourage used oil collection, and a selection of these is shown in the Table 3.

As well as incentives such as subsidies for collectors and provision of convenient collection sites, there can also be penalties and legal sanctions against improper disposal.

Re-refining has been considered to be an environmentally acceptable method of used oil disposal but it is currently not economically viable in most countries that do not provide subsidies.

Table 3. (Source: Industrial Lubrication and Tribology)

Incentive	Where in force
- Collector pays for used oil	Some developing countries (average U.S. $85)
- Government agencies must use a percentage of re-refined oils	U.S.
- Levy on lubricating oil sales which is paid to refiners	California
- Taxes and Subsidies	France, Italy, Finland, Spain, Germany
- Transport franchises	France
- Obligations on buyers	Austria, Germany, U.S.
- Restrictions on DIY sales	Austria (vendors to have disposal tanks)
- Public procurement	New Zealand
- Collection sites/curbside pick-up	Numerous countries
- Guaranteed price subsidy for gas from oil gasifiers	Italy
- Public education	Many countries

Germany is among the exceptions and has a strong re-refining industry in spite of losing a well-established financial advantage when the mineral oil tax was abolished on January 1, 1993. Previously re-refined oils were exempt from the mineral oil tax.

Other methods of disposal include incineration and use as fuel oil. In each case authorities have raised concerns about the potential for formation of dioxins if the conditions of combustion are not carefully controlled. Such concerns on the potential for dioxins to be formed by combustion of lubricants have arisen from time to time. The term "dioxins" is shorthand for a large family of 75 chemicals more accurately described as chlorinated dibenzo-p-dioxins. One of these chemicals (TCDD) is the most potent animal carcinogen ever tested. Dioxins can cause chloracne, as at the major incident at Seveso in 1976, but there is uncertainty about the carcinogenicity hazard presented to humans. Some authorities, notably the U.S. Environmental Protection Agency, consider TCDD a probable human carcinogen, and regulatory authorities all over the world have launched aggressive programs to identify and control the sources of dioxins and monitor their environmental distribution.

Dioxins could in principle be formed by combustion of any organic products containing chlorine. Automotive engine oils today do not contain any deliberately added chlorine-containing components, but several additive families may contain very small quantities of chlorine introduced from raw materials or catalysts as a by-product of manufacturing.

In the mid-1980s there was concern, particularly in Germany, over evidence that used oils collected for re-refining had been contaminated with polychloro-biphenyls (PCBs). PCBs (whose manufacture and use are now banned almost universally around the developed world) were used widely as transformer fluids, and form dioxins when subject to incomplete combustion. Unfortunately, once PCBs enter the re-refining chain they are very difficult to eliminate. A survey by the ATC[28,29] in 1986 gave results of a survey of chlorine and PCB contents of major additive types and 29 fresh and 37 used engine oils taken from the European market. None of the used oils showed any detectable level of PCBs. More recent studies[30,31] have shown that the trace amounts of chlorine in commercial engine lubricants do not contribute to vehicle dioxin and furan emissions, within the detection limits and test capabilities currently available.

Recently, more attention has focused on the potential use of cement kilns as environmentally acceptable means of used oil disposal. Cement kilns have the great advantages of operation at high temperature and of disposal of the metallic residue in the cement itself, thus avoiding the need for secondary disposal as would apply to the residues from re-refining. Recent papers[32,33] give evidence that the operation can meet legislated emissions standards for dioxins and heavy metals, at least in some countries. Another related approach to disposal which appears to offer environmental benefits is use in asphalt production, when the metals which are most difficult to dispose of are effectively encapsulated in asphalt, rather than concrete.

10.8 Transportation

In the worldwide shipping of any lubricant, the International Maritime Organization (IMO) requires classification, labeling, and choice of appropriate shipping vessels to be based on the classification of the components.

Automotive lubricants will be subject to national and international laws governing movement of goods but, in general, there are no special requirements relating to transport of automotive lubricants as distinct from other lubricants. In general they are considered to be non-hazardous and are not subject to hazardous substance regulations. There are specific regulations covering the construction of trucks which may be used for carrying lubricants but they are considered beyond the scope of this book.

10.9 Marketing Aspects

As "green" issues become increasingly popular with consumers, the motor and oil industries respond by finding ways to market their products as "environmentally friendly." Most of these claims are based on non-standard tests or non-legislated performance features.

Sometimes a product may be given an award of official recognition as with the German "Blue Angel" system which has been developed to promote environmentally friendly products.

An interesting OEM-led development has been the announcement by Mercedes-Benz at the end of 1991 of a system for evaluating passenger car motor oils by environmental standards.(34) Their initial assessment was based primarily on chemical analysis with positive or negative points being given for levels of particular elements (e.g., chlorine, zinc) and compounds (e.g., benzo-α-pyrene) and properties (e.g., biodegradability). Other features which they said should be recognized included the effects of engine oil on emissions, fuel consumption, catalyst life, oil recycling potential and oil drain period. Industry acceptance has been cautious, with several companies arguing for performance rather than chemical tests. However, the concept does give those companies with high scoring products the opportunity to advertise their benefits.

There is increasing interest in biodegradable engine lubricants.(35) Today legislation is primarily concerned with non-vehicular applications (outboard motors, chainsaws, etc.) but questions are being asked whether conventional automotive lubricants should be biodegradable. There is, however, no consensus that it makes economic or environmental sense for automotive engine oils to be biodegradable.

The attraction of being able to label a lubricant "environmentally friendly" has naturally encouraged oil marketers to seek advertising copy from biodegradation performance. As mentioned in Section 10.5, the test method CEC L-33-A-94 has been used (and possibly sometimes abused) in this connection.

References

1. "7th Amendment to EC Dangerous Substances Directive—Implications for the Petroleum Industry," CONCAWE report 54/93 and Directive 67/548/EEC with 7th amendment (92/32/EEC).
2. "Health Aspects of Lubricants," CONCAWE report 5/87.
3. Hewstone, R.K, "Used Lubricating Oils—Safety & Sense," *Oil Gas European Magazine*, 1/88.
4. "An Internationally Recognised Nomenclature System for Petroleum Additives," ATC Document 31, March 1991.
5. "Classification and User Labeling Information on Major Petroleum Additive Components," ATC Document 43, Revised February 1995.
6. "Classification and Labeling of Long Chain Calcium Alkaryl Sulfonates," ATC Document 42, May 1992.
7. "Implementation of the Classification and Labeling Requirements of the Dangerous Preparations Directive (88/379/EEC)," ATC Document 41, May 1992.
8. Linnett, S.L., *et al.*, "Aquatic Toxicity of Lubricant Additive Components," Technische Akademie Esslinger, 10th International Colloquium, Jan. 9-11, 1996.
9. European Community Directive 88/379/EEC.
10. Velasquez-Duhalt, R, "Environmental Impact of Used Motor Oil," *The Science of the Total Environment*, 79, pp 1-23, 1989.
11. Eyres, A.R. (on behalf of Institute of Petroleum Advisory Committee on Health), "Polycyclic Aromatic Hydrocarbon Contents of Used Metal Working Oils," *Petroleum Review*, 10, pp 32-35.

12. Results of toxicological studies, American Petroleum Institute, 1982.
13. "A Carcinogenesis Study of Used and Unused Diesel and Gasoline Engine Oils When Applied to Mouse Skin 1982," Institute of Petroleum SBER, 81.004, 1982.
14. "Precautionary Advice on the Handling of Used Engine Oils," CONCAWE Report 3/82.
15. "Guide to the Risks from Used Engine Oils," Society of Motor Manufacturers and Traders (UK), 1992 update.
16. Novick, N.J., Mehta, P.G., McGoldrick, P.B., "Assessment of the Biodegradability of Mineral Oil and Synthetic Ester Base Stocks, Using CO_2 Ultimate Biodegradability Tests and CEC-L-33-T-82," CEC/93/EL28, CEC Symposium, Birmingham, May 1993.
17. T.F. Bünemann and N.S. Battersby, CEC IL-24/PL-057, Final Report, 1994.
18. Exxon Research and Engineering Company, Unpublished Report, 1994.
19. CONCAWE, "Motor Vehicle Emission Regulations and Fuel Specifications," 1995 update.
20. Copan, W.G. and Haycock, R.F. (on behalf of ATC), "The World of Lubricant Additives and the Environment," CEC 93/SPO2, Presentation to CEC Symposium, Birmingham, May 1993.
21. Mayer, W.J., Lechman, W.C., and Hilden, D.L., "The Contribution of Engine Oil to Diesel Exhaust Particulate Emissions," SAE Paper No. 800256, Society of Automotive Engineers, Warrendale, Pa., 1980.
22. Cartellieri, W., and Tritthart, P., "Particulate Analysis of Light Duty Diesel Engines with Particular Reference to the Lube Oil Particulate Fraction," SAE Paper No. 840418, Society of Automotive Engineers, Warrendale, Pa., 1984.
23. U.K. Engine Emissions Consortium—Project D2, "Investigation of Lubricant Effects on Diesel Engine Emissions," information details proprietory to project group members.
24. Dowling, M., "The Impact of Oil Formulation on Diesel Engine Emissions," SAE Paper No. 922198, Society of Automotive Engineers, Warrendale, Pa., 1992.
25. "Used Oil Management in Selected Industrialised Countries," American Petroleum Institute Discussion Paper No. 64, January 1991.
26. *Industrial Lubrication and Tribology*, July/August 1994.

27. Donkelaar, P V, "Umweltverträglichkeit von Sumpf- und Verlustschmierung," *Tribologie und Schmiertechnik*, 36Jg 4/1989.
28. "Polychlorinated Biphenyls in Fresh and Used Engine Oils," ATC Document 19, July 1986.
29. Hewstone, R.K. and Spiess, G.T., "Handling, Reuse and Disposal of Used Lubricating Oils," *Petroleum Review*, March 1988.
30. Hutzinger, O., Essers, V., Hagenmaier H., *et al.*, Heft R 463 Informationstagung Motoren Frühjahr 1991, Würzburg 1991.
31. Hutzinger, O., Essers, V., Hagenmaier H., "Untersuchungen zur Emission halogenierter Dibenzodioxine und Dibenzofurane aus Verbrennungsmotoren," Forschungszentrum für Umwelt und Gesundheit GmbH 4/92, 1992.
32. McGrath, B., "Experience With Used Lubricating Oil as a Fuel for Cement Kilns," Kilnburn '92 Conference, Brisbane, Australia, September 1992.
33. "Used Oil Management: The Cement Kiln Option," Briefing Paper, Shell International Petroleum Company, February 1993.
34. *PARAMINS Post*, 9-4, p.9, February 1992.
35. Boehme, W., Hubmann, A., and Zeiner, W., " New Top Grade Engine Oil Optimised for Ecologically Sensitive Operational Areas," CEC/ 93/EL 30 CEC Symposium, Birmingham, U.K., 1993.

Chapter 11

The Future

11.1 The Influences for Change

Predicting the future is always difficult, but that is no reason not to make the attempt. We propose to look at the influences that are apparent at the present time and have the potential to change the type or level of automotive lubricants sales. We will look at various possible scenarios and indicate those we see as likely to develop and which imply changes to present patterns. In view of the size of the present vehicle population and the associated lubricant business, we caution against expecting dramatic shifts overnight. We will mainly be considering crankcase lubricants but will also consider other types of lubricants which may succeed those in use at the present time.

Our starting point is a conventional crankcase lubricant, made up with a series of additives such as dispersants, detergents, anti-oxidants, and anti-wear additives, and which in the majority of cases provides both diesel and gasoline engine performance, albeit with emphasis in one direction or another for some of the more specialized and expensive lubricants. Such lubricants have evolved over nearly a century, and a very strong influence will be required if this evolutionary development is not to continue. It is possible that environmental considerations may provide such an influence, but the strength of the pressures for changes to aid the environment varies from country to country and region to region and therefore could well result in at least an initial increase in diversity of products. While internationalization pressures are currently strong, these tend to succumb to local needs as soon as a strong local lobby develops.

These major influences that will affect the direction of the developments of conventional lubricant formulations are represented in Figure 94 as a series of pressures acting on current lubrication technology:

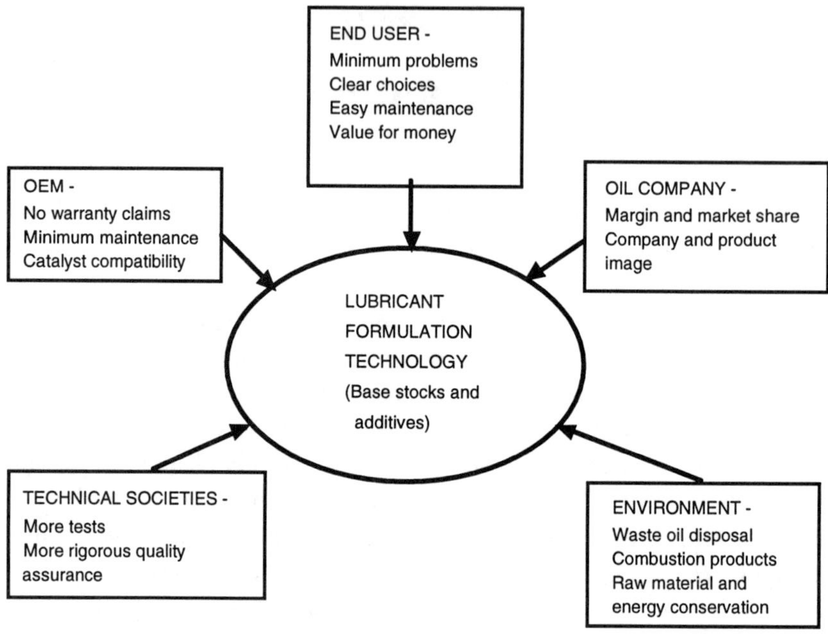

Fig. 94 Pressures for oil reformulation.

11.1.1 The End User

In Chapter 4 we mentioned the evidence that very few consumers know or understand anything about oil quality. They are influenced by factors such as OEM recommendations, brand advertising by oil companies, and price considerations. With the very low oil consumption of new modern vehicles many consumers probably make no decisions about the quality of the oil used in their engines, leaving this entirely to their local service garage. The choice of such a garage may be influenced by OEM affiliation, loyalty to an oil brand, or simple convenience. Warranty considerations will drive owners of new vehicles in the direction of OEM-sponsored service stations. Consumer organizations may also have an influence but this is felt to be minor.

The above considerations have been made with the domestic private vehicle owner and the small commercial operator in mind. Some comments on the large commercial fleet operators should perhaps also be added.[1] In the past, operators of large fleets such as bus companies, postal authorities, and in particular military authorities have had a very significant influence on the types of oils made available by the oil companies by setting their own technical requirements and purchasing oils which met those requirements. Although very significant in the past, the influence of such users, including the military authorities, is now considered to be slight. Strong recommendations from the OEMs and their issue of lists of approved lubricants have made it unnecessary for large users to run their own qualification programs. Any input from such consumers is likely to be via the technical authorities (see Section 11.1.4).

11.1.2 The Oil Companies

These are naturally striving to increase both margins and market share and will move in whichever direction appears likely to enable them to achieve this. On the one hand cost reductions can improve margins, whereas additional market benefits from new technology which give rise to validated market benefits can provide both higher market prices and higher market share. Different companies will find different ways to develop their products and may explore more than one direction with different brands. The oil companies will receive suggestions from and be assisted by the additive companies in developing new technology, but it is the oil companies who will make the final decisions on the products which are to be in the market. As the suppliers of the lubricants to the market, the potential ability of the oil companies to modify the demands from the OEMs, or from other areas of pressure, should not be underestimated.

It is also very much up to the oil companies to decide if there should be a wide range of oils marketed to suit all needs or if a restricted number of more multi-purpose oils should be available. Recently in Europe with a differentiated passenger car diesel market and increased promotion of fully synthetic oils there has been a significant increase in the number of passenger car motor oils offered by the major companies, and this trend seems likely to continue.

11.1.3 The Vehicle Manufacturer

The vehicle manufacturer, usually referred to as an Original Equipment Manufacturer or OEM, prescribes the quality of oil to be used in his vehicles under warranty conditions, and he hopes this will be used subsequently. This is true whether he manufactures the engines himself or purchases from another OEM, although in the latter case the engine manufacturer will have a significant influence.

The OEM is looking for reduced warranty claims, and also seeks to minimize the trouble the user will have in meeting the service requirements for the vehicle while meeting warranty conditions. He will also have to meet future legislation requirements, especially fuel economy and exhaust emission requirements, and if there are lubricant solutions which will enable him to attain these more easily, he will wish to use them. The move to SAE 5W-30 oils in the U.S. was mainly prompted by the legislated demands for Corporate Average Fuel Economy (CAFE) improvement.

In the passenger car sector the concept of "fill for life" crankcase lubricants has been actively considered as a means of sales promotion and avoiding waste oil disposal problems. The initial fill lubricant would be supplemented by top-up oil as required, and there would be no oil drains unless there were indications that the residual oil quality was grossly inadequate. In this connection the use of on-board diagnostics (OBD) will be desirable. Already several manufacturers have on-board systems that compute the time of oil changes from mileage run, the average speed and the number of starts which the vehicle has made. It is, however, much more difficult and expensive to develop on-board analytical systems that measure such properties as the viscosity and the residual base number of an oil which could have been influenced by engine malfunctions or modified by fuel dilution or top-up with an unsatisfactory lubricant. Technology for rapid analysis of small samples withdrawn from an engine already exists, but service stations are unlikely to be persuaded to invest in equipment or staff training to put this in place.

At the present time the attitude of the OEMs could perhaps be summarized as requiring more of the same type of engine protection that exists in current oils, but with increased attention to indirect environmental effects. While not necessarily identical to a marketed grade, the importance of factory-fill oils should be emphasized. These represent a substantial cost to the OEM and price is very important. However, the vehicle may run on the factory-fill oil for much of its warranty life, and certainly for the most critical period, so fully adequate quality is essential.

If new types of engines and/or vehicles are developed then their requirements would again be for minimum warranty problems. For new types of internal-combustion engines, the ability to use generally available lubricants could be considered important by the OEM, possibly leading to pressures for the development of new types of multi-purpose lubricants for both existing vehicles and those containing new types of engines.

11.1.4 The Technical Societies

In the past, technical societies such as the American "tripartite" and the European CCMC have really determined the qualities of oil marketed in their areas of influence. They would reach an equilibrium point between the pressures from the OEMs, the large users such as the military authorities, and the oil companies who had to produce the lubricants. An important concern in the past, for which they have been criticized, is that these bodies have in fact slowed down the rate of development of new oil qualities due to the time it took to obtain a consensus and develop the necessary test procedures and limits. It can be argued, however, that some inertia in the system is highly desirable. These societies or their successors should be capable of preventing a proliferation of minor improvements to existing types of lubricant, while developing a pattern for the introduction of new types of lubricant or radical new qualities for crankcase lubricants, in an ordered manner.

In the short term the technical societies will seek more harmonization, particularly international harmonization of lubricant qualities, to have more controls on lubricant quality in the field, and to produce a higher quality testing regime. ILSAC (see Section 6.3.5.2) will promote the advantages of internationalization of new specifications, but these may remain only as guidelines outside the U.S. ACEA has not joined ILSAC so far, and if it eventually does so is likely to insist that the conditions and different engine technologies in Europe require additions to ILSAC specifications. The existence of multiple quality levels in markets outside the U.S., and in particular in the developing countries, may also lead to local development of replacement specifications for obsolete API quality levels when these can no longer be run at test sites. A current example is the existence of the Japanese "JSE" sequences run in Japanese engines to define quality approximating API SE for the Asia/Pacific area. All the historical evidence is that simple adoption of the quality levels used in the U.S. cannot be done quickly or simply.

11.1.5 Environmental Pressures

These have the potential to be the greatest influence on the future composition of lubricants, although it is difficult at this time to be sure how rapidly pressures for change will develop. Environmental concerns can produce both primary effects (such as those from legislation limiting compositional or biological properties) and secondary effects resulting from changes in hardware or fuel quality which demand or permit changes in lubricant composition. The primary effects are more immediate but perhaps less likely, while secondary effects could in the long term be of great significance.

There is growing popular and governmental concern about the environment being encapsulated in legislation on matters such as recycling, safe waste disposal, biodegradability, and emissions to the atmosphere. There is a danger that substances considered harmful or as potential precursors of harmful substances may be restricted to unnecessarily low levels, regardless of how dangerous current levels are in fact. An example would be the restriction of the chlorine content of lubricants from a mere fear of dioxin production, without evidence that this happens in the field. The fitness for purpose of products must be remembered when compositional restrictions are applied, and more consideration must be given to "cradle to grave" analysis including global impacts, rather than focusing on simple compositional parameters.

Safe waste disposal is unlikely to have a direct impact on oil composition, although implementation of fill for life concepts to minimize user disposal problems could require stronger formulations. In the early 1990s there is much comment in the technical press, particularly in Europe and Japan, about biodegradable lubricants. These will certainly make rapid inroads into the market in certain sensitive areas such as two-stroke and industrial oils, but the case for biodegradable oils in the mainstream automotive market is much less clear. The technology is not developed, and even if it were, it might not necessarily provide a better-balanced answer to environmental protection than the application of more vigorous means of collection and safe disposal of used lubricants.

The impact of emissions legislation on lubricant composition should be considered as a secondary effect. Reduction of chlorine, phosphorus, zinc, and levels of other metals could be proposed in the interests of a better environment or improved catalyst life, although currently there is no agreement on whether a real problem exists or, if it does, how to tackle it. A secondary effect of even greater potential importance is a reduction of sulfur levels in diesel fuels to extremely low levels. The need for high TBN additives will be greatly reduced. But there are contrary forces at work. Reduction of NO_x in exhaust emissions to very low levels already requires the use of EGR in some Japanese HD diesel engines. This in turn puts more acids into the lubricant instead of the atmosphere and requires higher base numbers.[(2)]

The introduction of unleaded gasoline could have been expected to have a greater impact on the formulation of gasoline engine oils than has been so far, although the desire to lubricate both engines running on leaded fuel and also diesel engines has tended to act against the formulation of oils specifically for unleaded gasoline. As the use of leaded gasoline decreases and the use of low-sulfur diesel fuel increases, then one can foresee the introduction of lower ash crankcase oils with low detergency but high dispersancy and anti-oxidant properties. As metal detergency is reduced, it will probably become relatively easier to formulate with lower levels of ZDDP.

The above indicates the major pressures and constraints upon lubricant formulation technology. The direction in which it will ultimately move remains a matter for conjecture. In the past, change has almost entirely been evolutionary, with new additives or more of the old additives being used to improve the performance of lubricants in various well-defined areas. Evolution rather than revolution will probably remain the rule in the short term, but the environmental effects may prove ultimately able to bring about great change in the types and volumes of lubricants used. These could be from emissions/fuel effects as indicated above, or because the conventional, reciprocating, internal-combustion engine is gradually replaced by other types of prime mover. These possibilities will be discussed again later.

11.2 Predicting the Future

In 1964 one of us participated in a forecasting study of future transport developments in the U.K. The task was to predict technological changes to motor transport in the succeeding 10 years, with a 20-year long-term look ahead. After extensive discussions with the motor industry, Ricardo Consulting Engineers, and some academics, it became clear that 10 years was too short a time span for any changes of marketing significance to take place, and even after 20 years, penetration of new technology would be small even if development proceeded smoothly and continuously. The senior executives of the company found this difficult to accept in the light of continuous press comment on the great breakthroughs in automotive technology which were supposed to be taking place, at that time seen particularly as gas turbines for trucks and fuel cells for electric cars. However, a few figures on the replacement rate of old vehicles by new, and some discounting of the number of technological breakthroughs, eventually convinced them that this "non-prediction" was in fact the correct one. As it has turned out, there have been no major changes in the ensuing 30 years, developments being evolutionary and in the direction of higher specific engine power, longer oil drains, and more highly formulated oils. No major new type of prime mover has emerged, although more sophisticated engines and control systems and accessories such as turbochargers and exhaust catalyst systems have appeared.

We may perhaps now live in more stirring times, with environmental pressures and energy conservation being strong driving forces; however, the pool of existing vehicles is even larger, and increased new vehicle production increases the total pool as much as it eliminates the older vehicles. The more the total pool grows and the more the rate of conventional vehicle production rises, the greater is the inertia in the system to prevent a switch to new technologies such as electric and hybrid vehicles from having much short-term significance.

Having cautioned against assumption of rapid change, it may well be that the automotive industry has in fact finally arrived at a crossroad. There is a lot of interest in electric vehicles, although it can be argued that they contribute more to greenhouse gas emissions than internal-combustion vehicles if the ultimate power is derived from fossil fuels. The hybrid vehicle (using a low-power, low-emission engine to drive an electric generator) might be a more environmentally friendly option, but is likely to be expensive to produce. Cost/profitability considerations often oppose possible moves in the direction of greater reliability or lower environmental impact, and the strengths of such opposing forces need to be assessed. Developments in reciprocating engine technology, such as the increased use of two-stroke cycles for passenger car engines, might produce demands for alternative types of formulation, but oils already in the field will probably be used to lubricate the initially small numbers of such new types of engines. Once this is done, significant change or simplification becomes very difficult to achieve. Similar considerations will probably delay for a long time the introduction of the potentially interesting low TBN formulations discussed under Section 10.1, because the retention of some leaded gasoline and high-sulfur diesel fuel in some marketing areas will require general-purpose international lubricants to be similar to present-day types. An indication of the possible trends from the starting point of the current types of automotive lubricant is given in Figure 95, although the time scale and the proportions of eventual types of lubricant in the marketplace must be regarded as speculative.

In the following sections we will consider the various areas where change may take place, with an indication of our current thinking on the likelihood of these changes being implemented within a reasonable period of time.

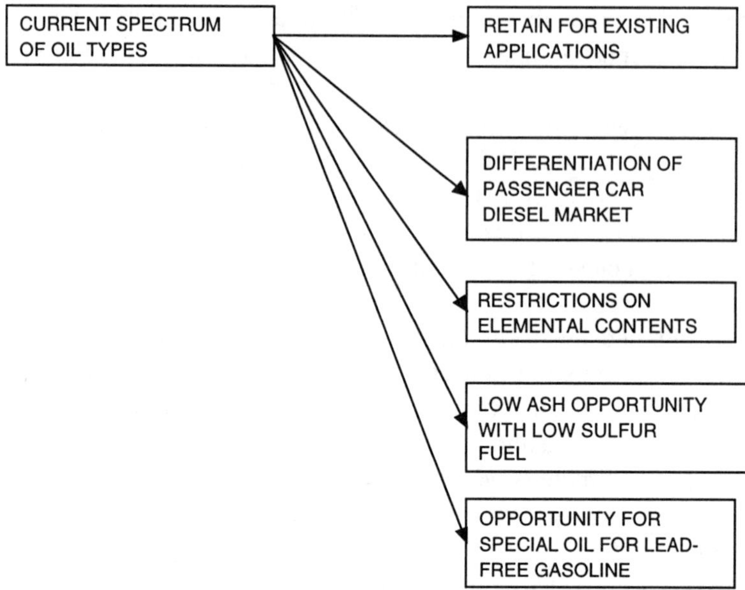

Fig. 95 Why there may be an increasing variety of oil types in the future.

11.3 Changes to Existing Types of Formulation

11.3.1 Alternative Base Stocks

Petroleum Derived Stocks

Use of unconventional hydro-isomerized or hydrocracked stocks will increase at the expense of the conventional broad spectrum extracted petroleum base stocks, this being driven by the need for low-volatility lubricants and by the tendency for recommendation of lower and lower viscosities in the interest of fuel economy. Catalytic dewaxing will be increasingly used as the means of achieving low pour points and startability performance in areas where low winter temperatures are encountered.

PAO and Ester Base Stocks

Growth in use of polyolefin base stocks will probably parallel the use of the unconventional petroleum-based stocks for companies or areas not producing the petroleum-derived materials. If biodegradability becomes a requirement then certain types of ester may become more widely used as base stocks, although not all esters are equally biodegradable and there is already concern about the significance of current types of biodegradability tests. Ester oils also usually have an inherent dispersancy property which could be useful in formulating possible new high-dispersancy/low-detergency oils if these become a feature in the marketplace. Other synthetic base stocks may come into prominence as a result of requirements arising in the future, but we cannot see any particular needs at this time other than those discussed above.

Vegetable Oils

Vegetable oils have already found some favor in terms of sustainable and biodegradable diesel fuel, and were the basis of most early lubricants. Rapeseed (colza) and sunflower oils appear to be the most likely candidates and could probably produce satisfactory crankcase oils of the present type. High oleic acid sunflower oils can be obtained by genetic selection as discussed by Lubrizol[(3)] and are said to have good oxidative stability and to be much more responsive to anti-oxidant treatment than rapeseed oils. With a need in many developed countries to find new outlets for farm products, vegetable oils could develop into a relatively low-cost source of biodegradable base stock for oils either similar to present types or for completely different types of lubricant. At the present time, however, little market penetration can be foreseen for products which have hitherto been regarded as expensive and of insufficient stability to formulate good lubricants.

Water

Water in the form of water/glycol mixtures has been proposed in the past for the lubrication of conventional types of internal-combustion engines. The advantages are the high thermal capacity for heat removal and the lack of a combustion product from the water itself. However, glycol can form gummy deposits and such a lubricant could probably not be inhibited to prevent rapid deterioration in engines of conventional metallurgy. In addition, the viscosity of such mixtures would probably be inadequate at high temperatures where there would also be a danger of water vaporization. At low temperatures the problem of phase separation of ice crystals from a water/glycol mixture would also have to be solved. Emulsion technology has advanced considerably with the development of emulsion hydraulic oils, and a biodegradable emulsion could be of interest in special applications if the low-temperature problems could be solved. Such an application could be hybrid power, for when the vehicle is out of use batteries are connected to an electricity supply which could also provide frost protection for the lubricant. The problems and the limited potential applications make it unlikely, however, that this inexpensive base stock will be utilized significantly, unless environmental developments take a dramatic and unexpected turn. Cost reduction would seem to be unlikely to provide a good reason to move to water-based lubricants.

11.3.2 Additive Technology

Invention of revolutionary new additive components for real breakthroughs in additive chemistry are, almost by definition, impossible to predict. However, real breakthroughs are very rare and most changes are by evolution of formulation technology and by discovery of component variations that perform the job more efficiently than previous types. In the past the initial discovery of detergent action in oils, the development of ashless dispersants, and the discovery of the anti-wear properties of ZDDP could be considered breakthroughs, although even in these cases there are elements of evolution as well as revolution. In the last 30 years the development of oil formulations has been almost entirely evolutionary with increased quantities of more efficient additives of the same general type as their predecessors being used in the development of new oil quality levels. In this context we are not considering changes such as switches from barium to calcium to magnesium sulfonates to be revolutionary. Perhaps changes in the types of polymer used for viscosity modifiers, with multifunctionality appearing, diminishing, and then re-appearing might be thought more fundamental, but we see the basic requirement of a high-molecular-weight, marginally oil-soluble polymer to be relatively unchanged.

Looking at the conventional type of crankcase oil which is in use today it looks unlikely that there will be any breakthroughs in additive or formulation technology that will affect the broad type of formulation of such oils. It is more likely that use of new base stocks, possibly for environmental reasons, will require changes to formulation technology, or that decisions to make lubricants more specialized with simplification of performance requirements for a multiplicity of different vehicle lubricants may require radical rebalancing of additive packages. Additive chemistry is sufficiently well developed that neither of these events is likely to require the development of radically new types of additives.

11.4 External Factors Influencing Oil Quality

11.4.1 New Hardware

As we have already pointed out, the introduction of new types of power plant in vehicles will initially have only a minor effect on the market for existing types of lubricant, but their needs will have to be catered to in the product mix and pressures such as those from environmental concerns may help to accelerate change. Of importance here is the acceptance or otherwise of the concept of one type of vehicle for running around in cities and towns, with a different and larger type of vehicle being used for long-distance driving. Experiments currently going on in some towns with small electric vehicles which may either be the property of the user or alternatively can be hired for single journeys, will help to indicate the willingness of the public to move to this type of in-town mobility. Personal Rapid Transport (PRT) systems that use "personal" vehicles on a public track network are as yet untested but may well offer an acceptable compromise between use of conventional public systems and private vehicles.[(4)] Apart from the PRT approach, one can basically see three types of vehicle being required, namely:

(a) The in-town runabout to back up any public transport system. Such a vehicle may be available on an hourly hire basis.

(b) The commuter car which need not have a range of more than 100 kilometres before refueling/recharging is required.

(c) The long-distance and recreational vehicle which is the family runabout and workhorse but is possibly not allowed into city centres.

Possible new power plants which may find their place in one or more of these categories include the following:

New two-stroke design

The possible emergence of new types of two-stroke engine for passenger cars was discussed in the section on two-stroke motor oils. These are at an early stage of development but it appears possible that one or more designs may well find their place on the road. A design incorporating a conventional sump would probably use oils similar to conventional present-day oils, but a once through total loss design or use of an external scavenging system could require a special oil probably of superior qualities to those presently used in two-stroke total loss motorcycle engines.

Electric vehicles

These are attracting interest because of their local zero air pollution, although the total emission of greenhouse gases when electricity generation is taken into consideration may well be greater than that of a conventional engine. Lubrication requirements would be minimal, consisting mainly of sealed-for-life grease systems, and therefore the effect of these vehicles would be to reduce the total requirements of crankcase oil rather than require a new type of oil. Significant penetration of the vehicle market by electric vehicles would reduce the sales of crankcase lubricants to levels where allocation of resources to develop new types could become a problem.

Hybrid vehicles

These are battery-electric vehicles but employ a small on-board engine to charge the batteries continuously. The most likely choice of engine would be a small diesel although gas turbines have also been suggested. One significant advantage of the hybrid vehicle over the purely battery-driven electric vehicle is the availability of waste heat for vehicle heating. Diesel drive would probably use a conventional type of oil, but use of gas turbine or other type of drive may require small quantities of a special oil.

Gas turbines
Considered in the 1960s (see Section 11.2) as a potential power source for future trucks, simple designs were found to be too fuel inefficient, while more complex designs were too expensive. Except for possible hybrid use (see above) they are unlikely to be resurrected as a major prime mover for land-based vehicles.

New transmissions
The variable cone, continuously variable transmission (CVT) has been revived with new metal belt designs and is now entering the market in the U.S., following launches in Asia/Pacific and Europe. Low-potency gear oils or engine oils are believed to be adequate lubricants at the present time, although the transmission could possibly benefit from development of specialized transmission oils.

A complete hydraulic drive to oil-mounted hydraulic motors has long been proposed as an ideal system for vehicles, but has never been adopted for over-the-road use. This system could permit use of a compact, efficient, constant-speed engine similar to that for the hybrid vehicle. Overall, it appears unlikely its use will extend out of the present area of application which is in certain off-the-road heavy equipment.

Hydraulic suspensions
The use of pressurized hydraulic systems for vehicle suspensions is growing, and increasing demands for initial fill, top-up, and replacement fluid can be expected. At the present time these fluids are not considered to be in the fill-for-life category, and therefore possibly represent an area of growth in the demand for automotive lubricants.

11.4.2 Hardware Problems

Historical evidence suggests that from time to time, in spite of improved OEM development procedures, lubricant-sensitive hardware problems will arise that have not been predicted. As in the past, it seems likely that competitive pressure, if nothing else, will force the oil industry and the additive industry to try to address the problems with oil-related solutions.

There will still be no easy way to establish whether such problems should be handled by technical societies or treated as the responsibility of individual OEMs.

11.4.3 Demands of Add-on Devices

We are thinking here of devices such as turbochargers and superchargers, catalytic systems, particulate traps, oxygen sensors, and afterburners, which may well have oil quality implications.

The widespread introduction of turbochargers to passenger vehicles in the early 1980s initially caused some problems owing to oil carbonization in the bearings. Partly by equipment redesign and partly by the use of well-inhibited oils, the problem has now been largely overcome but will have to be borne in mind if any radically new oil qualities are proposed. The use of turbochargers is expected to decline in favor of superchargers or more efficient, normally aspirated, multi-valve engines. Pressure charging will often be used to prolong the life or extend the range of an engine family.

Catalyst systems, whether the three-way catalyst in gasoline engines or the oxidation catalysts and/or possible de-NO_x catalysts for diesels, can suffer from poisoning from various additive elements in the lubricant, and ash from lubricant can also tend to cause blockage of both catalyst systems and the particulate traps which are designed primarily to remove carbon particles. Oils of much reduced ash content or even ashless oils would benefit all of these areas if they could provide adequate lubrication for the engine. To date, reaction to such needs has been very slow as evidenced by the fact that control of lubricant phosphorus levels to avoid possible catalyst poisoning has been under discussion for approximately 15 years but there is still no generally accepted limit for phosphorus content, or even agreement on the need for a limit. This is partly for the good reason that oil consumption is being reduced and catalyst technology is improving at the same time as exhaust system durability limits are tightening.

11.4.4 Alternative Fuels

Addition of methanol or ethanol or both to gasoline already takes place and there have been a few problems with some lubricants. However, satisfactory lubricants for gasohol type fuels can be considered to be similar to conventional types even if the components are somewhat rebalanced. A change to use of an 85+% methanol or ethanol or other synthetic oxygenated fuel will require a specialist lubricant, but at the present time there is no specification or approval system to cover such an oil. Most analysts do not see much future for alcohol alternative fuels in a free economy. Government subsidies or equivalent legislation would probably be needed. "Reformulated gasoline" as currently marketed in the U.S. is not considered sufficiently unconventional to be considered as an alternative fuel.

Gaseous fuels such as methane or hydrogen are possible but would require new designs of motor (possibly a turbine) and a different type of lubricant. Cost and storage difficulties militate against the use of such fuels, as do safety considerations.

In Europe considerable use is made of LPG (liquefied petroleum gas) as a passenger car fuel, and LPG refueling pumps are available at many service stations. It is used in conventional engines and conventional lubricants are used satisfactorily. Compressed natural gas (CNG) is receiving growing attention as a low-emissions fuel for buses, garbage collection vehicles, and other inner-city vehicles.

The possibility of moving to much lower ash levels or even ashless oils for use with lead-free gasoline and very low-sulfur diesel fuel has already been discussed under 11.1. The driving force for such a change could well come from the pressures discussed in the following sections.

11.4.5 Emission Effects

When oil passes into the combustion chamber in an engine it is burned, probably rather incompletely, and waste products including ash pass into the exhaust stream. Oil-derived deposits in the combustion chambers and on areas such as valve tulips can also cause changes in the emissions of an engine. At some time in the future it is anticipated that there will be a test requirement for motor oils related to the emission performance of standard engines using the test oil. One can speculate that oils which perform well would have low ash and polymer contents, but inevitably the situation will prove to be relatively complex and will need a considerable amount of development work.

11.4.6 Safety, Health and Environment

On testing, mineral base oils have generally been found to have very low orders of acute toxicity, probably largely due to the insoluble nature of the base stock. Some additives are, however, in themselves either irritant or toxic, and concern has been expressed about the presence of these in a commonly available material. Questions are particularly raised over the presence of ZDDP (which is an irritant) and pressures could develop for its removal from motor oils. Its presence is largely required by the pro-wear activity of the common detergents, and if these were reduced or replaced it might be possible to remove the ZDDP from the formulation. However, it is by no means certain that any other additives which would be added as a replacement would not have similar limitations.

Biodegradability is being seen by some as a desirable property for many types of motor oil, possibly for reasons which are more cosmetic than environmental. However, the introduction of biodegradable base stocks with conventional type additive treatments to provide a more environmentally friendly lubricant could result in the lubricant being more readily digested by mammals and therefore considerably more toxic than current oils.

Used gasoline engine oil is considered to be significantly more toxic, actually carcinogenic, than unused oil. This arises from the production of partially oxidized base oil and additive components in the hotter parts of an engine. In particular, polynuclear aromatics (PNA) are produced and build up in concentration while the oil is in use. The ultimate length of oil drain intervals and the concept of sealed-for-life crankcases could be affected by considerations of the maximum tolerable levels of PNA in the crankcase lubricant. Regulations on used oil handling and disposal could impact oil sales through supermarkets and high street stores to the do-it-yourself (DIY) market.

11.4.7 Oil Supply and Consumer Buying Habits

Despite the increase of newer cars on the roads, the number of older vehicles is not declining, as the scrapping rate is less than that of new production. The older vehicles tend to drift down toward younger owners who are short of money but have interest and ability to perform do-it-yourself routine maintenance. Unless retrospective emission or other legislation ultimately forces older vehicles off the roads, the market for do-it-yourself lubricants seems likely to continue at a significant fraction of the total volume of the passenger car oil sales volume. This is spread between service station sales, car part specialists, and, increasingly, from supermarkets. In Europe, with many marketed quality levels, it would seem most likely that the quality of oils sold to the DIY market would remain one or two levels below the premium grades. In the U.S., with more of a single quality market, the size of the DIY market may be less important. An additional factor there is the growing popularity of "while you wait" oil drain and replenishment facilities.

If sealed-for-life crankcases were to be developed, then one can postulate that the initial fill oil would be of special quality, but top-up oils could be quite different from the initial fill quality. Depending on the success of the development of on-board diagnostic and analytical systems, or alternatively the development of simple dipstick test kits, one can even postulate the possibility of several different types of top-up oil which would preferentially increase certain parameters, e.g., the dispersancy level, the alkalinity, or the viscosity modifier content. If this stage is ever reached, then the complexities of oil replenishment would require use of specialist facilities.

Used oil disposal requirements and/or concern about PNA levels might promote the development of such rapid drive-in oil change systems, but their service could be improved by a provision being made by the engine manufacturers for easy drain and replenishment of the oil. If recyclability comes to be considered as important, then *shorter* oil drain periods may be preferable, and such quick-change units would become a vital part of the recycling chain. If oil change periods did come down, the used oil could be more easily recycled, but the initial quality level would be different from that used today in a conventional oil of perhaps 10,000 kilometres plus oil drain interval.

Any changes such as those indicated above in the supplier practice and quality levels of crankcase oils would require careful development, standardization and promotion either by individual OEMs or, better, by the industry coordinating bodies. If this were done it would probably be better to specify oils by standardized designations rather than to confuse the ultimate end-user with too much information on the precise quality of the fluid being used. The new ILSAC mark is in fact intended to be just this—it indicates the latest approved quality and the mark does not change if the quality is increased.

Most of our thoughts above have been directed at the private passenger car market. However, in the case of both small and large fleet operators of trucks and other vehicles it is highly desirable that used oil is not routinely handled by mechanics, and therefore the concept of quick oil changes with special quality lubricants could also be applicable in this area. Fill for life, particularly for diesel engines, is probably less realistic, but the concept of specialized top-up oils alongside very long-life initial fill oils is probably a valid concept. Even more than in the case of the passenger cars this would depend on the availability of simple oil test kits either on-board or in the workshop.

11.5 Developments in Testing, Classifications, and Approvals

We believe that there is fundamental instability in the technical society structure at present and that more structural changes may emerge. Primarily because it is still extremely difficult to develop good, precise, field-correlated engine tests quickly and cheaply, there will continue to be a conflict between the OEMs who want the capability for rapid change, and the oil and additive industries who are more concerned about the cost and quality of tests.

The lubricants industry may continue in its present state, but there must be a chance that the motor industry (as threatened in the U.S. from time to time) will act independently of the technical societies. An alternative scenario is a refusal by the oil industry to develop the new qualities of oil that are requested. Some oil companies may also feel unduly constrained by the rigidity of the new rules covering how quality levels may be claimed, and seek to operate outside the existing industry structure.

Is there a way through the apparent impasse? We suggest that one route might be to limit severely the number and type of tests qualified for "industry control." The remaining tests might be those that best measured fundamental properties and were not strongly hardware related. There might be more scope for single-cylinder and rig-tests. This would leave individual OEMs to manage other tests outside the "core." Depending on the type and quality of the test, there might be freedom to accept a test result from many sources or more controls might be needed. For some expensive tests it might be more cost-effective for the industry to run all tests in a single laboratory (perhaps operated by the OEM) rather than spend millions of dollars on correlation programs.

The smaller the core of "critical" tests, the easier it is to respond rapidly to change, and for lubricant marketers to add their own special features to satisfy local market needs.

Some background to these views is given in the following sections:

11.5.1 Test Costs

With increasingly frequent reformulations and rapidly escalating costs of oil development and approval procedures, the cost of developing new formulations has risen markedly in the last decade. With extreme competition between the additive suppliers, the pricing of formulated packages to the oil companies has not risen correspondingly, and at the present time it seems doubtful if the crankcase oil additive package business provides a satisfactory return to the average additive supplier. The major suppliers are believed to be worried about the situation, and many of the smaller additive companies have already been sold or offered for sale. The ability of the remaining additive industry to absorb these cost increases without passing them on to the oil company and ultimately to the consumer must be in doubt. Despite the pressure from the technical societies for the development of ever higher-quality lubricants of the conventional type, it is likely that countervailing pressures will develop for oil companies to develop unique types of lubricant that perform well in a limited range of vehicles but which have significantly lower ingredient and approval costs compared to the current multi-purpose line of standardized oils.

Particularly in the case of the heavy-duty diesel sector, the number of different tests required for an approval has been increasing steadily. There are now several different specifications against which an HD oil can be qualified, and if an oil is to meet several of these specifications simultaneously then the number of tests is markedly greater than those required to demonstrate HD performance a decade ago.

Of even greater significance than the number of tests required is the cost of each individual test. The older tests in single-cylinder laboratory engines such as the Caterpillar 1G and 1G2 have been supplemented by tests in large multi-cylinder engines, which are not only expensive to obtain but consume large quantities of fuel and require special dynamometers and cooling systems to cater to their high power outputs. The table below gives an approximate comparison of the costs of testing an API CD oil versus a more recent CF-4 oil, on the basis of a single test for each engine type. The cost figures are approximations only but it can be seen that they have risen by around 500% on moving to the new oil quality.

HOW MULTI-CYLINDER TESTS HAVE INCREASED APPROVAL COSTS (1993)

ENGINE TEST	CYLINDERS	$k (US) approx. API CD	API CF-4
CRC L-38	1	6	6
CAT 1G2	1	18	
MACK T-6	6		56
MACK T-7	6		14
NTC-400	6		54
TOTAL COST		24	148

11.5.2 Quality Approval Procedures

The administrative procedures for obtaining approvals are becoming more complicated, with the adoption of the CMA Code of Practice worldwide (see Section 6.3.1), requirements in Europe for laboratories to be ISO 9000/EN45001 accredited, and the insistence of some OEMs such as Volkswagen that tests be run in independent laboratories of their choosing. In the Asia Pacific basin Japanese influence is continually growing and there is a possibility that a Japanese-inspired rigorous quality control system may emerge in that area. As mentioned earlier, they already have in place a system for defining API "SE"-type quality in Japanese engines.(5)

In general new requirements are for oils to meet certain pass/fail criteria set on a statistical basis with qualified oils normally producing first time passes in qualification tests run in any approved engine operating to the correct procedures. This is more difficult to achieve than the usual previous requirement of a single pass in an unspecified number of attempts. Given equal passing levels, the initial development of technology is liable to be both more time-consuming and expensive, and also the final oil will contain higher levels of additive treatment and be more expensive.

In the past, various individual OEMs and other authorities have granted additive "package approvals" on the basis of demonstrated performance of a standard additive in more than one base stock. This practice has given the additive companies the opportunity to minimize technical service costs to individual oil companies once the initial product development cost had been paid. The practice has reduced in the early 1990s, as organizations around the world have implemented tougher quality procedures. The new codes of practice such as for API SH approval have indicated certain possibilities for base stock changes, but we believe these will need modification in the light of experience. In this instance, the demands of guaranteed quality and saving in testing costs are in conflict, and we would not be surprised if some modified form of package approval concept, where demonstrated performance in several base stocks would be required, were not to re-emerge for wider application in the next few years.

11.5.3 Oil Quality Development

The Caterpillar Series 3 or API CD quality of oil represented the highest level of diesel quality in general automotive use for over 30 years. It was, and may still in some areas, be considered a satisfactory quality level for many diesel applications. In Japan, for example, a 1993 survey showed that 72% of diesel oils marketed were of API CD quality and only 6% of newer higher qualities, while 21% were of only API CC quality. The retention of API CC quality and even some API CB in both Europe and Japan after these specifications have been declared obsolete supports the suggestion that older qualities and particularly API CD may be around for some time.

In the last decade, however, engine manufacturers have begun to demand many new oil qualities as designs have changed to meet tougher emissions limits, which may require new or different additive treatments from those required simply to provide a high degree of detergency. Some of the requirements of different manufacturers seem to be incompatible,[2,6] and some manufacturers are not satisfied with the progress that has been made in attempting to meet their requirements; so at the present time the HD area has several alternative specifications (which the market requires to be met with as few oils as possible) and the manufacturers are seeking new and harder targets for the future. From their point of view it is only reasonable that development of oil qualities to suit their equipment should be performed in full-scale engines of the requisite type, and this is one reason why the cost of engine testing in the HD area has escalated so rapidly in recent years.

11.6 Future Crankcase Oils

Because of the huge market for existing vehicles, even dramatic changes in legislative demands, environmental pressures or new technology will make only a gradual impact on the total crankcase lubricants market.

We do not foresee changes that would convert the U.S. from being essentially a single-quality market or Europe, Japan, and other areas from being multiple-quality markets. The OEMs through the technical societies and trade associations will maintain a dominant influence on quality using API and ACEA classifications. Japanese manufacturers will have a growing influence, starting from their area of greatest market strength in Asia-Pacific.

In the interest of improved fuel economy, oil viscosity will generally fall with SAE 5W-30, already widely used in the U.S., being much more generally acceptable worldwide. Reduced engineering tolerances and new valve gear designs will permit the use of low-viscosity oils in modern engines without the onset of wear problems. To provide acceptable volatility such oils will use increasing quantities of unconventional base stocks, mainly of petroleum origin. Use of ester base stocks is unlikely to grow rapidly unless a requirement for oil biodegradability is introduced. Development of improved oil change procedures, together with effective collection and safe disposal of used oil, will probably make a requirement for biodegradability of oils for road vehicles unnecessary. Oil companies in their searches for market leadership will continue to find new ways to serve the customer. As new targets are established, new high-quality specialist oils will be developed, and in parallel new types of multi-purpose oil will offer new combinations of features.

Within the passenger car and light commercial vehicle sectors the average ash level of lubricants will decline slowly in the next 10 years or so, due to a reduced requirement for high TBN oils as lead-free gasoline and low-sulfur diesel fuel become the norm. Provided that relevant new engine test procedures are developed, we see no reason why ash levels in gasoline-engined vehicles should not decline slowly to zero if ZDDP can be removed cost-effectively. The concept of zero ash oils has long been seen as a desirable and marketable feature, but the last time it was tried in the passenger car market (in the 1950s) suitable technology was not available. It is believed that additive technology now exists to achieve this, but there is no set of standardized engine tests against which such oils could be qualified. The process of developing relevant engine tests operating on new fuel qualities with representative future oils will inevitably be slow and take time to achieve, but merits urgent consideration.

11.7 Other Automotive Lubricants

In the passenger car market the use of manual gearboxes will continue to decline. For commercial and residual passenger car usage, gear oils will tend to be of lower viscosity and in many cases multigraded, all in the interest of improved fuel economy. In most cases gear oils are effectively filled for life already, but this will become standard.

Automatic transmissions will show increased market penetration outside the U.S., and low-viscosity, long-life formulations will be sought by the OEMs. Fill-for-life concepts will gradually be introduced. For small passenger cars, the penetration of small CVTs is expected to increase, and development of a special lubricant will provide benefits of economy, transmission life, and possibly smoothness.

Special hydraulic oils will be increasingly required for active suspensions. Increased availability of mineral-oil-based hydraulic fluids for automotive use may lead to replacements of synthetic brake fluids by low-pour mineral fluids in the interest of reduced water take-up and seal compatibility. Low-pour-point, catalytically dewaxed base stocks could play an important role in the whole area of automotive hydraulics.

For front-wheel-drive vehicles improved greases will be required for the outboard end of constant velocity joints as these come closer to the higher temperature around the brake discs.

Greases will find extra uses in electric or other novel vehicles.

11.8 Conclusions

While most of the evidence suggests that the automotive lubricants market is a mature one and will follow slow and evolutionary change, there is enough external change to ensure that many chemists and engineers will be fully employed on the challenges of automotive lubrication for at least another generation.

We believe the pace of generating new specifications will increase rapidly in the next few years due to demands in the following areas:

a) Overcoming manufacturers' problems as they strive to improve engine emissions and fuel consumption.

b) Catering to the new fuels with comprehensive exclusive specifications for:
   - unleaded gasoline (not properly addressed so far)
   - low-sulfur diesel fuel
   - oxygenated special fuels
   - alcohol fuels
   - exotic fuels

c) New engine types, particularly new 2T designs for passenger cars.

Resource limitations suggest that only some of these will be able to be worked if the new standards for specification setting and rigorous testing are maintained.

There will be continuous realignment of oil grades with market needs. Cost pressures to rationalize grades will continue as new market opportunities arise. Oil companies will find new ways to get better market coverage with a different mix of products, but not necessarily a greater number of products.

In the passenger car sector the use of cracked/isomerized and/or catalytically dewaxed base stocks will grow rapidly in premium oils. Where second-line oils are common these and the commercial oils will continue to be formulated with mainly conventional extracted base stocks.

Development of lower additive treat oils for use with unleaded gasoline and very low-sulfur diesel will depress the additives market, and marketing of specialist rather than multi-purpose oils would bring further reductions. After the additives companies have made what cost savings they can, additive prices can be expected to rise to enable them to stay in business. More companies may well withdraw from the additives market.

The oil companies will also not have a smooth ride. There will be considerable costs in approving new oil qualities as demanded by the market or associated with the introduction of new lines of special oils. In the longer term, with reduction of average oil consumption per vehicle and the introduction of electric vehicles, the crankcase market will decline significantly in mature areas, although considerable growth will occur in the emerging automotive markets such as China and India.

Internationalization of specifications and oil qualities has many advantages, but the problems of defining oils that meet the demands of more than one region without being over-formulated for most applications are very great. It may be possible to operate a genuine worldwide ILSAC system, but we seriously doubt whether it is achievable with the current industry rules.

In case we appear too pessimistic, let us state that we feel the years ahead will be very interesting to those working with automotive lubricants, and while there will be many challenges, there will also be great rewards for those meeting them successfully.

References

1. "Fuels, Lubricants and the Fleet Operator," *PARAMINS Post*, Issue 9-4, p.12, February 1992.
2. Ohkawa, S., "Present and Future Lubricants for Construction Machines," Fuels and Lubricants Asia Conference, Singapore, January 1995.
3. *Lubrizol Newsline*, Vol. 11, No. 5, September 1993.
4. Gluck, S.J., Tauber, P., Schupp, B., Anderson, J.E., "Design and Commercialisation of the PRT 2000 Personal Rapid Transit System," Raytheon Company/Taxi 2000 Corporation, 3 November 1994.
5. Watanabe, S., "Japanese Engine Tests for Specifying Gasoline Engine Oils," Japan Lubricating Oil Society, Savant Conference, 1991.
6. Lavender, J., open letter to the oil and additive industries reproduced in *PARAMINS Post*, Issue 4/94, p.13.

Appendix 1

Glossary

A

Abrasion	Removal of material from a surface by scratching.
Acetylene	A reactive hydrocarbon characterized by the presence of a triple carbon-carbon bond. Particularly the lightest member, C_2H_2.
Acid	A hydrogen-containing substance that dissociates in water to produce hydrogen ions. It reacts with metals, oxides or bases to form salts.
Acid Number	A measure of acidity based on neutralization with alkali of known strength.
Acid Treating	A method of lubricating oil production by contacting with sulfuric acid.
Acronym	A short form of the name of an organization, etc., based on the initial letters of each word in the name.
Acrylic	Pertaining to the unsaturated acrylic acid CH_2:CH.COOH or its compounds.
Active Ingredient	The functional material within an additive, etc., which may also contain inert or carrier substances.
Acute Toxicity	The toxic effect resulting from a single exposure to a toxic substance.
Additive	A small amount of a chemical substance added to improve the properties of a material (such as a lubricant).
Adsorption	Adhesion of substances to a surface, such as some additives to metals.

Aerosol	A finely divided dispersion of a liquid or solid in a gas.
Alcohol	A class of organic chemical compounds characterized by the presence of a hydroxyl group, –OH , of general formula: $C_nH_{2n+1}OH$.
Aliphatic	Hydrocarbons which are of open chain type, without ring structures.
Alkali	A basic substance usually based on hydroxides or carbonates of alkali metals (Li, Na, K, etc.) or the alkaline earths (Ba, Ca, Mg, etc.) or the ammonium ion.
Alkyl	Denotes a monovalent radical derived from an aliphatic hydrocarbon, such as the methyl radical CH_3– derived from methane CH_4.
Alkylate	The product of a process in which alkyl groups are added to other molecules such as unsaturated hydrocarbons or benzene.
Amorphous	A non-crystalline substance with no discernible internal structure.
Amphoteric	A material which can behave either as an acid or a base depending on the type of substance it reacts with. An example is aluminum hydroxide.
Anhydrous	A substance that contains no water.
Apparent Viscosity	The viscosity of a liquid measured at a given shear rate for cases where the viscosity is shear rate dependent.
Aromatic	An unsaturated hydrocarbon characterized by the presence of one or more benzene rings.
Aryl	A radical or group derived from an aromatic molecule, such as the phenyl group C_6H_5– derived from benzene C_6H_6.
Ash	The non-combustible residue left after a substance is strongly heated in oxidizing conditions. The quantity of residue may depend on the time and conditions of heating.
Asperities	Microscopic projections from a surface, particularly on a sliding or bearing surface which result

	from machining operations or finishing treatment. They are the primary cause of friction if there is an insufficient film of lubricant between sliding surfaces.
Asphalt	Black or brown residual material from crude oil distillation, or the heavy material in crude oil. Also called bitumen when used as building or roofing material.
Asphaltines	The high-molecular-weight substances contained in asphalt and heavy residual base stocks which are insoluble in light paraffinic solvents.
Aspiration	The drawing of air at atmospheric pressure into a device such as an engine; also the drawing of a test liquid in droplet form into a flame or plasma for analytical purposes.
Atmosphere	The mixture of gases surrounding the earth. Also a unit of pressure equal to 760 mm of mercury, 14.7 psi, or 101.3 kiloPascals.
Atom	The smallest possible particle of a pure elemental substance.
Atomization	The conversion of a liquid into a spray of very fine droplets.
Automotive	Literally, self-propelled. Usually reserved to describe land-based vehicles such as cars (automobiles), buses, trucks, and off-highway vehicles.
Azeotrope	A liquid mixture of two or more components which boils at a temperature different to that of any of the different components.

B

Babbit	A relatively soft bearing metal used in early engine designs. It is an alloy of tin, copper, and antimony.
Bactericide	An additive used in many cutting oils to inhibit bacterial growth which causes degradation and unpleasant odors.

Barrel	A traditional petroleum unit of measure, equivalent to 42 U.S. gallons or 159 litres.
Base	An alkaline substance, that is, one that ionizes to produce hydroxyl ions, OH–, and reacts with acids to form salts plus water.
Base Number	A measure of the basicity (alkalinity) of a base which is obtained by reacting it with acid of known strength until neutralized.
Base Stock	The primary liquid constituent of a lubricant. Also the various separate components which may make up this liquid. Base stocks may be of petroleum fractions suitably refined, or synthetic.
Bearing	A component of a machine designed to support moving parts and minimize the friction between them. Rolling element bearings, such as ball or roller bearings, are more effective in reducing friction than plain bearings where the surfaces slide over each other.
Benzene	An important organic chemical consisting of a hexagonal ring of carbon atoms with a hydrogen atom attached to each. Many compounds with one or more benzene rings occur in petroleum, and are known as aromatic compounds.
Bevel Gear	A straight-toothed gear with the teeth cut on sloping faces and the gear shafts at an angle (normally a right angle).
Biodegradable	A material such as an oil or plastic is said to be biodegradable when it can be broken down by naturally occurring bacteria into simple substances which do not harm the environment.
Bitumen	A black material consisting mainly of petroleum residues, but which may contain other substances or be oxidized to make it more suitable for its principal uses in building products or for road building.
Bleeding	The separation of some of the liquid phase from a grease.

Blowby	The mixture of gases and chemicals that is blown past the piston rings of an internal-combustion engine by the combustion pressure, thereby contaminating the oil in the sump.
BMEP	Brake Mean Effective Pressure. An indicator for engine efficiency, based on the theoretical pressure to produce the same output from an engine under idealized circumstances as is found in practice.
Boiling Range	For a mixture of substances, such as a petroleum fraction, the temperature interval between the initial and final boiling points.
Bomb Oxidation	A test for the oxidation stability of a product obtained by sealing it in a closed container with oxygen under pressure. The drop in pressure of the oxygen is a measure of the amount of oxidation that has occurred.
Bottoms	In a distillation column, the material that does not distill over but remains at the bottom of the column. In crude oil distillation it is known as the residuum.
Boundary Lubrication	The condition when lubricant is present between sliding surfaces, but in insufficient quantity to provide an unbroken film to separate them completely. Resulting friction and wear can be alleviated by use of additives which are adsorbed onto the surfaces.
Brake	In engine testing, the device used to absorb power so that the test engine can be run at suitable speed and power levels.
Bright Stock	A high-viscosity base oil made from the bottoms of the vacuum distillation column, by de-asphalting and dewaxing.
Brookfield Viscosity	The viscosity or apparent viscosity of a product as measured by a Brookfield viscometer. A spindle is rotated in the product and the torque generated by fluid friction is read from a scale.
Bulk	Supply of product in large volumes rather than in packages. Common means of supply would be

	road or rail tank wagon, barge or ship. Supply in "container tanks" is usually called "semi-bulk."
Butadiene	A C_4 hydrocarbon containing two double bonds $H_2C{:}CH.CH{:}CH_2$. It polymerizes with itself or other substances to form rubbers.
Butane	Saturated C_4 hydrocarbon C_4H_{10}; can be straight or branched chain.
Butylene	Also called butene. A C_4 hydrocarbon with one double bond. The branched chain isobutylene is an important chemical raw material.
Butyl Rubber	A synthetic rubber obtained by polymerizing isobutylene with butadiene or isoprene (2-methyl butadiene).

C

Calibration	The process of setting up a measuring instrument by reference to standards of known value.
Cam	A rotating lobed disc used to open the valves in an engine.
Cam Follower	Also known as a tappet. The part in a pushrod-operated engine that rides on a cam to convert the cam rotation into a linear movement.
Carbon (deposit)	Solid black residue in piston grooves which can interfere with piston ring movement leading to wear and/or loss of power.
Carbon residue	The coke-like material left after strongly heating a lubricant base stock containing heavy or unstable fractions. Two alternative test methods, ASTM D 189 (Conradson coke) and ASTM D 524 (Ramsbottom Carbon residue) give different results. Not applicable to additive-containing oils.
Carbon Type	The distinction between paraffinic, naphthenic, and aromatic molecules. In relation to lubricant base stocks, the predominant type present.
Carbonyl	The divalent :CO radical found in aldehydes and organic acids and some other compounds.

Carboxylic Acid	The common organic acid, containing hydroxyl and carbonyl radicals. The radical is normally represented as –COOH.
Carburetor	The device in conventional gasoline engines that atomizes the fuel and mixes it with air in the correct proportions for combustion. Also contains a throttle plate to control the amount of mixture reaching the engine. Progressively being replaced by fuel injection systems.
Carcinogen	A substance that has been shown to cause cancer.
Catalyst	A substance that initiates or increases the rate of a chemical reaction, without itself being used up in the process.
Catalytic Converter	An emissions-control device containing catalysts that promote oxidation of hydrocarbons and carbon monoxide and reduction of nitrogen oxides, giving an exhaust containing mainly water, carbon dioxide, and nitrogen. The catalysts are poisoned by leaded gasoline and possibly by high levels of zinc and/or phosphorus derived from the lubricant.
Catalytic Cracker	A refinery unit that breaks down heavy petroleum molecules into simpler ones of greater utility.
Caustic	A highly alkaline substance such as sodium hydroxide.
Celsius	The temperature scale in which the freezing point of water is 0° and the boiling point 100°. Formerly known in English as Centigrade.
Centigrade	See Celsius.
Centipoise	A common unit of viscosity, equal to one hundredth of the fundamental viscosity unit, the Poise.
Centistoke	The unit of viscosity commonly used when a liquid falls through a capillary tube under its own weight. Related to the centipoise by the density of the fluid: Centistokes × Density = Centipoise.
Cetane No.	A measure of the ignition quality of a diesel fuel. A high cetane no. indicates a shorter lag between fuel injection and ignition.

Chromatography	An analytical technique whereby a complex substance is adsorbed on a solid or liquid substrate and progressively eluted by a flow of a substance (the eluant) in which the components of the substance under investigation are differentially soluble. The eluant can be a liquid or a gas. When the substrate is filter paper and the eluant a liquid, a chromatogram of colored bands can be developed by use of indicators. For gas chromatography, electronic detectors are normally used to indicate passage of the various components from the system.
Chronic Toxicity	The toxic effect resulting from prolonged exposure, possibly at low dose rates, to a substance which may be considered toxic.
Circulating System	A lubricating system in which oil is recirculated from a central sump to the parts requiring lubrication and then returned to the sump.
Clay Treatment	An older lubricating oil refining process in which the stream to be treated is contacted with activated clay to remove polar compounds and acids. Traces of water and solids are removed as the clay is filtered from the oil.
Cleveland Open Cup (COC)	A flash point test in which the surface of the sample is completely open to the atmosphere, and which is therefore relatively insensitive to small traces of volatile contaminants.
Cloud Point	As an oil is progressively cooled, the temperature at which wax starts to separate from the oil, producing a cloudy appearance.
Coefficient of Friction	A measured value indicating the degree of friction between two surfaces. Equal to the force required to slide one surface over the other divided by the force holding the surfaces together.
Cohesion	The molecular force holding a material together, thus the cohesion of certain greases and oils is important to avoid splatter.

Colloid	A suspension of microscopically fine particles in a liquid which do not settle out and are not easily filtered off. The liquid may well appear clear to the eye. Colloidal suspensions are sustained by ionic charges on the particles, and many lubricating oil additives (particularly detergents) consist of solid materials held in colloidal solution by surface active agents.
Combustion	Rapid oxidation or burning of a substance. Combustion of hydrocarbons under ideal conditions produces only water and carbon dioxide.
Combustion Chamber	The space between the piston and cylinder head in an internal-combustion engine where the charge of fuel plus air is burned to produce power.
Commercial Oils	The range of oils similar to or within the API "C" Service Categories of crankcase lubricants, and which are primarily for use in diesel engines.
Complex Soap	A grease composition in which the alkali is reacted both with a long chain acid and a short chain acid or an inorganic acid.
Compounded Oil	A lubricating oil that contains animal or vegetable oils or fats.
Compression-Ignition (CI)	Ignition of fuel by the heat generated in compressing the air charge, as in the diesel engine.
Consistency	The hardness or softness of a grease.
Co-polymer	A high-molecular-weight substance formed by the linking of two or more lower-molecular-weight monomers.
Corrosion	Chemical attack on a solid. In an engine, water causes corrosion (rust) on iron parts, and acids from fuel combustion or oil oxidation can cause corrosion of many different metals.
Cracking	Refining process in which large molecules are broken down into smaller molecules. Cracking takes place to some extent whenever high molecular material is heated strongly, but is accelerated by catalysts.

Crankcase Oil	Oil used for general lubrication in an engine where there is an oil sump below the crankshaft to which circulated oil returns. (The term can also be applied to lubrication of other reciprocating devices such as compressors, in which case the type of oil may be different.)
Crude Oil	Naturally occurring petroleum, before any refining or treatment.
Cut	A discrete fraction of a complex material such as a petroleum feedstock obtained by distillation.
Cutting Back	Dilution of a concentrated material with a solvent.
Cutting Oil	A lubricant used in machining operations for lubricating the tool in contact with the workpiece, and to remove heat. The fluid can be petroleum based, water based, or an emulsion of the two. The term "emulsifiable cutting oil" normally indicates a petroleum-based concentrate to which water is added to form an emulsion which is the actual cutting fluid.
Cyclic Hydrocarbon	A hydrocarbon containing ring structures, which can include the aromatic compounds containing benzene rings or more saturated or completely saturated rings. The latter are known as naphthenes. Rings are usually 5- or 6-membered, but larger rings are possible.
Cylinder Oil	A once-through lubricant injected into the ring zone of steam or large marine diesel engines, or of air compressors.

D

De-asphalting	The removal of heavy asphaltic material from a residual feedstock to produce a bright stock. A light paraffinic solvent is used to precipitate the asphalt then distilled off and recycled.
Demerits	A method of evaluating or rating engine parts after an engine test, in which deposits are rated on a

	scale with higher numbers signifying heavier deposits. Total weighted demerits is a sum composed of the demerits of several parts or areas which have been multiplied by laid-down weighting factors related to the importance of that area, thereby arriving at an overall assessment.
Demulsibility	The ability of an oil to separate from water on standing.
Demulsifier	An additive that promotes separation of oil and water from emulsions.
Density	An absolute property of a substance, the mass per unit volume of the material. It varies with the temperature of the material. The official S.I. unit is kg/m^3, but g/cm^3 and g/litre are widely used. Traditionally Specific Gravity, the ratio of the mass of a product to that of an equal volume of water at the same temperature, has been more widely used in the petroleum industry. Specific Gravity (S.G.) is dimensionless.
Dermatitis	Inflammation of the skin. Repeated contact with petroleum products can be a cause.
Detergent	In lubricants, an additive that reduces formation of piston deposits in engines. It will normally have acid-neutralizing properties and be capable of keeping finely divided solids in suspension. Most detergents are based on metallic soaps, and are known as overbased if they contain solid alkali in colloidal form.
Detergent-Inhibitor	Formerly a detergent which controlled bearing corrosion as well as deposits. This usually required control of oxidation, and the term is now taken to mean a detergent which also has antioxidant properties.
Dewaxing	Removal of wax from a base oil in order to reduce the pour point.
Dibasic Acid	An organic acid with two carboxylic acid groups in the same molecule.

Diester	The product of the reaction of a dibasic acid with one or more alcohols. Diesters are widely used as synthetic lubricants.
Diesel Engine	An internal-combustion engine in which the fuel is ignited by the heat generated in compressing the charge of air. The fuel is sprayed into the combustion chamber at high pressure by means of a diesel injector, and burns rapidly. The quantity of fuel is metered in to control power output.
Diluent	A solvent used to reduce the concentration of a substance. Lubricating oil additives frequently consist of major proportions of diluent oil containing active ingredient which, if not so diluted, would be difficult or impossible to handle and dissolve in base oil.
Dilution	In a crankcase oil, the contamination of the oil with fuel residues which reduce its viscosity. In analysis, quantitative dilution of a sample is often required in order to bring the range of constituents to be measured to the same level as those in a standard to which it is to be compared.
Dioxins	A chemical family of chlorinated aromatic compounds considered to be highly toxic and carcinogenic, at least to laboratory animals. Dioxins could be produced by incomplete combustion of petroleum products which contain chlorine.
Dispersant	An engine oil additive whose primary function is to hold in suspension solid and liquid contaminants, thereby passivating them and reducing engine deposits at the same time as sludge deposition is reduced. Most detergents have some dispersancy action. The ashless dispersants are polymer-based materials which are particularly useful in holding water in suspension in gasoline engine oils, as well as suspending solid particles. Dispersants of this type have little acid-neutralizing capability.

Dispersion	A suspension of small liquid or solid particles in a suspending medium. A colloidal dispersion is stable, but a dispersion does not have to be either colloidal or stable.
Distillation	The separation of a mixture of liquids of different boiling points by progressively raising the temperature. In a refinery distillation unit the temperature rises continuously from the top to the bottom of the column, and different fractions or cuts are drawn off at different heights.
Distillation Test	A laboratory test in which a petroleum sample is distilled in a standard apparatus. The Initial Boiling Point (IBP) is the temperature at which the first drop of liquid distills, and the Final Boiling Point (FBP) is the highest temperature reached.
Drag	The resistance to movement caused by oil viscosity.
Drawing (metals)	A metal shaping process involving simultaneous stretching and forming between dies.
Dry Lubrication	The situation when moving surfaces have no liquid lubricant between them.
Dry Sump	An engine design in which oil is not retained in a pan beneath the crankshaft thus permitting splash lubrication. There may be a remote sump from which oil is recirculated, or there may be a total loss system.
Drop(ping) Point	The temperature at which an oil commences to drip from a grease as it is progressively raised in temperature.
Dumbbell Blend	A mixture of two petroleum fractions with significantly different viscosities, volatilities, or other properties. Some properties of the mixture will appear as intermediate but others will be closer to the extreme values of one component.
Dynamic Viscosity	The absolute viscosity of a liquid as measured in a rotational instrument, as distinct from the kinematic viscosity where the liquid falls through a tube under its own weight.

E

Elastohydrodynamic Lubrication — The lubrication regime where highly loaded non-conforming surfaces undergo deformation to enlarge the contact area and the area of the oil film. At the same time the high pressures in the oil film lead to an increase in local oil viscosity. The phenomenon is particularly important in the case of rolling contact bearings.

Elastomer — A rubbery type of material.

Emissions — The principal regulated gaseous vehicle exhaust emissions that contribute to air pollution are unburned hydrocarbons, carbon monoxide, and nitrogen oxides. Solids (particulates) emitted primarily as visible smoke from diesel vehicles are also regulated. These pollutants can be greatly reduced with exhaust catalyst converters and particle traps. Of increasing concern is the emission of carbon dioxide which is an inevitable result of burning fuel. This contributes to the so-called "greenhouse effect" which may lead to global warming.

Empirical — In petroleum testing, the situation where the result of the test depends on the design and operation of the apparatus as much as on the material being tested. Thus comparison of materials must be made using identical apparatus in an identical manner, as there is no fundamental property being measured. Examples include the flash-point tests and pour point.

Emulsifier — An additive that promotes the formation of a stable emulsion.

Emulsion — A mixture of fine droplets of one fluid dispersed in another fluid. Emulsions can be of oil-in-water type or water-in-oil (invert emulsion). The size of the droplets of the dispersed phase determines

	whether the emulsion appears clear (transparent) or cloudy (milky emulsion).
Energy	The capacity to perform work. In an engine the stored chemical energy in the fuel is converted to thermal energy on combustion, and then to kinetic energy to propel the vehicle. The energy is not used up, but converted to low grade forms which are not readily usable, such as the heat of friction. The S.I. unit of energy is the Joule. Other units include the calorie and the kiloWatt-hour.
Engler Viscosity	An early empirical method of measuring viscosity which was used in continental Europe. Similar to the Saybolt method in the U.S.
Entrainment	The dispersion of a liquid or gas in a fluid without solution. Frequently a coarse unstable dispersion which does not amount to a foam or emulsion, arising from rapid flow of the carrier fluid in a situation where it can mix with the dispersed fluid.
Extreme Pressure	The lubrication regime where surfaces are sliding against each other under heavy load. The example for which the expression was coined is the teeth of the hypoid gears in an automotive rear axle. Extreme pressure additives react with the metal of the asperities under the high temperatures generated, producing compounds which shear readily without tearing the metal surface.
Erosion	The process of wearing away a surface by continued passage of a fluid, particularly when this contains suspended material.
Ester	An organic compound formed by the reaction of an acid (organic or inorganic) with an alcohol or other hydroxyl compound. Water is eliminated in the reaction. Several types of ester are useful as synthetic lubricants.
Ethane	A gaseous paraffin, C_2H_6.
Ethanol	Ethyl alcohol, C_2H_5OH, present in alcoholic drinks and mainly obtained by fermentation. Used as a fuel component in "gasohol."

Ethylene	The gaseous lowest olefin, C_2H_4, also called ethene. A basic petrochemical for polymer and other chemical production.
Evaporation	The conversion of a liquid into its vapor (gas) by the action of heat. Most liquids undergo some evaporation at ambient temperatures, although this may be undetectable.
Exhaust Gas Recirculation	Rerouting a proportion of exhaust gases to the inlet manifold of an internal-combustion engine. The oxygen content of the charge is reduced and fewer nitrogen oxides (NO_x) are produced.

F

Fahrenheit	The temperature scale in which the boiling point of water is 212° and the freezing point 32°.
Falex Test	A bench test for EP properties, using V-blocks around a rotating shaft.
Fat	An animal or vegetable material, consisting of the glyceryl esters of long chain organic acids (fatty acids).
Fatty Acid	A long chain monobasic carboxylic acid, $C_nH_{2n+1}COOH$. Fatty acids are obtained by hydrolysis of fats.
Fatty Oil	An oil of animal or vegetable origin.
Feedstock	The starting material for a refining or chemical process.
Film Strength	The property of a lubricant which enables it to lubricate under boundary conditions. Enhanced by addition of long chain polar molecules.
Flash Point	The lowest temperature at which the vapors from a heated product will ignite when exposed to the air under prescribed conditions. Flash Point tests are empirical and there is no correlation between results by different methods. There are two types

of test: Closed Cup (e.g., Pensky-Martens) in which the vapors are collected under a cover, and Open Cup (e.g., Cleveland Open Cup) in which the vapors are not trapped and vapor emission has to reach a certain level before a flash will occur. The open cup is much less sensitive to trace amounts of volatile matter.

Foam — A mixture of oil and air in a relatively stable form which may be caused by vigorous agitation in the presence of certain additive or impurities. Being compressible, it is a poor lubricant and will not operate hydraulic equipment properly. Excessive foaming can lead to loss of lubricant.

Fossil Fuel — Fuels such as coal and petroleum which have been formed by the action of heat and pressure on ancient vegetable or animal matter.

Four-ball Test — A mechanical tester used to evaluate the wear prevention properties of a lubricant, in which a ball- bearing is rotated against three other identical balls under various loads.

Four-square Tester — A device in which two pairs of gears are connected by two parallel shafts. By twisting one of the shafts (applying a torsional load) the loading on the teeth of the gears can be increased. One pair of gears is used to test the lubricating abilities of an oil or grease, and the other is made of very strong permanent gears. Examples include the Ryder, I.A.E., and FZG testers.

Four-stroke Cycle — A method of operating a reciprocating engine in which the power-producing stroke occurs once in every four strokes or piston movements, normally corresponding to two engine revolutions.

Fraction — A segregated part of crude oil or a petroleum product obtained by distillation.

Fuel Economy Oil — An engine lubricant which enables an engine or vehicle to use less fuel for a certain duty than a standard lubricant. Such oils are usually of lower than normal viscosity and may contain special anti-wear and/or lubricity additives.

Fuel Injection — The system of introducing fuel into an engine through a small nozzle under pressure. All pure diesel engines are dependent on fuel injection for their operation, with a separate injector for each cylinder. Gasoline engines can use single point injection (to inlet manifold or carburetor) or multi-point injection (to each cylinder).

Full Fluid Film (lubrication) — The situation where the surfaces in relative motion are separated by a continuous film of lubricant. The film may arise from hydrodynamic lubrication or be supplied by a hydrostatic lubrication system.

Furfural — A hydrocarbon liquid used as a solvent for aromatics, mercaptans, and polar compounds in the solvent refining of lubricating oil.

FZG Tester — A four-square gear oil tester of German origin (see list of Acronyms in Appendix 2).

G

Gas Oil — A heavy petroleum distillate, originally used to generate illuminating gas, now used as diesel fuel and a blending stock for fuel oil.

Gasohol — A blend of gasoline and anhydrous ethanol, usually in 90:10 ratio.

Gasoline — A wide and complex blend of natural and modified petroleum fractions, used as fuel in spark-ignition engines. For modern engines a high octane number is required to prevent engine knock. Lead alkyls were used for many years to increase octane number, but toxicity and catalyst-poisoning problems are increasingly leading to elimination of lead from gasoline. Increased quantities of high octane synthesized components are needed for blending unleaded gasoline.

Gas Turbine	A rotary engine in which air is compressed and directed to combustion chambers, where fuel is sprayed in and a continuous stream of high-pressure hot gases generated. These are directed at a turbine from which useful power can be obtained. The compressor can be driven by the same or a different turbine. Advantages over reciprocating engines include smooth output, compact size for a given power output, and a relative insensitivity to fuel quality.
Gears	Toothed machine parts for transmitting motion and power from one shaft to another. The shafts may be parallel or at an angle to each other (usually 90°). If one of a pair of gears is smaller than the other, it is known as the pinion, and the larger is called the gear. Differences in tooth design can provide quieter running, increased power handling, or accommodate different angles between the driving and driven shafts.
Gel	A semi-solid state of matter which arises from a suspension of solid particles with an affinity for each other in a liquid. On agitation the semi-solid material may re-liquefy. Gels frequently arise when a solution of a sparingly soluble substance is cooled, and some of this comes out of solution. Gels can also be formed from dispersion of certain finely divided but insoluble substances in a liquid.
Glyceride	An ester of glycerol (the simplest tri-hydric alcohol) with a fatty acid. Mixtures of glycerides are the main ingredients in natural fats and oils.
Graphite	A soft, naturally occurring form of carbon. The crystals are in the form of platelets which easily slide over each other, making it an effective solid lubricant. It can also be made synthetically.
Grease	A mixture of a lubricating liquid and a thickener, which has sufficient solidity to stay in place when undisturbed, but flows under the influence of motion to lubricate bearings, gears, etc. The

thickeners are commonly metal soaps of fatty acids, but solid thickeners and new chemical thickeners are being increasingly used.

H

Halogen	An element of the chemical family: Chlorine, Bromine, Fluorine, Iodine.
Heat Transfer Fluid	A liquid medium for taking heat from one part of a system and releasing it in another. It may be used for heating, when the source of heat may be remote from or unsuitable for the object to be heated, or it may be used for cooling hot objects when the heat is dissipated in a large thermal sink (cooling pond, river, etc.). An ideal heat transfer fluid has a high specific heat, is stable at the temperatures encountered, and is inexpensive (large volumes are often needed).
Heavy Duty (oil)	Originally a crankcase lubricant which contained detergent additives, used mainly in diesel engines but also in some large gasoline engines. More recently it is used to categorize lubricants containing high levels of detergency for specific application in large diesel engines.
Heavy Ends	The portions of a petroleum distillate fraction which are highest boiling, and therefore distill over last if the temperature is raised progressively.
Helical Gears	Gears cut with the teeth cut at angle to the center of rotation, so that the load is transferred progressively along the teeth with one tooth taking the load before the previous tooth has dropped out of mesh. This results in much quieter running than for straight-cut gears, where there is a jerky transfer of load from one tooth to another.
Heptane	A paraffinic hydrocarbon (or mixture of isomers) with a total of seven carbon atoms.

Hexane	A paraffinic hydrocarbon (or mixture of isomers) with a total of six carbon atoms.
Homogenization	The intimate mixing of insoluble material (solid or liquid) in a fluid by intense mechanical action or shearing. Greases and soluble-oil emulsions may require homogenization during manufacture.
Horsepower	An arbitrary unit of power equal to 33,000 foot-pounds/minute or 745.7 Watts. The horsepower rating of an engine depends on the conditions under which it is run, and whether accessories such as dynamos and pumps have their power consumption included in the total figure or not. Values therefore differ depending on the precise definition used.
Humidity	The amount of water vapor in the atmosphere. Absolute humidity is the actual amount of water in a given volume of air; relative humidity is the ratio of the actual amount of moisture to the maximum that the air could carry at a given temperature.
Hydrated	Containing water, as in certain minerals which contain water of crystallization. In greases, thickeners that have water incorporated.
Hydraulic	Originally, devices or systems using water as a working fluid. More usually today, systems for transferring power or movement by means of a hydraulic fluid which can be based on petroleum, synthetic oils, emulsions, or solutions of chemicals in water. The simplest devices consist of two linked cylinders containing pistons and filled with hydraulic fluid. Movement of one piston causes movement in the other. Systems can be continuously pressurized, and hydraulic motors can convert pressure into rotational movement.
Hydrocarbon	A chemical compound consisting of hydrogen and carbon atoms. Petroleum consists mainly of hydrocarbons. The existence of millions of different types of hydrocarbon is due to the tetrahedral

	arrangement of the four carbon valencies, and the ease with which carbon atoms link together to form chains or rings.
Hydrocracking	A refining process in which petroleum fractions are cracked at high pressure in a hydrogen stream, whereby a variety of lighter and more useful products is produced.
Hydrodynamic Lubrication	Full fluid film lubrication where the oil film is generated by the relative movement of the surfaces and the existence of an oil wedge.
Hydrofining	A hydrogen refining process. Such processes are used for improving the color and stability of products, particularly lubricating oil, and for removal of sulfur. The process is conducted at high temperatures in the presence of hydrogen and a catalyst.
Hydrogenation	The addition of hydrogen to unsaturated molecules, such as aromatics or those containing olefin linkages, to render them more stable. The process takes place in the presence of a catalyst, and can simply saturate existing material, or it can be severe as in the case of hydrocracking when chains are ruptured and hydrogen is added to the new chain ends.
Hydrolysis	The decomposition of certain compounds such as esters in the presence of water. Esters yield the original acid and alcohol when hydrolyzed.
Hydrolytic Stability	The relative ability of substances (such as ester lubricants) to resist hydrolysis.
Hydrometer	A device, consisting essentially of a tube weighted at the bottom, used to measure the specific gravity (density in relation to water) of petroleum products.
Hydrophilic	Compounds with an affinity for water.
Hydrophobic	Compounds that repel water.
Hydrostatic Lubrication	Full fluid film lubrication supplied by externally pressurized oil.

Hypoid Gears — Gears giving a right-angled drive with a pinion offset from the centerline of the gear. This gives a lowered driveline in rear-drive cars.

I

Immiscible — Not capable of being mixed to form a homogeneous whole: e.g., water and mineral oil.

Induction Period — A period of initial low-level activity in a chemical reaction before the reaction takes off at an ever increasing rate. An example is the induction period in petroleum oxidation reactions when oxidation proceeds at a constant slow rate for some time without seriously affecting the properties of the product.

Industrial Lubricant — Lubricants used either for lubrication of non-automotive industrial machinery or in the manufacturing processes, e.g., cutting and stamping oils.

Inhibitor — A chemical substance that prevents or slows down an undesirable reaction, such as oxidation of oil or rusting of components.

Inorganic — Compounds, and their chemistry, which do not contain hydrocarbons. Note: Some carbon compounds are considered inorganic (e.g., carbonates), and organic compounds are not necessarily derived from living material.

Insolubles — In petroleum and specifically lubricant testing the material which will not remain dissolved in a mixture of oil and a given solvent. (See Chapter 3 on testing.)

Internal-Combustion Engine — An engine driven directly by combustion gases rather than by a medium such as steam or the vapors of a volatile liquid. A consequence is the likely contamination of the lubricant by residues from fuel impurities or the results of incomplete combustion.

Invert Emulsion (or inverse emulsion)	Emulsions of water in oil rather than oil in water.
Ion	An atom or molecule that has either gained one or more electrons (making it negatively charged) or lost electrons (making it positively charged). Many substances split into ions of equal and opposite charge when dissolved in water.
Isomers	Large molecules, and particularly the hydrocarbons, can exist in varied structural forms. For example, octane has 18 different possible forms or isomers, all of which share the same molecular formula, C_8H_{18}.
Isoparaffin	A branched chain isomer of a paraffin molecule.
Isoprene	A hydrocarbon, C_5H_8, that contains two double-bonds and which can be polymerized to form polyisoprene rubber.
ISO Viscosity	The viscosity system set up by the International Organization for Standardization (ISO) for industrial lubricants. Viscosity bands are numbered in accordance with the viscosity in centistokes at 40°C at the center of each band. The bands are numbered 46/68/100/150/220/320/460/680/1000.

J

Jet Engine	A gas turbine for aircraft propulsion in which the sole function of the turbine is to drive the compressor and thrust is provided by the reaction to the jet efflux.
Jet Fuel	A kerosene-type fuel used for most aircraft gas turbines.
Joule	A unit of energy in the S.I. system, equal to the work done when a force of one Newton displaces an object by one metre.

Journal	The part of an axle or shaft which rotates within a bearing.

K

Kerosene	A light petroleum distillate, intermediate between gasoline and gas oil. Originally widely used as lamp oil, it is now mainly used as jet fuel and as a low-pollution diesel fuel in special situations.
KiloWatt-hour	The common electrical unit of work or energy based on kiloWatts consumed multiplied by hours run. One kiloWatt-hour (kWh) is equal to 3.6 megaJoules.
Kinematic Viscosity	The measurement of a liquid's resistance to flow under the force of gravity. It is equal to the absolute viscosity divided by the density of the fluid at the temperature of measurement.
Knock	Premature explosion of the fuel/air mixture in a spark-ignition engine such as a gasoline engine. High octane fuel reduces knock and prevents engine damage and power loss.

L

Lard (oil)	A natural fatty oil derived from pig fat.
Lacquer	A smooth deposit on engine parts ranging from yellow through brown to black and arising from polymerized fuel and lubricant decomposition products.
Land	The vertical portions of the sides of an engine piston between or above the ring grooves. (The portion below the lowest ring is called the piston skirt.) The land above the top ring is called the crown land.

Latent heat	The amount of heat required to achieve a change of state of a substance, i.e., of a liquid to a gas or a solid to a liquid. An equal amount is released when the reverse change takes place and a gas condenses or a liquid solidifies.
LD 50	A measure of toxicity, equal to the number of milligrams of a substance per kilogram of animal weight which causes the death of 50 percent of a population of test animals.
Lead Alkyl	Lead compounds such as tetraethyl lead or tetramethyl lead which act as octane improvers in gasoline. These compounds are now being phased out for reasons of exhaust catalyst deactivation and environmental pollution.
Lead Naphthenate	A lead soap often used in earlier times as a mild EP agent, but not now generally used for toxicity reasons. Lead oleate was also similarly used.
Light Ends	Low-boiling volatile materials in a petroleum fraction. They are often unwanted and undesirable, but in gasoline the proportion of light ends deliberately included are used to assist low-temperature starting.
Lime	Normally calcium oxide, but slaked lime contains hydroxide. Limestone, the source of most lime, is mainly calcium carbonate. Lime soap greases are calcium greases which may or may not have utilized lime as a raw material.
Linear Paraffins	Saturated hydrocarbons in which the carbon atoms form a continuous (unbranched) chain. Also called normal paraffins.
Lithium Grease	The most common type of grease today, based on lithium soaps.
LPG	Liquefied petroleum gas. Normally butane for leisure and small-scale domestic use, while liquid propane is used for industrial and larger scale use. In some areas mixtures may be available. A gasoline engine can be simply converted to run on

	LPG by replacing the carburetor with a throttle pedal-controlled metering device.
Lubrication	Reduction of friction and wear between rubbing surfaces by introducing a substance (a lubricant) between them. This can be a solid, a liquid, or a gas.
Lubricity	An ill-defined concept of ability to lubricate. It is usually considered to relate to film strength or the utility as a boundary lubricant. It is enhanced in petroleum lubricants by addition of polar compounds such as esters.
LWI	Load Wear Index. Originally called the Mean Hertz Load, it is a measure of the relative ability of a lubricant to prevent wear under the conditions of the Four-ball Test.

M

Mass Spectrometer	An analytical instrument for analyzing the hydrocarbon constituents of petroleum samples by accelerating ionized fragments in a magnetic field and analyzing the resultant spectrum photographically or by means of electronic detectors.
Mean Hertz Load	Now called Load Wear Index (see LWI).
Mercaptans	Malodorous sulfur compounds in crude oil, corresponding to alcohols but with the oxygen atoms replaced by sulfur. Synthetic mercaptans are used as stenching agents to aid detection of liquefied natural and petroleum gases.
Merit Rating	An arbitrary but carefully defined system for evaluating engine deposits in which cleaner parts are awarded higher numbers. A typical rating scheme would use a scale of 0 to 10, with 10 being absolutely clean. In many schemes even new parts do not achieve a rating of 10.

Methane	The lightest paraffin hydrocarbon, CH_4, a light, odorless, inflammable gas. It is the chief constituent of natural gas.
Methanol	Common name for methyl alcohol, CH_3OH. The simplest and lowest molecular weight alcohol.
Metric System	The international decimal system of weights and measures based on the metre and kilogram. (The S.I. system of units is based on an elaborated metric series of units.)
Microcrystalline Wax	Petroleum waxes derived from heavy petroleum fractions. They have higher melting points and viscosities than paraffin waxes and are usually darker in color.
Middle Distillate	Products such as kerosene and gas oil which are intermediate in boiling range between the fractions used for gasoline production and the bottoms of an atmospheric distillation column.
Military Specifications	Specifications set by the U.S. army or other armies for defining the required properties of products (including lubricants) which they may wish to purchase. The U.S. military was for many years in the forefront of lubricant specification development, but has recently moved to purchasing commercial products meeting industry standards of quality.
Mineral Oil	Petroleum oil as distinct from animal or vegetable oils.
Miscible	Capable of being mixed in any proportion without separation of phases.
Molybdenum Disulfide	MoS_2, a purplish-black solid lubricant which occurs naturally and can also be made synthetically. Often combined with greases to provide long-term low-friction lubrication for difficult applications.
Monomer	A single molecular substance of a type which can be polymerized to yield oligomers (a few molecules combined) or polymers (many molecules

	combined). A mixture of different monomers can be reacted together to form co-polymers.
Moped	Strictly, a motor-assisted pedal cycle, but the term is sometimes taken to include scooters and other small-horsepower two-wheeled vehicles.
Motor Oil	Engine oil used in motor vehicles.
Multigrade Oil	An engine or gear oil that meets more than one of the relevant SAE viscosity grade classifications, a summer grade with viscosity requirements at high temperatures and a winter grade with low-temperature viscosity requirements.
Multi-purpose Grease	A high-quality grease suitable for use in a variety of applications.
Multi-purpose Oil	An oil formulated to be suitable for a variety of applications. An example would be an oil suitable for both gasoline and diesel engines, and also satisfactory in some transmissions and hydraulic systems. Multi-purpose oils usually have limitations in relation to their performance in the most severe systems or conditions of operation, and compromises (such as to oil change recommendations) may be needed.

N

Naphtha	A light, relatively volatile, petroleum fraction obtained from primary distillation. Can be incorporated into gasoline, especially after reforming: can also be fractionated into solvents or used as feedstock for chemical production.
Naphthene	A cyclic saturated hydrocarbon. Cyclo-pentane and cyclo-hexane, with five and six carbons, respectively, in a ring, are the lightest stable naphthenes.
Naphthenic	A type of lubricating oil containing more molecules of a naphthenic type than paraffinic stocks. Naphthenic oils contain little wax and therefore

have low pour points. They also have good solvency power compared to paraffinic stocks, and tend to produce softer carbon residues if partially oxidized.

Narrow Cut — A petroleum fraction distilled to give a reduced distillation range, that is, with a less than average difference between the initial and final boiling points when subjected to laboratory distillation.

Neatsfoot Oil — A natural oil derived from animal hooves.

Neo-compounds — Compounds that contain the configuration of four carbon atoms linked tetrahedrally to a fifth, which therefore has no other atoms linked to it. Neo-esters have particularly valuable properties as synthetic lubricants due to their high thermal stability.

Neoprene — A type of synthetic rubber (chloroprene polymer) which has high resistance to oil and solvents. Has been widely used for gaskets and seals in automotive applications, but is being superseded by other types.

Neutralization Number — A measure of the acidity or alkalinity of an oil obtained by titrating it with either alkali or acid until neutral. (See Chapter 3.)

Newton — The unit of force in the S.I. system that is required to accelerate a mass of one kilogram one metre per second per second.

Newtonian — A fluid whose viscosity does not depend on the rate of flow (shear rate).

Niemann Test — An alternative name for the FZG gear oil test.

Nitrile Rubber — A type of synthetic rubber made by co-polymerizing butadiene and acrylonitrile. Also called Buna rubber. Resistant to oil, fuels and heat, it finds wide application in the automotive field.

Nitrogen Oxides — Air pollutants formed by high-temperature combustion in engines or furnaces consisting mainly of nitric oxide, NO, with some nitrogen dioxide, NO_2. Often abbreviated to NO_x or even NOX,

	nitrogen oxides are irritants and combine with hydrocarbon emissions in the presence of sunlight to form photochemical smog.
Noack Volatility	A method of measuring the volatility of lubricating oil, particularly popular in Europe.
Non-Newtonian	A fluid whose viscosity (apparent viscosity) depends on the shear rate at which the viscosity is measured.
Normal Force	The force exerted by one object on another in a direction at right angles to their mutual surface. If a block rests on an inclined plane, the normal force is not the weight of the block, but its component at 90 degrees to the plane. Friction is directly related to the normal force.
Normal Paraffin	A saturated hydrocarbon that contains no branching of the chain of carbon atoms. Also called linear paraffin.

O

Octane No.	A measurement of the anti-knock properties of gasoline, based on primary references of 100 for iso-octane and zero for normal heptane. It is measured in special test engines of variable compression ratio using secondary fuel standards that closely bracket the test fuel octane no. Alternative test procedures are used to indicate the Motor Octane No. or the Research Octane No. Octane No. can also be assessed on the road, and the Road Octane Requirement of a vehicle measured for different operating conditions.
Oiliness Agent	An additive, usually polar in nature, used to improve the lubricity of a mineral oil. Now usually called a boundary lubrication additive.
Olefin	A hydrocarbon that contains two carbon atoms linked by a double bond. The double bond is

relatively unstable, and olefins react readily with other compounds or polymerize to higher-molecular- weight, less unsaturated molecules.

Oleum — Effectively, very highly concentrated sulfuric acid. It can be used to refine lubricating oils (acid refining), producing "natural" sulfonates as a by-product.

Oligomer — A polymer that contains only a few of the monomer molecules.

Once-Through Lubrication — A system of lubrication where the lubricant is not collected or recycled. Even if the rate of lubrication is minimized, it can lead to greater pollution problems than with a return system.

Open Cup — The type of flash point test in which the volatile gases are free to disperse in the air until a high enough concentration is built up to ignite on passage of a flame.

Organic Compound — A chemical substance containing carbon and hydrogen and possibly other elements. While organic compounds are the basis of life forms, all organic compounds are not necessarily derived from living matter.

Otto Cycle — The operating process of a four-stroke reciprocating engine. The term is usually used to describe the gasoline-fueled, four-stroke engine as distinct from the diesel cycle, but some authorities consider the diesel to also be an Otto cycle engine.

Overheads — The light fractions removed from the top of a distillation column.

Oxidation — The combination of a substance with oxygen. This can be from the air or from an oxygen-containing chemical. Oxidation can be accelerated by heat, light, and metal or other catalysts. It can be slowed down (inhibited) by chemicals which interfere with the oxidizing agent, the catalyst, or one or more of the intermediate substances generated in an oxidation chain reaction. Most organic compounds oxidize via the formation of organic peroxides which catalyze further oxidation. Antioxidants can react with peroxides and halt the chain mechanism. The oxidation stability of hydrocarbons, with or without catalysts or oxidation inhibitors, is measured in a wide variety of different tests.

P

Paraffin — A saturated hydrocarbon (with no double or triple carbon-carbon bonds). It can be either straight or branched chain.

Paraffinic Oil — A lubricating oil containing a high proportion of paraffin molecules in its composition. When cooled, the straight chain paraffins come out of solution as wax, whose crystals tend to gel the oil. Paraffinic oils have high pour points, are relatively low in solvency compared to naphthenic oils, but have good viscosity/temperature characteristics.

Particulates — Solid airborne pollutants such as ash and smoke particles. Most pollution from particulates comes from natural or industrial processes, but diesel particulate emissions are of interest and are subject to controls.

Passivator — An additive which reacts with a metal surface, either to prevent it from acting as an oxidation catalyst, or to protect it from corrosion by lubricating oil or some other agent.

Pascal — The unit of pressure in the S.I. system. It is equal to a force of one Newton acting over an area of one square metre.

PCBs — Polychloro-biphenyls. A type of synthetic oil formerly used in some industrial applications such as transformers. PCBs are considered hazardous

	materials because of the capability to form dioxins when burned, and industrial use is banned in many countries. They can occur in used oil from industrial oil contamination, and there is evidence that they can be produced in an internal-combustion engine if the lubricant contains appreciable levels of chlorine.
PCV	Positive crankcase ventilation. The system whereby an engine crankcase is not vented to the air, but the blowby gases are sucked into the induction system and recycled through the engine. As the reactive gases are not lost to atmosphere, the use of PCV systems places extra demands on the lubricant.
Penetration	In grease testing the consistency (softness) is measured by a penetrometer, in which a cone is allowed to fall onto a prepared grease sample. The depth of penetration of the cone is called the penetration number.
P-M Flash	A flash point test result obtained with the Pensky-Martens closed flash point tester.
Pentane	A mixture of straight and branched chain paraffins with five carbon atoms. It is a light constituent of gasoline which aids starting due to its volatility.
Pentane Insols.	The insoluble material left behind when used lubricating oil is mixed with pentane and either centrifuged or filtered. As well as solid metallic, inorganic, or carbon particles, it also contains residues from oxidized oil which are insoluble in pentane.
Peroxide	A substance which contains two linked oxygen atoms. Inorganic peroxides include hydrogen and sodium peroxides, and many organic peroxides exist. Peroxides are relatively unstable, and act as powerful oxidizing agents. When oil is oxidized, peroxides are among the first substances formed, and go on to oxidize other oil molecules if not eliminated by a suitable inhibitor.

Petrochemical	Any chemical substance derived from crude oil or its products, or from natural gas. Some petrochemical products may be identical to others produced from other raw materials such as coal and producer gas.
Petrolatum	A semi-solid mixture consisting mainly of high-molecular-weight noncrystalline waxes, and produced in the dewaxing of heavy residua.
Petroleum	Crude oil and/or its products.
Petroil	A name given in some areas to a mixture of gasoline and lubricating oil used to lubricate small two-stroke engines.
Petrol	The popular British name for gasoline.
pH	The scale used for measuring acidity and alkalinity. It ranges from 0 (very acid) through 7.0 (neutral) to 14.0 (very alkaline).
Phenol	An aromatic compound with a hydroxyl group substituting one of the hydrogen atoms on the benzene ring. Phenols are reactive, mildly acidic materials used in the synthesis of many organic compounds.
Phenate	A metallic salt of a phenolic compound. Calcium and barium phenates made from alkyl-substituted phenols are well-known detergent additives, and magnesium and sodium phenates are also known.
Photometry	The measurement of light intensity; some analytical methods involve generation of light of specific wavelengths which characterize the elements present while their intensity indicates the concentration.
Pinion	In a pair of intermeshing gears, the smaller is known as the pinion.
Pipestill	A refinery distillation column, from which different fractions of the feed can be withdrawn at different heights. A temperature gradient runs up the pipe, which is hotter at the bottom from which heavy products are withdrawn, and cooler at the top where lighter products collect.

Piston	The moving element within the cylinder of a reciprocating engine. The energy of expanding gases drives down the piston and this is converted to rotational movement via a connecting rod and crankshaft. In a conventional orientation, the top of the piston is called the crown, and the sides will be grooved to take the piston rings which act as gas seals. Between the grooves are the lands, and between the top groove and the crown is the crown land. The area below the lowest groove is called the skirt. The piston pin or gudgeon pin on which the connecting rod pivots is held in the piston bosses.
Plain Bearing	A bearing in which the contacting surfaces are both smooth and contact is spread over a relatively large area, and there are no rolling elements such as balls or rollers.
Plasticizer	An organic compound added to rubbers or plastics to improve their flexibility. Leaching (removal) of plasticizer from seal materials by the solvent action of lubricants can lead to their premature failure.
PNA	Polynuclear aromatic compounds. Some occur naturally, and others can be formed in combustion processes. Some PNAs are known carcinogens.
Poise	A fundamental viscosity unit, equal to the force in dynes needed to move a one centimetre square surface past a parallel surface one centimetre away at a speed of one centimetre per second when the surfaces are separated by a liquid film.
Polar Compound	A chemical material whose molecules have one end electrically positive and the other electrically negative. Polar molecules are often strongly attracted to surfaces, and many additives are polar in nature.
Pollutants	Substances released into the environment which represent a hazard to man or to nature. Pollutants

	associated with the automotive industry include hydrocarbons, carbon monoxide, nitrogen oxides, sulfur dioxide and particulates into the atmosphere, and hydrocarbons into soil, groundwater and watercourses.
Polyester	A plastic or resin typically produced by polymerizing a mixture of a dibasic acid with a dihydric alcohol.
Polyglycols	Polymers of ethylene and/or propylene oxides, usually grown onto a suitable starting molecule. They are useful synthetic lubricants and hydraulic fluids.
Polyisoprene	A synthetic rubber, similar to natural rubber, made by polymerization of isoprene, C_5H_8.
Polymer	A substance made by polymerizing (linking together) a series of molecules of either one or more chemical types to form a higher-molecular-weight material. When more than one monomer (starting molecule) is involved, the material is usually referred to as a co-polymer. Unsaturated molecules polymerize readily to form either polymers or co-polymers.
Polyol Esters	Synthetic lubricants made by reacting fatty acids with polyhydric alcohols.
Pour Point	An empirical test of the gelling tendency of paraffinic oils at low temperatures. See ASTM test method D 97.
ppm	Parts per million (measure of concentration).
Pre-ignition	Ignition of the fuel-air mixture in a gasoline engine before the spark. It can be caused by glowing deposits in the combustion chamber, by overheating, a high compression ratio, or a combination of these factors. It can result in loss of power and engine damage.
Propane	The gaseous paraffin, C_3H_8. It can be liquefied as a form of LPG.

Propylene	The unsaturated olefin, C_3H_6. The monomer of polypropylene, it is used with ethylene to produce ethylene-propylene co-polymers.
psig	Pounds per square inch gauge. A measurement of pressure above that of atmospheric pressure.
Pushrod	Part of an engine valve mechanism transmitting the motion of a cam at the side of an engine to the rockers on top of the cylinder head, and hence to the valves.

Q

Quenching Oil	An oil used in metal hardening processes, in which parts are heated then plunged into liquid to produce rapid cooling. Oil has advantages over water, relating to a more controlled and slower cooling effect.

R

Radical	A group of atoms that remain intact during a chemical reaction, and thus behave as a single entity with a free valency. Examples are the methyl (CH_3–) and ethyl (C_2H_5–) radicals.
Raffinate	The refined stream of oil produced as a result of a refining process.
Rape Oil	Also known as colza oil, it is a vegetable oil obtained by crushing the seeds of the rapc (colza) plant.
Reduction	In organic chemistry, the opposite process to oxidation. A process in which a molecule becomes more electronegative. For example, the aldehyde group –CHO can be reduced to an alcohol –CH_2OH.
Refining	The process or combination of processes by which crude oil is converted into usable products.

Refrigerator Oil	The lubricant added to the working fluid in an expansion-type cooling unit which serves to lubricate the pump mechanism.
Re-refining	The process by which used oil can be processed to produce reclaimed stock with equivalent properties to the original oil. Less complete reclaiming processes will produce stocks for downgraded usage.
Residuum	The bottom fraction from a distillation column, containing the heaviest and most involatile material.
Resins	Solid or semi-solid materials containing carbon, hydrogen, and oxygen. Natural resins occur in plants and trees, especially pines. Synthetic resins are used in paints, in car body parts, in adhesives, and as rubber compounding agents. Examples are polyesters and acrylics. The term is also sometimes loosely applied to other solid polymers such as polystyrene and polyolefins.
Rheology	The study of the flow properties of semi-solid substances such as greases.
Rig-Test	A test to measure the performance properties of a product, using a relatively simple mechanical device rather than a full-scale piece of equipment. Rig-tests usually are run under severe and accelerated conditions and produce rapid results at relatively low cost. Their value is dependent on how well they have been correlated with real equipment and conditions.
Ring-Sticking	The seizure of a piston ring in its groove due to a build-up of varnish or lacquer. A stuck ring will not ride the cylinder walls and will result in power loss, increased blowby, and increased oil consumption.
Rotary Engine	An engine in which a major portion of the mechanism is in rotation, and from which a drive can be taken. The rotating portion can be the crankcase, the combustion chambers, or, as in the case of the

	Wankel engine, a rotating member equivalent to a piston.
Rust Inhibitor	An additive, normally of a polar type, that protects metal surfaces from rusting in the presence of water.
Rust Preventive	A compound used for coating the surface of steel to prevent rusting in storage. It may contain a rust inhibitor, but will also have a base of oil, wax, resin, or asphalt to act as a mechanical barrier to water and other harmful contaminants.

S

SAE Viscosity	The viscosity classification of a motor oil according to the system developed by the Society of Automotive Engineers and now in general use. "Winter" grades are defined by viscosity measurements at low temperatures and have "W" as a suffix, while "Summer" grades are defined by viscosity at 100°C and have no suffix. Multigrade oils meet both a winter and a summer definition and have designations such as SAE 10W-30, etc.
Salicylates	Salts of salicylic acid, a carboxylic acid based on benzene with a hydroxyl group adjacent to the carboxyl group. Salts of substituted salicylic acid (with side chain hydrocarbon groups on the benzene ring) are useful detergent additives.
Saponification	The process of producing soaps (metal salts of fatty acids) by heating a fatty acid or a fat with alkali.
Saybolt Viscosity	An early empirical measurement of viscosity used mainly in the U.S.
Scavenger	A component of leaded gasoline whose purpose is to react with lead compounds formed during combustion and convert them to a volatile form which passes out with the exhaust. Commonly

	used scavengers are ethylene dichloride and ethylene dibromide.
Scavenge Pump	In a circulating lubrication system, the pump which returns collected oil to the main reservoir for recycling.
Scoring	Scratching of sliding surfaces caused by wear debris. Scoring is seen as an intermediate level of distress for components such as gears, between scuffing and more severe grooving.
Scuffing	Localized matt areas on sliding surfaces caused by small-scale seizures and subsequent surface damage.
Seal Swell	The increase in volume of a seal component due to the action of the lubricant or other fluid. Practical tests for seal swelling tendency are often performed on test coupons of the same elastomer rather than on complete seals. Seal shrinkage is recorded as negative seal swell. A small degree of seal swell is often desirable to improve sealing.
Seizure	The welding together of surfaces sliding over each other, caused by frictionally generated heat. Seizure can stop the motion of the complete mechanism, or if on a smaller scale, can damage the sliding surfaces.
Series 3	A level of diesel detergency in a lubricant devised by the Caterpillar Tractor Company in 1955, superseding earlier levels of performance. It remains an important level of quality, and is related to API Service Category CD.
Shear Rate	A measure of the relative motion of adjacent layers in a fluid. Based on the assumption that the layers in intimate contact with surfaces move with those surfaces, the shear rate of a fluid between two surfaces in relative motion can be expressed as the relative rate of motion between the surfaces divided by their distance apart.

Shear Stress — The force needed to make such layers of liquid move relative to each other (see Shear Rate), and hence for the liquid to flow. For a so-called Newtonian fluid, the shear stress is proportional to the shear rate, where shear stress = shear rate × viscosity. For a non-Newtonian fluid, the ratio (and hence the viscosity) varies with shear rate.

S.I. Units — The international system of units of measurement, set up in 1960 and based on the metric system. As well as fundamental units it includes derived units such as the Newton for force, the Pascal for pressure, and the Joule for work.

Sidestream — A fraction taken from a side position and an intermediate temperature in a distillation column.

Silicones — Organo-siloxane polymers useful as stable synthetic lubricants. A minute addition of a higher-molecular-weight silicone to a mineral oil can reduce its tendency to foam.

Single-Grade — A motor oil which satisfies only one of the SAE viscosity classifications. This can be either a winter grade such as SAE 10W, or a summer grade such as SAE 30. In practice, most winter grades also meet a summer grade classification producing a multigrade oil such as SAE 10W-30.

Slack Wax — Wax of relatively high oil content obtained from lube oil dewaxing plants, suitable for a limited range of uses without further refining.

Slideway — The bed of a metal-shaping machine such as a lathe, along which the tool, turret, or other device slides to make contact with the workpiece.

Sludge — A black or dark brown semi-solid deposit in an engine which consists of an emulsion containing water and oil and also soot and other products of combustion. The consistency can vary from very soft to a firm gel, and on drying out it can leave hard carbonaceous deposits. Emulsion sludge is an off-white or pale brown mayonnaise of water

	and oil which occurs in the PCV or oil filler systems of an engine.
Soap	The salt of a fatty acid with an alkali or alkaline earth metal.
Solvent Extraction	A lube oil refining process in which a distillate fraction is mixed with a selective solvent such as sulfur dioxide, furfural or N-methyl pyrrolidone. Undesirable components such as aromatics and unsaturates dissolve in the solvent, while the main paraffinic components are immiscible and can be readily separated.
Solvent Neutral	A paraffinic base oil refined by solvent extraction and containing no free acidity.
Specific Gravity	The ratio of the mass of a given volume of product to the mass of an equal volume of pure water at the same temperature, which has to be stated when values of specific gravity are quoted.
Spindle Oil	Originally an inhibited low-viscosity oil for the lubrication of high-speed spindles of textile and other machinery, the term has come to indicate any low-viscosity oil.
Sperm Oil	An animal oil produced from spermaceti, a substance found in the head of the sperm whale, and originally a valuable and potent lubricant. Use of genuine sperm oil is now banned, and various synthetic sperm oil substitutes have been developed for gear oil and other applications.
Spectroscopy	An analytical technique in which radiation is split up into a spectrum, and relevant wavelengths are identified and possibly measured. The radiation can be of any wavelength, including visible light, x-ray, infrared, ultraviolet, microwave, etc.
Spur Gear	A gear with straight teeth cut parallel to the axis of rotation and used for transmitting motion between parallel shafts.
Stick-slip	Irregular noisy motion of sliding parts due to boundary lubrication and high static friction.

	Boundary lubrication additives can usually effect a cure.
Stoichiometric	The ratio of fuel to air (or any other two reacting substances) where the exact proportions for complete reaction of both, with none of each leftover, are present.
Straight Oil	A mineral oil containing no additives.
Stribeck Curve	Also called a ZN/P curve, this is a plot of friction against the parameter ZN/P where Z is the viscosity, N the speed of movement in a bearing, and P the pressure on the bearing. Named after a German scientist of the early 20th Century, this is one of the fundamental equations of bearing lubrication, and the curve illustrates the transition from boundary to full fluid film lubrication and then the rise of friction with increasing oil viscosity.
Strong Acid No.	An indication of mineral acid content, it is the amount of base reagent to titrate up to a pH value of 4.0 (neutral is pH = 7.0).
Strong Base No.	The amount of acid required to titrate strongly alkaline oils down to a pH of 11.0.
Styrene	A benzenoid chemical with an unsaturated ethylene side chain attached to the benzene ring, and which can be readily polymerized or co-polymerized with other unsaturated compounds to yield a range of plastics and rubbers.
Sulfonates	Hydrocarbons reacted with sulfur trioxide or oleum to produce salts containing the SO_2OM group, where M is a metal. Initially obtained as by-products from the acid refining of lubricants, the aromatic sulfonates are useful detergent additives. They are now produced from synthetic alkyl benzenes.
Supercharger	An air compressor used in the intake system of an engine whose purpose is to increase the amount of charge burned on each firing stroke. It is normally driven indirectly from the crankshaft and hence its

	efficiency is related to engine speed. (A turbo-charger is driven by a turbine in the exhaust stream, and is not solely dependent on engine revolutions for efficient operation.)
Surface-active	The ability of a compound to reduce the interfacial tension between water and oily substances, and hence to produce emulsions.
Surfactant	A surface active agent.
Synlube	Abbreviation of synthetic lubricant. A wide variety of chemical substances can be used as lubricants, with one or more advantages over conventional petroleum stocks. The main disincentive against such materials is their considerably higher cost.

T

Tag tester	A flash point test apparatus for low flash point liquids.
Talc	A white solid lubricant consisting of platelet crystals of a hydrated magnesium silicate.
Tappet	An alternative name for a cam follower, the cylindrical block which follows the motion of a cam to give a linear oscillatory motion.
TEL and TML	The lead-containing anti-knock compounds, tetraethyl lead and tetramethyl lead.
Terpolymer	A polymer made from a mixture of three different monomers.
Texture	A description of the physical appearance of a grease in standard terms: Brittle — has a tendency to crumble. Buttery — separates in short peaks with no fibers. Long Fiber — will string showing a fibrous appearance. Short Fiber — will not string but shows a fibrous structure.

	Stringy — strings without signs of fibers.
	Resilient — can be moderately compressed without permanent deformation.
Thermal Stability	Ability to resist high temperatures without decomposition. Not to be confused with oxidation stability where oxygen must be present and oxidation rather than decomposition products are formed.
Thio-	Compounds analogous to oxygen-containing compounds but with the oxygen replaced by sulfur. Thus alcohols have the formula R-OH and thioalcohols have the formula R-SH .
Thixotropic	The description given to materials such as grease that liquefy from a semi-solid state or show a reduction in viscosity when subjected to a shearing action.
Timken Test	A laboratory tester that measures the wear prevention and EP characteristics of oils or greases.
Titration	The addition of measured amounts of a chemical reagent to a sample of material with which it will react, in order to discover the concentration. The end-point when the reaction is complete is detected by a color indicator or an electronic device such as a pH meter. Thus the strength of an acid solution is measured by titrating it with measured amounts of alkali of known strength to a neutral end-point detected by a color change or a meter reading.
Toluene	Methyl benzene, C_7H_8.
Toluene Insols.	The insoluble material left behind when a used oil is mixed with toluene and filtered or centrifuged. Toluene will dissolve the degradation products of lube oxidation and any fuel residues, so only external contaminants and wear debris are detected.
Torque	A twisting force applied to a shaft.

Torque Converter — The device in an automatic transmission which takes the turning force of the engine and feeds it in a smooth controlled manner to the gears.

Torque Fluid — A special fluid of low viscosity and high stability used to transmit the drive through a torque converter.

Total Acid No. — A measure of the quantity of all acid components, weak and strong, in an oil sample. It is obtained by titrating with standard alkali to a pH of 7.0.

Total Base No. — A measure of total alkalinity obtained by titrating with acid to a pH of 7.0.

Tribology — The scientific study of friction and wear, usually with the object of minimizing their harmful effects on mechanisms and machinery.

Turbine — A device consisting of blades attached to a disc or rotor, which converts a flow of fluid into rotary motion. The fluid can be a gas or a liquid.

Turbocharger — A device for increasing the pressure and hence the mass of the charge of air or fuel and air entering an engine. A turbine in the stream of exhaust gases drives a compressor in the intake system.

Two-stroke — The firing cycle of an engine where every stroke which enlarges the combustion chamber volume is a power stroke. The following stroke serves both to exhaust the burned gases and compress a fresh charge. A two-stroke engine therefore has power supplied during each revolution of the engine, instead of every other revolution as in the case of a four-stroke engine.

U

Unconventional Base Stocks — Base stocks which have not been produced by the conventional vacuum distillation, solvent refining, solvent dewaxing, and clay or hydrogen-finishing

	of petroleum crude. Apart from synthetic stocks, there are unconventional petroleum stocks that are highly paraffinic and usually produced from a severe hydrocracking operation. Partly from the process characteristics and partly from their extreme paraffinicity, such stocks have narrower distillation ranges and lower volatility than conventional stocks of the same viscosity.
Unleaded Gasoline	Gasoline that does not contain lead compounds as octane improvers.
Unsaturated	Hydrocarbons that contain double or triple carbon-carbon linkages. Ethylenes, acetylenes, and aromatic compounds fall into this category, although aromatics are sometimes considered separately. Unsaturated compounds are reactive, and many ethylenic compounds are used as polymerization monomers or the starting chemicals for syntheses.
Unworked Penetration	The consistency of a grease measured in an undisturbed (unsoftened) state.

V

Vacuum Distillation	Distillation of petroleum under reduced pressure which enables heavy fractions to be vaporized at lower temperatures, thus reducing the amount of thermal decomposition.
Valve Train	The succession of parts in an engine that transmit the cam lift into valve opening. In a side camshaft, overhead valve engine, this consists of the camshaft, the cam followers (tappets), pushrods, the rockers and the valve stems.
Valve Recession	Also called valve beat-in or valve sink, this is a phenomenon which has occurred in some engines designed to run on leaded gasoline when these are run continuously on unleaded fuel. It is consid-

ered that lead residues act as valve seat lubricants and if hard alloy seat inserts are not used then seat wear occurs and the valve head sinks into the seat, with decrease of valve clearances and eventual failure to seal the combustion chamber.

Varnish — The first stage of tenacious deposition on engine parts such as pistons of the polymerized lube and fuel residues which progressively contaminate the oil. The deposits start as a mere iridescence and progress through yellow and orange to dark brown varnish-like layers. Heavy varnish eventually carbonizes and a layer of solid carbon will build up if the process is not stopped by use of suitable additives in the oil.

Viscometer — Sometimes called a viscosimeter, any device that can be used to measure the absolute, relative, or apparent viscosity of a fluid.

Viscosity Index — A measure of the viscosity/temperature behavior of an oil, based on an arbitrary scale of 0 to 100 with 0 being representative of a very naphthenic oil which thins down rapidly when heated, and 100 being representative of a paraffinic oil which thins less rapidly. See Section 3.2.1.

Viscosity Modifier — A polymeric additive that reduces the amount of thinning of an oil when it is heated. Originally called a Viscosity Index (V.I.) improver and later a Viscosity Improver, the term Viscosity Modifier is now generally preferred for this type of additive. The type of polymer used has undergone changes as the particular low-temperature targets for motor oils have been modified and elaborated.

Volatility — The evaporation tendency of a substance, measured as the amount which is lost when the substance is held at a given high temperature for a relatively long time. Some tests may take place under slightly reduced pressure, and some may direct streams of air over the sample. In most tests the volatile portions will be drawn off or blown away continuously. Results are empirical, depending on the design and operating conditions of the test.

Volumetric Efficiency — The ratio of the weight of air actually drawn into an engine under specified operating conditions to the weight of air which the engine cylinders could hold if each piston was stationary at the bottom of its stroke and the valves were closed. Volumetric efficiency decreases with engine speed due to the throttling effect of the induction system, and can be increased by supercharging or turbocharging.

W

Wax — High-molecular-weight hydrocarbons that separate from oil when its temperature is lowered. Paraffin wax separates from light paraffinic oils, and is crystalline. Microcrystalline waxes are present in heavy fractions and residua, and are darker with ill-defined crystal structure. Petrolatum is obtained from residua by propane precipitation and is of the microcrystalline type with a jelly-like consistency.

Weld Point — In the four-ball EP test, the loading at which the rotating ball either seizes to the other balls, or catastrophic scoring takes place.

White Oil — Petroleum oil refined to such an extent that all color is lost. Both medicinal and technical grades are available. Originally produced by severe acid treatment, white oils of both types are now more usually made by deep hydrogen refining.

Wide Cut — A petroleum fraction with a wide distillation range, and therefore containing more volatile material than a narrow-cut fraction of the same viscosity.

Worked Penetration	The consistency of a grease measured after undergoing a specified amount of agitation in a special device known as a "grease worker."
Worm Gear	A cylindrical gear with spiral flute(s), for 90° drives.

XYZ

Xylene	C_8H_{10}, a substituted benzene containing two methyl groups.
Yield Point	The minimum force needed to produce flow in a gelled substance.
ZDDP	Zinc dialkyl dithiophosphate, a powerful anti-oxidant and anti-wear additive.
ZN/P Curve	See Stribeck Curve.

Appendix 2

Common Automotive Acronyms

Organizations

AAMA	American Automobile Manufacturers Association
ACEA	Association des Constructeurs Européens d'Automobiles
AFNOR	Association Française de Normalisation (F)
AGMA	American Gear Manufacturers Association
API	American Petroleum Institute
ASLE	American Society of Lubrication Engineers
ASME	American Society of Mechanical Engineers
ASTM	American Society for Testing and Materials
ATC	Technical Committee of Petroleum Additive Manufacturers in Europe
ATIEL	Association Technique de I'Industrie Europèene des Lubrifiants
BSI	British Standards Institute
BTC	British Technical Council of the Motor and Petroleum Industries (UK CEC member)
CCMC	Comité des Constructeurs d'Automobiles du Marché Commun (Now replaced by ACEA)
CE/EC	Communauté Européene/European Community
CEC	Coordinating European Council for the development of Performance Tests for Transportation Fuels, Lubricants and Other Fluids
CEN	Centre Européen de Normalisation (European Committee for Standardisation)
CMA	Chemical Manufacturers Association (of USA)

CONCAWE	The Oil Companies European Organisation for Environmental and Health Protection. The acronym originally stood for "Conservation of Clean Air and Water in Europe."
CRC	Coordinating Research Council (USA)
CUNA	Commissione Tecnica di Unificazione Nell'Autoveicolo (Italian CEC member)
DGMK	Deutsche Gesellschaft für Mineralölwissenschaft und Kohlechemie (D)
DIN	Deutsches Institut für Normung (German Standards Office) or Deutsche Industrie Norm (German Standard)
DKA	Deutsche Koordinierungsausshuss (German CEC member)
EU	European Union
ELTC	Engine Lubricants Technical Committee (of CEC)
EMA	Engine Manufacturers Association (USA)
EOLCS	Engine Oil Licensing and Certification System (of API)
EPA	Environmental Protection Agency (USA)
FZG	Forschungstelle für Zahnräder und Getriebebau (German Gear Research Institute)
GFC	Groupement Française de Coordination (French CEC member)
IAE	Institution of Automobile Engineers (obsolete) (GB)
IFP	Institut Français du Pétrole
ILSAC	International Lubricant Standardization and Approval Committee
IP	Institute of Petroleum (UK)
ISO	International Organization for Standardization
JAMA	Japanese Automobile Manufacturers Association
JASO	Japanese Automobile Standards Organization
JSAE	Japanese Society of Automotive Engineers
LRI	Lubricants Review Institute (USA)
MVMA	Motor Vehicle Manufacturers Association (USA) (now replaced by AAMA)
NALSAS	North American Lubricant Standardization and Approval System (obsolete)
NLGI	National Lubricating Grease Association (USA)
NMMA	National Marine Manufacturers Association (USA)

NPRA	National Petroleum Refiners Association (USA)
OEM	Original Equipment Manufacturer (of cars, trucks, etc.)
SAE	Society of Automotive Engineers (USA)
TMC	Test Monitoring Center (of ASTM)

Products, Equipment, Testing

AI	Active Ingredient
AFR	Air/Fuel Ratio
ATF	Automatic Transmission Fluid
CAFE	Corporate Average Fuel Economy
CCS	Cold Cranking Simulator
CI	Compression Ignition
COC	Cleveland Open Cup (flash point)
CVT	Continuously Variable Transmission
DI	Detergent Inhibitor
DSC	Differential Scanning Calorimeter
EFI	Electronic Fuel Injection
EP	Extreme Pressure
FBP	Final Boiling Point
GC	Gas Chromatography
HD	Heavy Duty (vehicle or lubricant, usually diesel)
HT/HS	High Temperature/High Shear (Viscosity)
IBP	Initial Boiling Point
IR	Infrared
LOFI	Lube Oil Flow Improver
MOC	Mineral Oil Content
MTAC	Multiple Test Acceptance Criteria
MRV	Mini-Rotary Viscometer
PAH	Polyaromatic Hydrocarbon
PCB	Polychlorinated Biphenyl
PCD	Passenger Car Diesel
PM	Pensky-Martens (flash point)
PPD	Pour Point Depressant
RME	Rapeseed Methyl Ester
SAN	Strong Acid Number

SHPD	Super High Performance Diesel (oil)
SS/CO	Service Station and Commercial Oil
STUO	Super Tractor Universal Oil
2T or 2-T	Two-Stroke
TAN	Total Acid Number
TBN	Total Base Number
TBS	Tapered Bearing Simulator
VI	Viscosity Index (sometimes Viscosity Improver)
VM	Viscosity Modifier
WTD	Weighted Total Demerits (of engine test ratings)
ZDDP	Zinc Dialkyl Dithiophosphate

Appendix 3

Basic Petroleum Chemistry

Petroleum chemistry is a large branch of organic chemistry, which is essentially the study of compounds based on carbon. Organic chemistry is so-called because life on earth, and its residues, is based on carbon-containing substances. Hydrocarbon chemistry, based on materials containing just carbon and hydrogen, forms the major part of petroleum chemistry, although pure hydrocarbons are only a small part of the living world. Living substances contain other elements, particularly oxygen, nitrogen, sulfur, and some metals as well. As petroleum is a residue of long-dead animals and some plants, these other elements occur in crude petroleum and some products.

To understand chemistry, it is necessary to understand the basic structure of chemical compounds and how elements react, so this will be briefly summarized first.

Atoms and Molecules

All normal substances are composed of *molecules*, the smallest amount of substance that can exist. Molecules consist of *atoms* joined together, from a minimum of two to many thousands. An atom is the smallest amount of an *element* that can exist. There are about 100 elements including some man-made varieties produced in atomic piles and particle physics experiments. The lightest is hydrogen and the heaviest naturally occurring element is uranium.

The traditional though simplistic view of an atom is of a central, positively charged core, called the *nucleus*, and a surrounding cloud of negatively charged *electrons*. The composition of the nucleus defines the element, and is a mixture of positively charged protons, and neutrons which carry no charge. The *atomic number* of an element is the number of protons in the nucleus and the total of protons plus neutrons gives the *atomic weight*. (The number of neutrons can vary sometimes, giving different *isotopes* of the element.)

Chemical Reactions

The nature of the electron cloud determines how reactive the element is toward other elements. In some cases some electrons may move from one atom to another, leaving one positively charged and the other with a corresponding negative charge. The two atoms are therefore attracted to each other and there is a polar bond between them. This happens in the case of metals like sodium and halogens like chlorine:

$$\underset{\text{Metal}}{2Na} + \underset{\text{Gas}}{Cl_2} \longrightarrow \underset{\text{Salt}}{2 \times Na^+Cl^-}$$

(Sodium and Chlorine to Common Salt)

Note how different the product is from the starting elements. Elements that donate electrons are the metals and hydrogen; elements that readily accept electrons include the halogens and oxygen; and many elements are intermediate or can do either.

When elements like nitrogen, phosphorus, and sulfur are oxidized the resulting compounds are strongly acceptive.

When an electron acceptor is combined with hydrogen, an acid results:

$$H_2 + Cl_2 \longrightarrow \underset{\text{Hydrochloric acid}}{2 \times HCl}$$

(Hydrogen and Chlorine to Hydrochloric Acid)

$$H_2O + SO_3 \longrightarrow H_2SO_4$$

Sulphuric acid

(Water and Sulfur Trioxide to Sulfuric Acid)

The above type of polar or ionic bond is found in petroleum chemistry in relation to organic acids, like acetic acid (ethanoic acid) or the fatty acids. Some organic compounds containing oxygen, or oxygen and other elements like nitrogen, sulfur and phosphorus, can be persuaded to act like acids by the presence of a strong alkali. In particular this includes the phenols.

The other main mechanism by which atoms link together to form molecules is a mutual sharing of electrons in a so-called *covalent bond*, which usually results from effectively sharing two electrons. This is the most common type of bond in organic compounds.

The usual name for a bond is a *valency*. Elements exhibit characteristic valencies, whether ionic or covalent, although some elements can have different valency numbers in different circumstances. The normal valency states of oxygen, nitrogen, and carbon are shown below:

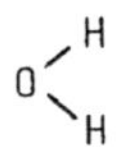

Divalent oxygen	Trivalent nitrogen	Tetravalent carbon
Water	Ammonia	Methane

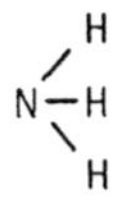

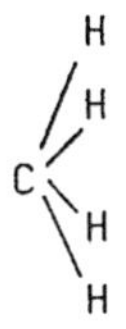

Structural Chemistry

An important property of the carbon atom is its great ability to bond with other carbon atoms, readily forming chains or ring structures. Some other elements do this to a lesser degree, but carbon is unique. Normally, the four valencies of carbon are assumed to point to the corners of a tetrahedron, with the carbon atom at the center. There is some flexibility in the bonds or valencies, however, and adjacent carbon atoms can link together with one, two, or three valencies. In pure hydrocarbons the other valencies around each carbon are occupied by hydrogen atoms. Oxygen, sulfur, and nitrogen also link easily to carbon, and when these form part of a ring structure we have a *heterocyclic* compound.

With the tetrahedral nature of the carbon valencies, depicting organic structures on a flat page is difficult. The normal convention is to show four bonds at 90° intervals whenever practicable. The following structures show typical molecules with their older common names, and the more internationally correct IUPAC official names in parentheses:

Paraffins (Alkanes)

```
   H              H  H             H  H  H
   |              |  |             |  |  |
H—C—H          H—C—C—H         H—C—C—C—H        ...........etc.
   |              |  |             |  |  |
   H              H  H             H  H  H

Methane          Ethane          Propane
```

With longer chains we have the possibility of branching:

```
                                                    H               H
                                                    |               |
  H  H  H  H  H  H  H  H                      H  H—C—H  H  H—C—H  H
  |  |  |  |  |  |  |  |                      |     |     |     |     |
H—C—C—C—C—C—C—C—C—H          OR        H—C ——— C ——— C ——— C ——— C—H
  |  |  |  |  |  |  |  |                      |     |     |     |     |
  H  H  H  H  H  H  H  H                      H  H—C—H  H     H     H
                                                    |
                                                    H

       n-octane                         iso-octane(2,2,4,trimethyl pentane)
```

The terminology, n- (for normal) means "straight chain," and iso- means "branched." The IUPAC name indicates the structure more exactly, based on identification of the longest discernible continuous chain.

The above are *saturated* compounds, that is, each carbon atom has four other atoms around it. These compounds are relatively stable. If two or more carbon atoms are joined by double bonds we have an *unsaturated* compound, which is less stable and reacts easily with other compounds.

<u>Olefins (Alkenes)</u>

```
                                                    H
                                                    |
                                                  H-C-H
H       H         H      H   H            H         |    H
 \     /           \     |   |             \        |    |
  C = C             C = C -  C - H          C = C -  C - H
 /     \           /         |             /             |
H       H         H          H            H              H
```

Ethylene (Ethene)

Propylene (Propene)

iso-butylene (2-methyl propene)

In the presence of a catalyst, unsaturated molecules can react with each other to form chains. This is *polymerization* (many molecules) or *oligomerization* (few molecules).

<u>Acetylenes (Alkynes)</u>

Triple bonds between carbon atoms are possible, with the resulting molecules being even more reactive, to the point of instability.

```
                                  H                      H    H
                                  |                      |    |
H - C≡C - H          H - C≡C - C - H          H - C≡C - C -  C - H
                                  |                      |    |
                                  H                      H    H
```

Acetylene (Ethyne)

Methyl acetylene (Propyne)

Ethyl acetylene (Butyne)

Note the use of terms based on the corresponding paraffin to describe the length of a side chain or radical (methyl, ethyl, propyl, butyl, etc.).

Cyclic Compounds

As well as chains, carbon atoms can also form ring structures. If these are saturated as in the case of cyclo-hexane the compounds are not very reactive, but the *aromatic* compounds based on the *benzene ring* are both reactive and found abundantly in natural products.

Cyclo-hexane

Benzene

Benzene has the equivalent of three double and three single bonds in the ring, but in fact the bonds are all equivalent, being smoothed out to a uniform structure (this also happens in straight chains if there are alternating or conjugated double and single bonds). This uniform bond structure is reflected in the newer way of depicting benzene rings in molecular diagrams:

Traditional Benzene ring

New Style Benzene ring

Oxygenated Compounds

Hydrocarbon chemistry is by definition concerned with molecules that contain only carbon and hydrogen. Organic chemistry, however, deals with all carbon-containing compounds, some containing a variety of other elements. Divalent oxygen is an important constituent of many commercial materials, being introduced by oxidation of hydrocarbons or reaction of unsaturated compounds with oxygen-containing molecules. Key types of such oxygenates are shown below:

Alcohol

Aldehyde

Carboxylic acid

Ketone

Ether

These diagrams introduce two new shorthand conventions. The hydroxyl radical is shown simply as –OH. (The aldehyde structure is frequently telescoped to –CHO and the carboxyl radical to –COOH.) The other convention is the use of R to indicate an alkyl group of any size or degree of branching.

An important reaction is that between alcohols and carboxylic acids to form esters:

$$R_1-OH + HO-\overset{O}{\overset{\|}{C}}-R_2 \rightleftharpoons R_1-O-\overset{O}{\overset{\|}{C}}-R_2 + H_2O$$

Alcohol Acid Ester Water

The bidirectional arrow means that the reaction is reversible. If an ester is heated with water, especially in the presence of acid or alkali, it undergoes hydrolysis to give the parent alcohol plus acid. Esters can also be made from alcohols and mineral acids.

Sulfur and Nitrogen

These are also commonly found in both naturally occurring and in synthetic organic materials. Thiophenes are ring compounds, and mercaptans are the sulfur versions of alcohols. Sulfides occur in certain additives, and can contain sulfur chains of up to eight atoms.

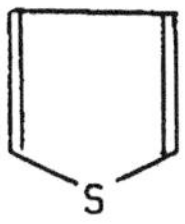

S

Thiophene

R—SH

Mercaptan
(Thiol)

R_1 S_n R_2

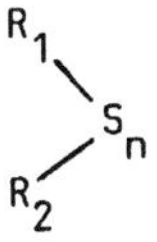

Polysulfide

Nitrogen occurs in pyrrole and pyridine rings, and in many other ways. Organic amines and their derivatives are important additive raw materials.

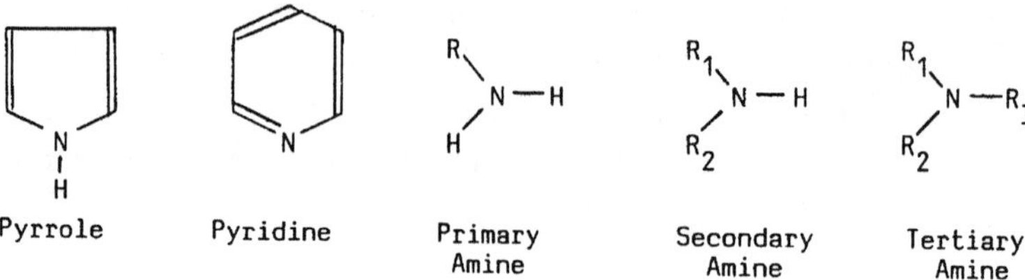

Organic nitrogen compounds are usually basic (alkaline) in behavior.

With the possibility of large molecules built up from carbon and hydrogen in many geometrical forms (isomers), and the introduction of unsaturation and other elements into molecules, it can be seen that the number of organic compounds verges on the infinite. Standard textbooks attempt to create order out of this chaos, and should be consulted for matters of detail.

Use of Catalysts

Finally, the importance of catalysis in commercial organic chemistry must be mentioned. Catalysts (materials that aid a reaction without being consumed in the process) can be used for breaking up large molecules (cracking), for reacting them with other materials (hydrogenation, sulfonation, etc.), or for building small molecules into larger ones (oligomerization or polymerization). Different catalysts and reaction conditions are of course used for different processes. Strong mineral acids are efficient catalysts for many reactions, which is why neutralization of acids in "blowby" in an engine is so important.

Appendix 4

The S.I. System of Units

The Système International of units was adopted in 1960 by 36 countries, including the U.S., at the 11th General Conference on Weights and Measures. It is a modernized version of the decimal system which is totally coherent and with which calculations and exchange of data should be simplified. General adoption has been slower than was hoped, but it is now the accepted system for all areas of technology.

The S.I. system consists of seven well-defined base units of measurement, two supplementary units of angular measurement, and many derived units resulting from mathematical combinations of the base and supplementary units.

Base Units

Quantity	Unit	Symbol
length	metre	m
mass	kilogram	kg
time	second	s
electric current	ampere	A
temperature	kelvin	K
amount of substance	mole	mol
luminous intensity	candela	cd

Supplementary Units

Quantity	Unit	Symbol
plane angle	radian	rad
solid angle	steradian	sr

Derived Units with Special Names (selected for interest)

Quantity	Unit	Symbol	Formula
frequency	hertz	Hz	1/s
force	newton	N	$kg{\cdot}m/s^2$
pressure	pascal	Pa	N/m^2
energy, work, heat	joule	J	N·m
power	watt	W	J/s
quantity of electricity	coulomb	C	A·s
electromotive force	volt	V	W/A
temperature	degree Celsius	°C	K(1)

(1) For temperature intervals a degree Celsius equals one kelvin. For the expression of thermodynamic temperature, Celsius temperature equals temperature in kelvins minus 274.15°.

A Selection of Other Derived Units

Quantity	Unit	Symbol
acceleration	metre per second squared	m/s^2
area	square metre	m^2
concentration	mole per cubic metre	mol/m^3
density of mass	kilogram per cubic metre	kg/m^3
heat capacity	joule per kelvin	J/K
moment of force	newton metre	N·m
power density	watt per square metre	W/m^2

Quantity	**Unit**	**Symbol**
specific heat capacity	joule per kilogram kelvin	J/(kg·K)
surface tension	newton per metre	N/m
thermal conductivity	watt per metre kelvin	W/(m·K)
velocity	metre per second	m/s
viscosity, dynamic	pascal second	Pa·s
viscosity, kinematic	square metre per second	m^2/s
volume	cubic metre	m^3

Multiplying Prefixes

These are used to avoid handling large numbers of digits or decimal places:

Prefix	**Multiplier**	**Symbol**
exa	10^{18}	E
peta	10^{15}	P
tera	10^{12}	T
giga	10^{9}	G
mega	10^{6}	M
kilo	10^{3}	k
hecto	10^{2}	h
deka	10	da
deci	10^{-1}	d
centi	10^{-2}	c
milli	10^{-3}	m
micro	10^{-6}	μ
nano	10^{-9}	n
pico	10^{-12}	p
femto	10^{-15}	f
atto	10^{-18}	a

Other Permitted Units in Common Use

Quantity	Unit	Symbol	S.I. Equivalent
time	minute	min	60 s
	hour	h	3600 s
	(day, week, month, year, etc.)		
angle	degree	°	π/180 rad
volume	litre	L	1 dm^3
mass	metric ton	t	10^3 kg
area	hectare	ha	10^4 m^2

Selected Conversion Factors from Older Units

To convert from:	to:	multiply by:
atmosphere (760 mm Hg)	pascal	$1.013\,25 \times 10^5$
Btu	joule	$1.055\,06 \times 10^3$
Btu/h	watt	$2.930\,71 \times 10^{-1}$
calorie	joule	4.186 80
centipoise	pascal second	1×10^{-3}
centistoke	square metre per second	1×10^{-6}
foot	metre	3.048×10^{-1}
square foot	square metre	$9.290\,30 \times 10^{-2}$
cubic foot	cubic metre	$2.831\,68 \times 10^{-2}$
foot-pound (ft·lbf)	joule	1.355 82
foot per second squared	metre per second squared	3.048×10^{-1}
gallon (US)	cubic metre	$3.785\,41 \times 10^{-3}$
gallon (UK)	cubic metre	$4.546\,1 \times 10^{-3}$
horsepower	watt	7.46×10^2
inch	metre	2.54×10^{-2}
inch of mercury, 60°F	pascal	$3.376\,85 \times 10$
pound-force, lbf	newton	4.448 22
pound (avoirdupois)	kilogram	$4.535\,92 \times 10^{-1}$
pounds/sq. inch	pascal	$6.894\,76 \times 10^3$
torr (mm Hg)	pascal	$1.333\,22 \times 10^2$
kilowatt-hour	joule	3.6×10^6

Units Used in Laboratory Practice

When the base units were originally selected, they were chosen to be of suitable magnitude for general and commercial use. Thus the kilogram was chosen as a base unit, rather than the gram. In the laboratory situation some of the base and derived units are unsuitably large, and sub-multiples are used for convenience.

- The gram (g) is used rather than the kilogram.
- The millipascal second (mPa·S), equal to a centipoise, is used instead of the pascal second.
- The millimetre squared per second (mm^2/s), equal to a centistoke, is used instead of the square metre per second.
- The correct S.I. small unit of volume is the cubic centimetre (cm^3). However, in laboratory analysis the millilitre (equal to a cubic centimetre), the litre (a cubic decimetre), and the microlitre are accepted practice, and permitted. The correct symbol for a millilitre is mL.

In contrast to the above cases, the kilowatt-hour (kW·h) is usually more convenient as a measure of energy than the watt-hour. One kilowatt-hour equals 3.6 megajoules (MJ).

More equivalents and a fuller explanation of the S.I. system will be found in the ASTM Standard Methods of Test books, under E 380.

Appendix 5

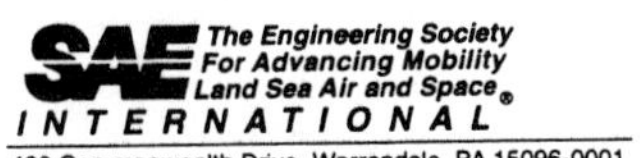

400 Commonwealth Drive, Warrendale, PA 15096-0001

SURFACE VEHICLE STANDARD

Submitted for recognition as an American National Standard

SAE J304	REV. MAR95
Issued 1942-01 Revised 1995-03	
Superseding J304 JUN93	

ENGINE OIL TESTS

(R) *1.* ***Scope***—The purpose of this SAE Standard is to describe test conditions and performance evaluation factors for both diesel and gasoline engine tests. Specifically, the tests described in this document are used to measure the engine performance requirements for engine oils described by the API Service Categories described in API Publication 1509, ASTM D 4485, SAE J183 and SAE J1423 standards, and U.S. military specifiications.

2. ***References***

2.1 Applicable Documents—The following publications form a part of this specification to the extent specified herein. The latest issue of SAE publications shall apply.

2.1.1 SAE PUBLICATIONS—Available from SAE, 400 Commonwealth Drive, Warrendale, PA 15096-0001.

2.1.1.1 SAE J183—Engine Oil Performance and Engine Service Classification (other than "Energy Conserving")

2.1.1.2 SAE J1423—Passenger Car and Light Duty Truck Energy-Conserving Engine Oil Classification

2.1.2 ASTM PUBLICATIONS—Available from ASTM, 1916 Race Street, Philadelphia, PA 19103-1187.

2.1.2.1 ASTM D 4485—Standard Specification for Performance of Engine Oils

2.1.2.2 ASTM Special Technical Publication 315I, Parts I, II, and III, "Multicylinder Test Sequences for Evaluating Automotive Engine Oils."

2.1.2.3 ASTM Special Technical Publication 509 and 509A, Parts I, II, and IV, "Single Cylinder Engine Tests for Evaluating Performance of Crankcase Lubricants."

2.1.2.4 Research reports containing the Sequence VI, Mack T-6, T-7, and T-8, Detroit Diesel 6V-53T, 6V-92TA, Caterpillar 1K, 1M-PC, 1N, and General Motors 6.2L procedures are available from ASTM. None of these procedures has been published as an ASTM Standard Test Method.

2.1.2.5 ASTM D 5290—Standard Test Method for Measurement of Oil Consumption, Piston Deposits, and Wear in a Heavy-Duty High-Speed Diesel Engine—NTC-400 Procedure

2.1.2.6 ASTM D 5119—Standard Test Method for Evaluation of Automotive Engine Oils in the CRC L-38 Spark-Ignition Engine

SAE J304 Revised MAR95

(R) *2.1.2.7 ASTM D 5302*—Standard Test Method for Evaluation of Automotive Engine Oils for Inhibition of Deposit Formation and Wear in a Spark-Ignition Internal Combustion Engine Fueled with Gasoline and Operated Under Low-Temperature, Light-Duty Conditions—Sequence VE Procedure

(R) *2.1.2.8* ASTM D 5533—Standard Test Method for Evaluation of Automotive Engine Oils in the Sequence IIIE, Spark-Ignition Engine

2.1.3 FEDERAL TEST METHOD STANDARDS—Available from General Services Administration, Business Service Center, Region 3, Seventh and D Streets SW, Washington, DC 20025.

2.1.3.1 FTM 354.1—Performance of Arctic Lubricating Oils in a Two-Cycle Diesel Engine Under Cyclic, Turbo-Charged Conditions

2.1.3.2 FTM 791—Lubricants, Liquid Fuels and Related Products; Methods of Testing

2.1.4 API PUBLICATIONS—Available from American Petroleum Institute, 1220 L Street NW, Washington, DC 20005.

2.1.4.1 API 1509—Engine Oil Licensing and Certification System

2.2 Related Publications

2.2.1 MILITARY SPECIFICATIONS AND COMMERCIAL ITEM DESCRIPTIONS—Available from Standardization Documents Order Desk, Building 4D, 700 Robbins Avenue, Philadelphia, PA 19111-5094.

MIL-L-2104F—Lubricating Oil—Internal Combustion Engine—Tactical Service
MIL-L-6082E—Lubricating Oil—Aircraft Reciprocating Engine
MIL-L-21260D—Lubricating Oil—Internal Combustion Engine—Preservative and Break-in
MIL-L-22851D—Lubricating Oil—Aircraft Piston Engine (Ashless Dispersant)
MIL-L-46167B—Lubricating Oil—Internal Combustion Engine—Arctic Service
CID A-A-52039—Lubricating Oil—Automotive Engine—API Service SG
CID A-A-52306—Lubricating Oil—Heavy-Duty Diesel Engine

(R) ***3. Test Applicability***—Engine tests have been used for many years to determine performance characteristics of engine oils, and the development of new engine tests is an ongoing process. This document describes a number of current tests, as well as many tests that are considered by ASTM to be obsolete. Obsolete tests are so defined because specified fuels, reference oils, and/or engine parts are no longer available, or because the tests are not monitored by the test developer or ASTM. This document includes many API engine oil categories based on such obsolete tests, as well as other categories based on currently available tests. As of this revision, the only included API service categories based on current tests are the SG, SH, CD, CD-II, CE, CF, CF-2, CF-4, CG-4, Energy Conserving, and Energy Conserving II categories. Similarly, this document contains both obsolete and current U.S. military specifications, and as of this revision the only current U.S. military specifications included are the MIL-L-2104F, MIL-L-6082E, MIL-L-21260D, MIL-L-22851D, and MIL-L-46167B specifications.

(R) ***4. Diesel Engine Tests***—Many tests for piston ring sticking, piston ring and cylinder wear, valve train wear, oil consumption, oil thickening, and general deposit accumulation have been developed using single-cylinder and multicylinder compression ignition (diesel) engines. Current single-cylinder tests in use include the Caterpillar 1G2, 1K, IM-PC, and IN. Current multicylinder tests include the Detroit Diesel 6V-53T, 6V-92TA, Mack T-6, T-7, T-8, Cummins NTC-400, and General Motors 6.2L. (See Tables 1 and 2.)

SAE J304 Revised MAR95

(R) ***5. Gasoline Engine Tests***—Lubricant performance is evaluated in gasoline engine tests in terms of rust and corrosion, sludge, varnish, piston ring zone deposits and condition, wear, bearing corrosion, and fuel efficiency. For those tests listed in SAE J183 and J1423, Table 3 gives the test conditions and Table 4 gives procedure references, evaluation factors, and the API performance categories and U.S. military specifications in which the tests are included.

(R) ***6. Military Specifications and CIDs***—MIL-L specifications have historically been used as both primary performance specifications, including engine testing, as well as a part of the procurement process via Qualified Products Lists. Recently, there has been some movement to adopt a CID (Commercial Item Description) system in place of the MIL-L Specification/QPL system. The first example of this was the adoption of CID A-A-52039 for API Service Category SG oils, and the concurrent cancellation of MIL-L-46152E and the corresponding QPL-46152-14. CID A-A-52039 defines oil requirements in terms of API Service Category SG, SAE viscosity grades, and various other bench tests. A similar example was the adoption of CID A-A-52306 for heavy-duty diesel oils used in wheeled vehicles (only) which formerly would have required MIL-L-2104 products. With the exception of L-38 stay-in-grade viscosity tests, these CIDs do not specify engine test requirements per se, other than via API categories. For this reason, CID information is not listed in this document in the tables concerned with test applications, Table 2 and 4.

7. Notes

7.1 Marginal Indica—The (R) is for the convenience of the user in locating areas where technical revisions have been made to the previous issue of the report. If the symbol is next to the report title, it indicates a complete revision of the report.

PREPARED BY THE SAE FUELS AND LUBRICANTS TECHNICAL COMMITTEE 2—
HEAVY-DUTY TYPE ENGINE OILS

SAE J304 Revised MAR95

(R) TABLE 1A—DIESEL ENGINE TEST CONDITIONS

	Engine Displacement cm^3	Engine Displacement in^3	Test Duration, h total	Test Duration, h each phase	Oil Change Period, h	Engine Speed, rpm	Fuel Rate, kW	Fuel Rate, btu/min	Load bmep, kPa	Load bmep, psi
L-1[a]	3 400	208	480		120	1000	51.8	2 950	520	75
Modified L-1[a]	3 400	208	480		120	1000	51.8	2 950	520	75
1D[a]	3 400	208	480		120	1200	98.4	5 600	925	134
1G[a]	2 200	134	480		120	1800	102.8	5 850	972	141
1G2	2 200	134	480		120	1800	102.8	5 850	972	141
1H[a]	2 200	134	480		120	1800	87.0	4 950	758	110
1H2[a]	2 200	134	480		120	1800	87.0	4 950	758	110
6V-53T	5 220	318	240	0.5 2.0 0.5 2.0	120	1000 2800 1000 2200	N/S 670 N/S 549	N/S 38 142 N/S 31 296	172 917 172 1034	25 133 25 150
6V-92TA	9 046	552	100	8.0 8.0	None	1200 2300	≥653 ≥1130	≥37 326 ≥64 265	2062 1076	299 156
T-6	11 000	672	600	4.0 4.0 4.0	None	1400 1800 2100	523 628 697	29 756 35 762 39 641	1531 1420 1262	222 206 183
T-7	11 000	672	150		None	1200	514	29 235	1655	240
T-8	12 000	728	250		None	1800	800	45 692	1459	212
NTC-400	14 000	855	200		None	2100	865	49 237	1289	187
1K	2 400	149	252		None	2100	140.7[b]	7 995	1240	180
1M-PC	2 200	134	120		None	1800	102.8	5 850	972	141
1N	2 400	149	252		None	2100	140.7[b]	7 995	1 240	180
6.2L	6 200	379	50		None	1000	106.1	6 033	620	90

[a] This test is obsolete and is included for historical purposes only.
[b] Within the test this parameter is expressed in metric units, and the specified fuel rate is 8430 kJ/min.
N/S = not specified

SAE J304 Revised MAR95

(R) TABLE 1B—DIESEL ENGINE TEST CONDITIONS

	Air-to-Engine Temperature °C	Air-to-Engine Temperature °F	Air-to-Engine Pressure Abs., kPa	Air-to-Engine Pressure Abs., Inch Hg	Water Outlet Temperature °C	Water Outlet Temperature °F	Oil-to-Bearing Temperature °C	Oil-to-Bearing Temperature °F	Fuel Sulfur, % mass
L-1[a]	Room, not over 38	Room, not over 100	Atmospheric	Atmospheric	80.8	177.5	64.2	147.5	0.35 min
Modified L-1[a]	Room, not over 38	Room, not over 100	Atmospheric	Atmospheric	80.8	177.5	64.2	147.5	1.0 ± 0.05
1D[a]	93	200	150.3	44.5	93.3	200	79.4	175	1.0 ± 0.05
1G[a]	124	255	179	53	87.8	190	96.1	205	0.35 min[b]
1G2	124	255	179	53	87.8	190	96.1	205	0.37-0.43[b]
1H[a]	77	170	135	40	71.1	160	82.2	180	0.35 min[b]
1H2[a]	77	170	135	40	71.1	160	82.2	180	0.37-0.43[b]
6V-53T	N/S 143 N/S 116	N/S 290 N/S 240	N/S 240 N/S 190	N/S 71 N/S 56	76.7 76.7 76.7 76.7	170 170 170 170	79.4 100 79.4 95	175 212 175 203	0.37-0.43[b] 0.37-0.43[b] 0.37-0.43[b] 0.37-0.43[b]
6V-92TA	35 35	95 95	 97	Not Specified 29	84 84	172-190 172-190	102 111	216 232	0.1-0.4 0.1-0.4
T-6	66 85 99	150 185 210	98 139 156	29 41 46	87.8 87.8 87.8	190 190 190	112.8 112.8 112.8	235 max 235 max 235 max	0.1-0.3 0.1-0.3 0.1-0.3
T-7	113	135	102	30	85.0	185	112.8	235 max	0.40 max
T-8	43	109	186-199	55-59	85	185	100-107	212-225	0.03-0.05
NTC-400	143	290	142-169	42-50	85.0	185	121.1	250	0.40
1K	127	260	240	71	93	199.4	107	224.6	0.38-0.42[b]
1M-PC	124	255	179	53	87.8	190	96	205	0.38-0.42[b]
1N	127	260	240	71	93	200	107	225	0.03-0.05
6.2L	32	90	97	29	120	248	120	248	0.03-0.05

[a] This test is obsolete and is included for historical purposes only.
[b] Sulfur must be of "Natural Origin."
N/S = not specified

(R) TABLE 2—DIESEL ENGINE TEST REFERENCES, PERFORMANCE EVALUATION FACTORS, AND APPLICATIONS

Test	Procedure Reference See Ref. 2.1.3.1, 2.1.3.2, 2.1.2.3, 2.1.2.4, 2.1.2.5	Evaluation Deposits	Evaluation Wear	Evaluation Oil Consumption	Evaluation Other	Test Application (See Ref. 2.1.1.1, 2.1.2.1)
L-1[a]	FTMS 791A - FTM 332	X				CA, MIL-L-2104
Modified L-1[a]	FTMS 791A - FTM 345	X				CB, SC, SD, MIL-L-2104A
ID[a]	FTMS 791B - FTM 340.3	X				CD, MIL-L-2104C, MIL-L-45199B
1G[a]	FTMS 791B - FTM 341.4	X				CD, MIL-L-2104C, MIL-L-45199B
1G2[c]	ASTM STP 509A (PART 1), FTMS 791C FTM 341.4	X				CD, CD-II, CE, MIL-L-2104E, MIL-L-21260D, MIL-L-46167B
1H[a]	FTMS 791B - FTM 346.2	X				CC, SD, MIL-L-46152, MIL-L-21260B, MIL-L-46167, MIL-L-2104B
1H2[a,c]	ASTM STP 509A (PART II), FTMS 791C FTM 346.2	X				CC, MIL-L-46152D, MIL-L-21260C, MIL-L-46167A, MIL-L-46152E
6V-53T	ASTM Research Report RR:D02:1222[b]	X	X			CD-II, MIL-L-2104E, MIL-L-21260D, MIL-L-46167B
6V-92TA	ASTM Research Report RR:D02:1319	X	X			MIL-L-2104F, CF-2
T-6	ASTM Research Report RR:D02:1219	X	X	X	Oil Thickening	CE, CF-4
T-7	ASTM Research Report RR:D02:1220				Oil Thickening	CE, CF-4, MIL-L-2104F
T-8	ASTM Research Report RR:D02:1324			X	Oil Thickening, Filter Plugging	CG-4
NTC-400	ASTM D 5290	X	X	X		CE, CF-4
1K	ASTM Research Report RR:D02:1273	X		X		CF-4, MIL-L-2104F
1M-PC	ASTM Research Report RR:D02:1320	X				CF, CF-2
1N	ASTM Research Report RR:D02:1321	X		X		CG-4
6.2L	ATM Research Report RR:D02:1323		X			CG-4

[a] This test is obsolete and is included for historical purposes only.

[b] FTMS 791C - FTM 354.1 (for MIL-L-46167B) is the same as ASTM RR:D-2-1222 except FTMS 791C - FTM 354.1 test conditions call for 21.7% less load at max power and torque modes, 675 rpm and water outlet temperature of 100 °F at idle mode.

[c] The 1H2 (or alternatively, the 1G2) was originally used for the SG Category, but this requirement was dropped in 1992.

TABLE 3—GASOLINE ENGINE TEST CONDITIONS

Test	Engine Type	Engine Displacement cm^3	Engine Displacement In^3	Time Total, h	Time Each Phase	Air-Fuel Ratio	Load kW	Load bhp	Fuel Flow kg/h	Fuel Flow lb/h	Speed, rpm	Temperature Coolant °C	Temperature Coolant °F	Temperature Oil °C	Temperature Oil °F
L-4[a]	L6	3540	216	36		14.5	22.4	30			3150	93.3	200	137.8	280
L-38	Single	700	42.5	40		14.0	Not Controlled	Not Controlled	2.04/ 2.27	4.5- 5.0	3150	93.3	200	143.3	290
LTD[a]	Single	700	42.5	180	3 h	14.5	Not Controlled	Not Controlled	2.13	4.7	1800	48.9	120	Not Controlled	Not Controlled
					1 h	14.5			2.13	4.7	1800	93.3	200		
Modified LTD[a]	Single	700	42.5	180	3 h	15.25	Not Controlled	Not Controlled	2.13	4.7	1800	48.9	120	Not Controlled	Not Controlled
					1 h	15.25			2.13	4.7	1800	93.3	200		
Falcon[a]	L6	2790	170	55	45 min	Max Vac.	None	None			500	46.1	115	48.9	120
					2 h	15.5	23.04	30.9			2500	51.7	125	79.4	175
Sequence 1[a]	V8	6460	394	5	10 min	14.0	1.5	2			2500	35.0	95	48.9	120
Sequence II[a]	V8	6460	394	30	3 h	14.0	18.6	25			1500	35.0	95	48.9	120
Sequence IIA[a]	V8	6460	394	22	20 h	12.0	18.6	25			1500	35.0	95	48.9	120
					2 h	12.0	18.6	25			1500	48.9	120	48.9	120
Sequence IIB[a]	V8	6960	425	24	20 h	13.0	18.6	25			1500	40.6	105	48.9	120
					2 h	13.0	18.6	25			1500	48.9	120	48.9	120
					2 h	16.5	74.6	100			3600	93.3	200	135.0	275
Sequence IIC[a]	V8	6960	425	32	28 h	13.0	18.6	25			1500	43.3	110	48.9	120
					2h	13.0	18.6	25			1500	48.9	120	48.9	120
					2h	16.5	74.6	100			3600	93.3	200	126.7	260
Sequence IID	V8	5740	350	32	28 h	13.0	18.6	25			1500	43.3	110	48.9	120
					2 h	13.0	18.6	25			1500	43.3	110	48.9	120
					2 h	16.5	74.6	100			3600	93.3	200	126.7	260
Sequence III[a]	V8	6460	394	40		15.0	63.4	85			3400	93.3	200	129.4	265
Sequence IIIA[a]	V8	6460	394	40		16.5	63.4	85			3400	93.3	200	129.4	265
Sequence IIIB	V8	6960	425	56	7 h	16.5	74.6	100			3600	65.6	150	93.3	200
					7 h	16.5	74.6	100			3600	93.3	200	135.0	275
Sequence IIIC[a]	V8	6960	425	64		16.5	74.6	100			3000	118.3	245	148.9	300
Sequence IIID[a]	V8	5740	350	64		16.5	74.6	100			3000	118.3	245	148.9	300
Sequence IIIE	V6	3800	231	64		16.5	50.6	67.8			3000	115.0	239	149.0	300.2
Sequence IV[a]	V8	5920	361	24	2 h	Not Controlled	None	None			3600	82.2	180	104.4	220
					2 h						0	12.8	55	Not Controlled	Not Controlled

SAE J304 Revised MAR95

TABLE 3—GASOLINE ENGINE TEST CONDITIONS (CONTINUED)

Test	Engine Type	Engine Displacement cm³	Engine Displacement in³	Time Total, h	Time Each Phase	Air-Fuel Ratio	Load kW	Load bhp	Fuel Flow kg/h	Fuel Flow lb/h	Speed, rpm	Temperature Coolant °C	Temperature Coolant °F	Temperature Oil °C	Temperature Oil °F
Sequence V[a]	V8	6030	368	192	45 min	9.5	None	None			500	46.1	115	48.9	120
					2 h	15.5	78.3	105			2500	51.7	125	79.4	175
					75 min	15.5	78.3	105			2500	76.7	170	96.1	205
Sequence VB[a]	V8	4740	289	192	45 min	9.5	None	None			500	46.1	115	48.9	120
					2 h	15.5	64.58	86.6			2500	51.7	125	79.4	175
					75 min	15.5	64.58	86.6			2500	77.2	171	93.9	201
Sequence VC[a]	V8	4950	302	192	2 h	b	64.58	86.6			2500	57.2	135	79.4	175
					75 min		64.58	86.6			2500	76.7	170	93.3	200
					45 min		1.5	2			500	46.1	115	48.9	120
Sequence V-D[a]	OHC4	2290	140	192	2 h	b	24.98	33.5			2500	57.2	135	79.4	175
					75 min		24.98	3.5			2500	68.3	155	86.1	187
					45 min		0.75	1.0			750	48.9	120	48.9	120
Sequence VE	OHC4	2290	140	288	2 h	b	24.98	33.5			2500	51.7	125	68.3	155
					75 min		24.98	33.5			2500	85.0	185	98.9	210
					45 min		0.75	1.0			750	46.1	115	46.1	115
Sequence VI					13 h							54.0	130	66.0	150
	V6	3800	231	65	32 h	14.1	5.96	8.0			1500	93.0	200	107.0	225
					20 h							110.0	230	135.0	275

[a] This test is obsolete and is included for historical purposes only.

[b] Equivalence ratio is controlled by monitoring carbon monoxide and oxygen concentrations.

(R) TABLE 4—GASOLINE ENGINE TEST REFERENCES, PERFORMANCE EVALUATION FACTORS, AND APPLICATIONS

Test	Procedure Reference (See Ref. 2.1.2.2, 2.1.2.4, 2.1.2.6, 2.1.3.2)	Evaluation Rust and Corrosion	Evaluation Sludge	Evaluation Varnish	Evaluation Wear	Evaluation Other	Test Application (See Ref.2.1.1.1, 2.1.1.2, 2.1.2.1)
L-4[a]	FTMS 791A-FTM 3402 (2.1.3.2)			X		Bearing Weight Loss	CA, CB, SB
L-38	ASTM D 5119(2.1.2.6)			X		Bearing Weight Loss	CA, CB, CC, CD, CD-11, SG, MIL-L-45199B, MIL-L-2104F
L-38	ASTM D5119 (2.1.2.6)					Bearing Weight Loss	SB, SC, SD, SE, SF, SH, CE, CF, CF-4, CG-4, MIL-L-2104 (through E), MIL-L-6082E, MIL-L-21260D, MIL-L-22851D, MIL-L-46152 (through E), MIL-L-46167B
LTD[a]	FTMS 791B-FTM 348.1 (2.1.3.2)		X	X			CC, MIL-L-2104B
Modified LTD[a]	FTMS 791B-FTM 348.2 (2.1.3.2)		X	X			CC
Falcon[a]	Ford Motor Company FLTM BJ 11-2	X					SD
Sequence 1[a]	ASTM STP 315, 315A				X		MS (Obsolete)
Sequence II[a]	ASTM STP 315, 315A	X					MS (Obsolete)
Sequence IIA[a]	ASTM STP 315B, 315C	X					CC, SC, MIL-L-2104B
Sequence IIB[a]	ASTM STP 315D, 315E	X					CC, SD, SE, MIL-L-2104C, MIL-L-46152
Sequence IIC[a]	ASTM STP 315F, 315G	X					SE, CC, MIL-L-2104C, MIL-L-46152
Sequence IID	ASTM STP 315I, Part I (2.1.2.2)	X				Lifter Sticking	SE, SF, SG, SH, CC, MIL-L-2104E, MIL-L-46152D, MIL-L-21260D, MIL-L-46167B, MIL-L-46152E
Sequence III[a]	ASTM STP 315, 315A		X	X			MS (Obsolete)
Sequence IIIA[a]	ASTM STP 315B, 315C		X	X	X		SC
Sequence IIIB[a]	ASTM STP 315D		X	X	X		SD
Sequence IIIC[a]	ASTM STP 315E, 315F, 315G		X	X	X	Oil Thickening	SE, MIL-L-46152
Sequence IIID[a]	ASTM STP 315H, Part II (2.1.2.2)		X	X	X	Oil Thickening	SE, SF, MIL-L-45152C
Sequence IIID[a]	ASTM STP 315H, Part II (2.1.2.2)				X		MIL-L-2104D, MIL-L-46167A, MIL-L-21260C
Sequence IIIE	ASTM D 5533		X	X	X	Oil Thickening, Ring Sticking	SG, SH, MIL-L-46152D, MIL-L-46152E, MIL-L-2104F
Sequence IIIE	ASTM D 5533					Oil Thickening	CG-4
Sequence IIIE	ASTM Research Report RR:D02:1225				X		MIL-L-2104E, MIL-L-46167B, MIL-L-21260D
Sequence IV[a]	ASTM STP 315, 315A, B, C, D				X		MS (Obsolete) , SB, SC, SD
Sequence V[a]	ASTM STP 315, 315A, B		X	X			MS (Obsolete), SC
Sequence VB[a]	ASTM STP 315C, D		X	X			SD
Sequence VC[a]	ASTM STP 315E, F, G		X	X			SE, CC, MIL-L-2104C, MIL-L-46152A
Sequence V-D[a]	ASTM STP 315H, Part III (2.1.2.2)		X	X	X		SE, SF, MIL-L-46152C, MIL-L-2104D, MIL-L-46167A
Sequence V-D[a]	ASTM STP 315H, Part III (2.1.2.2)		X	X			CC, MIL-L-2104C, MIL-L-21260C
Sequence VE	ASTM D 5302		X	X	X		SG, SH, MIL-L-46167B, MIL-L-46152D, MIL-L-2104E MIL-L-21260D, MIL-L-46152E
Sequence VI	ASTM Research Report RR:D02:1204					Fuel Efficiency	Energy Conserving, Energy Conserving II, MIL-L-46152E

[a] This test is obsolete and is included for historical purposes only.

[b] For MIL-L-6082E and MIL-L-22851D, the oil temperature is 135 °C (275 °F).

Appendix 6

The Precision of Laboratory Tests

The following tables give examples of the values of repeatability (same operator, same laboratory) and reproducibility (different operators and laboratories) obtained by the ASTM as a result of cooperative test programs. In normal circumstances two test results on the same sample should not differ by more than these values in 19 cases out of 20.

Common tests used in the evaluation of automotive lubricants are cited. Values for other tests will be found in the relevant ASTM test method.

Tests are grouped according to type. The reduction in precision on going from physical tests to chemical tests to bench rig-tests to engine tests is apparent.

ASTM No.	Description	Repeatability	Reproducibility
Physical Tests			
D 445	Kinematic viscosity	0.35%	0.70% (of the value)
D 2893	Brookfield viscosity		
	at 10,000 cP max	320 cP	800 cP
	at 20,000 cP max	840 cP	1880 cP
	at 50,000 cP max	2800 cP	5900 cP
	at 100,000 cP max	6700 cP	11,800 cP
	at 200,000 cP max	16,600 cP	23,200 cP
	at 400,000 cP max	54,800 cP	65,600 cP

ASTM No.	Description	Repeatability	Reproducibility
Physical Tests			
D 4683	TBS viscometer	0.96%	2.59%
D 4684	Mini Rotary Viscometer		
	at –15°C	4.2%	8.4%
	at –20°C	7.3%	12.1%
	at –25°C	11.7%	17.5%
	at –30°C	9.3%	18.4%
	at –35°C	20.7%	21.5%
D 97	Pour Point	2.87°C	6.43°C
D97	Flash Point (P-M)	5°C	10°C
D 892	Foaming Tendency		
	Seq. I, II	23%	47%
	Seq. III	37%	110%
Chemical Tests			
D 664	Neutralization No.		
	Manual	7 mg	20 mg
	Automatic	6 mg	28 mg
D 2896	Base Number		
	New oils	3%	7%
	Used oils with back titration	24%	32%

ASTM No.	Description	Repeatability	Reproducibility
D4628	Atomic Absorption Analysis		
	Calcium at 0.01%	0.001%	0.004%
	Calcium at 0.30%	0.010%	0.035%
	Magnesium at 0.01%	0.001%	0.003%
	Magnesium at 0.30%	0.010%	0.032%
	Zinc at 0.01%	0.001%	0.002%
	Zinc at 0.20%	0.008%	0.018%
	(e.g., for 0.01% of calcium, repeatability is 0.001/0.01 = 10% of value)		
D 4927	X-ray Fluorescence Analysis (internal standard method)		
	Calcium at 0.10%	0.005%	0.018%
	Phosphorus at 0.10%	0.007%	0.011%
	Sulfur at 0.10%	0.004%	0.020%
	Zinc at 0.10%	0.002%	0.005%
D 4951	ICP Spectroscopic Analysis		
	Example: Zinc 0.001 to 1.0%	3.1% of value	10.0% of value

Bench Rig-Tests

ASTM No.	Description	Repeatability	Reproducibility
D 2783	Four-ball EP Test, LWI	17%	44%
D 4172	Four-ball Wear Prevention	0.12 mm	0.28 mm (difference in scar diameter)
D 217	Grease Penetration		
	unworked	8	19 penetration units
	worked	7	20

ASTM No.	Description	Repeatability	Reproducibility
Engine Test			
D 5302	Sequence VE Test		
		average sludge	107.7% of demerit level*
		rocker cover sludge	131.1% of demerit level*
		average varnish	1.19 merits in 10 max
		piston skirt varnish	0.78 merits
		average cam wear	159.4%
		maximum cam wear	169.4%

* 10 minus merit level

No repeatability estimate is available due to scarcity of data for same oil on same stand in same laboratory with same operator.

Appendix 7

INTERNATIONAL

400 Commonwealth Drive, Warrendale, PA 15096-0001

SURFACE VEHICLE STANDARD

Submitted for recognition as an American National Standard

SAE J183	REV. JUN91
Issued 1970-06 Revised 1991-06-17	
Superseding J183 JUN90	

ENGINE OIL PERFORMANCE AND ENGINE SERVICE CLASSIFICATION (OTHER THAN "ENERGY CONSERVING")

1. ***Scope***—The scope of this SAE Standard is to outline the joint engine oil classification efforts of API, ASTM, and SAE. The designation, status, and descriptions of the categories are presented, as well as the test techniques and primary performance criteria.

(R) *2.* ***References***

2.1 Applicable Documents—The following publications form a part of this specification to the extent specified herein. The latest issue of SAE publications shall apply.

2.1.1 SAE Publications—Available from SAE, 400 Commonwealth Drive, Warrendale, PA 15096-0001.

SAE J300—Engine Oil Viscosity Classification

SAE J303—Internal Combustion Engine Service Classifications

SAE J304—Engine Oil Tests

SAE J1146—The Automotive Lubricant Performance and Service Classification Maintenance Procedure

SAE J1423—Passenger Car and Light-Duty Truck Energy-Conserving Engine Oil Classification

2.1.2 API Publication—Available from American Petroleum Institute, 1220 L Street N.W., Washington, DC 20005.

API Publication 1509—Engine Service Classification System and Guide to Crankcase Oil Selection

2.1.3 ASTM Publications—Available from ASTM, 1916 Race Street, Philadelphia, PA 19103.

ASTM D 4485—Standard Performance Specification for Automotive Engine Oils

ASTM E 178—Recommended Practice for Dealing with Outlying Observations

ASTM RR:D-2-1194—Research Report: Measurement of Lubricating Oil Performance with a Multicylinder Diesel Laboratory Engine Procedure

ASTM RR:D-2-1219—Supporting Data for D 4485, Performance Specification for Automotive Engine Oils (Multicylinder Engine Test Procedure for the Evaluation of Lubricants—Mack T-6)

ASTM RR:D-2-1220—Supporting Data for D 4485, Performance Specification for Automotive Engine Oils (Multicylinder Engine Test Procedure for the Evaluation of Lubricants—Mack T-7)

ASTM RR:D-2-1222—Supporting Data for D 4885, Performance Specification for Automotive Engine Oils (Test Method for Measurement of Lubricating Oil Performance in Two-Stroke Cycle Turbo-Supercharged Diesel Engines)

ASTM RR:D-2-1273—Caterpillar 1K Test ASTM Research Report

SAE J183 Revised JUN91

ASTM STP 315—Multicylinder Test Sequences for Evaluating Automotive Engine Oils

ASTM STP 509—Single Cylinder Engine Tests for Evaluating Performance of Crankcase Lubricants

2.1.4 CRC Publication—Available from Coordinating Research Council, 219 Parimeter Center Parkway, Atlanta, GA 30346.

CRC Modified Supplemental Diesel Engine Rating Manual No. 15

2.1.5 Federal Publications—Available from The Standardization Documents Order Desk, Building 4D, 700 Robbins Avenue, Philadelphia, PA 19111-5094.

Federal Test Method Standard 791—Lubricants, Liquid Fuels and Related Products; Methods of Testing

3. API Engine Service Classification System—Prior to 1947, automotive engine oils were classified by SAE J300 in terms of viscosity only. In order to permit the recommendation of oils by classes which would include factors other than viscosity, the American Petroleum Institute adopted in 1947 a system which divided crankcase oils into three classes depending on the properties of the oil and the operating conditions under which it was intended to be used. In this system, crankcase oils were classified as: Regular Type, Premium Type, or Heavy-Duty Type. Generally, the Regular Type oils were straight mineral oils, Premium Type contained oxidation inhibitors, and Heavy-Duty Type contained oxidation inhibitors plus detergent-dispersant additives.

(R) These early service classifications did not recognize that diesel and gasoline engines might have different engine oil requirements or that the requirements for either type of engine are influenced significantly by the characteristics of the fuel burned and operating conditions, especially cold weather "start and stop" operation. Consequently, the API developed a new classification system based on the severity of engine service. This system was developed in 1952 and revised in 1955.

(R) The API Engine Service Classification System described and classified, in general terms, the service conditions under which engines were operated. It included three categories for gasoline engines (ML, MM, and MS) and three for diesel engines (DG, DM, and DS). In 1960 the system was again modified and performance definitions were added to define the service categories by the adoption of the ASTM Multicylinder Sequence Tests (Sequence I, II, III, IV, and V). Detail regarding these categories was given in SAE J303, Internal Combustion Engine Service Classifications, which last appeared in the 1971 SAE Handbook.

Because the API Engine Service Classification lacked precise technical definitions of quality, gasoline and diesel engine oils were described using combinations of the API Service Classification, and individual company and military specifications. Supplementary quality definitions were found necessary and these supplemental definitions were incorporated into individual company specifications. This practice encouraged the development of special lubricants acceptable to only one equipment manufacturer. Also, the performance level indicated by each category changed periodically and thus it became necessary to include supplementary definitions in communications regarding engine oil.

It became apparent that more effective means must be found to communicate engine oil performance and engine service classification information between the automotive equipment manufacturer, the petroleum industry, and the customer. Accordingly, in 1969 and 1970, the API, ASTM, and SAE cooperated in establishing the present classification as a joint effort to provide these means. An additional performance and service classification (SAE J1423) was established in 1983 to characterize the "energy-conserving" features of engine oils. By these classifications, engine oils can be more precisely defined and selected according to their performance characteristics than heretofore, and they can be more easily related to the type of usage for which each is intended.

SAE J183 Revised JUN91

3.1 The primary responsibility of the cooperating organizations in establishment and administration of the classification is:

3.1.1 SAE—Evaluation of categories suggested and promulgation of the categories to be included. (See SAE J1146.)

3.1.2 ASTM—Establishment of test methods and performance limits, that is, development of a technical language describing the categories to engine builders and oil formulators. (See ASTM Standard D 4485.)

3.1.3 API—Identification of the categories and description of their service use, that is, elucidation of the technical language to consumers. (See API Publication 1509.)

(R) **3.2** Tables 1 and 2, which summarize the Engine Oil Performance and Engine Service Classification, were prepared in cooperation with API and ASTM. This classification system is divided into two major categories, "S" and "C." The "S" category denotes automotive gasoline engine services and is comprised of categories SA, SB, SC, SD, SE, SF, and SG. The "C" category denotes commercial diesel engine services and is comprised of categories CA, CB, CC, CD, CD-II, CE, and CF-4. In some instances an engine manufacturer may specify an engine oil that meets more than one engine service category, such as SG-CD. Oils that meet more than one service category may be so designated.

The categories shown in Tables 1 and 2 include all engine oils which have been, or currently are, marketed in substantial volume for use in passenger cars, gasoline and diesel powered trucks, and gasoline and diesel powered off-highway equipment. Some of the categories are not currently recommended by manufacturers for use in their equipment. It should be noted that some individual recommendations may also include compositional or proprietary considerations in a product choice. Should any doubt arise in the application of this category, the engine builder and oil supplier should be consulted.

3.3 The letter designations of categories are labeled either NO TESTS REQUIRED, OBSOLETE TEST TECHNIQUES, or ACTIVE TEST TECHNIQUES. Categories with letter designations labeled NO TESTS REQUIRED or OBSOLETE TEST TECHNIQUES continue to be shown for historical purposes, but should be used only if explicitly recommended by the equipment builder.

3.3.1 NO TEST REQUIRED is the status designation of category SA for which no performance definition exists.

3.3.2 OBSOLETE TEST TECHNIQUES is the status of an engine oil category that includes performance tests where engine parts and/or test fuel and/or reference oils are no longer generally available, and the tests are no longer being monitored by the test developer or ASTM. An oil's performance in a category with this status cannot be evaluated. Letter designations currently labeled OBSOLETE TEST TECHNIQUES include SB, SC, SD, SE, SF, CA, CB, and CC.

(R) 3.3.3 ACTIVE TEST TECHNIQUES is the status of an engine oil category that involves engine parts, test fuels, and reference oils that are currently available, and for which the tests are being monitored by the ASTM Test Monitoring Center and/or the appropriate ASTM surveillance panel. Letter designations currently labeled ACTIVE TEST TECHNIQUES include SG, CD, CD-II, CE, and CF-4.

SAE J183 Revised JUN91

3.4 This document will be reviewed in its entirety on an annual basis to insure conformance with the requirements of the automotive and petroleum industries and the consumer.

The piston zone definitions listed in Figure 1 apply:

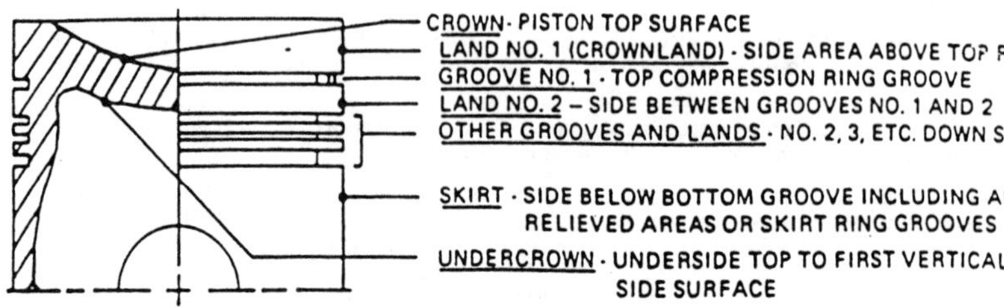

FIGURE 1—CRC PISTON ZONE DEFINITIONS

(R) ***4. Notes***

4.1 Marginal Indicia—The (R) is for the convenience of the user in locating areas where technical revisions have been made to the previous issue of the report. If the symbol is next to the report title, it indicates a complete revision of the report.

PREPARED BY THE SAE FUELS & LUBRICANTS TECHNICAL COMMITTEE 2—
HEAVY-DUTY TYPE ENGINE OILS

SAE J183 Revised JUN91

(R) TABLE 1—DESIGNATION, STATUS, AND DESCRIPTIONS OF CATEGORIES

API Letter Designation	Status	API Engine Service Description	ASTM Engine Oil Description
SA	No Test Required	**Formerly for Utility Gasoline and Diesel Engine Service (Obsolete)** The category SA denotes service typical of older engines operated under such mild conditions that the protection afforded by compounded oils is not required. This category has no performance requirements, and oils in this category should not be used in any engine unless specifically recommended by the equipment manufacturer.	No performance requirements have been established for this category since it describes oil containing no performance additives.
SB	Obsolete Test Techniques	**Minimum Duty Gasoline Engine Service (Obsolete)** The category SB denotes service typical of older engines operated under such mild conditions that only minimum protection afforded by compounding is desired. Oils designed for this service have been used since the 1930's and provide mild antiscuff capability and resistance to oil oxidation and bearing corrosion. They should not be used in any engine unless specifically recommended by the equipment manufacturer.	(The test procedure used to define this category is obsolete.) Oil meeting the performance requirements of the following gasoline engine tests: The Sequence IV test has been correlated with vehicles used in consumer service prior to 1958, particularly with regard to valve train scuffing. Either the L-4 or the L-38 test provides a measurement of copper-lead bearing weight loss under high-temperature conditions.
SC	Obsolete Test Techniques	**1964 Gasoline Engine Warranty Maintenance Service (Obsolete)** The category SC denotes service typical of gasoline engines in 1964 through 1967 models of passenger cars and some trucks operating under engine manufacturers' warranties in effect during those model years. Oils designed for this service provide control of high- and low-temperature deposits, wear, rust, and corrosion in gasoline engines.	(The test procedure used to define this category is obsolete.) Oil meeting the performance requirements measured in the following gasoline and diesel engine tests: The IIA and IIIA gasoline engine tests are run in series on the same oil charge. The IIA has been correlated with vehicles used in short-trip service with regard to rusting, and the IIIA has been correlated with vehicles used in high-temperature service, primarily with regard to valve train wear. These correlations were developed with vehicles in use prior to 1958. The Sequence IV gasoline engine test has been correlated with vehicles used in consumer service prior to 1958, particularly with regard to valve train scuffing. The Sequence V test has been correlated with vehicles used in stop-and-go service prior to 1957, primarily with regard to sludge, varnish and valve tip wear. The L-38 gasoline engine test requirement provides a measurement of copper-lead bearing weight loss under high-temperature operating conditions. The L-1 (1.0% mass fuel sulfur) diesel engine test requirement provides a measurement of high-temperature deposits.

SAE J183 Revised JUN91

(R) TABLE 1—DESIGNATION, STATUS, AND DESCRIPTIONS OF CATEGORIES (CONTINUED)

API Letter Designation	Status	API Engine Service Description	ASTM Engine Oil Description
SD	Obsolete Test Techniques	**1968 Gasoline Engine Warranty Maintenance Service (Obsolete)** The category SD denotes service typical of gasoline engines in 1968 through 1970 models of passenger cars and some trucks operating under engine manufacturers' warranties in effect during those model years. This category may also apply to certain 1971 and/or later models as specified (or recommended) in the owner's manuals. Oils designed for this service provide more protection against high- and low-temperature engine deposits, wear, rust, and corrosion in gasoline engines than oils that are satisfactory for API Engine Service Category SC and may be used when API Engine Service Category SC is recommended.	(The test procedure used to define this category is obsolete.) Oil meeting the performance requirements measured in the following gasoline and diesel engine tests: The IIB and IIIB gasoline engine tests are run in series on the same oil charge. The IIB has been correlated with vehicles used in short-trip service with regard to rusting, and the IIIB has been correlated with vehicles used in high-temperature service, primarily with regard to valve train wear. These correlations were developed with vehicles in use prior to 1968. The Sequence IV gasoline engine test has been correlated with vehicles used in consumer service prior to 1958, particularly with regard to valve train scuffing. The Sequence VB test has been correlated with vehicles used in stop-and-go service prior to 1965, primarily with regard to sludge, varnish, and valve tip wear. The L-38 gasoline engine test requirement provides a measurement of copper-lead bearing weight loss under high-temperature operating conditions. The L-1 (1.0% mass fuel sulfur) or 1H diesel engine test requirements are alternatives that provide a measurement of high-temperature deposits. The Falcon gasoline engine test has been used to provide additional performance requirements with regard to rusting.
SE See footnote a.	Obsolete Test Techniques	**1972 Gasoline Engine Warranty Maintenance Service (Obsolete Starting in 1989)** The category SE denotes service typical of gasoline engines in passenger cars and some trucks beginning with 1972 and certain 1971 through 1979 models operating under engine manufacturers' warranties. Oils designed for this service provide more protection against oil oxidation, high-temperature engine deposits, rust, and corrosion in gasoline engines than oils that are satisfactory for API Engine Service Categories SD or SC and may be used when either of these categories is recommended.	Oil meeting the performance requirements measured in the following gasoline engine tests: The IIB, IIC and IID gasoline engine tests were correlated with vehicles used in short-term service prior to 1968, 1971, and 1978, respectively - particularly with regard to rusting. The IIIC and IIID gasoline engine tests were correlated with vehicles used in high-temperature service prior to 1971 and 1978, respectively - particularly with regard to oil thickening and valve train wear. The VC and V-D gasoline engine tests were correlated with vehicles used in stop-and-go service prior to 1971 and 1978, respectively - particularly with regard to varnish and sludge. The L-38 gasoline engine test requirement provides a measurement of copper-lead bearing weight loss under high-temperature operating conditions.

SAE J183 Revised JUN91

(R) TABLE 1—DESIGNATION, STATUS, AND DESCRIPTIONS OF CATEGORIES (CONTINUED)

API Letter Designation	Status	API Engine Service Description	ASTM Engine Oil Description
SF See footnote a.	Obsolete Test Techniques	**1980 Gasoline Engine Warranty Maintenance Service** The category SF denotes service typical of gasoline engines in passenger cars and some trucks beginning with the 1980 through 1988 model years operating under engine manufacturers' recommended maintenance procedures. Oils developed for this service provide increased oxidation stability and improved antiwear performance relative to oils that meet the minimum requirements for API Service Category SE. These oils also provide protection against engine deposits, rust, and corrosion. Oils meeting API Service Category SF may be used when API Service Categories SE, SD, or SC are recommended.	Oil meeting the performance requirements measured in the following gasoline engine tests: The IID gasoline engine test has been correlated with vehicles used in short-trip service prior to 1978, particularly with regard to rusting. The IIID gasoline engine test has been correlated with vehicles used in high-temperature service prior to 1978, particularly with regard to oil thickening and valve train wear. The V-D gasoline engine test has been correlated with vehicles used in stop-and-go service prior to 1978, particularly with regard to varnish, sludge, and valve train wear. The L-38 gasoline engine test requirement provides a measurement of copper-lead bearing weight loss under high-temperature operating conditions. NOTE (added in 1988): Many automobile manufacturers' warranty requirements for 1986-1988 gasoline engines include CC or CD requirements in addition to SF.
SG See footnote a.	Active Test Techniques	**1989 Gasoline Engine Warranty Maintenance Service** The category SG denotes service typical of present gasoline engines in passenger cars, vans, and light-duty trucks operating under manufacturers' recommended maintenance procedures. Category SG quality oils include the performance properties of API Service Category CC. (Certain manufacturers of gasoline engines require oils also meeting the higher diesel engine Category CD.) Oils developed for this service provide improved control of engine deposits, oil oxidation, and engine wear relative to oils developed for previous categories. These oils also provide protection against rust and corrosion. Oils meeting API Service Category SG may be used when API Service Categories SF, SE, SF/CC, or SE/CC are recommended.	Oil meeting the performance requirements measured in the following gasoline and diesel engine tests: The IID gasoline engine test has been correlated with vehicles used in short-trip service prior to 1978, particularly with regard to rusting. The IIIE gasoline engine test has been correlated with vehicles used in high-temperature service prior to 1988, particularly with regard to oil thickening and valve train wear. The VE gasoline engine test has been correlated with vehicles used in stop-and-go service prior to 1988, particularly with regard to sludge and valve train wear. The L-38 gasoline engine test requirement provides a measurement of copper-lead bearing weight loss and piston varnish under high-temperature operating conditions. The 1H2 diesel engine test requirement provides a measurement of high-temperature piston deposits.
CA	Obsolete Test Techniques	**Diesel Engine Service (Obsolete)** Service typical of diesel engines operated in mild to moderate duty with high-quality fuels and occasionally has included gasoline engines in mild service. Oils designed for this service provide protection from bearing corrosion and from ring belt deposits in some naturally aspirated diesel engines when using fuels of such quality that they impose no unusual requirements for wear and deposit protection. They were widely used in the late 1940s and 1950s but should not be used in any engine unless specifically recommended by the equipment manufacturer.	(The test procedure used to define this category is obsolete.) Oil meeting the performance requirements measured in the following diesel and gasoline engine tests: The L-1 (0.4% mass fuel sulfur) naturally-aspirated diesel engine test provides a measurement of piston deposits. The L-38 (or alternatively the L-4) gasoline engine test requirement provides a measurement of copper-lead bearing weight loss and piston varnish under high-temperature operating conditions.

SAE J183 Revised JUN91

(R) TABLE 1—DESIGNATION, STATUS, AND DESCRIPTIONS OF CATEGORIES (CONTINUED)

API Letter Designation	Status	API Engine Service Description	ASTM Engine Oil Description
CB	Obsolete Test Techniques	**Diesel Engine Service (Obsolete)** Service typical of diesel engines operated in mild to moderate duty, but with lower quality fuels which necessitate more protection from wear and deposits. Occasionally has included gasoline engines in mild service. Oils designed for this service were introduced in 1949. Such oils provide necessary protection from bearing corrosion and from high temperature deposits in normally aspirated diesel engines with higher sulfur fuels.	(The test procedure used to define this category is obsolete.) Oil meeting the performance requirements measured in the following diesel and gasoline engine tests: The L-1 (1% mass fuel sulfur) naturally-aspirated diesel engine test provides a measurement of piston deposits. The L-38 (or alternatively the L-4) gasoline engine test requirement provides a measurement of copper-lead bearing weight loss and piston varnish under high-temperature operating conditions.
CC	Obsolete Test Techniques	**Diesel Engine Service** The category CC denotes service typical of certain naturally aspirated, turbocharged, or supercharged diesel engines operated in moderate- to severe-duty service and certain heavy-duty gasoline engines. Oils designed for this service provide protection from high-temperature deposits and bearing corrosion in these diesel engines and also from rust, corrosion, and low-temperature deposits in gasoline engines. These oils were introduced in 1961.	Oil meeting the performance requirements measured in the following diesel and gasoline engine tests: The 1H2 diesel engine test has been correlated with indirect injection engines used in moderate-duty operation, particularly with regard to piston and ring groove deposits. The L-38 gasoline engine test requirement provides a measurement of copper-lead bearing weight loss and piston varnish under high-temperature operating conditions. The Modified LTD gasoline engine test provides a measurement of sludge and varnish. The IID gasoline engine test has been correlated with vehicles used in short-trip service prior to 1978, particularly with regard to rusting.
CD	Active Test Techniques	**Diesel Engine Service** The category CD denotes service typical of certain naturally aspirated, turbocharged, or supercharged diesel engines where highly effective control of wear and deposits is vital or when using fuels of a wide quality range, including high sulfur fuels. Oils designed for this service were introduced in 1955 and provide protection from bearing corrosion and from high-temperature deposits in these diesel engines.	Oil meeting the performance requirements measured in the following diesel and gasoline engine tests: The 1G2 diesel engine test has been correlated with indirect injection engines used in heavy-duty operation, particularly with regard to piston and ring groove deposits. The L-38 gasoline engine test requirement provides a measurement of copper-lead bearing weight loss and piston varnish under high-temperature operating conditions.
CD-II	Active Test Techniques	**Severe Duty Two-Stroke Cycle Diesel Engine Service** Service typical of two-stroke cycle diesel engines requiring highly effective control over wear and deposits. Oils designed for this service also meet all performance requirements of API Service Category CD.	Oil meeting the performance requirements measured in the following diesel and gasoline engine tests: The 1G2 diesel engine test has been correlated with indirect injection engines used in heavy-duty operation, particularly with regard to piston and ring groove deposits. The 6V-53T diesel engine test has been correlated with vehicles equipped with two-stroke cycle diesel engines in high-speed operation prior to 1985, particularly with regard to ring and liner distress. The L-38 gasoline engine test requirement provides a measurement of copper-lead bearing weight loss and piston varnish under high-temperature operating conditions.

SAE J183 Revised JUN91

(R) TABLE 1—DESIGNATION, STATUS, AND DESCRIPTIONS OF CATEGORIES (CONTINUED)

	API Letter Designation	Status	API Engine Service Description	ASTM Engine Oil Description
	CE	Active Test Techniques	**1983 Diesel Engine Service** Service typical of certain turbocharged or supercharged heavy-duty diesel engines manufactured since 1983 and operated under both low-speed, high-load and high-speed, high-load conditions. Oils designed for this service may also be used when API Engine Service Category CD is recommended for diesel engines.	Oil meeting the performance requirements of the following diesel and gasoline engine tests: The 1G2 diesel engine test has been correlated with indirect injection engines used in heavy-duty service, particularly with regard to piston and ring groove deposits. The T-6, T-7, and NTC-400 are direct injection diesel engine tests. The T-6 has been correlated with vehicles equipped with engines used in high-speed operation prior to 1980, particularly with regard to deposits, oil consumption, and ring wear. The T-7 test has been correlated with vehicles equipped with engines used in lugging operation prior to 1984, particularly with regard to oil thickening. The NTC-400 diesel engine test has been correlated with vehicles equipped with engines in highway operation prior to 1983, particularly with regard to oil consumption control, deposits, and wear. The L-38 gasoline engine test requirement provides a measurement of copper-lead bearing weight loss under high-temperature operating conditions.
(R)	CF-4		This category was adopted in 1990 and describes oils for use in high-speed, four-stroke-cycle, diesel engines. API CF-4 oils exceed the requirements of the CE category providing improved control of oil consumption and piston deposits. These oils should be used in place of CE oils. They are particularly suited for on-highway, heavy-duty truck applications. When combined with the appropriate "S" category, for example, SG, they can also be used in gasoline and diesel-powered personal vehicles such as automobiles, light trucks, and vans when recommended by the vehicle or engine manufacturer. Based on L-38, Cat 1K, Mack T-6 and T-7, Cummins NTC 400 (revised).	Oil meeting the performance requirements in the following diesel and gasoline engine tests: The 1K diesel engine test, which has been correlated with direct injection engines used in heavy-duty service prior to 1990, particularly with regard to piston and ring groove deposits. The T-6 diesel engine test which has been correlated with vehicles equipped with engines used in high-speed operation prior to 1980, particularly with regard to deposits, oil consumption, and ring wear. The T-7 diesel engine test which has been correlated with vehicles equipped with engines operated largely under lugging conditions prior to 1984, particularly with regard to oil thickening. The NTC-400 diesel engine test which has been correlated with vehicles equipped with engines in highway operation prior to 1983, particularly with regard to oil consumption control, deposits, and wear. The L-38 gasoline engine test which is used to measure copper-lead bearing weight loss.

a. Oils identified as SE, SF, and/or SG, with or without any CC or CD category designation, may be intended for engine/vehicle emission control systems containing catalysts and oxygen sensors. To minimize emission control system deterioration, it may be advantageous to utilize formulations of controlled phosphorus concentration and alkaline earth metal-to-phosphorus ratios. Formulations so developed must not compromise engine durability.

No standard test is currently available for evaluating the effect of engine oil formulations on emission control systems. However, further informational guidance on this subject may be found in SAE Fuels and Lubricants Technical Subcommittee 1, Engine Oil/Catalyst and Oxygen Sensor Compatibility Task Force Status Report, dated October 1985.

SAE J183 Revised JUN91

TABLE 2—TEST TECHNIQUES AND PRIMARY PERFORMANCE CRITERIA

API Letter Designation	Test Techniques[b, t]	Primary Performance Criteria[b]	Primary Performance Criteria[b]	Primary Performance Criteria[b]	Primary Performance Criteria[b]
SA	None	None			
			L-4		L-38
SB	L-4[f] or L-38[c]	Bearing weight loss, mg, max	500		500
	Sequence IV[f]	Cam scuffing		None	
		Lifter scuff rating, max		2	
			IIA	IIIA	
SC	Sequences IIA[f] and IIIA[f]	Avg rust rating, min	8.2	—	
		Cam and lifter scuffing	—	None	
		Avg cam plus lifter wear, mm (in) max	—	0.064 (0.0025)	
		Avg sludge rating, min	—	9.5	
		Avg varnish rating, min	—	9.7	
	Sequence IV[f]	Cam scuffing		None	
		Lifter scuff rating, max		2	
	Sequence V[f]	Total engine sludge rating, min		40	
		Avg piston skirt varnish rating, min		7.0	
		Total engine varnish rating, min		35	
		Avg intake valve tip wear, mm (in) max		0.051 (0.0020)	
		Ring sticking		None	
		Oil ring clogging, %, max		20	
		Oil screen plugging, %, max		20	
	L-38[d]	Bearing weight loss, mg, max		50	
			IIB	IIIB	
SD	Sequences IIB[f] and IIIB[f]	Avg rust rating, min	8.8	—	
		Cam and lifter scuffing	—	None	
		Avg cam plus lifter wear, mm (in) max	—	0.076 (0.0030)	
		Avg sludge rating, min	—	9.6	
		Avg varnish rating, min	—	9.6	
	Sequence IV[f]	Cam scuffing		None	
		Lifter scuff rating, max		1	
	Sequence VB[f]	Total engine sludge rating, min		42.5	
		Avg piston skirt varnish rating, min		8.0	
		Total engine varnish rating, min		37.5	
		Avg intake valve tip wear, mm (in) max		0.038 (0.0015)	
		Oil ring clogging, %, max		5	
		Oil screen plugging, %, max		5	
	L-38[d]	Bearing weight loss, mg, max		40	
			L-1		1H
	L-1 (0.95% min sulfur fuel) or 1H[d,f,g]	Groove No. 1 (top) carbon fill, % vol, max	25		30
		Groove No. 2 lacquer coverage, % area, max	—		50
		Groove No. 2 and below	Essentially clean		—
		Land No. 3 and below	—		Essentially clean
	Falcon[d,f]	Avg engine rust rating, min		9.0	
			IIB	IIC	IID
SE	Sequence IIB[f], IIC[f], or IID	Avg engine rust rating, min	8.9	8.4	8.5
		Number stuck lifters	None	None	None

SAE J183 Revised JUN91

TABLE 2—TEST TECHNIQUES AND PRIMARY PERFORMANCE CRITERIA (CONTINUED)

API Letter Designation	Test Techniques[b, t]	Primary Performance Criteria[b]	Primary Performance Criteria[b]	Primary Performance Criteria[b]	Primary Performance Criteria[b]
			IIIC		IIID
	Sequence IIIC[f] or IIID[f]	Viscosity increase at 37.78 °C (100 °F) and 40 test h, %, max	400		400
		Viscosity increase at 40 °C and 40 test h, %, max	—		375
		Avg engine ratings at 64 test h			
		Avg sludge rating, min	9.2		9.2
		Avg piston skirt varnish rating, min	9.3		9.1
		Avg oil ring land deposit rating, min	6.0		4.0
See footnote a.		Ring sticking	None		None
		Lifter sticking	None		None
		Scuffing and wear at 64 test h			
		Cam or lifter scuffing	None		None
		Cam plus lifter wear, mm (in)			
		Average	0.025 (0.0010)		0.102 (0.0040)[e]
		Maximum	0.051 (0.0020)		0.254 (0.0100)[e]
				VC	V-D
	Sequence VC[f] or V-D[f]	Avg engine sludge rating, min		8.7	9.2
		Avg piston skirt varnish rating, min		7.9	6.4
		Avg engine varnish rating, min		8.0	6.3
		Oil screen clogging, %, max		5	10.0
		Oil ring clogging, %, max		5	10.0
		Compression ring sticking		None	None
		Cam wear, mm (in)			
		Average, max		—	Rate and Report[h]
		Maximum, max		—	Rate and Report[h]
	L-38	Bearing weight loss, mg, max		40	
SF	Sequence IID	Avg engine rust rating, min		8.5	
		Number stuck lifters		None	
	Sequence IIID[f]	Viscosity increase at 40 °C (64 test h)		375	
		Avg sludge rating, min		9.2	
		Avg piston skirt varnish rating, min		9.2	
		Avg oil ring land deposit rating, min		4.8	
		Ring sticking		None	
		Lifter sticking		None	
		Scuffing and wear			
		Cam and lifter scuffing		None	
See footnote a.		Cam plus lifter wear, mm (in)			
		Average, max		0.102 (0.0040)	
		Maximum, max		0.203 (0.0080)	
	Sequence V-D[f]	Avg engine sludge rating, min		9.4	
		Avg piston skirt varnish rating, min		6.7	
		Avg engine varnish rating, min		6.6	
		Oil ring clogging, %, max		10.0	
		Oil screen clogging, %, max		7.5	
		Compression ring sticking		None	
		Cam wear, mm (in)			
		Average, max		0.025 (0.0010)	
		Maximum, max		0.064 (0.0025)	
	L-38	Bearing weight loss, mg, max		40	
SG	Sequence IID	Avg engine rust rating, min		8.5	
		Number stuck lifters		None	

SAE J183 Revised JUN91

(R) TABLE 2—TEST TECHNIQUES AND PRIMARY PERFORMANCE CRITERIA (CONTINUED)

API Letter Designation	Test Techniques[b, t]	Primary Performance Criteria[b]	Primary Performance Criteria[b]	Primary Performance Criteria[b]	Primary Performance Criteria[b]
	Sequence IIIE	Viscosity increase @ 40 °C			
		(64 test h), % Max		375	
		Avg sludge rating, min		9.2	
		Avg piston skirt varnish rating, min		8.9	
		Avg oil ring land deposit rating, min		3.5	
		Lifter sticking		None	
		Scuffing and wear (64 test h)			
		Cam or lifter scuffing		None	
		Cam plus lifter wear, mm (in)			
		Average, max		0.030 (0.0012)	
		Maximum, max		0.064 (0.0025)	
		Ring sticking		Rate and Report	
	Sequence VE	Avg engine sludge rating, min		9.0	
		Cam cover sludge rating, min		7.0	
See footnote a.		Avg piston skirt varnish rating, min		6.5	
		Average engine varnish rating, min		5.0	
		Oil ring clogging, % max		15.0	
		Oil screen clogging, % max		20.0	
		Compression ring sticking (hot stuck)		None	
		Cam wear, mm (in)			
		Average, max		0.13 (0.005)	
		Maximum, max		0.38 (0.015)	
	IH2[s]	Groove No. 1 (top) carbon fill, % vol, max		45	
		Weighted total demerits, max		140	
		Ring side clearance			
		Loss, mm (in), max[m]		0.013 (0.0005)	
	L-38	Bearing weight loss, mg, max		40	
		Piston skirt varnish rating, min		9.0	
			L-4		L-38
CA	L-4[f] or L-38[c]	Bearing weight loss, mg, max	120-135		50
		Piston skirt varnish rating, min	9.0		9.0
	L-1 (0.35% min sulfur fuel)[f,g]	Groove No. 1 (top) carbon fill, % vol, max		25	
		Groove No. 2 and below		Essentially clean	
CB	L-4[f] or L-38[c]	Same as CA			
	L-1 (0.95% min sulfur fuel)[f,g]	Same as CA, except groove No. 1 (top) carbon fill, % vol, max		30	
CC	L-38	Bearing weight loss, mg, max		50	
		Piston skirt varnish rating, min		9.0	
			LTD		Modified LTD
	LTD[f] or Modified LTD[c,f,j,k]	Piston skirt varnish rating, min	7.5		7.5
		Total engine varnish rating, min	—		42
		Total engine sludge rating, min	35		42
		Oil ring plugging, %, max	25		10
		Oil screen clogging, %, max	25		10
			IIA	IIB IIC	IID
	Sequence IIA[f], IIIB[c,f], IIC[f], or IID	Avg engine rust rating, min	8.2	8.2 7.6	7.7

SAE J183 Revised JUN91

TABLE 2—TEST TECHNIQUES AND PRIMARY PERFORMANCE CRITERIA (CONTINUED)

API Letter Designation	Test Techniques[b, t]	Primary Performance Criteria[b]	Primary Performance Criteria[b]	Primary Performance Criteria[b]	Primary Performance Criteria[b]
			1H		IH2
	1H[f,g] or 1H2[g]	Groove No. 1 (top) carbon fill, % vol, max	30		45
		Groove No. 2 lacquer coverage, % area, max	50		—
		Land No. 3 and below	Essentially clean		—
		Weighted total demerits, max	—		140
		Ring side clearance Loss, mm (in) max[m]	—		0.013 (0.0005)
CD	1D[f,g,j]	Groove No. 1 (top) carbon fill, % vol, max		75	
		Groove No. 2 and below		Essentially clean	
	1G[f,g] or 1G2[g]	Groove No. 1 (top) carbon fill, % vol, max	60		80
		Land No. 2 carbon and lacquer coverage, % area, max	50		—
		Groove No. 2 carbon and lacquer coverage, % area, max	30		—
		Land No. 3 and below	Essentially clean		—
		Weighted total demerits, max	—		300
		Ring side clearance Loss, mm (in) max[m]	—		0.013 (0.0005)
	L-38	Bearing weight loss, mg, max		50	
		Piston skirt varnish rating, min		9.0	
CD-II	1G2[g]	Groove No. 1 (top) carbon fill, % volume, max		80	
		Weighted total demerits, max		300	
		Ring side clearance Loss, mm (in) max[m]		0.013 (0.0005)	
	L-38	Bearing weight loss, mg, max		50	
		Piston skirt varnish rating, min		9.0	
	6V-53T[r]	Piston area			
		Weighted total demerits, avg, max		400	
		Hot stuck rings		None	
		Face distress, Nos. 2 & 3 rings demerits, avg, max		13.0	
		Liner and head area			
		Liner distress, avg % area, max		12.0	
		Valve distress		None	
(R) CE	1G2[g]	Groove No. 1 (top) carbon fill, % vol, max		80	
		Weighted total demerits, max		300	
		Ring side clearance Loss, mm (in), max[m]		0.013 (0.0005)	
	L-38	Bearing weight loss, mg, max		50	
	T-6[n]	Merit rating, min		90[p]	
	T-7[n]	Average rate of viscosity increase during last 50 h, cSt @ 100 C/h, max		0.040	

SAE J183 Revised JUN91

(R) TABLE 2—TEST TECHNIQUES AND PRIMARY PERFORMANCE CRITERIA (CONTINUED)

API Letter Designation	Test Techniques[b, t]	Primary Performance Criteria[b]	Primary Performance Criteria[b]	Primary Performance Criteria[b]	Primary Performance Criteria[b]
	NTC-400[q]	Oil consumption		Candidate oil consumption second order regression curve must fall completely below the published mean plus one standard deviation curve for the applicable reference oil.[w]	
		Camshaft roller follower pin wear average, max, mm (in)		0.051 (0.002)	
		Crownland (top land) deposits, % area covered with heavy carbon, average, max		25	
		Piston deposits, third ring land, total CRC demerits for all six pistons, max		40	
(R) CF-4	1K[v]	Weighted Demerits (WDK)[u], max		Candidate oil WDK must be less than or equal to the published test limits based on the applicable reference oil for one, two, or three test runs.[w]	
		Groove No. 1 (top) carbon fill (TGF)[u], % volume, max		Candidate oil TGF must be less than or equal to the published test limits based on the applicable reference oil for one, two, or three test runs.[w]	
		Top land heavy carbon (TLHC)[u], % max		Candidate oil TLHC must be less than or equal to the published test limits based on the applicable reference oil for one, two, or three test runs.[w]	
		Oil consumption, average, kg/kw-h, max		0.0005	
		Piston ring sticking		None	
		Piston, ring and liner scuffing		None	
	L-38	Bearing weight loss, mg, max		50	
	T-6[n]	Merit rating, min[p]		90	
	T-7[n]	Average rate of kinematic viscosity increase during last 50 h, cSt @ 100 °C/h, max		0.040	
	NTC-400[q]	Oil consumption		Candidate oil consumption second order regression curve must fall completely below the published mean curve for the applicable reference oil.[w]	
		Camshaft roller follower pin wear, mm (in), average, max		0.051 (0.002)	
		Crownland (top land) deposits, area covered with heavy carbon, %, average, max		15	

a. Oils identified as SE, SF, and/or SG, with or without any CC or CD category designation, may be intended for engine/vehicle emission control systems containing catalysts and oxygen sensors. To minimize emission control system deterioration, it may be advantageous to utilize formulations of controlled phosphorus concentration and alkaline earth metal-to-phosphorus ratios. Formulations so developed must not compromise engine durability.

No standard test is currently available for evaluating the effect of engine oil formulations on emission control systems. However, further informational guidance on this subject may be found in SAE Fuels and Lubricants Technical Subcommittee 1, Engine Oil/Catalyst and Oxygen Sensor Compatibility Task Force Status Report, dated October 1985.

b. Detail regarding many of these test techniques, including a description of their objectives or significance, may be found in SAE J304, ASTM STP 315 and 509, and Federal Test Method Standard 791. The dimensionless numbers listed under Primary Performance Criteria refer to arbitrary rating scales as follows:

(1) When a maximum rating is quoted as 2, the reference is to a scale of 1 to 6 with 1 being perfect.

(2) When minimum ratings are quoted at about 4 to 9.7, the reference is to a scale where 10 is "clean." A list of appropriate CRC rating manuals may be obtained from Coordinating Research Council, Inc. 219 Perimeter Center Parkway, Atlanta, GA 30346.

(3) When minimum ratings are quoted at about 35 to 45, the reference is to a scale where 50 is "clean."

SAE J183 Revised JUN91

TABLE 2—TEST TECHNIQUES AND PRIMARY PERFORMANCE CRITERIA (CONTINUED)

c. Test conditions or performance requirements changed since originally promulgated.
d. This test technique has also been used in this evaluation.
e. Because of the change in test techniques, no meaningful correlation can be established between IIIC and IIID wear data. These recommended values are intended to assure adequate wear protection.
f. This test is obsolete; engine parts and/or test fuel, and/or reference oils are no longer generally available and the test is no longer being monitored by the test developer or ASTM.
g. Refer to Figure 1 for nomenclature of piston zones.
h. Because Sequence VC Test contained no valve train wear measurement requirements, no meaningful relationship could be established with V-D tests. However, for informational guidance in development of SE quality oils, cam wear values of 0.051 mm (0.0020 in) average and 0.102 mm (0.0040 in) max should be considered.
j. An acceptably referenced test must be immediately preceded or followed in the same engine stand by an REO 191 test which gives a total engine sludge rating within ±1.28 standard deviation of the historical average for this reference oil.
k. An oil can also satisfy this requirement by meeting or exceeding the Sequence VC or V-D deposit limits of the SE category.
l. As of December 1980, it was determined that the 1D test is no longer necessary to describe API category CD since meeting the performance limits of the 1G or 1G2 test virtually assured that the performance limits of the 1D test are met.
m. This refers to losses in the piston groove and ring side clearances.
n. Test technique T-6 may be found in ASTM Research Report RR:D-2–1219, and test technique T-7 may be found in ASTM Research Report RR:D-2–1220.
p. Requires greater than zero merits on all individual ratings. Merit rating method may be found in ASTM Research Report RR:D-2-1219.
(R) q. Test technique NTC-400 may be found in ASTM Research Report RR:D-2-1194.
(R) r. Test technique 6V-53T may be found in ASTM Research Report RR:D-2-1222. Current information is available from ASTM Section D.02.BO.02.
s. Passing 1G2 test results (TGF = 80% max and WTD = 300 max, using CRC manual No. 15) will be accepted in place of 1H2 results for SG.
t. Test results are valid only if the tests are run in calibrated stands monitored by the ASTM Test Monitoring Center.
(R) u. See CRC Modified Supplement Diesel Engine Rating Manual No. 15.
(R) v. Test technique 1K may be found in ASTM Research Report RR:D-2-1273.
(R) w. Applicable NTC-400 reference oil consumption curves and Caterpillar 1K reference oil one, two, and three test limits are published twice per year by the ASTM Test Monitoring Center. Copies of these data may be obtained by contacting the Center at ASTM TMC, 4400 Fifth Avenue, Pittsburgh, PA 15213. Applicable reference oil data to which candidate oil data are to be compared are included with each engine test report.

Certain test limits for some of the API C Categories are statistically derived on the basis of the performance of selected reference engine oils used in the monitoring of these tests by the ASTM Test Monitoring Center. These include the oil consumption limit for the NTC-400 test and the weighted demerits (WDK), top groove fill (TGF), and top land heavy carbon (TLHC) limits for the 1K test.

The ASTM Test Monitoring Center publishes reference oil test data every six months along with specific limits for each of the parameters and the time period for which the specific limits apply. Copies of these data may be obtained by contacting: ASTM TMC, 4400 Fifth Avenue, Pittsburgh, PA 15213 (phone 412-268-3315) (fax 412-268-6899).

The application of the 1K test in determining the performance limits for the API CF-4 Category allow the running of multiple tests, if necessary. Limits are published by the TMC against which the results of the first 1K test in a program are compared. In addition, limits are also published for two-test and three-test programs.

In applying the limits for two-test and three-test programs, the results for the WDK, TGF, TLHC, and average oil consumption of the two or three tests are averaged and compared to the numerical limits published by the TMC or shown in Table 2.

In a three-test program, allowance is made for excluding one of the tests as an outlier. The basis for determining whether a test result is an outlier is ASTM E 178. In applying E 178 to the 1K test, each parameter is considered individually. If one parameter on one of the first three tests is more than the limits shown in the list for Limits for 1K Test Outlier Determinations (W.1), then that test may be considered an outlier and a fourth test run. In determining the results of the three-test program, the results of the outlier test are not used in calculating the average results which are compared to the published limits.

W.1 Limits for 1K Test Outlier Determinations:

Parameter	Outlier Limit[1]
Weighted Demerits (WDK), min:	Mean + 92
Top Groove Fill (TGF), min:	Mean + 22
Top Land Heavy Carbon (TLHC), min:	Mean + 6

[1] The means used in these limits are the means of the individual parameters for the first three 1K tests in a program. The constants are three times the standard deviations of each parameter from the original 30-test matrix data base on reference oil TMC 809 rounded to the nearest whole number.

400 Commonwealth Drive, Warrendale, PA 15096-0001

SURFACE VEHICLE INFORMATION REPORT

Submitted for recognition as an American National Standard

SAE J2227 REV. AUG95

Issued 1991-06
Revised 1995-08

Superseding J2227 JUN94

INTERNATIONAL TESTS AND SPECIFICATIONS FOR AUTOMOTIVE ENGINE OILS

Foreword—Engine and laboratory tests are utilized to determine the performance of engine oils. The API, ASTM, and SAE have established engine tests and classifications to describe engine oil performance. Such tests are included in SAE J183, Engine Oil Performance and Engine Service Classification (Other Than "Energy Conserving"), SAE J304, Engine Oil Tests, and SAE J300, Engine Oil Viscosity Classification. Additionally a test to characterize the energy-conserving characteristics of engine oils is described in SAE J1423. Engine and laboratory tests apart from those described in these SAE documents are also established in Europe and Japan. The purpose of this document is to summarize the respective international tests and specifications utilized to characterize the performance of service fill automotive engine oils outside of North America. Since specifications are likely to change frequently, it is recommended that all specifications be confirmed with the appropriate manufacturer or Technical Society at the time that critical usage is contemplated.

(R) ***1. Scope***—This SAE Information Report lists engine and laboratory tests for service fill engine oils which are associated with specifications and classifications established outside of North America. These specifications and classifications include those developed prior to June 1, 1995, by International Technical Societies as well as individual original equipment manufacturers. The information contained within this report applies to engine oils utilized in gasoline- and diesel-powered automotive vehicles.

2. References

2.1 Applicable Documents—The following publications form a part of this SAE Information Report to the extent specified herein. The latest issue of SAE publications shall apply.

2.1.1 SAE PUBLICATIONS—Available from SAE, 400 Commonwealth Drive, Warrendale, PA 15096-0001.

SAE J183—Engine Oil Performance and Engine Service Classification (Other Than "Energy-Conserving")
SAE J300—Engine Oil Viscosity Classification
SAE J304—Engine Oil Tests
SAE J1423—Classification of Energy-Conserving Engine Oil for Passenger Cars, Vans, and Light-Duty Trucks

2.1.2 API PUBLICATION—Available from the American Petroleum Institute, 1220 L Street, Northwest, Washington, DC 20005.

API Publication 1509—Engine Oil Licensing and Certification System

SAE J2227 Revised AUG95

2.1.3 CEC PUBLICATIONS—Available from the Coordinating European Council, Madou Plaza - 25th Floor, Place Madou 1, B-1030 Brussels, Belgium.

CEC Annual Report
CEC Catalogue of Methods

2.1.4 CCMC PUBLICATIONS—Available from ACEA (Association of the European Automobile Manufacturers), 211 rue du Noyer, B-1040, Bruxelles.

CCMC Ref: FL/28/83—CCMC European Oil Sequence for Service Fill Engine Oils for Gasoline Engines—Classes G-1, G-2, and G-3
CCMC Ref: FL/37/84—CCMC European Oil Sequence for Service Fill Oils for Diesel Engines—Classes D-1, D-2, D-3, and PD-1
CCMC Ref: FL/29/89—CCMC European Oil Sequence for Service Fill Oils for Gasoline Engines—Classes G-4 and G-5
CCMC Ref: FL/19/91—CCMC European Oil Sequence for Service Fill Oils for Diesel Engines—Classes D-4, D-5, and PD-2
CCMC Ref: FL/20/91—Evolution of the CCMC Engine Lubricant Sequences

2.1.5 INTERNATIONAL LUBRICANT STANDARDIZATION AND APPROVAL COMMITTEE STANDARD—Available from the American Automobile Manufacturers Association, 7430 Second Avenue, Suite 300, Detroit, Michigan 48202.

International Lubricant Standardization and Approval Committee (ILSAC) Standard for Passenger Car Engine Oils

2.1.6 JASO PUBLICATIONS—Available from the Society of Automotive Engineers of Japan Inc., 10-2 Goban-cho, Chiyoda-ku, Tokyo 102, Japan. (M-333-93 presently available in Japanese only.)

JASO Standard M328-91—Valve Train Wear Test Procedure for Evaluating Automobile Gasoline Engine Oils
JASO Standard M331-91—Low and Medium Temperature Detergency Test Procedure for Evaluating Automobile Gasoline Engine Oils
JASO Standard M333-93—High Temperature Oxidation Stability Test Procedure for Evaluating Automobile Gasoline Engine Oils
JASO Standard M336-90—High Temperature and High Load Detergency Test Procedure for Evaluating Automobile Diesel Engine Oils

2.1.7 MILITARY SPECIFICATIONS—Available from Standardization Document Order Desk, 700 Robbins Avenue, Building #4, Section D, Philadelphia, PA 19111-5094.

U.S. Military Specification MIL-L-2104C[1]
U.S. Military Specification MIL-L-46152A[2]

(R) [1] MIL-L-2104E and earlier editions are obsolete. Revision MIL-L-2104F and Commercial Item Description A-A-52306 currently in use.
(R) [2] MIL-L-46152 is obsolete and replaced by Commercial Item Description A-A-52039.

SAE J2227 Revised AUG95

(R) 2.1.8 MANUFACTURER PUBLICATIONS—Available from the respective European Original Equipment Manufacturers.

MAN Specifications 270, 271, QC-13-017
Mercedes-Benz Betriebsstoff—Vorschriften
MTU Oil Type 1 and 2
MWM Deutz Motor Technical Circular 0199-3002
MWM Deutz Motor Technical Circular 0199-2090
Perkins Engine Specification P.M.S. S.1.01—1983, TSD 3187-1990, PS No. 7294/SB019
Rover Group Specifications RES.22.OL.G-4, RES.22.OL.PD-2 RES.22.OL.D-5
Scania Specifications 8/2—8405 and 0-840330
Volkswagen Specification VW 500 00, 501 01, and 505 00
Volvo Drain Specification (VDS and VDS-2)

(R) 3. ***European CCMC Sequences***—CCMC Oil Sequences were established by the Committee of Common Market Automobile Constructors to ensure that suitable lubricants are available to meet the minimum requirements of European vehicles. CCMC was dissolved at the end of 1990. ACEA (Association of the European Automobile Manufacturers), the new association of the European automobile manufacturers, formed in February 1991, has decided to retain the CCMC oil sequences and their original designation for a transitional period. ACEA is engaged in a broad range of activities including safety and environmental concerns and any regulations which have a direct impact on the European automobile industry. ACEA members are all the European motor vehicle manufacturers including Ford Europe, GM Europe, Scania, and Volvo. The ACEA is in the process of developing new sequences for engine oil.

The CCMC Oil Sequences define laboratory tests and engine tests which lubricants must satisfy to achieve the minimum performance requirements established by European manufacturers. These sequences are divided into two main categories for both gasoline (G Sequences) and diesel (D Sequences) engines. The categories are divided further into sub-groups which more precisely define the performance for specific applications.

3.1 **CCMC G-1**—Defines normal viscosity oils utilized in gasoline engines including SAE 10W, 15W, and 20W multigrades. CCMC G-1 is obsolete as of April 1989.

3.2 **CCMC G-2**—Defines normal viscosity oils utilized in gasoline engines which provide high protection and include SAE 10W, 15W, and 20W multigrades. CCMC G-2 is obsolete as of January 1990 and has been replaced by CCMC G-4.

3.3 **CCMC G-3**—Defines low viscosity oils utilized in gasoline engines which provide high protection and include SAE 5W-30, 10W-30, 5W-40, and 10W-40 grades. CCMC G-3 is obsolete as of January 1990 and has been replaced by CCMC G-5.

(R) 3.4 **CCMC G-4**—Defines oils utilized in gasoline engines which provide high protection and include SAE 10W, 15W, and 20W multigrades. CCMC G-4 replaces CCMC G-2 as of January 1990, reflecting engine oils which offer improved wear and deposit control protection and better evaporative loss characteristics. CCMC G-4 also adds requirements for oil seal elastomer compatibility, foaming tendency, valve train scuffing, and black sludge. ACEA sequences will replace CCMC G-4 in 1996 for marketed oils.

(R) 3.5 **CCMC G-5**—Defines oils utilized in gasoline engines which provide high protection and include SAE 5W and SAE 10W multigrades. CCMC G-5 replaces CCMC G-3 as of January 1990, reflecting engine oils which offer improved wear and deposit control protection and better evaporative loss and shear stability characteristics. CCMC G-5 also adds requirements for oil seal compatibility, foaming tendency, valve train scuffing and black sludge. ACEA sequences will replace CCMC G-5 in 1996 for marketed oils.

3.6 **CCMC D-1**—Defines engine oils utilized in naturally aspirated commercial diesel engines in light-duty operation. CCMC D-1 is obsolete as of April 1989.

SAE J2227 Revised AUG95

3.7 **CCMC D-2**—Defines engine oils utilized in naturally aspirated and turbocharged commercial diesel engines in heavy-duty operation. CCMC D-2 is obsolete as of January 1990 and has been replaced by CCMC D-4.

3.8 **CCMC D-3**—Defines engine oils utilized in naturally aspirated and turbocharged commercial diesel engines in extra heavy-duty operation. CCMC D-3 is obsolete as of January 1, 1990, and has been replaced by CCMC D-5.

(R) 3.9 **CCMC D-4**—Defines engine oils utilized in naturally aspirated and turbocharged commercial diesel engines in heavy-duty operation. CCMC D-4 replaces CCMC D-2 as of January 1990, reflecting engine oils which offer improved wear and deposit control protection and better evaporation loss characteristics. CCMC D-4 also adds requirements for oil elastomer compatibility, low-speed oil thickening, and foaming tendency. ACEA sequences will replace CCMC D-4 in 1996 for marketed oils.

(R) 3.10 **CCMC D-5**—Defines engine oils utilized in naturally aspirated and turbocharged commercial diesel engines in extra heavy-duty operation. CCMC D-5 replaces CCMC D-3 as of January 1990, reflecting engine oils which offer improved wear and deposit control protection and better shear stability and evaporative loss characteristics. CCMC D-5 also adds requirements for oil elastomer compatibility, low-speed oil thickening, and foaming tendency. ACEA sequences will replace CCMC D-5 in 1996 for marketed oils.

(R) 3.11 **CCMC PD-1**—Defines engine oils utilized in passenger car naturally aspirated and turbocharged diesel engines. CCMC PD-1 is obsolete as of January 1990 and has been replaced by CCMC PD-2.

(R) 3.12 **CCMC PD-2**—Defines engine oils utilized in passenger cars naturally aspirated and turbocharged diesel engines. CCMC PD-2 replaces CCMC PD-1 as of January 1990 reflecting engine oils which offer improved wear and deposit control protection and better evaporative loss characteristics. CCMC PD-2 also adds requirements for oil elastomer compatibility and foaming tendency. ACEA sequences will replace CCMC PD-2 in 1996 for marketed oils.

4. ***European Original Equipment Manufacturer Specifications***—In addition to the performance requirements set by the CCMC Sequences, original equipment manufacturers in Europe have developed individual specifications for engine oils which may demand additional laboratory and engine performance testing. These specifications are based upon engine type, service, and application.

4.1 **MAN 269**—Defines minimum laboratory and engine test requirements for naturally aspirated engines of the Nuremberg and Brunswick design. Quality level is MIL-L-46152A and covers SAE 20W-20, 20W-30, and SAE 30 grades without viscosity index improver. This specification is obsolete as of July 1, 1990.

4.2 **MAN 270**—Defines minimum laboratory and engine test requirements for naturally aspirated and turbocharged diesel engines both for stationary equipment and for vehicles. Quality level required is MIL-L-2104C/MIL-L-46152A, CCMC D-4, API CD/SE covering monograde oils without VI improvers.

4.3 **MAN 271**—Defines minimum laboratory and engine test requirements for naturally aspirated and turbocharged diesel engines both for stationary equipment and for vehicles. Quality level required is MIL-L-2104C/MIL-L-46152A, CCMC D-4, API CD/SE and covers multigrade oils (SAE 10W-40, 15W-40, and 20W-50).

4.4 **MAN QC-13-017**—Defines minimum laboratory and engine test requirements for super high-performance turbocharged diesel engine oils for turbocharged and non-turbocharged engines whenever a higher performance level than MAN 270 or MAN 271 is required. Satisfactory performance in a MAN 500 Hour Engine Test is required.

4.5 **MERCEDES-BENZ 226.0**—Describes single-grade engine oils approved for commercial vehicles which are equipped with non-turbocharged diesel engines.

SAE J2227 Revised AUG95

4.6 **MERCEDES-BENZ 227.0**—Describes single-grade engine oils approved for passenger cars and commercial vehicles equipped with turbocharged and non-turbocharged diesel engines. For commercial vehicles, attention must be paid to oil drain intervals.

(R) 4.7 **MERCEDES-BENZ 227.1**—Describes multigrade engine oils approved for commercial vehicles equipped with turbocharged and non-turbocharged engines. Attention must be paid to oil drain intervals. Quality Level is equivalent to CCMC D-4, but with more requirements.

4.8 **MERCEDES-BENZ 228.0**—Describes single-grade engine oils approved for all Mercedes-Benz diesel engines including turbocharged commercial vehicle diesel engines operating with increased oil drain intervals.

4.9 **MERCEDES-BENZ 228.1**—Describes multigrade engine oils approved for all Mercedes-Benz diesel engines including turbocharged commercial diesel engines operating with increased oil drain intervals. Quality level is equivalent to CCMC D-4 plus more stringent requirements for bore polishing and cylinder wear.

(R) 4.10 **MERCEDES-BENZ 228.2**—Describes single-grade engine oils as in the case of Mercedes-Benz 228.0 which are also suited in given commercial vehicles for longer drain intervals compared to 228.0.

(R) 4.11 **MERCEDES-BENZ 228.3**—Describes multigrade engine oils as in the case of Mercedes-Benz 228.1 which are also suited in given commercial vehicles for longer drain intervals. Quality level is equivalent to CCMC D-5.

Performance of 227.0 and 227.1, 228.0 and 228.1, 228.2 and 228.3 is equivalent. 228.3 is suitable for the longest drain intervals and has the first recommendation.

(R) 4.12 **MTU OIL TYPE 1 AND TYPE 2**—Defines the performance of both single- and multigrade engine oils recommended for MTU diesel engines. Quality level required is Mercedes-Benz 227.0, 227.1 (Type 1 Oils) or 228.2, 228.3 (Type 2 Oils).

4.13 **MWM DEUTZ TR 0199-3002**—Defines the performance of both single and multigrade engine oils recommended for all Deutz diesel engines and/or small-size Deutz MWM engines. Quality level required is API CC, CD, CE, and CCMC D-4, D-5.

4.14 **MWM DEUTZ TR 0199-2090**—Defines the performance of both single and multigrade engine oils recommended for big-size Deutz MWM engines. Quality level required is API CC, CD, CE, and CCMC D-4, D-5.

4.15 **PERKINS P.M.S. S1.01—1983, TSD 3187—1990, PS No. 7294/SB019**—Defines the performance of both single- and multigrade oils utilized in Perkins naturally aspirated and turbocharged diesel engines. P.M.S. S1.01-1983 applies to Peterborough engines, TSD 3187-1990, Shrewsbury engines, and PS No. 729H/SB019 Gardner Engines.

4.16 **ROVER GROUP—RES.22.OL.G-4**—Applies to service fill engine oils for use in gasoline engines for passenger cars, light commercial vehicles, and 4 x 4 dual purpose cross-country vehicles (up to 4 tons) in all worldwide markets.

4.17 **ROVER GROUP—RES.22.OL.PD-2**—Applies to service fill engine oils for use in turbocharged and naturally aspirated diesel engines for passenger cars, light commercial vehicles, and 4 x 4 dual purpose cross-country vehicles (up to 4 tons) in all worldwide markets.

4.18 **ROVER GROUP RES.22.OL.D-5**—Applies to service fill engine oils for use in highly rated turbocharged diesel engines for passenger cars, light commercial vehicles, and 4 x 4 dual purpose cross-country vehicles (up to 4 tons) in all worldwide markets.

SAE J2227 Revised AUG95

4.19 Scania Specification—Defines the performance of engine oils for naturally aspirated and turbocharged diesel engines to be used in maintenance programs.

4.20 VOLKSWAGEN, VW 500 00—Defines laboratory and engine test requirements for service fill engine oils in VW, Audi, and Seat gasoline and naturally aspirated diesel engines. Quality level is CCMC G-5 plus additional performance requirements for evaporative loss, seal compatibility, piston cleanliness, black sludge, and cam and tappet wear.

4.21 VOLKSWAGEN, VW 501 01—Defines laboratory and engine test requirements for service fill engine oils in VW, Audi, and Seat gasoline and naturally aspirated diesel engines. Quality level is CCMC G-4 plus additional requirements for seal compatibility, piston cleanliness, black sludge, and cam and tappet wear.

4.22 VOLKSWAGEN VW 505 00—Defines laboratory and engine test requirements for service fill engine oils in VW/Audi turbocharged passenger car and commercial diesel engines and Seat diesel engines with exhaust-driven supercharger with and without boost intercooling. Quality level is CCMC PD-2 plus additional requirements for piston cleanliness, seal compatibility, and cam and tappet wear.

(R) **4.23 VOLVO DRAIN SPECIFICATION (VDS)**—Defines the performance of SAE 15W-40 and SAE 10W-30 engine oils intended for turbocharged engines running under extended drain conditions. Volvo VDS-2 specification was introduced in 1992 and will be recommended for all Volvo truck engines meeting the 1996 European emission requirements. VDS-2 covers SAE 5W-30, 5W-40, 10W-30, 10W-40, and 15W-40 grades.

5. European Military Specifications—Laboratory and engine tests which are utilized to define the CCMC sequences are also incorporated in the development of various Military specifications throughout Europe.

6. International Lubricant Standardization and Approval Committee Standard—The American Automobile Manufacturers Association (AAMA) and the Japan Automobile Manufacturers Association, Inc. (JAMA), through an organization called the International Lubricant Standardization and Approval Committee (ILSAC), jointly developed and approved a minimum performance standard for gasoline-fueled passenger car engine oils. This standard (ILSAC GF-1), which became official in October of 1990, includes the performance requirements and chemical and physical properties of those engine oils that vehicle manufacturers may deem necessary for satisfactory equipment life and performance. Included within the standard are both engine and bench test requirements, as well as additional requirements for fuel efficiency, catalyst compatibility (phosphorus content), and low-temperature viscosity.

7. Japanese Classifications and Specifications—Japanese vehicle manufacturers, in general, rely on the API Classification System to recommend engine oils for service fill applications. Additionally, "in-house" procedures are also required by many manufacturers. The Japanese Automobile Standards Organizations (JASO) which is comprised of automobile and truck manufacturers, oil and additive companies, and government authorities, has worked to unify the engine oil evaluation procedures in Japan. Four test procedures are currently established. Three of these procedures address lubricant performance in gasoline engines. The fourth procedure evaluates diesel engine oil performance.

7.1 JASO Gasoline Engine Test Procedures

7.1.1 JASO M328-91—Specifies the test procedure for the evaluation of the wear resistance of valve trains of automobile gasoline engine oils.

7.1.2 JASO M333-93—Specifies the test procedure for the evaluation of the high-temperature oxidation stability of lubricating oils for automobile gasoline engines.

7.1.3 JASO M331-91—Specifies the test procedure for the evaluation of the low- and medium-temperature detergency of lubricating oils for automobile gasoline engines.

SAE J2227 Revised AUG95

7.2 JASO Diesel Engine Test Procedure

7.2.1 JASO M336-90—Specifies the test procedure for the evaluation of the high-temperature and high-load detergency of lubricating oils for automobile diesel engines.

8. European Engine Tests

(R) **8.1 Gasoline Engine Tests**—A number of gasoline engine tests have been developed by the European CEC and original equipment manufacturers to evaluate the ability of engine oils to prevent piston deposits, sludge, varnish, rust and corrosion, and wear. These tests include procedures established by Volkswagen and Mercedes-Benz as well as tests which utilize Peugeot and Petter engines. Engine test conditions for these procedures are presented in Table 1. Table 2 provides procedure reference, performance evaluation factors, and test applications.

(R) **8.2 Diesel Engine Tests**—Many diesel engine test procedures are established by the European CEC and original equipment manufacturers to evaluate diesel engine oil performance. Included in these are tests which utilize engines developed by Volkswagen and Mercedes-Benz as well as procedures incorporating Petter and an MWM diesel engine. Engine test conditions for these procedures are provided in Table 3. Table 4 presents the respective procedure references, performance evaluation factors, and test applications.

9. ILSAC Engine Tests—The following engine tests are included within the ILSAC GF-1 performance standard.

9.1 ASTM Sequence IID Test—Specifies the test procedure which measures the ability of engine oil to protect valve train and oil pump components against rust and corrosion deposits.

9.2 ASTM Sequence IIIE Test—Specifies the test procedure which evaluates the ability of engine oil to minimize high-temperature oxidation and thickening, sludge and varnish deposits, and wear.

9.3 ASTM Sequence VE Test—Specifies the test procedure which evaluates the ability of engine oil to protect against sludge, varnish deposits, and wear.

9.4 L-38 Test Method—Specifies the test procedure which evaluates the ability of engine oil to protect against copper/lead bearing weight loss and piston deposit formation. The test can also be utilized to evaluate engine oil shear stability.

9.5 ASTM Sequence VI Test—Specifies the test procedure which evaluates the ability of engine oils to provide engine fuel efficiency improvement.

The ASTM Sequence VI test is included as an additional requirement to API SH. Performance requirements for all other tests listed are identical to API SH. Fleet testing may also be requested. Test conditions and performance evaluation factors for ILSAC GF-1 engine tests can be referenced within SAE J183 and SAE J304.

10. Japanese Engine Tests

10.1 Gasoline Engine Tests—The engine test procedures developed within Japan for gasoline engine oil performance are referenced by JASO procedures M328-91, M333-93, and M331-91. These include tests developed by both Nissan and Toyota for valve train wear, high-temperature oxidation stability, and low- and medium-temperature detergency. Engine test conditions are presented in Table 5. Table 6 provides procedure reference, performance evaluation factors, and test applications.

10.2 Diesel Engine Tests—The engine test procedure for diesel engine oil performance is referenced by JASO procedure M336-90. This procedure utilizes the Nissan engine and evaluates automobile diesel engine oil detergency under high-temperature and high-load conditions. Engine test conditions are presented in Table 7. Table 8 provides procedure reference, performance evaluation factors, and test application.

SAE J2227 Revised AUG95

11. Laboratory Test Procedures—In addition to the previously discussed engine test procedures, a series of laboratory tests are also employed as part of the CCMC and ILSAC performance criteria as well as original equipment manufacturers' specifications in Europe. The tests include evaluation for engine oil shear stability, high-temperature/high-shear viscosity, volatility, foaming tendency, cam and tappet wear, seal compatibility, engine oil filterability, flash point, and homogeneity/miscibility. Table 9 provides a list of important procedures for the evaluation of both gasoline and diesel engine oils along with the respective applications.

12. Notes

12.1 Marginal Indicia—The (R) is for the convenience of the user in locating areas where technical revisions have been made to the previous issue of the report. If the symbol is next to the report title, it indicates a complete revision of the report.

PREPARED BY THE SAE FUELS AND LUBRICANTS TECHNICAL COMMITTEE 1—AUTOMOTIVE ENGINE OILS

SAE J2227 Revised AUG95

(R) TABLE 1—EUROPEAN GASOLINE ENGINE TEST CONDITIONS

Manufacturer	Engine No.	Engine Displacement cm^3	Cylinders	Fuel	Test Duration Hours	Test Duration Stages	Engine[5] Speed rpm	Oil Temp °C	Coolant Temp °C	Fuel Cons. kg/h	Fuel Cons. L/h	Air/Fuel Ratio	Engine[5] Load NM	Engine Load kW	Procedure[1]
Peugeot (PSA)	TU3M	1360	4	Unleaded CEC RF 83-A-91	100	40 h 60 h	1500 3000	40 100	45 90	1.5 4.0		[2]	10 35		CEC-L-38-A-94
VW	1302	1285	4	Premium Fuel 0.15 g/l Lead Max Approved to DIN 51600	50		4200	100			15.0	[3]		31.5	DKA 6/79
Petter	W-1	468	1	Leaded CEC RF 80	36		1500	130	147 inlet 150 outlet		1.59	11.7-12.1		2.5	CEC-L-02-A-78
Mercedes Benz	M111	2000	4	Special fuel batch to batch approval	224	48 h 1 h 75 h 100 h	Alternating 750-1950 1500-6000 3850 3750 780-5500	45 max 100-140 123 37-130	–4 to 40 98 31-97				Alternating 0-31.5 95-194 WOT 0-WOT		CEC-L-53-T-95
Peugeot	TU3M	1360	4	RF-83-A-91	96	11 h 50 min 10 min	WOT Oil make up	150	110 max	16.5					CEC-L-55-T-95

(1) CEC Procedures available from the Coordinating European Council, Madou Plaza - 25th Floor, Place Madou 1, B-1030 Brussels, Belgium
(2) As per manufacturer's recommendation: 0.8 to 1.5% Vol CO
(3) As per manufacturer's recommendation: 2% ± 1% Vol CO
(4) As per manufacturer's recommendation: 0.1 to 1.0% Vol CO
(5) WOT = Wide Open Throttle

(R) TABLE 2—EUROPEAN GASOLINE ENGINE TEST PERFORMANCE EVALUATION FACTORS

Test	Procedure[1]	Piston Deposits	Sludge	Varnish	Wear	Other	Test Application
Peugeot TU3M	CEC-L-38-A-94				X		CCMC G-4, G-5, ROVER RES.22.OL.G-4, ACEA
VW 1302	DKA 6/79	X		X	X	Oil Consumption	VW 500 00, 501 01
Petter W-1	CEC-L-02-A-78				X	Bearing Corrosion Viscosity Increase	CCMC, G-4, G-5, ROVER RES.22.OL.G-4
Mercedes Benz M111	CEC-L-53-T-95		X				VW 500 00, 501 01, ACEA
Peugeot TU3M	CEC-L-55-T-95	X				Ring Sticking Viscosity Increase	ACEA

(1) CEC procedures available from the Coordinating European Council, Madou Plaza - 25th Floor, Place Madou 1, B-1030 Brussels, Belgium

SAE J2227 Revised AUG95

(R) TABLE 3—EUROPEAN DIESEL ENGINE TEST CONDITIONS

Test Manufacturer	Test Engine No.	Engine Displacement cm^3	Test Duration Hours	Test Duration Stages	Oil Change Period Hours	Engine Speed rpm	Fuel Rate mg^3/Stroke (Unless Noted)	Engine Power (kW)	Air-to-Engine Temperature °C	Air-to-Engine Pressure	Coolant Outlet Temperature °C	Oil-to-Bearing Temperature °C	Fuel Sulfur % Mass	Procedure[1]
Mercedes-Benz	OM 364A	3972	300[2]	1.5 h 0.5 h 0.5 h 50 h	None	2600 1500 1000 2600	63 65.5 mg/stroke 67 70 mg/stroke 64 67 mg/stroke 63 65.5 mg/stroke	85 58 34 85	max 150 record record max 150	1150 mbar	98 record record 98	122 record record 122	0.25-0.30 [3]	CEC-L-42-A-92
MWM	KD 12E	850	50	...	None	2200	3.1 kg/h	10.7	25-35	Atmospheric	110	110	1.0	CEC-L-12-U-93
Petter	AVB	553	100	...	None	2250	3.7 kg/h	14.7	75	1280 mbar	100	90	1.0	CEC-L-24-A-78
Volkswagen TC Intercooled	1431	1600	50	...	None	4500	31.1 mg/stroke	55 min	50	670 mbar	90	130	0.3	CEC-L-46-T-93
Volkswagen naturally aspirated diesel	1435	1588	50	...	None	4800	20 + 1 mg/stroke	40	28	Atmospheric	90	128	0.3	VW PV 1431
Peugeot (PSA)	XUDII ATE	2088	75	2 min 27 min	None	1000 4300	... 47.5	0 >80	... 60	Atmospheric 710 mbar	... 100	... 110	CEC RF-90-A-92	CEC-L-56-T-95
Mercedes Benz	OM602A	2497	200	14 stages	None	Alternating 0-5000	...	Minimum to maximum	30	800-825 mbar boost during stage 10 and 11	20-92	...	...	CEC-L-51-T-95

(1) CEC procedures available from the Coordinating European Council, Madou Plaza · 25th Floor, Place Madou 1, B-1030 Brussels, Belgium
(2) 3 major 100 h stages consisting of 20 x 1-1/2 h cycles + 50 h steady state
(3) CEC-RF 90-A-81

(R) TABLE 4—EUROPEAN DIESEL ENGINE TEST PERFORMANCE EVALUATION FACTORS

Test	Procedure Reference[1]	Piston Deposits	Rust and Corrosion	Sludge	Varnish	Wear	Other	Test Application
Mercedes-Benz OM 364A	CEC-L-42-A-92	X		X	X	X	Oil Cons. Bore Polishing	M-B 227 and 228, CCMC D-4, D-5, Rover[2], MTU
Peugeot (PSA) XUDII ATE	CEC-L-56-T-95	X					Viscosity Increase	ACEA
Mercedes Benz OM 602A	CEC-L-51-T-95	X		X		X	Viscosity Increase, Bore Polishing	ACEA
MWM KD12E	CEC-L-12-A-76	X			X			MAN 270/271, CCMC D-1, D-2, MTU, Perkins
Petter AVB	CEC-L-24-A-78	X		X	X			European Military
Volkswagen TC Intercooled	CEC-L-46-T-93	X			X		Ring Sticking	VW 505 00
Volkswagen	PV 1435	X			X		Ring Sticking	VW 500 00, VW 501 01

(1) CEC procedures available from the Coordinating European Council, Madou Plaza - 25th Floor, Place Madou 1, B-1030 Brussels, Belgium
(2) RES.22.OL.D-5

SAE J2227 Revised AUG95

TABLE 5—JAPANESE GASOLINE ENGINE TEST CONDITIONS

Manufacturer	Engine No.	Engine Displacement cm^3	Test Duration Type	Test Duration Hours	Test Duration Stages	Engine Speed rpm	Oil Temp °C	Outlet Coolant Temp °C	Air/Fuel Ratio	Engine(1) Load	Spring Load	JASO(2) Procedure
Toyota	3A	1452	4 OHC	200		1000	60-65			0	standard	M328-91
Nissan	VG20E	1998	6 OHC	300(3)	24 min	800	50	42		19.6		M331-91
				200(4)	24 min	1800	96	85		98.1		
					12 min	3500	117	97		93.1		
Toyota	1G-FE	1988	6 OHC	96(3) 48(4)		4800	149	120	14.5	58.8		M333-93

(1) In units of N·m unless otherwise indicated
(2) Available from the Society of Automotive Engineers of Japan, Inc., 10-2 Goban-cho, Chiyoda-ku, Tokyo 102, Japan
(3) High-grade oils
(4) Regular-grade oils

TABLE 6—JAPANESE GASOLINE ENGINE TEST PERFORMANCE EVALUATION FACTORS

Test	Procedure Reference(1)	Evaluation Piston Rings	Evaluation Rust & Corrosion	Evaluation Sludge	Evaluation Varnish	Evaluation Wear	Evaluation Other	Test Application(2)
Toyota 3A	M328-91					X		API SE, SG
Nissan VG20E	M331-91	X		X	X	X		API SE, SG
Toyota 1G-FE	M333-93	X		X	X	X	Viscosity Increase	API SE, SG

(1) Available from the Society of Automotive Engineers of Japan, Inc., 10-2 Goban-cho, Chiyoda-ku, Tokyo 102, Japan
(2) JASO Engine Test Procedures are applied in Japan to API performance categories. JASO procedures are not part of the official service category description established by the American Petroleum Institute, but are utilized in Japan to supplement API performance testing.

SAE J2227 Revised AUG95

TABLE 7—JAPANESE DIESEL ENGINE TEST CONDITIONS

Manufacturer	Engine No.	Engine Displacement cm^3	Test Duration Hours	Test Duration Stages	Oil Change Period	Engine Speed	Fuel Rate	Engine Load	Air to Engine Temp °C	Spring Load	Coolant Outlet Temp °C	Oil Temp °C	Fuel Sulfur % Mass	JASO[1] Procedure
Nissan Diesel	SD 22	2164	50[2] 100[3]		Addition every 5 h	4000	36 mm^3/ Stroke Cyl.	Above 55 P.S.	25-30		80	120	0.4-0.5	M336-90

(1) Available from the Society of Automotive Engineers of Japan, Inc., Goban-cho, Chiyoda-ku, Tokyo 102, Japan
(2) API CC Level
(3) API CD Level

TABLE 8—JAPANESE DIESEL ENGINE TEST PERFORMANCE EVALUATION FACTORS

Test	Procedure Reference[1]	Evaluation Piston Rings	Evaluation Rust & Corrosion	Evaluation Sludge	Evaluation Varnish	Evaluation Wear	Evaluation Other	Test Application[2]
Nissan SD 22	M336-90	X		X	X	X		API CC, CD

(1) Available from the Society of Automotive Engineers of Japan, Inc., 10-2 Goban-cho, Chiyoda-ku, Tokyo 102, Japan
(2) JASO Engine Test Procedures are applied in Japan to API performance categories. JASO procedures are not part of the official service category description established by the American Petroleum Institute, but are utilized in Japan to supplement API performance testing.

SAE J2227 Revised AUG95

(R) TABLE 9—LABORATORY TEST PROCEDURES

Test	Reference[1]	Description	Applicable Specifications
Shear Stability	CEC-L-14-A-93	Evaluates the shear stability of polymer containing lubricating oils utilizing a diesel fuel injector rig. Shear stability is defined as a permanent percent drop in kinematic viscosity at 100 °C.	CCMC, Volkswagen, Rover, Mercedes-Benz, ACEA
	L-38 Test Method ASTM STP 509A Part IV	Evaluates the shear stability of engine oils utilizing a single-cylinder engine test. Fuel stripped kinematic viscosity is determined after 10 h test time.	ILSAC GF-1
NOACK Volatility	CEC-L-40-A-93 DIN 51581	Evaluates the volatility or evaporative loss tendency of lubricating oils. Volatility is expressed as a percentage of loss in mass after 1 h at 250 °C.	CCMC, MAN, MTU, Rover, Volkswagen, Mercedes-Benz, ILSAC GF-1, ACEA
	ASTM D 2887	Evaluates the boiling point range distribution of petroleum products.	ILSAC GF-1
High-Temperature/ High-Shear Viscosity	CEC-L-36-A-90	Evaluates the dynamic viscosity of lubricating oils at 150 °C and 10^6 S^{-1} shear rate.	CCMC, Mercedes-Benz, Volkswagen, Rover, ILSAC GF-1, ACEA
	ASTM D 4683		ILSAC GF-1
Foaming Tendency	ASTM D 892	Evaluates the foaming characteristics of lubricating oils at specified temperatures.	CCMC G-4, G-5, PD-2, D-4, D-5, ILSAC GF-1, ROVER, ACEA
Oil/Elastomer Compatibility	CEC-L-39-T-87	Evaluates the degree of compatibility of lubricating oils and cured elastomers used in the car industry. Elastomer materials include acrylics, nitriles, fluorinated, and silicones.	CCMC G-4, G-5, PD-2, D-4, D-5, ROVER, ACEA
Cam and Tappet	P-VW 5106	Evaluates the ability of lubricating oils to prevent cam and tappet wear.	VW 505 00, 501 01, 500 00
Filterability	GM 9099P	Evaluates the tendency of an oil to form a precipitate which can plug the oil filter.	ILSAC GF-1
Flash Point	ASTM D 92 ASTM D 93	Evaluates the flash point of oils utilizing an open (ASTM D 92) or closed (ASTM D 93) cup apparatus.	ILSAC GF-1
Homogeneity and Miscibility	Federal Test Method Standard 791B, Method 3470	Evaluates if an oil is and will remain homogeneous and if it is miscible with certain standard reference oils after being submitted to a prescribed cycle of temperature ranges.	ILSAC GF-1

(1) - CEC procedures available from the Coordinating European Council, Madou Plaza - 25th Floor, Place Madou 1, B-1030 Brussels, Belgium
- DIN procedures available from DIN, Postfach 1107, D-1000 Berlin 30, W. Germany
- ASTM procedures available from ASTM, 1916 Race Street, Philadelphia, PA 19103-1187
- P-VW 5106 available from Volkswagen, Wolfsburg, Germany
- GM 9099P available from General Motors Corporation, CPE-Engineering Standards W-3, Warren, MI 48090-9010
- Federal Test Method 791B available from the General Services Administration, Business Service Center, Region 3, Seventh and D Street, SW, Washington, DC 20025

Appendix 8

Approximate Engine and Rig-Test Prices - 1994

(Representative of typical inter-company charges for contracted tests)

USA TESTS GENERAL

Test Description	Typical Cost (US $)	Duration (Hrs)	Oil Required (Litres)
Labeco L38	6400	40	10
Sequence 2D Rust	11900	32	20
Sequence 2D Rust Screener	8200	32	25
Sequence 3E	16800	64	20
Sequence 5E Sludge	21000	288	20
ASTM Fuel Economy	11500	3 days	20
Cummins NTC 400	38700	200	210
Mack T6	59900	600	210
Mack T7	14800	150	135
Mack T8	28600	250	180
Detroit Diesel 6V92TA	46300	100	210
TO4 - Sequence 1220	2000	8	40
TO4 - Bronze Friction Retention	4600	140	40

Test Description	Typical Cost (US $)	Duration (Hrs)	Oil Required (Litres)
Caterpillar 1G2	19200	480	80
Caterpillar 1K	20300	252	40
Caterpillar Micro Oxidation	1300	2	1
FZG ASTM 4998	1800	3 × 20 hrs	4
GM 6.2 Litre Wear	13700	50	45
Caterpillar 1N	19100	252	40
Caterpillar 1MPC	12000	120	40
L38 Stay in Grade	5000	10	12
Cummins L10 High Soot Test	20800	100	95

EUROPEAN TESTS GENERAL

Test Description	Typical Cost (US $)	Duration (Hrs)	Oil Required (Litres)
Petter AVB 50 HR	4800	50	10
Petter AVB 100 HR	5400	100	10
MWMB (KD 12E)	3600	50	10
MAN D-2866 KF (QC 13-017)	110000	500	205
M102E Sludge	30200	185	45
OM616 Combi	26400	206	65
OM364A	46200	300	100
OM602A	40900	200	50
Fuel Injected Peugeot TU3 100 HR	11900	100	10

Test Description	Typical Cost (US $)	Duration (Hrs)	Oil Required (Litres)
Peugeot XUD 7TE Endurance	27700	320	65
MIII Sludge Test	-	300	45
Peugeot XUD II ATE Soot Loading	20000	75	50
Saab Turbo Deposits	15000	167 (500 cycles)	10
VW 1302 Wear	11500	50	15
VW Naturally Aspirated Diesel	12600	50	25
VW Intercooled Turbo Diesel	13300	50	25
Petter W1	4600	36	10
VW Cam & Tappet Rig	2000	16	10
Peugeot TU3 High Temp Deposits	14200	96	20
MAN Single Cylinder	23700	200	40
OM441LA	77300	400	200

TRANSMISSION/MISCELLANEOUS

Test Description	Typical Cost (US $)	Duration (Hrs)	Oil Required (Litres)
C4 Seals	1600	70	5
C4 Bench	3000	70	12
C4 Graphite Friction	2200	23	4
C4 Paper Friction	2900	42	4

Test Description	Typical Cost (US $)	Duration (Hrs)	Oil Required (Litres)
C4 Thermal Oxidation	8300	300	20
C4 Wear (Vickers Pump)	3600	100	20
GM Clutch Plate Test (Dexron III)	4400	100	19
GM Band Clutch Friction (Dexron™ III)	4400	100	19
GM Oxidation Test 11000 (Dexron™ III)	300	19	
GM 20,000 Cycle Test (Dexron™ III)	23000	-	19
GM Vehicle Test (Dexron™ III)	3000	-	19
Ford 15,000 Cycle Clutch Friction (Mercon™)	4600	-	4
Ford 20,000 Cycle Test (Mercon™)	5700	-	4
ABOT (Mercon™)	2900	300	4
4L60 Cycle Test (Mercon™)	22600	-	
Ford Shift Feel Test (Mercon™)	9500	-	19
Volkswagen Seals	300	282	1.5
Daimler Benz Seals	200	168	2
MAN Seals	300	168	2
CCMC Seals	500	168	2

Test Description	Typical Cost (US $)	Duration (Hrs)	Oil Required (Litres)
Nissan SD22 Detergency	24000	100	40
Toyota Valve Train Wear	4500	200	20
Toyota High Temp Oxidation	18000	48	20
Nissan Low/Medium Temp Detergency	19000	200	20

2T

Test Description	Typical Cost (US $)	Duration (Hrs)	Oil Required (Litres)
Susuki SX 800 Blocking	3100	Variable	20
Susuki SX 800 Smoke	1000	2	1
Honda DIO Lubricity	1900	3	1
Honda DIO Detergency	2200	1	1
Honda DIO Severe Detergency	2400	3	1
Husqvarna Chainsaw	1200	5	1
Yamaha Y350M	5900	20	4
Yamaha Y350M2	5900	20	4
Yamaha 50CC Tightening (Lubricity)	3000	8	4
Yamaha 50CC - 50hr Preignition	6700	50	4
Yamaha 50CC - 100hr Preignition	9000	100	12
Yamaha 50CC Tightening (Lubricity)	3100	20	4

Test Description	Typical Cost (US $)	Duration (Hrs)	Oil Required (Litres)
W2 Identification (IR & INSP)	600	70	1
W2 Bench Tests	800	70	1
OMC 40H Detergency (TC-WII)	23400	98	20
OMC 70HP Detergency (TC-W3)	28400	98	60
Mercury 15HP (TCW3)	13400	100	12

Appendix 9

400 Commonwealth Drive, Warrendale, PA 15096-0001

SURFACE VEHICLE STANDARD

Submitted for recognition as an American National Standard

SAE J300	REV. DEC95
Issued 1911-06 Revised 1995-12 Superseding J300 MAR93	

ENGINE OIL VISCOSITY CLASSIFICATION

1. ***Scope***—This SAE Standard defines the limits for a classification of engine lubricating oils in rheological terms only. Other oil characteristics are not considered or included.

2. ***References***

2.1 Applicable Documents—The following publications form a part of this specification to the extent specified herein. The latest issue of ASTM and SAE publications shall apply.

2.1.1 SAE Publications—Available from SAE, 400 Commonwealth Drive, Warrendale, PA 15096-0001.

SAE J510—Lubricants for Two-Stroke-Cycle Engines
SAE J1536—Two-Stroke-Cycle Engine Oil Miscibility/Fluidity Classification

2.1.2 ASTM Publications—Available from ASTM, 1916 Race Street, Philadelphia, PA 19103-1187.

ASTM D 97—Standard Test Method for Pour Point of Petroleum Oils
ASTM D 445—Standard Test Method for Kinematic Viscosity of Transparent and Opaque Liquids (and the Calculation of Dynamic Viscosity)
ASTM D 2500—Standard Test Method for Cloud Point of Petroleum Oils
ASTM D 3244—Standard Practice for Utilization of Test Data to Determine Conformance With Specifications
ASTM D 3829—Standard Test Method for Predicting the Borderline Pumping Temperature of Engine Oil
ASTM D 4683—Standard Test Method for Measuring Viscosity at High Temperature and High-Shear Rate by Tapered Bearing Simulator
ASTM D 4684—Standard Test Method for Determination of Yield Stress and Apparent Viscosity of Engine Oils at Low Temperature
ASTM D 4741—Standard Test Method for Measuring Viscosity at High Temperature and High-Shear Rate by Tapered-Plug Viscometer
ASTM D 5133—Standard Test Method for Low Temperature, Low Shear Rate, Viscosity/Temperature Dependence of Lubricating Oils Using a Temperature-Scanning Technique
ASTM D 5293—Standard Test Method for Apparent Viscosity of Engine Oils Between –30 and –5 °C Using the Cold-Cranking Simulator
ASTM Data Series DS 62—The Relationship Between High-Temperature Oil Rheology and Engine Operation—A Status Report
ASTM STP 1068—High-Temperature, High-Shear Oil Viscosity—Measurement and Relationship to Engine Operation
ASTM STP 1143—Low-Temperature Lubricant Rheology: Measurement and Relevance to Engine Operation

SAE J300 Revised DEC95

2.1.3 OTHER PUBLICATIONS

CEC L-36-A-90—The Measurement of Lubricant Dynamic Viscosity Under Conditions of High Shear Using the Ravenfield Viscometer

CRC Report No. 409—Evaluation of Laboratory Viscometers for Predicting Cranking Characteristics of Engine Oils at 0 °F and -20 °F, April 1968

Hodges and Rodgers, "Some New Aspects of Pour Depressant Treated Oils," Oil and Gas Journal, p. 89, October 4, 1947

McNab, Rodgers, Michaels, and Hodges, "The Pour Stability Characteristics of Winter Grade Motor Oils," Quarterly Transactions, Society of Automotive Engineers, Inc., Vol. 2, No. 1, p. 34, January 1948

3. ***Significance and Use***—The limits specified in Table 1 are intended for use by engine manufacturers in determining the engine oil viscosity grades to be used in their engines, and by oil marketers in formulating, manufacturing, and labeling their products. Oil marketers are expected to distribute only products which are within the relevant specifications in Table 1.

Disputes between laboratories as to whether a product conforms with any specification in Table 1 shall be resolved by application of the procedures described in ASTM D 3244. For this purpose, all specifications in Table 1 are *critical specifications* to which conformance based on reproducibility of the prescribed test method is required. The product shall be considered to be in conformance if the Assigned Test Value (ATV) is within the specification.

Two series of viscosity grades are defined in Table 1: (a) those containing the letter W and (b) those without the letter W. Single viscosity-grade oils ("single-grades") with the letter W are defined by maximum low-temperature cranking and pumping viscosities, and a minimum kinematic viscosity at 100 °C. Single-grade oils without the letter W are based on a set of minimum and maximum kinematic viscosities at 100 °C, and a minimum high-shear viscosity at 150 °C and 10^6 s^{-1}. Multiviscosity-grade oils ("multigrades") are defined by both of the following criteria:

(1) Maximum low-temperature cranking and pumping viscosities corresponding to one of the W grades, and

(2) Maximum and minimum kinematic viscosities at 100 °C and a minimum high-shear viscosity at 150 °C and 10^6 s^{-1} corresponding to one of the non-W grades.

4. ***Low-Temperature Test Methods***—The low-temperature cranking viscosity is measured according to the procedure described in ASTM D 5293 and is reported in centipoise (mPa·s). Viscosities measured by this method have been found to correlate with engine speeds developed during low-temperature cranking.

The pumping viscosity is a measure of an oil's ability to flow to the engine oil pump and provide adequate oil pressure during the initial stages of operation. The pumping viscosity is measured in centipoise (mPa·s) according to the procedure in ASTM D 4684. This procedure uses the Mini-Rotary Viscometer to measure either the existence of yield stress or the viscosity in the absence of measured yield stress after the sample has been cooled through a prescribed slow cool (so-called TP1) cycle. This cooling cycle has predicted as failures several SAE 10W-30 and SAE 10W-40 engine oils which are known to have suffered pumping failures in the field after short-term (two days or less) cooling. These field failures are believed to be the result of the oil forming a gel structure that results in excessive yield stress and/or viscosity of the engine oil. The significance of the ASTM D 4684 method is projected from the preceding SAE 10W-30 and SAE 10W-40 data.

SAE J300 Revised DEC95

(R) TABLE 1—SAE VISCOSITY GRADES FOR ENGINE OILS[1]

SAE Viscosity Grade	Low-Temperature (°C) Cranking Viscosity[2], cP Max	Low-Temperature (°C) Pumping Viscosity[3], cP Max With No Yield Stress	Kinematic Viscosity[4] (cSt) at 100 °C Min	Kinematic Viscosity[4] (cSt) at 100 °C Max	High-Shear Viscosity[5] (cP) at 150 °C and 10^6 s^{-1} Min
0W	3250 at −30	60 000 at −40	3.8	—	—
5W	3500 at −25	60 000 at −35	3.8	—	—
10W	3500 at −20	60 000 at −30	4.1	—	—
15W	3500 at −15	60 000 at −25	5.6	—	—
20W	4500 at −10	60 000 at −20	5.6	—	—
25W	6000 at −5	60 000 at −15	9.3	—	—
20	—	—	5.6	<9.3	2.6
30	—	—	9.3	<12.5	2.9
40	—	—	12.5	<16.3	2.9 (0W-40, 5W-40, and 10W-40 grades)
40	—	—	12.5	<16.3	3.7 (15W-40, 20W-40, 25W-40, 40 grades)
50	—	—	16.3	<21.9	3.7
60	—	—	21.9	<26.1	3.7

NOTE—1 cP = 1 mPa·s; 1 cSt = $1 mm^2/s$

[1] All values are critical specifications as defined by ASTM D 3244 (see text, Section 3).
[2] ASTM D 5293
[3] ASTM D 4684 (see also Appendix B and text, Section 4.1): Note that the presence of any yield stress detectable by this method constitutes a failure regardless of viscosity.
[4] ASTM D 445
[5] ASTM D 4683, CEC L-36-A-90 (ASTM D 4741)

SAE J300 Revised DEC95

Limited test work has shown that in a few specific instances, stable pour point (Appendix B, Test Method for Stable Pour Point of Engine Oils), borderline pumping temperature (ASTM D 3829), and/or Scanning Brookfield method (ASTM D 5133) can provide additional information regarding low-temperature performance. It is suggested that these tests be conducted when formulating new engine oils, or when there are significant changes in base oil or additive components of existing products.

Because engine pumping, cranking, and starting are all important at low temperatures, the selection of an oil for winter operation should consider both the viscosity required for successful oil flow, as well as that for cranking and starting, at the lowest ambient temperature expected.

5. ***High-Temperature Test Methods***—Kinematic viscosity at 100 °C is measured according to ASTM D 445, and the results are reported in centistokes (mm^2/s). Kinematic viscosities have been related to certain forms of oil consumption and have been traditionally used as a guide in selecting oil viscosity for use under normal engine operating temperatures. Also, kinematic viscosities are widely used in specifying oils for applications other than in automotive engines.

   High-shear viscosity measured at 150 °C and 10^6 s^{-1} and reported in centipoise (mPa·s) is widely accepted as a rheological parameter which is relevant to high-temperature engine performance. In particular, it is generally believed to be indicative of the effective oil viscosity in high-shear components of an internal combustion engine (for example, within the journal bearings and between the rings and cylinder walls) under severe operating conditions. While the specific temperature and shear rate conditions experienced by an oil in a particular application depend on mechanical design and operating parameters, the measurement conditions specified in Table 1 are representative of a wide range of engine operating conditions.

   Many commercial engine oils contain polymeric additives for a variety of purposes, one of the most important of which is viscosity modification. Specifically, the use of such additives in creating multigrade oils is commonplace. However, oils containing a significant polymeric additive concentration, whether for viscosity modification or another lubricant function, are generally characterized by having a non-Newtonian, "shear thinning" viscosity (i.e., a viscosity which decreases with increasing shear rate).

   To insure that polymer-containing oils do not create a situation in which the viscosity of the oil decreases to less than a specified limit, minimum values of high-shear viscosity are assigned to each of the non-W viscosity grades in Table 1. A special situation exists regarding the SAE 40 grade. Historically, SAE 0W-40, 5W-40, and 10W-40 oils have been used primarily in light-duty engines. These multigrade SAE 40 oils must meet a minimum high-temperature, high-shear viscosity limit of 2.9 cP.

   In contrast, SAE 15W-40, 20W-40, 25W-40, and 40 oils have typically been used in heavy-duty engines. The manufacturers of such engines have required high-shear viscosity limits consistent with good engine durability in high-load, severe service applications. Thus, SAE 15W-40, 20W-40, 25W-40, and single-grade 40 oils must meet a minimum high-temperature, high-shear viscosity limit of 3.7 cP.

   Acceptable methods for measuring high-temperature, high-shear rate viscosities are ASTM D 4683 and CEC L-36-A-90 (or equivalent method ASTM D 4741).

6. ***Labeling***—In properly describing the viscosity grade of an engine oil according to this document, the letters "SAE" must precede the grade number designation. In addition, for multigrade oil formulations this document requires that the W grade precede the non-W grade, and that the two grades be separated by a hyphen (i.e., SAE 10W-30). Other forms of punctuation or separation are not acceptable.

SAE J300 Revised DEC95

Most oils will meet the viscosity requirements of at least one of the W grades. Nevertheless, consistent with historic practice, any Newtonian oil may be labeled as a single-grade oil (either with or without a W). Oils which are formulated with polymeric viscosity index improvers for the purpose of making them multiviscosity-grade products are non-Newtonian and must be labeled with the appropriate multiviscosity grade (both W and high-temperature grade). Since each W grade is defined on the basis of maximum cranking and pumping viscosities as well as minimum kinematic viscosities at 100 °C, it is possible for an oil to satisfy the requirements of more than one W grade. In labeling either a W grade or a multiviscosity grade oil, only the lowest W grade satisfied may be referred to on the label. Thus, an oil meeting the requirements for SAE grades 10W, 15W, 20W, 25W, and 30 must be referred to as an SAE 10W-30 grade only.

The intent of the low-temperature portion of SAE J300 is to insure that if oil viscosity is sufficiently low for an engine to crank, the viscosity must also be low enough that the oil will flow after the engine starts. Accordingly, the cranking viscosity is the primary criterion for establishing the W grade. Specifically, an oil must meet the pumping viscosity requirement of the lowest W grade satisfied by the cranking viscosity. If the W grade defined by the pumping viscosity is higher than the lowest grade satisfied by the cranking viscosity, the oil does not meet the requirements of this document and is, therefore, inappropriate for use.

Similarly, the intent of the kinematic viscosity limits for each W grade is to insure that the viscosities of these oils are high enough at engine operating temperatures to provide adequate protection. Thus, if the kinematic viscosity at 100 °C does not meet the requirements of the lowest W grade satisfied by the cranking viscosity, then the oil does not meet the requirements of this document and is, therefore, inappropriate for use.

Some engine oils are prediluted, usually to assist in mixing with fuel when used in certain two-stroke-cycle engines. If any viscosity grade in SAE J300 is used to describe a prediluted engine oil, the grade indicated should relate to the viscosity of the oil in its undiluted state. In displaying SAE J300 viscosity grades of prediluted oils, containers should indicate that the SAE grade applies to the oil in its undiluted state.

More accurately, the rheological properties of two-stroke-cycle engine oils should be identified using the terminology and grades described in SAE J1536. Further information on prediluted oils is also provided in SAE J1510.

7. Notes

7.1 Marginal Indicia—The (R) is for the convenience of the user in locating areas where technical revisions have been made to the previous issue of the report. If the symbol is next to the report title, it indicates a complete revision of the report.

PREPARED BY THE SAE FUELS AND LUBRICANTS TECHNICAL COMMITTEE 1—
PASSENGER CAR TYPE ENGINE OILS

SAE J300 Revised DEC95

APPENDIX A
TEST METHOD FOR APPARENT VISCOSITY OF ENGINE OILS BETWEEN –40 AND 0 °C USING THE COLD CRANKING SIMULATOR

The test procedure formally described in this appendix has been standardized in ASTM D 5293—Standard Test Method for Apparent Viscosity of Engine Oils Between –30 and –5 °C Using the Cold-Cranking Simulator.

SAE J300 Revised DEC95

APPENDIX B
TEST METHOD FOR STABLE POUR POINT OF ENGINE OILS

Use ASTM D 4684 for the Determination of the Pumpability Viscosity Requirements in Table 1.

When formulating new engine oils or when there are significant changes in base oils or additives, the following stable pour point test method is suggested to check the characteristics of formulated engine oils using as formulation guidelines the previously established limits of –35 °C, max, for SAE 5W oils and –30 °C, max, for SAE 10W oils.

B.1 Scope

B.1.1 The test for stable pour point is primarily intended for use with engine lubricating oils. The potential for applicability to other lubricants is unknown.

B.2 Summary of Method

B.2.1 After preliminary warming, the sample is subjected to a controlled temperature/time cycle over five and one-half to seven days. The cycle was originally established to reproduce pour instability or reversion which has occurred during storage of oils in moderately cold cyclic ambient conditions. More recent work has shown relevance to engine oil pumpability failure. Oils exhibiting pour reversion are essentially "solid" resulting from wax gel formation, at temperatures significantly above their ASTM D 97 pour points.

NOTE—Refer to: McNab, Rodgers, Michaels, and Hodgers, "The Pour Stability Characteristics of Winter Grade Motor Oils," Quarterly Transaction, Society of Automotive Engineers, Inc., Vol. 2, No. 1, p. 34, January 1948; Hodges and Rodgers, "Some New Aspects of Pour Depressant Treated Oils," Oil and Gas Journal, October 4, 1947, p. 89.

B.3 Definition

B.3.1 Pour Stability Temperature—That specified temperature at which oil remains fluid on completion of an established temperature/time cycle.

B.3.2 Stable Pour Point—The lowest temperature at which oil remains fluid when subjected to the specified temperature/time cycle.

B.4 Apparatus

B.4.1 Test Jar—Identical to ASTM D 97 and D 2500 pour point/cloud point test jar.

B.4.2 Thermometer—ASTM E 1 6C with temperature range of +20 to –80 °C.

B.4.3 Cork or Rubber Stoppers—To fit test jar.

B.4.4 Any equipment suitable to heat sample uniformly to precondition test samples.

B.4.5 Cooling Bath—Low temperature with controller to follow temperature/time cycles from +15 to –45 °C. Spacing between test jars is to be about 15 mm with jars suspended so that cooling medium circulates around bottom and sides of jar.

B.4.6 Temperature Recorder—Two channels to record temperatures of bath and sample.

SAE J300 Revised DEC95

B.5 Procedure

B.5.1 Adjust cooling bath temperature to 15 °C with one temperature sensing bulb in the cooling bath medium.

B.5.2 Prepare two temperature measurement samples as follows:

B.5.2.1 Select a sample oil which is known to be fluid to at least –45 °C.

B.5.2.2 Fill each of two test jars with approximately 60 mL of selected oil sample. Identify these bottles as "Temperature Measurement Sample."

B.5.2.3 Prepare cork stopper to accommodate the standardized calibrated ASTM thermometer.

B.5.2.4 Insert stopper and thermometer into one jar so that thermometer immersion line is visible but not more than 3 mm above top of stopper. Place jar in center of cooling bath.

B.5.2.5 Prepare cork stopper to accommodate recorded temperature sensing bulb.

B.5.2.6 Insert stopper and one temperature sensing bulb in the second jar and position the bulb approximately 7 mm into the control oil sample. Place jar in center of cooling bath next to jar with thermometer.

B.5.2.7 Place the other temperature sensing bulb in cooling bath medium adjacent to the two control sample bottles.

B.5.3 Prepare samples of test oils using clean, dust-free test jars. Fill jar with about 60 mL of test oil.

B.5.4 Pretreat the test oil samples.

B.5.4.1 Heat sample in such a way as to maintain oil temperature at 80 °C for 2 h.

B.5.4.2 After allowing sample to cool to room temperature, stopper the jar with a clean, solid cork or rubber stopper.

B.5.4.3 Place test sample jars in cooling bath adjacent to control sample jars. All samples must be at same level if liquid bath is used.

B.5.5 Prepare bath for cyclic temperature test.

B.5.5.1 Temperature of bath should be 15 °C. Check thermometer and recorded temperature of temperature measurement sample.

B.5.5.2 If liquid bath is used, adjust level in bath to slightly above sample level in test jars.

B.5.5.3 Initiate the temperature cycle as indicated in Tables B1 and B2.

B.5.6 During the final cool down, check proper temperature control each day as follows:

B.5.6.1 Read the "Temperature Measurement Sample" thermometer. Return this sample to the center of the bath.

B.5.6.2 Compare this temperature with the recorded temperature.

B.5.6.3 Determine whether a correction is required in the reading of recorded temperature. Estimate the correct time to make the first pour stability determination at the correct thermometer temperature (±1 °C).

SAE J300 Revised DEC95

TABLE B1—CYCLE C, SOFT METRICATION

Total Time, h	Time, h	Direction and Temp., °C
0	0	Set at 15 °C
15	15	Down to −22 °C
17	2	Down to −23 °C
19	2	Up to −21 °C
21	2	Up to −18 °C
26	5	Up to −14 °C
31	5	Up to −12 °C
34	3	Up to −11 °C
50	16	At −11 °C
60	10	Up to 0 °C
62	2	At 0 °C
63	1	Down to −1 °C
66	3	Down to −3 °C
69	3	Down to −4 °C
73	4	Down to −5 °C
91	18	At −5 °C
94	3	Down to −6 °C
96	2	Down to −7 °C
168	72	Down to −41 °C

SAE J300 Revised DEC95

TABLE B2—CYCLE C, READING TIMES ON FINAL DROP

Approx. h to Test	Temperature, °C
91	-5
98	-8
106	–12
113	–15
119	–18
126	–21
132	–24
138	–27
145	–30
152	-33
158	–36
164	–39
168	–41

B.5.7 The stable pour point is determined during the final cool down in the temperature/time cycle as follows:

B.5.7.1 At the sample temperature of –12 °C, carefully remove the test jar vertically from the bath and carefully tilt only enough to ascertain whether the oil surface moves and is "fluid." If movement is detected while tilting, return the bottle to a vertical position and carefully replace in bath. Total time for this operation shall be less than 3 s. Use care in handling jars in and out of bath. Shaking can cause a change in the onset or rate of gelation. Handle jars by cork end only. If frosting occurs, wipe with a rag to prevent heating of sample.

B.5.7.2 If no movement of the oil is detected when the jar is tilted to 90 degrees (horizontal) for 5 s, the sample is "solid." Record the reading of the temperature measurement sample thermometer.

B.5.7.3 For oils which remain fluid, repeat step B.5.7.1 at successively lower temperatures, in 3 °C increments, until no movement of the oil is detected and the oil is "solid" by B.5.7.2, or until temperature cycle is complete.

B.6 Report

B.6.1 Report stable pour point as 3 °C higher than the temperature recorded in B.5.7.2. If the sample is still fluid at –41 °C, report stable pour point as less than or equal to –41 °C.

Appendix 10

THE ILSAC* MINIMUM PERFORMANCE STANDARD FOR PASSENGER CAR ENGINE OILS

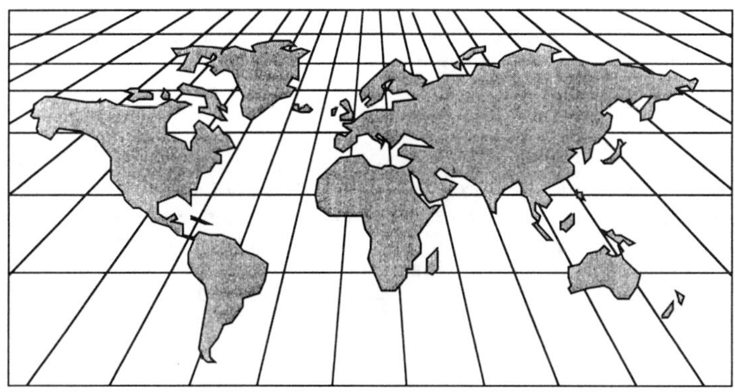

*International Lubricant Standardization and Approval Committee

Officially Revised By:

Japan Automobile Manufacturers Association, Inc.
and
Motor Vehicle Manufacturers Association of the United States, Inc.

October 12, 1992

THE ILSAC MINIMUM PERFORMANCE STANDARD FOR PASSENGER CAR ENGINE OILS GF-1

INTRODUCTION

The Motor Vehicle Manufacturers Association of the United States, Inc. (MVMA) and the Japan Automobile Manufacturers Association, Inc. (JAMA), through an organization called the International Lubricant Standardization and Approval Committee (ILSAC), jointly developed and approved a minimum performance standard for gasoline-fueled passenger car engine oils.

This standard includes only the performance requirements and chemical and physical properties of those engine oils that vehicle manufacturers may deem necessary for satisfactory equipment life and performance. It is the oil marketer's responsibility to be aware of and comply with all applicable legal and regulatory requirements on substance use restrictions, labeling, and health and safety information, and to conduct its business in a manner which represents minimum risk to consumers and the environment.

This ILSAC minimum performance standard, including all of the additional requirements outlined in section four, comprises the latest standard for passenger car engine oils. Diesel engine oils are not covered in this specification, but may be the topic of future discussions between ILSAC and groups representing diesel engine builders.

SUMMARY

The ILSAC standard is composed of five parts. The first section on viscosity uses the Society of Automotive Engineers (SAE) Engine Oil Viscosity Classification, SAEJ300. The second section encompasses the American Petroleum Institute (API) SH performance requirements. The third section contains specifications for bench test performance parameters, such as volatility, foaming tendency, high-temperature, high-shear rate viscosity, filterability, etc. The fourth section contains additional requirements including fuel efficiency, catalyst compatibility, and low-temperature viscosity. Key reference documents are listed in the final section.

The truest evaluation of an engine oil product is satisfactory performance in a variety of vehicle fleet tests which simulate the full range of customer driving conditions. The engine sequence tests listed in this document have been specified instead of fleet testing to minimize testing time and costs. This simplification of test requirements is only possible because the specified engine sequence tests have been correlated to a variety of vehicle tests.

The correlations between engine sequence tests and fleet tests are judged valid based only on the range of base oils, refining processes and additive technologies which have demonstrated satisfactory performance in widespread use at the time this standard was first issued October 22, 1990 and revised October 12, 1992. The introduction of base oils, refining processes or additive technologies which constitute a significant departure from existing practice would require supporting fleet test data and appropriate ASTM engine tests to validate the correlation between the fleet tests and engine sequence tests for that different base oil, refining process, or additive technology. This fleet testing would be in addition to the other requirements listed in this specification.

It is the responsibility of any individual or organization introducing a new technology which they claim will provide equivalent or better performance to ensure their engine test results still correlate with customer field service. Also, the marketer must ensure there is no adverse effect to vehicle components or emission control systems. No marketer can claim to be acting in a reasonable and prudent manner if the marketer knowingly uses a new technology based only on the results of engine sequence testing without verifying suitability in vehicle fleet testing which simulates the full range of customer operation.

MINIMUM PERFORMANCE STANDARD

The revised ILSAC GF-1 minimum performance standard is shown in Table 1.

Section 1

The first section of the standard deals with viscosity. It utilizes the most widely-accepted definition of viscosity, SAE Standard J300 (1). Table 1 specifies the latest revision of this document, in order to keep the ILSAC standard current.

Section 2

The second section of the standard defines ASTM engine tests and corresponding requirements used to define API SH category engine oil performance (2,3). The American Society for Testing and Materials (ASTM) Sequence IID test is used to define the low-temperature rust and corrosion protection provided by engine oils. High-temperature valve train wear, oil thickening, and deposits are evaluated in the ASTM Sequence IIIE test. Low- to medium-temperature sludge and wear are determined in the ASTM Sequence VE test. The L-38 test method defines the bearing corrosion protection provided by engine oils. The 1H2 or 1G2 test which defined piston cleanliness was dropped from the October 22, 1990 version of this standard because of concern over interpretation of test results. A replacement test is being sought to evaluate high temperature deposit formation.

Section 3

The bench test requirements are outlined in Section 3. High-temperature, high-shear-rate viscosity provides an estimate of bearing oil film thickness and, thus, relates to bearing life (4). A value of 2.9 mPa•s at 150°C and 1 million sec^{-1} is considered by MVMA and JAMA members to provide adequate assurance of bearing durability in passenger car engines.

Volatility, as measured by either the Noack or ASTM simulated distillation method, is included in the standard because volatility has been shown to correlate with oil consumption in the field (5, 6). The values were selected to provide acceptable oil economy in the field. The higher allowable volatility values specified for the lighter viscosity grade oils are an acknowledgement of the difficulties encountered with existing refining equipment when manufacturing the lighter base stocks necessary for such oils. There is a real need to improve this limit over time, and base oil manufacturers should make plans to modify equipment to satisfy future requirements which will likely be more stringent.

A filterability test is incorporated in the standard in order to ensure the water tolerance of oils under low-temperature conditions. The limits in the General Motors Engine Oil Filterability Test (GM 9099P) correspond to GM's and Ford's initial fill requirements. ASTM has been requested to standardize this test and to consider having the ASTM Test Monitoring Center handle distribution of reference oils and filter paper. This would provide worldwide availability of the test method and test materials.

ASTM Foam Test (D 892) limits similar to Ford and General Motors'initial-fill and U.S. military specifications are incorporated in the ILSAC standard to ensure that foaming will not be a problem in current and future engines, which tend to run at higher speeds and sometimes incorporate balance shafts, both of which can promote foaming. The Sequence IV portion of this test, although not formally part of the ASTM procedure yet, is believed to correlate better with foaming under high-speed engine operating conditions. The intent of including the Sequence IV portion of this test as a report-only item is to gather data on this procedure so that, after it has been included in the D 892 standard, it can be added to the ILSAC standard with an appropriate maximum acceptable limit.

Two alternative flash point methods are also included in the standard, primarily to cover safety and materials handling concerns.

A shear stability requirement for the 10-hour oil sample from the L-38 test to remain within the original SAE viscosity grade is also included. An investigation into alternative shear stability methods will be conducted for possible use in future standards.

ILSAC(GF-1) 12Oct92

Requirements for homogeneity and miscibility are included in the standard primarily as quality control checks, to ensure that the oil is blended properly (i.e., that the additives have not settled out).

Section 4

Section 4 of the ILSAC standard incorporates additional requirements. All three of the additional requirements listed in Section 4 must be met in order for an oil to satisfy the licensing requirements of the ILSAC certification mark in the American Petroleum Institute Engine Oil Licensing and Certification System (EOLCS). The fuel efficiency requirement is extremely important, since use of engine oils providing at least a 2.7% fuel economy improvement in the ASTM Sequence VI test could provide substantial fuel savings.

No currently-acceptable standard test exists for determining the catalyst poisoning effect of engine oils. In the absence of such a test, and since it has been shown that engine-oil-derived phosphorus poisons emission control devices (7), it is believed prudent to limit the phosphorus content of the engine oil to 0.12 mass percent maximum.

The last portion of Section 4 of the standard deals with the low-temperature viscosity of engine oils, as defined by SAE J300 (1). The low-temperature viscometric properties of multiviscosity grade engine oils are important as they relate to cold starting performance in gasoline-fueled passenger cars.

Section 5

Section 5 of the standard references procedures for conducting the tests included in the standard.

REFERENCES

1. SAE Standard, Engine Oil Viscosity Classification - SAE J300, (latest edition) SAE Handbook.

2. API Publication 1509, "Engine Service Classification and Guide to Crankcase Oil Selection," 12th Edition, January 1993, American Petroleum Institute.

3. ASTM Standard D 4485, "Standard Performance Specification For Automotive Engine Oils," ASTM (latest edition).

4. J. A. Spearot, C. K. Murphy, A. K. Deysarkar, "Interpreting Experimental Bearing Oil Film Thickness Data," SAE Paper No. 892151.

5. F. E. Didot, E. Green, R. H. Johnson, "Volatility and Oil Consumption of SAE 5W-30 Engine Oil," SAE Paper No. 872126.

6. L. R. Carey, D. C. Roberts, H. Shaub, "Factors Influencing Engine Oil Consumption in Today's Automotive Engines," SAE Paper No. 892159.

7. SAE Fuels and Lubricants Technical Committee 1 Engine Oil/Catalyst and Oxygen Sensor Compatibility Task Force Status Report, October 1985.

ILSAC(GF-1) 12Oct92

TABLE 1.

ILSAC PASSENGER CAR ENGINE OIL MINIMUM PERFORMANCE STANDARD

1. VISCOSITY REQUIREMENTS (A)
as defined by the latest revision
of SAE Standard J300

2. ENGINE TEST REQUIREMENTS - ASTM D 4485 (B)

2a. Engine Rusting

	ASTM Sequence IID Test (C)
Average Rust Rating	8.5 minimum
Stuck Lifters	none

2b. Wear and Oil Thickening

	ASTM Sequence IIIE Test (D)
Increase in Viscosity at 40°C	375% maximum
Piston Skirt Varnish	8.9 minimum
Ring Land Deposits	3.5 minimum
Average Engine Sludge	9.2 minimum
Stuck Piston Rings	no oil related
Cam and Lifter Wear	
Average, μm	30 maximum
Maximum, μm	64 maximum
Oil Consumption, l	5.1 maximum

2c. Sludge and Wear

	ASTM Sequence VE Test (E)
Average Engine Sludge	9.0 minimum
Rocker Cover Sludge	7.0 minimum
Average Engine Varnish	5.0 minimum
Piston Skirt Varnish	6.5 minimum
Cam Wear,	
Average, μm	130 maximum
Maximum, μm	380 maximum
Oil Ring Clogging	15% maximum
Oil Screen Clogging	20% maximum
Hot Stuck Rings	none

ILSAC(GF-1) 12Oct92

2d. Bearing Corrosion

	Test Method L-38 (F)
Bearing Weight Loss, mg	40 maximum
Piston Skirt Varnish	9.0 minimum

3. BENCH TEST REQUIREMENTS

3a. HTHS Viscosity at 150°C and 10^6 s^{-1} (ASTM D 4683, ASTM D 4741, or CEC L-36-A-90)(B)

For all viscosity grades: 2.9 mPa•s minimum

3b. Volatility Either:

	Sim. Dis.(ASTM D 2887)(B) or	Evap. Loss (CEC L-40-7-87)(G)
SAE 0W- and 5W- multi-viscosity grades:	20% maximum at 371°C	25% maximum, 1 h at 250°C
All other SAE viscosity grades:	17% maximum at 371°C	20% maximum, 1 h at 250°C

3c. Filterability

GM EOFT (GM 9099P) (H)
50% maximum flow reduction

3d. Foaming Tendency ASTM D 892 (Option A) (B)

	Foaming	Settling*
Sequence I:	10 ml maximum	0 ml maximum
Sequence II:	50 ml maximum	0 ml maximum
Sequence III:	10 ml maximum	0 ml maximum
Sequence IV:	report	report

* settling determined after 5 min, except Sequence IV after 5 seconds
NOTE: Sequence IV test conditions are the same as Sequence I, except 150°C and 200 ml/min flow rate

3e. Flash Point

ASTM D 93 (ISO 2719) (B) or	ASTM D 92 (B)
185°C minimum	200°C minimum

ILSAC(GF-1) 12Oct92

3f. Shear Stability

L-38 test - 10-hour stripped viscosity - must remain in original SAE viscosity grade

3g. Homogeneity and Miscibility

Federal Test Method 791B, Method 3470 (I)
Shall remain homogeneous and, when mixed with SAE reference oils, shall remain miscible.

4. ADDITIONAL REQUIREMENTS

4a. Fuel Efficiency

ASTM Sequence VI Test (J)
Improvement (EFEI): 2.7% minimum

4b. Catalyst Compatibility

Phosphorus content: 0.12 (% mass) maximum

4c. Low-Temperature Viscosity (A)

Viscosity, mPa•s, at Temperature, °C, maximum	
Cranking	Pumping
3500 at -20	30,000 at -25

5. APPLICABLE DOCUMENTS

A. SAE Standard, Engine Oil Viscosity Classification - SAE J300, SAE Handbook.

B. ASTM Annual Book of Standards, Section 5, Petroleum Products and Lubricants, ASTM, 1992.

C. ASTM STP 315H, "Multicyclinder Test Sequences for Evaluating Automotive Engine Oils," Part 1, Sequence IID Test, American Society for Testing and Materials (ASTM), September 15, 1979.

D. ASTM Research Report RR-D:2-1225, "Sequence IIIE Test", ASTM, 1988.

ILSAC(GF-1) 12Oct92

E. ASTM Research Report RR-D:2-1226, "Sequence VE Test", ASTM, 1988.

F. ASTM STP 509A, American Society for Testing and Materials.

G. Coordinating European Council (CEC) L-40-7-87.

H. General Motors Engineering Standard GM 9099P, "Engine Oil Filterability Test (EOFT)," May 1980. ASTM was formally requested (letter of 9/6/90) to develop this procedure into a standard.

I. Federal Test Method 791B, Method 3470.

J. ASTM Research Report RR-D: 2-1204, "Standard Dynamometer Test Methods for Measuring the Energy Conserving Quality of Engine Oils," ASTM, August 1985

ILSAC(GF-1) 12Oct92

CCMC Sequences (obsolete mid-1996)

CCMC European Oil Sequence for Service-Fill Oils for Gasoline Engines
Classes G4 and G5

FL/19/91
April 1991

This sequence defines the minimum quality level of a product for presentation to CCMC Members. Performance parameters other than those covered by the following tests or more stringent limits may be indicated by individual member companies.

1. Laboratory Tests				
			Values	
Properties	**Test Method**	**Unit**	**G-4**	**G-5**
1.1. VISCOSITY CHARACTERISTICS according to SAE J300 JUN87			only SAE grades 10W-X 15W-X 20W-X	only SAE grades 5W-X 10W-X
1.2. SHEAR STABILITY Viscosity after 30 cycles measured at 100°C	Bosch Injector CEC L-14-A-78	mm^2/s	XW-30 ≥ 9 XW-40 ≥ 12 XW-50 ≥ 14	Stay in grade after shear twist
1.3. HIGH SHEAR RATE HIGH TEMP. VISCOSITY T = 150°C Shear rate: $10^6 s^{-1}$	CEC L-36-T-84	mPa·s	≥ 3.5	≥ 3.5
1.4. EVAPORATIVE LOSS Max weight loss after 1 h at 250°C	CEC L-40-T-87	%	10W-X ≤ 15 15W-X ≤ 13 20W-X ≤ 13	≤ 13
1.5. OIL ELASTOMER COMPATIBILITY Max. variation of characteristics after immersion for 7 days in fresh oil without preaging	CEC L-39-T-87		ELASTOMER TYPE RE1 (Fluoro-elastomer) / RE2 (ACM) / RE3 (Silicone) / RE4 (NBR)	
Hardness DIDC		points	0/+5 / −5/+5 / −25/0 / −5/+5	
Tensile strength		%	−50/0 / −15/+10 / −30/+10 / −20/0	
Elongation rupture		%	−60/0 / −35/+10 / −20/+10 / −50/0	
Volume variation		%	0/+5 / −5/+5 / 0/+30 / −5/+5	
1.6. FOAMING TENDENCY (tendency-stability)	ASTM D 892 without option A	mL	Sequence I (24°C): 10-nil Sequence II (94°C): 50-nil Sequence III (24°C): 10-nil	

2. Engine Tests						
					Values	
	Requirements	Properties	Test Method	Unit of Measure	G-4	G-5
HIGH TEMPERATURE TESTS	2.1. CORROSION L38	Bearing weight loss max.	ASTM STP 509A	mg	40	40
	OR		or			
	PETTER W1	Bearing weight loss max.	CEC L-02-A-78	mg	25	25
	2.2. HIGH TEMPERATURE OXIDATION	Viscosity increase at 40°C max.	ASTM Research Report D-2: 1225	%	300	200
		Piston skirt varnish min.		merit	8.9	8.9
		Ring land varnish min.			3.5	3.5
		Sludge min.			9.2	9.2
		Ring sticking			none	none
		Lifter sticking			none	none
		Cam and lifter wear average		µm	< 30	< 30
		(maximum)		µm	<60	< 60
	2.3. HIGH TEMPERATURE DEPOSITS	Ring sticking		merit	Test procedure under development (PL-39) In the meantime additional requirements under 2.6	
		Piston skirt varnish min.	(1)	merit		
		Oil thickening				
		Oil consumption				
	RING STICK					
LOW AND HIGH TEMPERATURE SLUDGE & WEAR	2.4. LOW TEMPERATURE SLUDGE	Engine sludge (average min.)	ASTM Research Report D-2: 1126	merit	9	9
		Piston skirt varnish min.		merit	6.5	6.5
		Average engine varnish min.			5	5
		Oil ring clogging		%	<15	<15
		Comp. ring sticking			none	none
		Oil screen clogging		%	< 20	<20
		Avg. cam wear max.		µm	130	130
		Max. cam wear max.		µm	380	380
	2.5. VALVE TRAIN SCUFFING WEAR (TU3)	Cam wear aver.	CEC L-38-T-87 (100 H)	µm	< 15	< 15
		max. wear		µm	< 20	< 20
		Pad merit (aver. of 8 pad min.)			7.5	7.5
	2.6. BLACK SLUDGE M102E	Average engine sludge merit min.	CEC L-41-T-88	merit	> RL140	> RL140
		Ring sticking		merit	≥ RL140	≥ RL139
		Piston deposit		merit	≥ RL140	≥ RL139
		Cam wear		µm	≤ RL139	≤ RL139
		Follower wear		µm	≤ RL139	≤ RL139
	2.7. PRE-IGNITION 132	Hours without pre-ignition min.	CEC L-34-U-82	hours	80	80
	2.8. RUST TEST		ASTM II D STP 315 H-PART I	merit	8.5	8.5

(1) The present Cortina test is not adequate for today's engines. CEC is asked to develop immediately an adequate test procedure.

CCMC European Oil Sequence for Service-Fill Oils for Diesel Engines Classes D4, D5 and PD2

FL/20/91
April 1991

This sequence defines the minimum quality level of a product for presentation to CCMC Members. Performance parameters other than those covered by the following tests or more stringent limits may be indicated by individual member companies.

1. Laboratory Tests					
			Values		
			Pass. cars	Industrial vehicles	
Properties	Test Method	Unit	PD-2	D-4	D-5
1.1. VISCOSITY According to SAE J300 JUN87		-	XW-30 XW-40 XW-50	20W-20 30 40 XW-30 XW-40 XW-50	20W-20 30 40 XW-30 XW-40 XW-50
1.2. SHEAR STABILITY Viscosity after 30 cycles measured at 100°C	Bosch Injector CEC L-14-A-78	mm^2/s	XW-30 ≥ 9 XW-40 ≥ 12 XW-50 ≥ 14	XW-30 ≥ 9 XW-40 ≥ 12 XW-50 ≥ 14 no requirements for single grades	
1.3. HIGH SHEAR RATE HIGH TEMP. VISCOSITY T = 150°C Shear rate: 10^6 s^{-1}	CEC L-36-T-84	mPa·s	≥ 3.5	≥ 3.5 (≥ 3.3 for 20W-20 & 30)	
1.4. EVAPORATIVE LOSS Max weight loss after 1 h at 250°C	CEC L-40-T-87	%	10W-X ≤ 15 ≤ 13 for all others	10W-X ≤ 15 ≤ 13 for all others	10W-X ≤ 15 ≤ 13 for all others
1.5. SULFATED ASH	ASTM D 874	% (weight)	≤ 1.8	≤ 2.0	
1.6. OIL ELASTOMER COMPATIBILITY Max. variation of characteristics after immersion for 7 days in fresh oil without preaging	CEC L-39-T-87		ELASTOMER TYPE		
			RE1 (Fluoro-elastomer)	RE2 (ACM)	RE3 (Sili-cone) / RE4 (NBR)
Hardness DIDC		points	0/+5	–5/+5	–25/0 / –5/+5
Tensile strength		%	–50/0	–15/+10	–30/+10 / –20/0
Elongation rupture		%	–60/0	–35/+10	–20/+10 / –50/0
Volume variation		%	0/+5	–5/+5	0/+30 / –5/+5
1.7. FOAMING TENDENCY (tendency-stability)	ASTM D 892 without option A	mL	Sequence I (24°C): 10-nil Sequence II (94°C): 50-nil Sequence III (24°C): 10-nil		

2. Engine Tests

					Values		
					Pass. cars	**Industrial vehicles**	
	Requirements	**Properties**	**Test Method**	**Unit of Measure**	**PD-2**	**D-4**	**D-5**
HIGH TEMPERATURE TESTS	2.1. RING STICKING and PISTON CLEANLINESS VW 1.6 TC Diesel	RING STICKING PISTON CLEANLINESS	CEC L-35-T-84	Merit	≥ RL148		
	2.2. BORE POLISHING and PISTON CLEANLINESS OM 364 A	BORE POLISHING PISTON CLEANLINESS	CEC L-42-T-89	% Merit		≤ 16 ≥ 24	≤ 2 ≥ 38
LOW AND MEDIUM TEMPERATURE SLUDGE WEAR	2.3. LOW SPEED OIL THICKENING TEST (1)	Viscosity Increase	Mack T-7 ASTM RRD2-1220	cSt/h	-	< 0.04	
	2.4. LOW TEMPERATURE OIL THICKENING		to be developed (OM-602 A PL-38)		limit to be deter-mined	-	-
	2.5. RUST IID	AV/ENGINE RUST min.	ASTM 315 H Part I	Merit	8.5	8.1	8.1
	2.6. WEAR (2) OM 616	Cam Wear: average (max.) Cyl. Wear: average (max.)	CEC L-17-A-78	µm	10 (20) 5 (12)	10 (20) 10 (24)	10 (20) 5 (12)

(1) A suitable European test should be developed.

(2) 2.6 will be replaced by 2.4.

European
Automobile
Manufacturers
Association

ACEA EUROPEAN OIL SEQUENCES

1996

SERVICE FILL OILS FOR

GASOLINE ENGINES

LIGHT DUTY DIESEL ENGINES

HEAVY DUTY DIESEL ENGINES

Laboratory tests for gasoline engine oils;
Engine tests for gasoline engine oils;
Laboratory tests for light duty diesel engine oils;
Engine tests for light duty diesel engine oils;
Laboratory tests for heavy duty diesel engine oils;
Engine tests for heavy duty diesel engine oils.

ACEA
Rue du Noyer 211
B-1040 Bruxelles
Tel (32 2) 732 55 50
Fax (32 2) 732 60 01
(32 2) 732 42 67
TVA BE 444 072 631
SGB 210-0069404-04

FL/52/95
14 December 1995

ACEA	ACEA EUROPEAN OIL SEQUENCES FOR SERVICE-FILL OILS	DECEMBER 1995 Sheet 2 of 8

This document details the ACEA European Oil Sequences for Service-fill Oils for Gasoline engines, for Light Duty Diesel engines, and for Heavy Duty Diesel engines. These sequences define the minimum quality level of a product for presentation to ACEA members. Performance parameters other than those covered by the tests shown or more stringent limits may be indicated by individual member companies.

These sequences replace the CCMC sequences as a means of defining engine lubricant quality from 1 January 1996.

CONDITIONS FOR USE OF PERFORMANCE CLAIMS AGAINST THE ACEA OIL SEQUENCES

ACEA requires that any claim for oil performance to meet these sequences must be based on credible data and controlled tests in accredited test laboratories.

All engine performance testing used to support a claim of compliance with the ACEA sequences must be generated according to the European Engine Lubricants Quality Management System (EELQMS). This system, which is described in the ATIEL Code of Practice[1], addresses product development testing and product performance documentation, and involves the registration of all candidate and reference oil testing and defines the compliance process. First allowable use of ACEA performance claims will be from 1 January 1996. Oil marketers must demonstrate commitment to comply with the ATIEL Code of Practice from this date. Compliance with ATIEL Code of Practice will be mandatory for any claim of meeting ACEA sequences from 1 January 1997.

The marketer of an oil claiming to meet ACEA performance requirements is responsible for all aspects of product liability.

---

[1] The ATIEL Code of Practice is available from ATIEL (Association Technique de l'Industrie Européenne des Lubrifiants), Madou Plaza, 25th floor, Place Madou 1, B -1030 Brussels, Belgium.

ACEA	ACEA EUROPEAN OIL SEQUENCE FOR SERVICE-FILL OILS FOR GASOLINE ENGINES	December 1995 Sheet 3 of 8

This sequence defines the minimum quality level of a product for presentation to ACEA members.
Performance parameters other than those covered by the tests shown or more stringent limits may be indicated by individual member companies.

REQUIREMENTS	TEST METHOD	PROPERTIES	UNIT	LIMITS			
				ACEA : A1-96	ACEA : A2-96	ACEA : A3-96	
1. LABORATORY TESTS							
1.1 Viscosity Grades		SAE J300 March 1993		No restriction except as defined by shear stability and HT/HS requirements. Manufacturers may indicate specific viscosity requirements related to ambient temperature.			
1.2 Shear Stability	CEC-L-14-A-88 (Bosch Injector)	Viscosity after 30 cycles measured at 100°C.	mm^2/s	xW-20 : stay in grade xW-30 ≥ 8.6 xW-40 ≥ 12.0	 xW-30 ≥9.0 xW-40 ≥12.0 xW-50 ≥15.0	stay in grade	
1.3 Viscosity High Temperature High Shear Rate	CEC-L-36-A-90 (Ravenfield)	Viscosity at 150°C and 10^6 s^{-1} shear rate	mPa.s	min 2.9 max. 3.5	>3.5	>3.5	
1.4 Evaporative Loss	CEC-L-40-A-93 (Noack)	Max. weight loss after 1 h at 250°C	%	≤15	≤15 for 10W-x or lower. ≤13 for others	≤13	
1.5 Sulfated Ash	ASTM D874		% m/m	≤1.5			
1.6 Oil Elastomer Compatibility	CEC-L-39-X-95	Max. variation of characteristics after immersion for 7 days in fresh oil without pre-ageing		Elastomer type RE1	RE2	RE3	RE4
		Hardness DIDC	points	0/+5	-5/+5	-25/0	-5/+5
		Tensile strength	%	-50/0	-15/+10	-30/+10	-20/0
		Elongation at rupture	%	-60/0	-35/+10	-20/+10	-50/0
		Volume variation	%	0/+5	-5/+5	0/+30	-5/+5
1.7 Foaming Tendency (1)	ASTM D892 without option A	Tendency - stability	ml	Sequence I (24°C) 10 - nil Sequence II (94°C) 50 - nil Sequence III (24°C) 10 - nil			

(1) A foaming test at high temperatures is needed.

ACEA	ACEA EUROPEAN OIL SEQUENCE FOR SERVICE-FILL OILS FOR GASOLINE ENGINES	December 1995 Sheet 4 of 8

This sequence defines the minimum quality level of a product for presentation to ACEA members.
Performance parameters other than those covered by the tests shown or more stringent limits may be indicated by individual member companies.

REQUIREMENTS	TEST METHOD	PROPERTIES	UNIT	LIMITS		
				ACEA : A1-96	ACEA : A2-96	ACEA : A3-96
2. ENGINE TESTS						
2.1 High Temperature Oxidation	ASTM 315H Part II (Sequence IIIE)	Viscosity increase at 40°C	%	≤ 200	≤ 200	≤ 100
		Piston skirt varnish	merit	≥ 8.9	≥ 8.9	≥ 8.9
		Ring land deposits	merit	≥3.5	≥3.5	≥3.5
	Under protocol & requirements for API SH	Average sludge	merit	≥9.2	≥9.2	≥9.2
		Ring sticking		none	none	none
		Lifter sticking		none	none	none
		Cam & Lifter wear, ave.	µm	≤ 30	≤ 30	≤ 30
		Cam & Lifter wear, max.	µm	≤ 60	≤ 60	≤ 60
		Oil consumption	l	≤ 5.1	≤ 5.1	≤ 5.1
2.2 High Temperature Deposits Ring Sticking Oil Thickening	CEC-L-55-T-95 (TU3M)	Ring sticking (each part)	merit	≥ 9.0	≥ 9.0	≥ 9.0
		Piston varnish (7 elements)	merit	≥ 60	≥ 60	≥ 65
		Viscosity increase at 40°C	%	≤ 40	≤ 40	≤ 40
2.3 Low Temperature Sludge	ASTM 315H Part III (Sequence VE)	Average engine sludge	merit	≥ 9.0	≥ 9.0	≥ 9.0
		Cam cover sludge	merit	≥ 7.0	≥ 7.0	≥ 7.0
		Ave. piston skirt varnish	merit	≥ 6.5	≥ 6.5	≥ 6.5
	Under protocol & requirements for API SH	Average engine varnish	merit	≥ 5.0	≥ 5.0	≥ 5.0
		Comp. ring (hot stuck)		none	none	none
		Oil screen clogging	%	≤ 20	≤ 20	≤ 20
		Cam wear, average	µm	≤ 130	≤ 130	≤ 130
		Cam wear, max.	µm	≤ 380	≤ 380	≤ 380
2.4 Valve Train Scuffing Wear	CEC-L-38-A-94 (TU3M)	Cam wear, average	µm	≤ 10	≤ 10	≤ 10
		Cam wear, max.	µm	≤ 15	≤ 15	≤ 15
		Pad merit (ave. of 8 pads)	merit	≥ 7.5	≥ 7.5	≥ 7.5
2.5 Black Sludge	CEC-L-53-T-95 (M111)	Engine sludge, average	merit	≥ RL140	≥ RL140	≥ RL140
2.6 Fuel Economy (1)	CEC-L-54-X-94 (M111)	Fuel economy vs. Reference Oil RL191 (15W-40)	%			

(1) To be implemented when test achieves 'T' status and limits are defined.
Limits for A1 are expected to be significantly better than reference oil, and A3 to be no worse than reference oil.

Note: Lubricant performance related to fuel economy and emissions are growing issues and should be addressed by CEC-test procedures.

ACEA	ACEA EUROPEAN OIL SEQUENCE FOR SERVICE-FILL OILS FOR LIGHT DUTY DIESEL ENGINES	December 1995 Sheet 5 of 8

This sequence defines the minimum quality level of a product for presentation to ACEA members.
Performance parameters other than those covered by the tests shown or more stringent limits may be indicated by individual member companies.

REQUIREMENTS	TEST METHOD	PROPERTIES	UNIT	LIMITS			
				ACEA : B1-96	ACEA : B2-96	ACEA : B3-96	
1. LABORATORY TESTS							
1.1 Viscosity Grades		SAE J300 March 1993		No restriction except as defined by shear stability and HT/HS requirements. Manufacturers may indicate specific viscosity requirements related to ambient temperature.			
1.2 Shear Stability	CEC-L-14-A-88 (Bosch Injector)	Viscosity after 30 cycles measured at 100°C.	mm^2/s	xW-20 : stay in grade xW-30 ≥ 8.6 xW-40 ≥ 12.0	xW-30 ≥9.0 xW-40 ≥12.0 xW-50 ≥15.0	stay in grade	
1.3 Viscosity High Temperature High Shear Rate	CEC-L-36-A-90 (Ravenfield)	Viscosity at 150°C and 10^6 s^{-1} shear rate	mPa.s	min 2.9 max. 3.5	>3.5	>3.5	
1.4 Evaporative Loss	CEC-L-40-A-93 (Noack)	Max. weight loss after 1 h at 250°C	%	≤15	≤15 for 10W-x or lower. ≤13 for others	≤13	
1.5 Sulfated Ash	ASTM D874		% m/m	≤1.8			
1.6 Oil Elastomer Compatibility	CEC-L-39-X-95	Max. variation of characteristics after immersion for 7 days in fresh oil without pre-ageing		Elastomer type			
				RE1	RE2	RE3	RE4
		Hardness DIDC	points	0/+5	-5/+5	-25/0	-5/+5
		Tensile strength	%	-50/0	-15/+10	-30/+10	-20/0
		Elongation rupture	%	-60/0	-35/+10	-20/+10	-50/0
		Volume variation	%	0/+5	-5/+5	0/+30	-5/+5
1.7 Foaming Tendency (1)	ASTM D892 without option A	Tendency - stability	ml	Sequence I (24°C) 10 - nil Sequence II (94°C) 50 - nil Sequence III (24°C) 10 - nil			

(1) A foaming test at high temperatures is needed.

ACEA	ACEA EUROPEAN OIL SEQUENCE FOR SERVICE-FILL OILS FOR LIGHT DUTY DIESEL ENGINES	December 1995 Sheet 6 of 8

This sequence defines the minimum quality level of a product for presentation to ACEA members.
Performance parameters other than those covered by the tests shown or more stringent limits may be indicated by individual member companies.

REQUIREMENTS	TEST METHOD	PROPERTIES	UNIT	LIMITS		
				ACEA : B1-96	ACEA : B2-96	ACEA : B3-96
2. ENGINE TESTS						
2.1 Ring Sticking and Piston Cleanliness	CEC L-46-T-93 (VW 1.6 TC D)	Ring sticking	merit	≥ RL 148	≥ RL 148	≥ RL 148
		Piston cleanliness	merit	≥ RL 148	≥ RL 148	≥ RL 148
2.2 Medium Temperature Dispersivity	CEC-L-56-T-95 (XUD11ATE)	Viscosity increase at 100°C and 3% soot	%	≤ 200	≤ 200	≤ 125
		Piston merit (5 elements)	merit	≥ 43	≥ 43	≥ 46
2.3 Wear	CEC-L-51-T-95 (OM602A)	Cam wear	µm	≤ RL 148	≤ RL 148	≤ 50% of RL 148
		Viscosity increase at 40°C	%	Rate & report	Rate & report	Rate & report
		Bore polishing	%	Rate & report	Rate & report	Rate & report
		Piston cleanliness	merit	Rate & report	Rate & report	Rate & report
		Average engine sludge	merit	Rate & report	Rate & report	Rate & report
		Cylinder wear	µm	Rate & report	Rate & report	Rate & report
		Oil consumption	kg/test	Rate & report	Rate & report	Rate & report

Note: Lubricant performance related to fuel economy and emissions are growing issues and should be addressed by CEC-test procedures.

ACEA	ACEA EUROPEAN OIL SEQUENCE FOR SERVICE-FILL OILS FOR HEAVY DUTY DIESEL ENGINES	December 1995 Sheet 7 of 8

This sequence defines the minimum quality level of a product for presentation to ACEA members.
Performance parameters other than those covered by the tests shown or more stringent limits may be indicated by individual member companies.

REQUIREMENTS	TEST METHOD	PROPERTIES	UNIT	LIMITS				
				ACEA : E1-96	**ACEA : E2-96**	**ACEA : E3-96**		1)
1. LABORATORY TESTS								
1.1 Viscosity Grades		SAE J300 March 1993		No restriction except as defined by shear stability and HT/HS requirements. Manufacturers may indicate specific viscosity requirements related to ambient temperature.				
1.2 Shear Stability	CEC-L-14-A-88 (Bosch Injector)	Viscosity after 30 cycles measured at 100°C.	mm^2/s	xW-30 ≥ 9.0 xW-40 ≥12.0 xW-50 ≥15.0 No requirements for single grades				
1.3 Viscosity High Temperature High Shear Rate	CEC-L-36-A-90 (Ravenfield)	Viscosity at 150°C and 10^6 s^{-1} shear rate	mPa.s	>3.5				
1.4 Evaporative Loss	CEC-L-40-A-93 (Noack)	Max. weight loss after 1 h at 250°C	%	≤13				
1.5 Sulfated Ash	ASTM D874		% m/m	≤2.0				
1.6 Oil Elastomer Compatibility	CEC-L-39-X-95	Max. variation of characteristics after immersion for 7 days in fresh oil without pre-ageing		Elastomer type				
				RE1	RE2	RE3	RE4	
		Hardness DIDC	points	0/+5	-5/+5	-25/0	-5/+5	
		Tensile strength	%	-50/0	-15/+10	-30/+10	-20/0	
		Elongation rupture	%	-60/0	-35/+10	-20/+10	-50/0	
		Volume variation	%	0/+5	-5/+5	0/+30	-5/+5	
1.7 Foaming Tendency (2)	ASTM D892 without option A	Tendency - stability	ml	Sequence I (24°C) 10 - nil Sequence II (94°C) 50 - nil Sequence III (24°C) 10 - nil				
1.8 Turbocharger / Intercooler Deposits (3)								

(1) A new additional test sequence defining higher oil performance is under development. Engine performance will be based on OM 441 LA test (CEC L-52-X-94)
(2) A foaming test at high temperatures is needed.
(3) Test procedure and limits are urgently needed.

ACEA	ACEA EUROPEAN OIL SEQUENCE FOR SERVICE-FILL OILS FOR HEAVY DUTY DIESEL ENGINES	December 1995 Sheet 8 of 8

This sequence defines the minimum quality level of a product for presentation to ACEA members.
Performance parameters other than those covered by the tests shown or more stringent limits may be indicated by individual member companies.

REQUIREMENTS	TEST METHOD	PROPERTIES	UNIT	LIMITS			
				ACEA : E1-96	ACEA : E2-96	ACEA : E3-96	1)
2. ENGINE TESTS							
2.1 Bore Polishing / Piston Cleanliness	CEC L-42-A-92 (OM 364 A)	Bore polishing	%	≤ 14.0	≤ 8.0	≤ 2.5	
		Piston cleanliness	merit	≥ 24	≥ 31	≥ 35	
		Average cylinder wear	μm	≤ 8	≤ 7	≤ 6	
		Sludge	merit	≥ 9.0	≥ 9.0	≥ 9.5	
		Oil consumption	kg/test	≤ 25.0	≤ 18.0	≤ 12.0	
2.2 Wear	CEC-L-51-T-95 (OM602A)	Cam wear	μm	≤ RL 148	≤ RL 148	≤ RL 148	
		Viscosity increase at 40°C	%	Rate & report	Rate & report	Rate & report	
		Bore polishing	%	Rate & report	Rate & report	Rate & report	
		Piston cleanliness	merit	Rate & report	Rate & report	Rate & report	
		Average engine sludge	merit	Rate & report	Rate & report	Rate & report	
		Cylinder wear	μm	Rate & report	Rate & report	Rate & report	
		Oil consumption	kg/test	Rate & report	Rate & report	Rate & report	
2.3 Soot in Oil	ASTM D 44 85 (Mack T-8)	Viscosity increase at 3.8% soot :					
		1st test	cSt	——	Rate & report	≤ 11.5	
		2 test ave	cSt	——	Rate & report	≤ 12.5	
		3 test ave.	cSt	——	Rate & report	≤ 13.0	
		Filter plugging, Diff. pressure	kPa	——	Rate & report	≤ 138	
		Oil consumption.	g/kWh	——	Rate & report	≤ 0.304	

(1) A new additional test sequence defining higher oil performance is under development.
Engine performance will be based on OM 441 LA test (CEC L-52-X-94)

Note: Lubricant performance related to fuel economy and emissions are growing issues and should be addressed by CEC-test procedures.

Appendix 11

INTERNATIONAL

400 Commonwealth Drive, Warrendale, PA 15096-0001

SURFACE VEHICLE STANDARD

Submitted for recognition as an American National Standard

SAE J306 | REAF. OCT91

Issued 1924-02
Reaffirmed 1991-10

Reaffirming J306

AXLE AND MANUAL TRANSMISSION LUBRICANT VISCOSITY CLASSIFICATION

Foreword—This reaffirmed document has been changed only to reflect the new SAE Technical Standards Board format.

1. ***Scope***—This SAE Standard is intended for equipment manufacturers in defining and recommending axle and manual transmission lubricants, for oil marketers in labeling such lubricants with respect to viscosity, and for users in following their owner's manual recommendations. The SAE viscosity grades shown in Table 1 constitute a classification for axle and transmission lubricants in terms of viscosity only; the change in viscosity with use, or other gear lubricant qualities, are not considered.

TABLE 1 - AXLE AND MANUAL TRANSMISSION LUBRICANT VISCOSITY CLASSIFICATION

SAE Viscosity Grade	Maximum Temperature for Viscosity of 150 000 cP[1] °C	Viscosity at 100 °C cSt[2] Minimum	Viscosity at 100 °C cSt[2] Maximum
70W	-55[3]	4.1	—
75W	-40	4.1	—
80W	-26	7.0	—
85W	-12	11.0	—
90	—	13.5	<24.0
140	—	24.0	<41.0
250	—	41.0	—

1 Centipoise (cP) is the customary absolute viscosity unit and is numerically equal to the corresponding SI unit of millipascal-second (mPa·s).
2 Centistokes (cSt) is the customary kinematic viscosity unit and is numerically equal to the corresponding SI unit of square millimeter per second (mm^2/s).
3 The precision of ASTM Method D 2983 has not been established for determinations made at temperatures below −40 °C; consequently, this fact should be realized in any producer-consumer relationship. It is expected that ASTM will shortly undertake work in the range down to −55 °C for ASTM D 2983.

2. ***References***

2.1 Applicable Documents—The following publications form a part of this specification to the extent specified herein. The latest issue of SAE publications shall apply.

2.1.1 ASTM PUBLICATIONS—Available from ASTM, 1916 Race Street, Philadelphia, PA 19103.

ASTM D 445—Method of Test for Viscosity of Transparent and Opaque Liquids

ASTM D 2983—Method of Test for Apparent Viscosity at Low Temperature Using the Brookfield Viscometer

Printed in U.S.A.

SAE J306 Reaffirmed OCT91

3. Axle and transmission lubricant SAE viscosity grades should not be confused with engine oil SAE viscosity grades. (Compare Table 1 in this report with Table 1 in SAE J300.) A gear lubricant and an engine oil having the same viscosity will have widely different SAE viscosity grade designations as defined in the two viscosity classifications. For instance, an SAE 80W gear lubricant can have the same viscosity characteristics as an SAE 20W-20 engine oil; and SAE 90 gear lubricant viscosity can be similar to that of an SAE 40 or SAE 50 engine oil.

This classification is based on the lubricant viscosity measured at both high and low temperatures. The high-temperature values are determined according to ASTM D 445, with the results reported in centistokes (cSt).[2] The low-temperature values are determined according to ASTM D 2983 and these results are reported in centipoises (cP).[1] These two viscosity units are related as follows in Equation 1:

$$\frac{\text{cP}}{\text{Density, g/cm}^3} = \text{cSt} \qquad \text{(Eq.1)}$$

NOTE—[1] and [2] refer to footnotes in Table 1.

Density is measured at the test temperature.

This relationship is valid for Newtonian fluids; it is an approximation for non-Newtonian fluids.

A multiviscosity graded lubricant, such as SAE 80W-90, meets both the low- and high-temperature requirements shown in Table 1. That is, it conforms to the SAE 80W requirement at low temperature and is in the range provided for SAE 90 at high temperature.

The selection of an axle or transmission lubricant should be based on the lowest and highest service temperatures. The multiviscosity graded lubricants may be satisfactory at both temperature extremes. The 150 000 cP viscosity value used for the definition of low-temperature properties is based on a series of tests in a specific rear axle design. These tests have shown that pinion bearing failure has occurred at viscosities higher than 150 000 cP and the Brookfield method was shown to give adequate precision at this viscosity level. However, it should be pointed out that other axle designs may tolerate higher viscosities or fail at lower viscosities. The Brookfield low-temperature viscosity curves for several gear lubricants, made with conventional petroleum base stocks, are shown in a viscosity-temperature chart in Figure 1. It must be recognized that some gear lubricants can have viscosity-temperature relationships different than those shown in this chart.

Other applications may require considerably different Brookfield viscosity limits. For example, experience has indicated that, for satisfactory ease of shifting, many manual transmissions require a lubricant viscosity not exceeding 20 000 cP at the shifting temperature.

In recommending axle and transmission lubricants by SAE viscosity grade, the following temperatures are suggested as a uniform practice: -40 °C, -26 °C, and -12 °C.

PREPARED BY THE SAE FUELS AND LUBRICANTS TECHNICAL COMMITTEE 3—GEAR LUBRICANTS

SAE J306 Reaffirmed OCT91

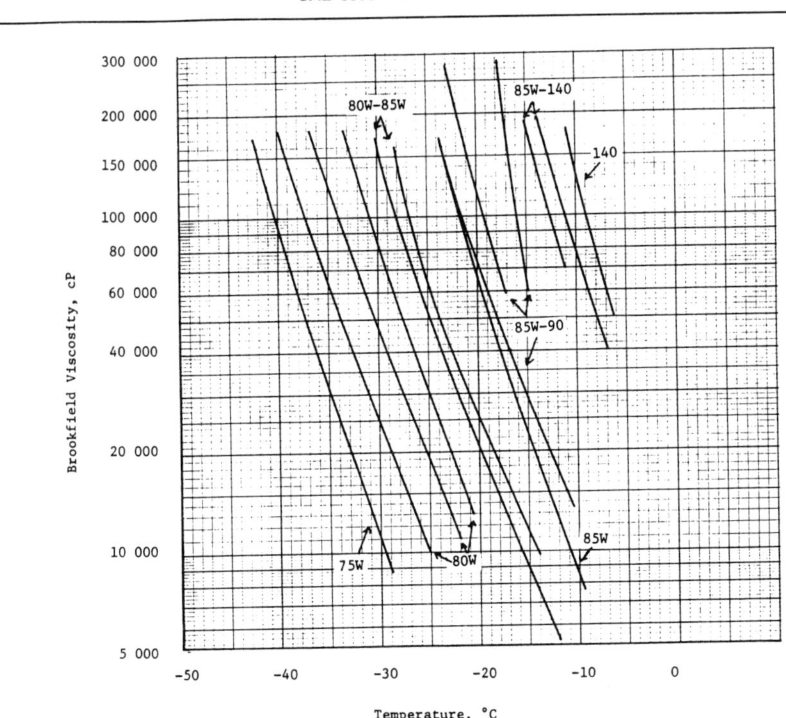

FIGURE 1—BROOKFIELD VISCOSITY VERSUS TEMPERATURE FOR TYPICAL GEAR LUBRICANTS (SAE VISCOSITY GRADES INDICATED)

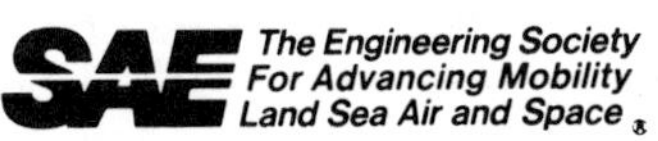

400 COMMONWEALTH DRIVE, WARRENDALE, PA 15096

FUELS AND LUBRICANTS REPORT

SAE J308

Issued February 1924
Revised June 1989
Superseding J308 JUN87

AXLE AND MANUAL TRANSMISSION LUBRICANTS

1. SCOPE:

This SAE Information Report was prepared by the SAE Fuels and Lubricants Technical Committee for two purposes: 1. to assist the users of automotive equipment in the selection of axle[1] and manual transmission lubricants for field use, and 2. to promote a uniform practice for use by marketers of lubricants and by equipment builders in identifying and recommending these lubricants by a service designation.

2. FOREWORD:

In 1943, the U.S. Army Ordnance Department (currently U.S. Army Belvoir Research, Development and Engineering Center) began qualifying gear lubricants against U.S. Army Specification 2-105. This specification has gone through several revisions and is now identified as MIL-L-2105D. The American Petroleum Institute recognizes gear lubricants meeting this latter specification as API Service GL-5 (API GL-5).

In 1977, the U.S. Army terminated direct sponsorship of the qualification process and contracted with SAE to: (1) perform the reviewing activity, and (2) make recommendations relative to the acceptance of candidate products under the military gear lubricant specification. In accordance with its contract with SAE, the U.S. Army retains sole responsibility for approving and qualifying products to its specification.

Following termination of the U.S. Army sponsorship, the SAE Board of Directors established a Lubricants Review Institute (LRI), which in turn has established an LRI Gear Lubricant Review Committee. This committee developed procedures for submitting candidate lubricants for review as well as procedures for reviewing such lubricants. The LRI activities are reviewed by SAE Legal Counsel to ensure compliance with applicable federal and state laws. The LRI Gear Lubricant Review Procedures can be obtained from SAE headquarters in Warrendale, PA.

[1]Axle(s) in this document are defined as drive axles incorporating reduction gearing and/or differential gears.

2. (Continued):

Performance Characteristics--In axles and manual transmissions, gears and bearings of different designs are employed under a variety of service conditions. Therefore, the selection of a lubricant involves careful consideration of the performance characteristics required.

3. LOAD-CARRYING CAPACITY:

One of the most important gear lubricant performance characteristics is load-carrying capacity. Some gears are operated under such loads and speeds that the very low load-carrying capacity of untreated oil[2] is adequate. However, most gears require lubricants of greater load-carrying capacity, which is provided through the use of additives.

Gear lubricants compounded to achieve increased load-carrying capacity are referred to as "extreme pressure" (EP) lubricants. However, when this term is applied to a gear lubricant, it means only that the load-carrying capacity of the lubricant is greater than that of untreated oil, with no distinction as to how much greater it may be. Therefore, to differentiate among EP lubricants of various load-carrying capacities, it is necessary to classify them further. The Coordinating Research Council (CRC) and ASTM have developed tests and the American Petroleum Institute (API) has assigned performance designations to aid in this classification.

The following designations from API Publication 1560, Lubricant Service Designation for Automotive Manual Transmissions and Axles, November, 1981, have been amended with the objective of improving user understanding of intended lubricant application.

API GL-1 Designates the type of service characteristic of manual transmissions operating under such mild conditions of low unit pressures and minimum sliding velocities, that untreated oil may be used satisfactorily. Oxidation and rust inhibitors, defoamers, and pour depressants may be used to improve the characteristics of lubricants intended for this service. Frictional modifiers and extreme pressure additives shall not be utilized.

Ø Due to speeds and loads involved, untreated oil is generally not a satisfactory lubricant for many passenger car manual transmissions[3]. For some truck and tractor manual transmissions, untreated oils may be used successfully. In all cases, the transmission manufacturers' specific lubricant recommendations should be followed.

API GL-2 Designates the type of service characteristic of automotive type worm-gear axles operating under such conditions of load, temperature, and sliding velocities, that lubricants satisfactory for API GL-1 service will not suffice.

[2]Untreated oil is defined as either refined petroleum or synthetic lubricant base oil containing no supplemental performance additives.

[3]Automatic or semiautomatic transmissions, fluid couplings, torque converters, and tractor hydraulic systems usually require special lubricants. For the proper lubricant to be used, consult the manufacturer or lubricant supplier.

3. (Continued):

Products suited for this type of service contain antiwear or very mild extreme-pressure agents which provide protection for worm gears.

API GL-3 Designates the type of service characteristic of manual transmissions and spiral-bevel axles operating under mild to moderate to severe conditions of speed and load. These service conditions require a lubricant having load-carrying capacities greater than those that will satisfy API GL-1 service, but below the requirements of lubricants satisfying the API GL-4 service.

Gear lubricants designated for API GL-3 service are not intended for hypoid gear applications.

API GL-4 Designates the type of service characteristic of spiral-bevel and hypoid[4] gears in automotive axles operated under moderate speeds and loads. These oils may be used in selected manual transmission and transaxle applications. (User should consult axle/transmission manufacturers' specific lubricant recommendations).

While this service designation is still used commercially to describe lubricants, some test equipment used for performance verification is no longer available. ASTM is investigating the possibility of redefining service designation API GL-4 using modern test equipment.

API GL-5 designates the type of service characteristic of gears, particularly hypoids in automotive axles operated under high-speed and/or low-speed, high-torque conditions. Lubricants qualified under U.S. Military specification MIL-L-2105D (formerly MIL-L-2105C) satisfy the requirements of the API GL-5 service designation. Details of the API GL-5 performance tests are contained in ASTM Publication STP-512A, Laboratory Performance Tests for Automotive Gear Lubricants Intended for API GL-5 Service.

API GL-6 designates the type of service characteristic of gears designed with a very high pinion offset. Such designs typically require (gear) score protection in excess of that provided by API GL-5 gear oils. A shift to more modest pinion offsets coupled with the obsolescence of original API GL-6 test equipment and procedure has greatly diminished the commercial need for API GL-6 gear lubricants.

3.1 Reference Oils: The current reference oils required for each of the API GL-5 tests are listed below:

Performance Test	Reference Oil	
	High Level	Low Level
L-33	RGO 125	RGO 122
L-37	RGO 105	RGO 103
L-42	RGO 110	RGO 108
L-60	RGO 4668	RGO 4669

[4]Frictional requirements for axles equipped with limited slip differentials are normally defined by the axle manufacturer.

3.1 (Continued):

These reference oils are available from the ASTM Test Monitoring Center, 4400 Fifth Avenue, Pittsburgh, PA 15213.

New reference oils for the L-37 test are being developed by ASTM. The timing for their adoption is indefinite.

4. VISCOSITY AND VISCOSITY IMPROVERS:

Refer to SAE J306 for axle and manual transmission lubricant viscosity classification information.

5. STABILITY AND OXIDATION RESISTANCE:

Factors affecting stability and oxidation characteristics while the lubricant is in service include ambient temperature, duty cycle, length of service, and the effects of contamination. Even when lubricants are stored (prior to use), care should be exercised to ensure that they are not exposed to extremes in temperature and are kept free of contaminants. These precautions are intended to ensure optimum lubricant life.

For automotive axles and transmissions in mild service, the temperature of the lubricant may not be sufficiently high to cause oxidation. However, for vehicles in severe conditions of service such as passenger cars pulling trailers or for trucks or buses in service where high temperatures occur, oxidation resistance is an important factor. Accordingly, only oils with a high degree of oxidation resistance should be used in these more severe applications. The vehicle operator should consult the manufacturer's service guide for drain and refill recommendations.

6. FOAMING:

Excessive foaming may interfere with proper lubrication of gear and bearing surfaces and, consequently, should be avoided. Further, foaming can cause leakage via normal venting passages, thereby, reducing lubricant sump volume. Defoamers are used to minimize this potential lubricant problem.

7. CHEMICAL ACTIVITY OR CORROSION:

In order to obtain gear lubricants with adequate load-carrying capacity, chemical additives are usually employed as compounding ingredients. Gear lubricant additives are generally designed to reduce or control corrosion of both ferrous and copper-containing metals. The darkening of brightly polished copper alloys or other metal surfaces in contact with the lubricant does not necessarily indicate that the lubricant will cause harmful corrosion in service. Corrosion may be minimized by the choice of a lubricant containing the proper combination of chemical additives, particularly those stable in the presence of water.

8. SEAL COMPATIBILITY:

While the primary function of a gear lubricant is to protect gears and bearings, consideration must be given to the effect of a lubricant on elastomers or other seal materials used in the design of the component. Simple immersion tests such as those described in ASTM D 471 may be used to establish the relative compatibility of the lubricant and the seal material. Successful performance on such tests does not automatically ensure satisfactory seal performance under field service conditions. ASTM is currently investigating tests suitable for this purpose.

9. MIXING GEAR LUBRICANTS:

As a general practice, the mixing of lubricants should be avoided. Mixing gear lubricants with even small amounts of other types of lubricants can result in antagonistic reactions between the additive chemicals in the mixture. Such reactions may result in a significant loss of gear protection.

As a general practice, the mixing of MIL-L-2105D approved lubricants as in a top-up situation should not impair lubricant performance. MIL-L-2105D lubricants are required to demonstrate satisfactory storage stability when mixed with previously qualified gear lubricants as a condition of the MIL-L-2105D approval process.

The phi (Ø) symbol is for the convenience of the user in locating areas where technical revisions have been made to the previous issue of the report. If the symbol is next to the report title, it indicates a complete revision of the report.

Appendix 12

Automatic Transmission Fluid Specifications and Approvals

Most transmission and vehicle manufacturers do not publish full ATF specifications, but evaluate potentially suitable fluids with their own procedures in the laboratory and on the road. The most complete specifications are issued by General Motors and Ford in the U.S., and even here the final checking and assessment is performed by the OEMs for oil approvals.

The GM and Ford specifications and approvals issued by them are widely used around the world as a starting point for other OEMs. With very few exceptions, fluids are required to be friction-modified, so DEXRON™ or recently qualified Ford fluids are specified. The most recent edition of these basic specifications does not necessarily have to be followed, and there can be a lag between issue of a new U.S. specification and its adoption as the new baseline standard elsewhere.

Transmission manufacturers have differing friction materials and seals incorporated in their units, so friction and elastomer compatibility tests are specific to each manufacturer or even individual types of transmission. Oxidation and other durability tests are also usually performed in full-sized transmissions specific to each OEM.

Viscometric requirements differ, but this is principally due to some manufacturers having more severe requirements for low-temperature performance than others, rather than specification of a distinct level of viscosity. Thus most specifications could in theory be met by a single fluid of high overall performance. Practically, approval costs and the reluctance of many OEMs to approve a large number of fluids means that ATFs have tended to be regionally specific.

The specifications or approval requirements of key manufacturers are listed below. Because friction, elastomer compatibility and transmission tests are material and manufacturer specific, and in many cases outside the U.S. cannot be performed by other laboratories, we have not given the complex details of these tests. In many cases such details are not published and manufacturer requirements prior to conducting their own in-house tests amount only to evidence of other approvals (usually for DEXRON™ II or later) and viscometrics.

U.S. Passenger Car Transmissions

A new DEXRON™ III specification from GM and a new Ford MERCON™ specification have recently been issued. Chrysler's published requirements have not been updated for some time. They and other U.S. OEMs accept either DEXRON™ III or MERCON™ fluids as service-fill in their transmissions. These two dominating specifications are summarized below:

Specification	**DEXRON™ III**	**MERCON™**
Viscometrics (new fluid)		
Visc. @ 100°C, cSt	-	6.8 min
Brookfield visc., cP		
at –20°C	1500 max	1500 max
at –40°C	20,000 max	20,000 max
Other tests on new fluid		
Flash Point, °C, (D92)	170 min	177 min
Fire Point, °C, (D92)	185 min	-
Cu corrosion (D130)	1b max	1b max
Fe corrosion (D1748)	none	none
Vane pump wear (D2882)	15 mg max	10 mg max

Specification	DEXRON™ III	MERCON™
Foam tests:	GM method nil @ 95°C 5 mL max @ 135°C (break time 15 secs max)	Foam stability 200 mL/min @ 100°C 100 mL after 5 seconds nil after 60 seconds
Friction tests	[Tests on SAE No.2 machine of given clutch and band materials for torque under set conditions and endurance characteristics]	
Transmission tests	4L60 oxidation 4L60 cycling	ABOT oxidation 4L60 as DEXRON™ III cycling
Vehicle tests	Performed by GM	Performed by Ford
Compositional requirements	none	none
Elastomer compatibility	Various tests	Various tests

U.S. Heavy-Duty and Off-Road Transmissions

Fluid specifications are dominated by the Allison C-4 and Caterpillar TO-2 specifications. (Caterpillar also has their TO-4 specification for special situations which requires special base stocks.) The requirements for heavy-duty transmissions are summarized in SAE J1285 which is reproduced here:

INTERNATIONAL
400 Commonwealth Drive, Warrendale, PA 15096-0001

SURFACE VEHICLE RECOMMENDED PRACTICE

Submitted for recognition as an American National Standard

SAE J1285 | REAF. JAN85

Issued 1980-02
Reaffirmed 1985-01

Superseding J1285 FEB80

POWERSHIFT TRANSMISSION FLUID CLASSIFICATION

This SAE Recommended Practice was prepared by the SAE Fuels and Lubricants Technical Committee.

(a) to assist the designers and users of heavy-duty transmissions in the selection of powershift transmission fluids for field use and
(b) to promote a uniform practice for use by marketers of lubricants and equipment builders in identifying and recommending these fluids by type.

This classification is designed for fluids used in heavy-duty truck, bus, earthmoving, and marine transmissions or steering clutches. The fluids must perform the following five functions:

1. Transmit hydrodynamic energy in a torque converter.
2. Transmit hydrostatic energy in hydraulic circuits.
3. Lubricate bearings, bushings, gears, and moving parts.
4. Provide proper frictional properties in lubricated bands and clutches.
5. Provide heat transfer medium for liquid- or air-cooled systems to maintain suitable operational temperature range.

SIGNIFICANCE AND METHOD OF MEASURING POWERSHIFT TRANSMISSION FLUID PROPERTIES

1. ***Viscosity***—Viscosity controls the efficiency of the hydraulic control and torque converter systems. Viscosity grades and recommended temperature ranges are chosen by the individual manufacturer. Both Newtonian and non-Newtonian fluids are used. The viscosity grades are classified by SAE J300d. All other properties of the fluids are covered by this classification.

Powershift transmission fluids must flow readily at low temperatures to oil screens and through inlet tubes. This property is evaluated by the Brookfield Viscometer test, ASTM D 2983, and by the Pour Point test, ASTM D 97.

2. ***Foaming Characteristics***—Suppression of the foaming tendency of fluids in a powershift transmission is essential to proper operation. Foaming of the transmission fluid can produce erratic pump, converter, and hydraulic control response; and frequently results in fluid loss through the breather or filler tube. Measurement of foaming tendency and foam stability is used for evaluating fluid suitability. The technique for measuring this property in a powershift transmission fluid is the ASTM D 892 Foaming Characteristics test.

SAE J1285 Reaffirmed JAN85

3. ***Fluid/Seal Compatibility***—Compatibility of the powershift transmission fluid with elastomeric seal materials must be established during fluid formulation. Accepted design procedure involves use of a reference elastomer to determine seal swell, shrink, and hardening tendencies in a candidate fluid. Seal materials must be selected to meet transmission performance requirement with the established fluid formulation. Bench test procedure such as ASTM D 471, Method of Test for Change in Properties of Elastomeric Vulcanizates Resulting from Immersion in Liquids, and ASTM D 2240, Method of Test for Indentation of Rubber and Plastics by Means of a Durometer, are of value for screening purposes.

4. ***Rust Protection***—The bench test used to evaluate the rust-preventative properties of powershift transmission fluids is ASTM D 1748, Method Test for Rust Protection by Metal Preservatives in the Humidity Cabinet.

5. ***Wear Resistance***—Powershift transmission fluids must inhibit scoring and wear of rubbing surfaces in the transmission. There is no transmission test available which correlates with wear in field service. A power steering pump test is used for evaluation of the wear performance of powershift transmission fluids.

6. ***Oxidation Stability***—Fluids used in powershift transmissions must be capable of operating at temperatures up to 150°C. Introduction of air, through normal transmission breathing, results in severe oxidizing conditions which change many new fluid characteristics. Some effects which oxidation can produce are:

   (a) Alteration of frictional characteristics which result in excessive clutch and band slippage, producing high localized clutch temperatures, which in turn make oxidizing conditions more severe.
   (b) Acids or peroxides formed in fluid oxidation which may be corrosive to bushing and thrust washer materials, and detrimental to elastomeric seal materials and the composition clutch plates.
   (c) Viscosity increases great enough to degrade transmission operation.
   (d) Sludge which can plug hydraulic controls and fluid passages.
   (e) Varnish formation which can lead to control valve or governor sticking and ultimate transmission failure.
   (f) Oxidation products which can reduce antifoamant effectiveness.

7. ***Friction Retention***—Matching fluid-friction properly with clutch and band materials is a fundamental design consideration in all currently produced powershift transmissions.

   Fluid friction characteristics are important in automatic transmissions that utilize lubricated clutches to change gear ratios. Extensive performance and durability testing is performed in actual transmissions and bench friction test apparatus. No single friction test can evaluate the requirements of different clutch plate friction materials. The performance of powershift transmission fluids with bronze faced friction plates is evaluated using a 15 000 cycle test procedure and the performance with graphite faced plates is evaluated with a second 5500 cycle procedure. Both procedures are conducted in the SAE No. 2 friction machine.

References

1. C. R. Potter and R. H. Schaefer, "Development of Type C-3 Torque Fluid for Heavy-Duty Power Shift Transmission." Paper 770513 presented at the SAE Earthmoving Industry Conference, Peoria, April 1977.
2. J. A. McLain, "Oil Friction Retention Measured by Caterpillar Oil Test No. TO-2." SAE Transactions, Vol. 86 (1977), Paper 770512.

SAE J1285 Reaffirmed JAN85

TABLE 1—TEST TECHNIQUES AND PERFORMANCE CRITERIA

Test Technique	Limits	Test Method
1. **Brookfield Viscosity** at -18°C, cSt	2500 max for 5W and 10W oils report result for other SAE grades	ASTM D 2983
2. **Pour Point, °C**	-30 max for SAE 5W or 10W oils -25 max for SAE 15W oils -15 max for SAE 30 or 40 oils	ASTM D 97
3. **Foaming Characteristics**		
Sequence 1, cm^3	25/0 max	ASTM D 892
Sequence 2, cm^3	50/0 max	
Sequence 3, cm^3	25/0 max	
4. **Seal Compatibility**		[a]
(a) Total Immersion Test[b] (Nitrile Rubber)		
Volume Change, %	0 to +5 for SAE 5W and 10W oils -1 to +5 for other SAE grades	
Hardness Change, Points	0 to ±5	
(b) Dip Cycle Test[b] (Polyacrylate)		
Volume Change, %	0 to +10	
Hardness Change, Points	0 to +5	
(c) Tip Cycle Test[b] (Silicone)		
Volume Change, %	0 to +5	
Hardness Change, Points	0 to -10	
5. **Rust Resistance**	No rust ("no more than 3 random spots 1 mm or less in diameter at least 6 mm away from edge of panel")	[a]
Other Properties		
6. **Wear Test**	Pump Cam Ring shall show grinding pattern for 360 deg and be free from scuffing, scoring, or chatter marks.	[a]
7. **Oxidation Test**	—No significant varnish or sludge on transmission parts —No blackening or flaking of copper containing parts —Oil shall not gain more than 15% in viscosity at 100°C	[a]
8. **Friction Retention Tests**		
(a) Test A—(5500 cycles)	Slip Time—0.85 s max Dynamic Torque—102 N·m min Decrease in Dynamic Torque—40 N·m max	[a]
(b) Test B—(15 000 cycles)	Stopping Time Increase—15% max Wear—Bronze disc 0.25 mm max Steel Plate 0.1 mm max	[c]

[a] Detroit Diesel Allison Division Transmission Engineering Specifications Serial No. TES 122.
[b] Nominal values which are adjusted for each elastomer batch.
[c] Caterpillar Engineering Specification No. TO-2.

PREPARED BY THE SAE FUELS AND LUBRICANTS TECHNICAL COMMITTEE

For tractors, companies have their own Tractor Hydraulic Fluid (THF) specifications covering use in transmissions as well as in hydraulic actuators. The requirements of the key manufacturers are summarized below:

Test		J I Case/ International Harvester MS 1207	Ford M2C134D	John Deere J 20 D	Massey-Ferguson M 1143
Kin Visc @ 100°C (cSt)		6.2 min	9.0 min	7.0 min	13.5 max
App Visc @ –118°C (cP)		-	4000 max	-	4000 max
App Visc @ –120°C (cP)		3500 max	-	1500 max	-
App Visc @ –130°C (cP)		15,000 max	-	-	-
App Visc @ –140°C (cP)		-	-	20,000 max	-
V.I.		95-115	-	-	-
Visc Stab. (heating)		-	+10% max	-	below +25%
Visc Stab. (shear)		-	–16% max	5.0 cSt min	9.0 cSt min
Flash Point D92, °C		195 min	190 min	150 min	above 200
Pour Point, °C		–37 max	–37 max	–45 max	below –34
S. Ash %		1.15-1.30	-	-	-
Metals %		Specified	-	-	-
Foam mL	I	50/below 10	20/0	25/0	50/0
	II	50/below 10	50/0	50/0	50/0
	III	50/below 10	20/0	25/0	50/0
Cu corrn.		(No corrosion in special	2b max	-	1a max
Rusting		multi-metal test)	No rust	No rust	No rust

Water sensitivity
Wear / EP performance
Elastomer compatibility
Transmission tests
Field tests

(Manufacturers have own special tests for these properties)

Approval Procedures Outside the U.S.

These almost invariably start with a requirement that a candidate fluid is already approved against either a DEXRON™ or a Ford ATF specification. At the present time DEXRON™ III or Ford MERCON™ approvals have become a common requirement, but until recently the required approval was often not necessarily the latest U.S. variant.

Transmission and vehicle manufacturers often have specific viscosity requirements in addition to those in the U.S. specification. If these are also met, then the OEM has to be persuaded that there is a need for a new approval. He will then test the fluid exhaustively in his own compatibility, friction, oxidation and full-scale transmission and field tests. Few of these tests are published and generally cannot be performed outside the manufacturer's own laboratories, although there are some exceptions for screening purposes. Candidate fluids are almost invariably evaluated against the OEM's reference fluid(s) rather than required to meet laid-down specification limits.

European Passenger Car Transmissions

The European manufacturers include Mercedes-Benz, Volkswagen and ZF in Germany, and Renault in France. GM and Ford also have transmission plants in France.

Some passenger car manufacturers also have their own ATF specifications for use in bought-in transmissions. An example is Volvo who has severe low-temperature requirements and who will approve previously qualified DEXRON™ or Ford fluids provided they also meet their more restrictive specifications. An indication of individual testing requirements is given below:

Mercedes-Benz	DEXRON™ approval Kin. visc. at 100°C normally 7.0 cSt but 5.5 cSt min. Satisfactory elastomer and initial friction tests by applicant Friction and wear rig-tests by M-B Cycling tests in several types of transmission for wear, engagement and general behavior Oxidation test in cycled transmission Field tests in various vehicles for up to 150,000 km (some may be done by the applicant) Trailer-towing tests in different models

ZF	DEXRON™ approval Friction tests FZG testing Cycling tests Comprehensive field tests
Volkswagen	DEXRON™ approval Comprehensive in-house testing
Renault	DEXRON™ approval Visc @ 40°C 33-37 cSt Visc @ 100°C 6-8 cSt Visc @ –10°C below 1000 cP Visc @ –40°C 80,000 cP max Pour Point –36°C max In-house performance evaluation
Volvo	Ford 33G approval or DEXRON™ approval V.I. 140 min or 130 min Visc @ 100°C 7.0 cSt min Pour Point –40°C max Testing by applicant (NB: Different types of fluid for different purchased transmissions)

European Bus, Heavy-Duty and Off-Road Transmissions

The important European manufacturers are Mercedes-Benz, Voith, Renk and ZF in Germany, and Volvo and Scania in Sweden. Caterpillar and Allison also have plants in Europe. Approval procedures are similar to those for passenger car transmissions.

Mercedes-Benz	DEXRON™ approval DKA friction tests Clutch friction and wear tests Field tests (90,000 km in at least 5 buses)

Voith	DEXRON™ approval Friction tests FZG test Seal tests Oxidation test Field test 60,000 km
Renk	DEXRON™ approval Friction tests Field testing (minimum 30,000 km)
ZF	DEXRON™ approval Friction tests FZG test Field tests
Volvo	DEXRON™, Allison or Ford approvals depending on transmission Physical parameters Corrosion tests (Testing for Volvo approvals is principally by the applicant)

Japanese ATF Approval Requirements

DEXRON™ approval plus specific viscosity limits:

	Toyota	**Nissan**	**Honda**	**Mazda**	**Mitsubishi**
Visc @ 100°C cSt min	7.5	7.3	7.5	7.0	7.5
Visc @ –30°C cP max	-	-	-	3500	-
Visc @ –40°C cP max	20,000	20,000	20,000	-	-
Shear loss % max	15	10	10	5	15

Fluids are subject to intensive in-house investigation for approvals.

Appendix 13

The Engineering Society For Advancing Mobility Land Sea Air and Space®
INTERNATIONAL
400 Commonwealth Drive, Warrendale, PA 15096-0001

SURFACE VEHICLE RECOMMENDED PRACTICE

Submitted for recognition as an American National Standard

SAE J310 | REV. JUN93

Issued 1951-09
Revised 1993-06-30

Superseding J310 APR90

(R) AUTOMOTIVE LUBRICATING GREASES

1. ***Scope***—This SAE Recommended Practice was developed by SAE, and the section "Standard Classification and Specification for Service Greases" cooperatively with ASTM, and NLGI. It is intended to assist those concerned with the design of automotive components, and with the selection and marketing of greases for the lubrication of certain of those components on passenger cars, trucks, and buses. The information contained herein will be helpful in understanding the terms related to properties, designations, and service applications of automotive greases.

2. ***References***

2.1 **Applicable Documents**—The following publications form a part of this specification to the extent herein. The latest issue of SAE publications shall apply.

2.1.1 SAE PUBLICATION—Available from SAE, 400 Commonwealth Drive, Warrendale, PA 15096-0001.

SAE J1146—Automotive Lubricant Performance and Service Classification and Maintenance Procedure

2.1.2 ASTM PUBLICATIONS—Available from ASTM, 1916 Race Street, Philadelphia, PA 19103-1187.

ASTM D 128—Analysis of Lubricating Grease
ASTM D 217—Cone Penetration of Lubricating Grease
ASTM D 566—Dropping Point of Lubricating Grease
ASTM D 942—Oxidation Stability of Lubricating Greases by the Oxygen Bomb Method
ASTM D 972—Evaporation Loss of Lubricating Greases and Oils
ASTM D 1092—Apparent Viscosity of Lubricating Greases
ASTM D 1263—Leakage Tendencies of Automotive Wheel Bearing Greases
ASTM D 1264—Water Washout Characteristics of Lubricating Greases
ASTM D 1403—Cone Penetration of Lubricating Grease Using One-Quarter and One-Half Scale Cone Equipment
ASTM D 1478—Low-Temperature Torque of Ball Bearing Greases
ASTM D 1742—Oil Separation from Lubricating Grease During Storage
ASTM D 1743—Corrosion Preventive Properties of Lubricating Greases
ASTM D 1831—Roll Stability of Lubricating Grease
ASTM D 2265—Dropping Point of Lubricating Grease Over Wide-Temperature Range
ASTM D 2266—Wear Preventive Characteristics of Lubricating Grease (Four-Ball Method)

Printed in U.S.A.
90-1202B EG

SAE J310 Revised JUN93

ASTM D 2509—Measurement of Load-Carrying Capacity of Lubricating Grease (Timken Method)
ASTM D 2595—Evaporation Loss of Lubricating Greases Over Wide-Temperature Range
ASTM D 2596—Measurement of Extreme-Pressure Properties of Lubricating Grease (Four-Ball Method)
ASTM D 3527—Life Performance of Automotive Wheel Bearing Grease
ASTM D 4170—Fretting Wear Protection by Lubricating Greases
ASTM D 4289—Compatibility of Lubricating Grease with Elastomers
ASTM D 4290—Leakage Tendencies of Automotive Wheel Bearing Grease Under Accelerated Conditions
ASTM D 4693—Low-Temperature Torque of Greased-Lubricated Wheel Bearings
ASTM D 4950—Standard Classification and Specification for Automotive Service Greases

2.1.3 NGLI PUBLICATIONS—Available from NGLI, 4635 Wyandotte Street, Kansas City, MO 64112.

NGLI Recommended Practice for Lubricating Passenger Car Wheel Bearings
NGLI Recommended Practice for Lubricating Passenger Car Ball Joint Front Suspensions
NGLI Recommended Practice for Grease Lubricated Truck Wheel Bearings

3. ***Definition of Lubricating Grease***—A lubricating grease is a solid to semifluid mixture of a liquid lubricant and a thickening agent. Additives to impart special properties or performance characteristics may be incorporated. The liquid component may be a mineral (petroleum) oil or a synthetic liquid; the thickener may be a metallic soap or soaps or a nonsoap substance such as an organophilic modified clay, a urea compound, carbon black, or other material. The viscosity of the fluid, the thickener concentration, and the chemical nature of the thickener may vary widely. The properties of the finished grease are influenced by the manufacturing process as well as by the materials used.

4. ***Basic Performance Requirements***—Greases are most often used instead of fluids where a lubricant is required to maintain its original position in a mechanism, especially where opportunities for frequent relubrication may be limited or economically unjustifiable. This requirement may be due to the physical configuration of the mechanism, the type of motion, the type of sealing, or to the need for the lubricant to perform all or part of any sealing function in the prevention of lubricant loss or the entrance of contaminants. Because of their essentially solid nature, greases do not perform the cooling and cleaning functions associated with the use of a fluid lubricant. With these exceptions, greases are expected to accomplish all other functions of fluid lubricants.

A satisfactory grease for a given application is expected to:

a. Provide adequate lubrication to reduce friction and to prevent harmful wear of bearing components

b. Protect against corrosion.

c. Act as a seal to prevent entry of dirt and water.

d. Resist leakage, dripping, or undesirable throw off from the lubricated surfaces.

e. Resist objectionable change in structure or consistency with mechanical working (in the bearing) during prolonged service.

f. Not stiffen excessively to cause undue resistance to motion in cold weather.

g. Have physical characteristics suitable for the method of application.

h. Be compatible with elastomer seals and other materials of construction in the lubricated portion of the mechanism.

i. Tolerate some degree of contamination, such as moisture, without loss of significant characteristics.

SAE J310 Revised JUN93

j. Have suitable oxidation and thermal stability for the intended application.

*5. **Properties of Greases***

5.1 **Consistency**—A measure of relative hardness. This property is commonly expressed in terms of the ASTM penetration or NLGI consistency number. The ASTM penetration is a numerical statement of the actual penetration of the grease sample, in tenths of a millimeter, by a standard test cone under stated conditions. The higher the penetration value, the softer the grease. The National Lubricating Grease Institute classifies greases according to their ASTM penetration as shown in Table 1.

TABLE 1—NLGI CONSISTENCY NUMBER

NLGI Consistency No.	ASTM Worked (60 Strokes) Penetration at 25 °C (77 °F) tenths of a millimeter[1]	NLGI Consistency No.	ASTM Worked (60 Strokes) Penetration at 25 °C (77 °F) tenths of a millimeter[1]
000	445 to 475	3	220 to 250
00	400 to 430	4	175 to 205
0	355 to 385	5	130 to 160
1	310 to 340	6	85 to 115
2	265 to 295		

[1] ASTM D 217 Cone Penetration of Lubricating Grease.

The consistency of a grease is an important factor in its ability to lubricate, seal, and remain in place, and to the methods and ease by which it can be dispensed and applied. Most automotive greases are in the NLGI No. 1, 2, or 3 range, that is, ranging from soft to medium consistency.

5.2 **Texture and Structure**—The appearance and feel of greases. A grease may be described as smooth, buttery, fibrous, long- or short-fibered, stringy, tacky, etc. These characteristics are influenced by the viscosity of the fluid, type of thickener, proportion of each of these components, presence of certain additives, and process of manufacture. There are no standard test methods for quantitative definitions of these properties. Texture and structure are factors in the adhesiveness and ease of handling of a grease.

5.3 **Structural Stability**—The ability of a grease to retain its as-manufactured consistency and texture despite age, temperature, mechanical working, and other influences, or its ability to return to its original state when a transient influence is removed.

5.4 **Mechanical Stability**—The resistance of a grease to permanent changes in consistency due to the continuous application of shearing forces.

The stability of a grease is important to its ability to provide adequate lubrication and sealing and to remain properly in place during use.

5.5 **Apparent Viscosity**—The ratio of shear stress to rate of shear at a stated temperature and shear rate. Grease is by nature a plastic material. Therefore, the usual concept of viscosity valid for simple fluids (that is, internal resistance to flow) is not entirely applicable. The ratio of shear stress to shear rate varies as the shear rate changes. The apparent viscosity of most greases decreases with increase of either temperature or shear rate. Apparent viscosity greatly influences the ease of handling and dispensing a grease.

SAE J310 Revised JUN93

5.6 Dropping Point—The temperature at which the grease generally passes from a plastic solid to a liquid state, and flows through an orifice under standard test conditions. The dropping point is incorrectly regarded by some as establishing the maximum temperature for acceptable use. Performance at high temperature also depends on other factors such as duration of exposure, evaporation resistance, and design of the lubricated mechanism.

5.7 Oxidation Resistance—The resistance to chemical deterioration in storage and in service caused by exposure to air. It depends basically on the stability of the individual grease components, and can be improved by use of antioxidants. Oxidation resistance is important wherever long storage or service life is required or where high temperatures prevail even for short periods.

5.8 Protection Against Friction and Wear—A protection greatly influenced by the viscosity and type of the fluid component and by grease structural and consistency characteristics. This performance characteristic can be altered by use of additives.

5.9 Protection Against Corrosion—A protection of ferrous components achieved primarily by the inclusion of suitable additives in the grease. The effectiveness of the protection is influenced also by the chemical and physical properties, such as interactions with other additives, consistency and base oil viscosity (both of which will determine how effectively the grease will seal out corrosive and other undesirable material), and the interaction with water. The effect of water on the grease can be significant. Some greases are water resistant or waterproof, which means that they resist the washing effect of water and do not absorb it to any significant extent. Other greases can absorb varying amounts of water without appreciable damage to their structure or consistency, and may provide better rust protection than waterproof greases which can permit the accumulation of free water in bearings.

5.10 Bleeding or Oil Separation—The separation of liquid lubricant from a grease. Slight bleeding is regarded as desirable by some as indicative of good lubricating ability in rolling element bearings.

5.11 Color—A superficial grease property without performance significance.

Of the previous properties, oxidation resistance, protection against friction and wear, protection against corrosion, and structural stability are probably of most importance in automotive service as far as actual performance in bearings is concerned.

There is, of course, the problem of getting grease to the bearings to be lubricated. Certain terms, by no means of strict, rigid interpretation, are used to describe the factors involved: feedability, pumpability, and dispensability.

5.12 Feedability or Slumpability—The ability to flow to the suction of the grease-dispensing equipment or mechanism to be lubricated.

5.13 Pumpability—The ability to flow through the grease-dispensing lines at a satisfactory rate, without the necessity of using excessively high pressure.

5.14 Dispensability—The ease with which a grease may be transferred from its container to the point of application. For practical purposes, it is a combination of feedability and pumpability.

SAE J310 Revised JUN93

6. ***Grease Testing***—Many of the previous grease properties are determined by tests which have been standardized or otherwise accorded industry recognition. These, in conjunction with simulated performance tests, permit some approximate judgment for the proper selection of greases for a given application. They are, however, not considered to be replacements for, or equivalent to, longtime service tests.

Table 2 shows some of the more important tests identified as to sponsor, title, and purpose.

TABLE 2—GREASE TESTS

Test Designation	Test Purpose
ASTM D 128, Analysis of Lubricating Grease	Quantitative determination of specified constituents, such as soap, unsaponifiable matter (mineral oil), water, free alkali, free fatty acid, glycerine, and insolubles. NOTE—This procedure has a supplementary method useful for greases containing nonsoap thickeners or synthetic fluids.
ASTM D 217, Cone Penetration of Lubricating Grease	Measurement of consistency.
ASTM D 566, Dropping Point of Lubricating Grease	Determination of temperature at which grease generally passes from plastic to liquid state; not regarded as indicative of service suitability; limited to dropping points up to 260 °C (500 °F). (In this test, some greases may release oil before the grease flows which is defined as their dropping points.)
ASTM D 942, Oxidation Stability of Lubricating Greases by the Oxygen Bomb Method	Determination of resistance to oxidation under static conditions in a sealed system at elevated temperatures, not indicative of the stability of greases under dynamic service conditions, nor the stability of greases stored in containers for long periods, nor the stability of films of grease on machine parts.
ASTM D 972, Evaporation Loss of Lubricating Greases and Oils	Evaluation of weight loss by evaporation at temperatures up to 150 °C (300 °F).
ASTM D 1092, Apparent Viscosity of Lubricating Greases	Determination of apparent viscosity in temperature range of -54 to 38 °C (-65 to 100 °F); results relatable to ease of handling and dispensing.
ASTM D 1263, Leakage Tendencies of Automotive Wheel Bearing Greases	Evaluation of leakage tendencies from an unsealed wheel bearing assembly, run for 6 h at 104 °C (220 °F); permits screening candidate greases; not a replacement for longtime service tests. NOTE—Replaced by ASTM D 4290 in many updated specifications.
ASTM D 1264, Water Washout Characteristics of Lubricating Greases	Evaluation of resistance to water washout from rotating bearings at 38 °C (100 °F) and at 80 °C (175 °F) under prescribed conditions; not a replacement for actual service tests; not suitable for fibrous greases.
ASTM D 1403, Cone Penetration of Lubricating Grease Using One-Quarter and One-Half Scale Cone Equipment	Essentially same as ASTM D 217, using reduced-size apparatus to evaluate small grease samples, but limited to greases of NLGI No. 0 to 4 consistency.

SAE J310 Revised JUN93

TABLE 2—GREASE TESTS (CONTINUED)

Test Designation	Test Purpose
ASTM D 1478, Low-Temperature Torque of Ball Bearing Greases	Determination of the extent to which a grease retards the rotation of a slow-speed ball bearing when subjected to temperatures below -18 °C (0 °F). This method was developed using a test temperature of -54 °C (-65 °F) and greases with extremely low torque characteristics. Although higher test temperatures are commonly used, the precision statements may not apply to temperatures other than -54 °C (-65 °F) or to greases with torque characteristics different from those used to establish precision. NOTE—ASTM D 4693 is better suited for greases having higher torque chacteristics.
ASTM D 1742, Oil Separation from Lubricating Grease During Storage	Determination of tendency of oil constituent to separate from parent grease while in containers; suitable for NLGI No. 1 or harder greases; results are indicative of oil separation in containers, but not of oil separation under dynamic service conditions.
ASTM D 1743, Corrosion Preventive Properties of Lubricating Greases	Determination of surface damage due to corrosion, such as pitting, etching, rusting, or black stains on raceways and rollers of tapered roller bearings which have been run-in and stored for a prescribed period at 52 °C (125 °F) and 100% relative humidity.
ASTM D 1831, Roll Stability of Lubricating Grease	Determination of changes to consistency after working in tester for 2 h at room temperature. Although test significance has not been determined, changes in worked penetration of a grease after rolling are believed to be an indication of shear stability under low shear conditions.
ASTM D 2265, Dropping Point of Lubricating Grease Over Wide-Temperature Range	Same purpose as ASTM D 566 but ASTM D 2265 is valid for temperatures up to 330 °C (625 °F).
ASTM D 2266, Wear Preventive Characteristics of Lubricating Grease (Four-Ball Method)	Determination of wear preventive characteristics of grease when a rotating loaded steel ball slides against three similar stationary steel balls, measured by wear-scar diameters on stationary balls after completion of test; not indicative of results in actual service, and cannot distinguish between extreme-pressure (EP) and nonextreme-pressure (non-EP) greases.
ASTM D 2509, Measurement of Load-Carrying Capacity of Lubricating Grease (Timken Method)	Determination of load-carrying ability of lubricating greases by Timken Lubricant and Wear Tester. In this device, a rectangular steel test block is forced against a rotating steel ring. Scar width and surface conditions are noted. Method differentiates between lubricants of various extreme-pressure levels; not a replacement for actual service tests.
ASTM D 2595, Evaporation Loss of Lubricating Greases Over Wide-Temperature Range	Evaluation of weight loss by evaporation at temperatures between 93 and 316 °C (200 and 600 °F).
ASTM D 2596, Measurement of Extreme-Pressure Properties of Lubricating Grease (Four-Ball Method)	Evaluation of load-carrying properties at high loads. Determines: a. Load-wear index (formerly mean-Hertz load) b. Weld point by Four-Ball EP Tester
ASTM D 3527, Life Performance of Automotive Wheel Bearing Grease	Evaluation of the high-temperature life performance of wheel bearing grease.

SAE J310 Revised JUN93

TABLE 2—GREASE TESTS (CONTINUED)

Test Designation	Test Purpose
ASTM D 4170, Fretting Wear Protection by Lubricating Greases[1]	Evaluation of fretting wear protection characteristic by measuring mass loss of ball thrust bearings oscillated under load; correlates with fretting protection performance of greases in wheel bearings of passenger cars shipped long distances.
ASTM D 4289, Compatibility of Lubricating Grease with Elastomers	Determination of hardness and volume changes in elastomers caused by contact with lubricating grease at elevated temperatures.
ASTM D 4290, Leakage Tendencies of Automotive Wheel Bearing Grease Under Accelerated Conditions	Evaluation of leakage tendency of a grease from unsealed wheel bearings run 20 h at 1000 rpm and thrust loaded to 111 N (25 lb force). Unlike ASTM D 1263, this method, which is conducted at a higher temperature, 160 °C (320 °F), differentiates among wheel bearing greases having distinctly different high-temperature leakage characteristics.
ASTM D 4693, Low-Temperature Torque of Greased-Lubricated Wheel Bearings	Determination of the viscous resistance of a grease in a wheel bearing assembly rotated at low speed in a low-temperature environment; used to evaluate both wheel bearing and chassis greases for performance in low-temperature service. NOTE—Greases having torque characteristics that permit evaluation in both ASTM D 1478 and ASTM D 4693 will not give the same torque values in the two tests because of differences in bearings and test apparatus.

[1] NLGI Spokesman, August 1983, page 156.

7. ***Designation of Greases***—Greases are commonly classified and designated according to chemical composition, such as lithium-soap grease; by broad type of usage, such as antifriction bearing grease or multipurpose grease; by specific properties such as high-temperature grease; by special additives, such as extreme-pressure grease or graphite grease; and by specific applications, such as automotive-wheel-bearing grease. SAE recognizes the following designations for greases used in servicing passenger cars, trucks, and buses according to their specific applications.

7.1 **Wheel Bearing Grease**—Designates lubricating greases of such composition, structure, and consistency as to be suitable for longtime use in antifriction wheel bearings. The properties and composition of greases used in ball-type wheel bearings can be significantly different than those used in tapered roller-type wheel bearings. Generally, ball-type wheel bearings used in modern automotive vehicles are not serviceable.

NOTE—Generally, these greases resist the deteriorating effects of temperature and the separating effects of centrifugal action. They have good antirust properties. They should not exhibit oil-soap separation or excessive softening which could result in leakage that could lead to braking failure.

7.2 **Universal Joint Grease**—Designates lubricating greases of such composition, structure, and consistency as to be suitable for the lubrication of those types of automotive universal joints requiring grease lubrication.

NOTE—In many cases, the service relubrication of universal joints can be satisfied with NLGI categories LA or LB (Table 3). However, some designs, such as constant velocity joints, or some types of service, may require special greases. Manufacturers' recommendations or lubrication charts should be consulted.

SAE J310 Revised JUN93

TABLE 3—SUMMARY OF NLGI AND ASTM DESIGNATION, DESCRIPTION, AND PERFORMANCE REQUIREMENTS FOR AUTOMOTIVE SERVICE GREASES

NLGI Letter Designation	NLGI Service Description	ASTM D 4950 Performance Description	ASTM D 4950 Performance Requirements
Chassis Service			
LA	Service typical of chassis components and universal joints in passenger cars, trucks, and other vehicles under mild duty only. Mild duty will be encountered in vehicles operated with frequent relubrication in noncritical applications.	The grease shall satisfactorily lubricate chassis components and universal joints where frequent relubrication is practiced (at intervals 3200 km or 2000 miles or less for passenger cars). During its service life, the grease shall resist oxidation and consistency degradation while protecting the chassis components and universal joints from corrosion and wear under lightly loaded conditions. NLGI 2 consistency greases are commonly recommended, but other grades may also be recommended.	Conform to requirements of Table 4.
LB	Service typical of chassis components and universal joints in passenger cars, trucks, and other vehicles under mild to severe duty. Severe duty will be encountered in vehicles operated under conditions which may include prolonged relubrication intervals, or high loads, severe vibration, exposure to water or other contaminants, etc.	The grease shall satisfactorily lubricate chassis components and universal joints at temperatures as low as -40 °C (-40 °F) and at temperatures as high as 120 °C (248 °F) over prolonged relubrication intervals (more than 3200 km or 2000 miles for passenger cars). During its service life, the grease shall resist oxidation and consistency degradation while protecting the chassis components and universal joints from corrosion and wear even when aqueous contamination and heavily loaded conditions occur. NLGI 2 consistency greases are commonly recommended, but other grades may also be recommended.	Conform to requirements of Table 4.
Wheel Bearing Service			
GA	Service typical of wheel bearings operating in passenger cars, trucks, and other vehicles under mild duty. Mild duty will be encountered in vehicles operated with frequent relubrication in noncritical applications.	The grease shall satisfactorily lubricate wheel bearings over a limited temperature range. Many products of this type are limited to bearing temperatures of -20 to 70 °C (-4 to 158 °F). No additional performance requirements are specified for these greases.	Conform to requirements of Table 5.

TABLE 3—SUMMARY OF NLGI AND ASTM DESIGNATION, DESCRIPTION, AND PERFORMANCE REQUIREMENTS FOR AUTOMOTIVE SERVICE GREASES (CONTINUED)

NLGI Letter Designation	NLGI Service Description	ASTM D 4950 Performance Description	ASTM D 4950 Performance Requirements
GB	Service typical of wheel bearings operating in passenger cars, trucks, and other vehicles under mild to moderate duty. Moderate duty will be encountered in most vehicles operated under normal urban, highway, and off-highway service.	The grease shall satisfactorily lubricate wheel bearings over a wide temperature range. The bearing temperatures may range down to 40 °C (-40 °F), with frequent excursions to 120 °C (320 °F). During its service life, the grease shall resist oxidation, evaporation, and consistency degradation while protecting the bearings from corrosion and wear. NLGI 2 consistency greases are commonly recommended, but NLGI 1 or 3 grades may also be recommended.	Conform to requirements of Table 5.
GC	Service typical of wheel bearings operating in passenger cars, trucks, and other vehicles under mild to severe duty. Severe duty will be encountered in certain vehicles operated under conditions resulting in high bearing temperatures. This includes vehicles operated under frequent stop-and-go service (buses, taxis, urban police cars, etc.), or under severe braking service (trailer towing, heavy loading, mountain driving, etc.).	The grease shall satisfactorily lubricate wheel bearings over a wide temperature range. The bearing temperatures may range down to -40 °C (-40 °F), with frequent excursions to 160 °C (320 °F) and occasional excursions to 200 °C (392 °F). During its service life, the grease shall resist oxidation, evaporation, and consistency degradation while protecting the bearings from corrosion and wear. NLGI 2 consistency greases are commonly recommended, but NLGI 1 or 3 grades may also be recommended.	Conform to requirements of Table 5.

7.3 Chassis Grease—Designates lubricating greases of proper consistency to be applied at periodic intervals in accordance with equipment manufacturers' recommendations, with grease guns through grease fittings, into the various parts of automotive chassis requiring grease lubrication.

NOTE—When no means are provided for periodic relubrication, the ability of a grease to retain its performance characteristics over long intervals of time and service becomes critical. This applies to seals as well because only seals in good condition can effectively bar the entrance of water, dirt, and other contaminants, and minimize loss of grease by leakage.

SAE J310 Revised JUN93

TABLE 4—"L" CHASSIS GREASE CATEGORIES

Category	Test	Property	Acceptance Limit
LA	D 217	Consistency, worked penetration, mm/10	220 to 340[1]
	D 566 or D 2265	Dropping Point, °C, min	80
	D 2266	Wear Protection, scar diameter, mm, max	0.9
	D 4289	Elastomer CR Compatibility:	
		volume change, %	0 to 30
		hardness change, Durometer-A points	0 to -10
LB	D 217	Consistency, worked penetration, mm/10	220 to 340[1]
	D 566 or D 2265	Dropping Point, °C, min	150
	D 2266	Wear Protection, scar diameter, mm, max	0.6
	D 4289	Elastomer CR Compatibility:	
		volume change, %	0 to 30
		hardness change, Durometer-A points	0 to -10
	D 1742	Oil Separation, mass %, max	10
	D 1743	Rust Protection, rating, max	Pass
	D 2596	EP Performance:	
		load wear index, kgf, min	30
		weld point, kgf, min	200
	D 4170	Fretting Protection, mass loss, mg, max	10[2]
	D 4693	Low-Temperature Performance, torque at -40 °C, N·m, max	15.5

[1] Vehicle manufacturer's requirement may be more restrictive; grease containers should display NLGI Consistency Number as well as category designation.

[2] The fretting wear requirement is significant in passenger car and light-duty truck service, but it has not been shown to be significant in heavy-duty truck applications.

SAE J310 Revised JUN93

TABLE 5—"G" WHEEL BEARING GREASE CATEGORIES

Category	Test	Property	Acceptance Limit
GA	D 217	Consistency, worked penetration, mm/10	220 to 340[1]
	D 566 or D 2265	Dropping Point, °C, min	80
	D 4693	Low-Temperature Performance, torque at -20 °C, N·m, max	15.5
GB	D 217	Consistency, worked penetration, mm/10	220 to 340[1]
	D 566 or D 2265	Dropping Point, °C, min	175
	D 4693	Low-Temperature Performance, torque at -40 °C, N·m, max	15.5
	D 1264	Water Resistance at 80 °C, %, max	15
	D 1742	Oil Separation, mass %, max	10
	D 1743	Rust Protection, rating, max	Pass
	D 2266	Wear Protection, scar diameter, mm, max	0.9
	D 3527	High-Temperature Life, hours, min	40
	D 4289	Elastomer NBR-L Compatibility:	
		volume change, %	-5 to +30
		hardness change, Durometer-A points	+2 to -15
	D 4290	Leakage Tendencies, g, max	24
GC	D 217	Consistency, worked penetration, mm/10	220 to 340[1]
	D 566 or D 2265	Dropping Point, °C, min	220
	D 4693	Low-Temperature Performance, torque at -40 °C, N·m, max	15.5
	D 1264	Water Resistance at 80 °C, %, max	15
	D 1742	Oil Separation, mass %, max	6
	D 1743	Rust Protection, rating, max	Pass
	D 2266	Wear Protection, scar diameter, mm, max	0.9
	D 3527	High-Temperature Life, hours, min	80
	D 4289	Elastomer NBR-L Compatibility:	
		volume change, %	-5 to +30
		hardness change, Durometer-A points	+2 to -15
	D 4290	Leakage Tendencies, g, max	10
	D 2596	EP Performance:	
		load wear index, kgf, min	30
		weld point, kgf, min	200

[1] Vehicle manufacturer's requirement may be more restrictive; grease containers should display NLGI Consistency Number as well as category designation.

SAE J310 Revised JUN93

7.4 **Multipurpose Grease**—Designates lubricating greases of such composition, structure, and consistency to meet the performance requirements for chassis grease (more than 3200 km (2000 mile) service life), wheel bearing grease, universal joint grease, and other automotive uses of a miscellaneous nature, such as fifth-wheel service.

NOTE—Some chassis lubricants are satisfactory as multipurpose greases. The grease manufacturer should be consulted as to the multipurpose qualities of his product. Greases designated NLGI GC-LB are multipurpose greases by definition.

7.5 **Extreme Pressure or EP**—Not a designation by usage, but is applied to greases with high load-carrying capacity, determined usually by the Timken method or the Four-Ball EP Test or similar. In some cases, the EP property results from a surface-active additive that imparts antiwear or antiseize properties beyond the capabilities of the usual fluid, thickener, or other finely dispersed lubricating solids in the grease. Extreme-pressure or wear-reducing properties may be incorporated in any of the usage types, most frequently those designated as multipurpose.

7.6 **Greases for Other Vehicle Needs**—Automotive equipment may require special greases not as yet designated by SAE. Examples of such applications are speedometer cables and brake adjustors.

8. ***Grease Application***—Automotive greases are applied by hand packing, by hand- and power-operated pressure guns, and by hand- and power-operated central systems fitted to individual vehicles. In wheel-bearing lubrication, a bearing packing device should be used, as it is more effective, faster, and less wasteful of grease than hand packing. Mixing of different types of greases in wheel bearings should be avoided to preclude excessive thinning and leakage.

The prime consideration in applying greases is that of cleanliness: of containers and dispensing and pumping equipment and in the removal of surface grease and dirt accumulation from application points such as plugs and grease gun fittings.

Excessive dispensing pressures and pumping rates are to be avoided. They tend to cause seal deformation and rupture and are wasteful of lubricant.

Automotive servicing literature is voluminous on the subject of grease lubrication. Important sources are vehicle manufacturers' service bulletins, oil company bulletins and lubrication charts, and trade organization manuals, such as NGLI publications: "Recommended Practice for Lubricating Passenger Car Wheel Bearings," "Recommended Practice for Lubricating Passenger Car Ball Joint Front Suspensions," and "Recommended Practice for Grease Lubricated Truck Wheel Bearings."

9. ***Grease Properties as Related to Types of Service***—Service requirements determine the relative importance of the aforementioned grease properties for each kind of application and set the level of performance needed. Table 6 is a generalized summary of the grease properties of primary importance in the several fields of automotive use previously discussed. Certain properties, such as texture or structure, consistency, and apparent viscosity, are not included in the summary, because it is assumed they will be appropriate to the purposes of the individual grease types.

SAE J310 Revised JUN93

TABLE 6—RELATIVE IMPORTANCE OF LUBRICATING GREASE PROPERTIES FOR AUTOMOTIVE USES SHOWN[1]

Property	Wheel Bearings	Universal Joints	Chassis	Multipurpose Applications
Structural Stability (inc. Mechanical Stability)	H	M	H	H
High Dropping Point (High-Temp. Service)	H	M	M	H
Oxidation Resistance	H	M	H	H
Protection Against Friction and Wear	M	H	H	H
Protection Against Corrosion	M	M	H	M
Protection Against Washout	M	M	H	M

[1] H = Highest, M = Moderate.

10. ***Standard Classification and Specification for Service Greases***—After years of cooperative effort, SAE, NLGI, and ASTM developed a system for the designation, description, classification, and specification of greases for service relubrication. This system has been accepted by both the grease and automotive industries. It was first published in 1989 as ASTM D 4950. ASTM D 4950 is a grease specification expressly intended for service applications. Specifications for factory-fill grease are generally more restrictive and often contain additional performance requirements described by nonstandard tests. However, there is nothing in ASTM D 4950 to preclude its use by equipment manufacturers to describe initial-fill greases. The pertinent requirements are summarized in Table 3.

Automotive service greases are classified into two groups, those suitable for chassis relubrication (including universal joints), and those suitable for the relubrication of serviceable-type wheel bearings. These are further separated into performance categories: two Chassis Grease categories, LA and LB, and three Wheel Bearing Grease categories, GA, GB, and GC. Tables 4 and 5 list the requirements (Acceptability Limits) for the respective categories. These two tables do not constitute all of the requirements of ASTM D 4950. ASTM D 4950 also includes specific descriptions of the service applications and performance requirements, which are included in the standard to ensure the selection of greases suited to the intended application. For quick comparisons, a qualitative guide to the specified requirements is shown in Figure 1.

NLGI has developed a symbol (10.1) to be used to identify greases that conform to the requirements of ASTM D 4950. For greases meeting the requirements of both a Chassis Grease category and a Wheel Bearing category, **multipurpose grease** nomenclature may be used, provided the appropriate NLGI Designation for each group is included. This latter provision is essential to avoid confusion with commercial, nonautomotive, "multipurpose" greases.

ASTM D 4950 and the NLGI service descriptions and symbols were generated and will be maintained and expanded in accordance with SAE J1146.

10.1 **NLGI Symbol for Automotive Service Greases**—To provide a means of readily highlighting greases of the classification that meet the highest performance requirements, NLGI has proposed the use of a standardized identifying symbol. The symbol shall be used only with greases meeting the highest automotive grease performance requirements, i.e., categories GC or LB, or both, as illustrated in Figure 2. The NLGI symbol shall not be used with categories GA, GB, or LA. NLGI licenses grease packagers and marketers to use the symbol, whose size may be varied to meet packaging and labeling needs. It is the responsibility of the packager to verify that the product conforms to the requirements of the categories used in the symbol.

SAE J310 Revised JUN93

D 4950 AUTOMOTIVE SERVICE GREASE REQUIREMENTS

		SERVICE GREASE CATEGORIES				
		CHASSIS		WHEEL BEARING		
TEST	DESCRIPTION	LA	LB	GA	GB	GC
D 217	Penetration	✓	✓	✓	✓	✓
D 566	Dropping Point	✓	✓	✓	✓	✓
D 1264	Water Washout	–	–	–	✓	✓
D 1742	Oil Separation	–	✓	–	✓	✓
D 1743	Rust Protection	–	✓	–	✓	✓
D 2266	4-Ball Wear	✓	✓	–	✓	✓
D 2596	4-Ball EP	–	✓	–	–	✓
D 3527	High-Temperature Life	–	–	–	✓	✓
D 4170	Fretting Wear	–	✓	–	–	–
D 4289	Elastomer Compatibily	✓	✓	–	✓	✓
D 4290	Leakage	–	–	–	✓	✓
D 4693	Low-Temperature Torque	–	✓	✓	✓	✓

FIGURE 1—GUIDE TO REQUIREMENTS FOR GREASE CATEGORIES

FIGURE 2—NLGI SYMBOL FOR LABELING AUTOMOTIVE LUBRICATING GREASES

SAE J310 Revised JUN93

11. Notes

11.1 Marginal Indicia—The (R) is for the convenience of the user in locating areas where technical revisions have been made to the previous issue of the report. If the symbol is next to the report title, it indicates a complete revision of the report.

PREPARED BY THE SAE FUELS AND LUBRICANTS TECHNICAL COMMITTEE 4—LUBRICATING GREASES

Appendix 14

Classifications and Specifications for Two-Stroke Oils

(a) The CEC/API Classification

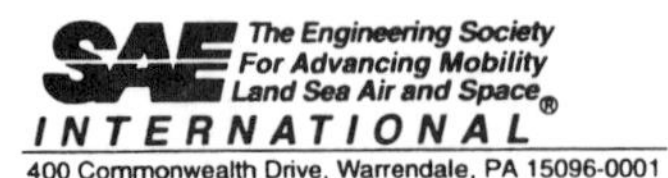

400 Commonwealth Drive, Warrendale, PA 15096-0001

SURFACE VEHICLE STANDARD

Submitted for recognition as an American National Standard

SAE J2116 | REV. JUN93

Issued 1990-10-19
Revised 1993-06-22

Superseding J2116 OCT90

TWO-STROKE-CYCLE GASOLINE ENGINE LUBRICANTS PERFORMANCE AND SERVICE CLASSIFICATION

*1. **Scope***—This SAE Standard was prepared by Technical Committee 6, Small Engine Lubricants, of SAE Fuels and Lubricants Division. The intent is to improve communications among engine manufacturers, engine users, and lubricant marketers in describing lubricant performance characteristics. The key objective is to ensure that a correct lubricant is used in each two-stroke-cycle engine.

1.1 **Background**—SAE J1510 previewed the cooperative effort of SAE, ASTM, API, and CEC in developing a universal classification for engine performance. SAE J1510 provides a great deal of information on the properties of two-stroke-cycle lubricants.

SAE J1536 is a classification in rheological terms only. SAE J1536 is a companion classification to SAE J2116. By use of both SAE J1536 and SAE J2116, any lubricant can be classified in terms of both rheology and engine performance.

*2. **References***

2.1 **Applicable Documents**—The following publications form a part of this specification to the extent specified herein. The latest issue of SAE publications shall apply.

2.1.1 SAE PUBLICATIONS—Available from SAE, 400 Commonwealth Drive, Warrendale, PA 15096-0001.

SAE J1510—Lubricants for Two-Stroke-Cycle Gasoline Engines
SAE J1536—Two-Stroke-Cycle Oil Miscibility/Fluidity Classification

2.1.2 ASTM PUBLICATIONS—Available from ASTM, 1916 Race Street, Philadelphia, PA 19103-1187.

ASTM D 4681-87—Specification for Lubricants for Two-Stroke-Cycle Gasoline Engines (TSC-4)
ASTM D 4857-88—Test Method for Determination of the Ability of Lubricants to Minimize Ring Sticking and Piston Deposits in Two-Stroke-Cycle Gasoline Engines Other Than Outboards
ASTM D 4858-88—Test Method for Determination of the Tendency of Lubricants to Promote Preignition in Two-Stroke-Cycle Gasoline Engines
ASTM D 4859-88—Specification for Lubricants for Two-Stroke-Cycle Spark-Ignition Gasoline Engines - TC
ASTM D 4863-88—Test Method for Determination of Lubricity of Two-Stroke-Cycle Gasoline Engine Lubricants

Printed in U.S.A.
90-1202B EG

SAE J2116 Revised JUN93

2.1.3 NATIONAL MARINE MANUFACTURERS PUBLICATION—Available from NMMA, 401 North Michigan, Chicago, IL 60611.

TC-W (312-84)

3. ***Performance Characteristics***—There are a number of engine test rating areas which are indicative of the contribution of a lubricant to the proper performance and durability of a two-stroke-cycle engine. In each category within this classification, the relevant rating areas are given numerical limits which permit assignment of a pass or fail to the performance of a lubricant. These areas include:

a. Ring sticking
b. Varnish (which may include piston skirts, lands, and undercrowns)
c. Preignition
d. Scuffing
e. Exhaust system blockage

Table 1 relates these performance characteristics to the critical lubrication requirements of each of the four Performance and Service categories. To assist in understanding the purpose of each category, normal engine service applications are also provided.

(R) 4. ***Performance Criteria***—Table 2 summarizes the ASTM standard test methods, test engines, primary performance criteria, and status for each category.

5. ***Notes***

5.1 Marginal Indicia—The (R) is for the convenience of the user in locating areas where technical revisions have been made to the previous issue of the report. If the symbol is next to the report title, it indicates a complete revision of the report.

PREPARED BY THE SAE FUELS AND LUBRICANTS TECHNICAL
COMMITTEE 6—SMALL ENGINE LUBRICANTS

SAE J2116 Revised JUN93

TABLE 1—PERFORMANCE AND SERVICE CLASSIFICATION TWO-STROKE-CYCLE GASOLINE ENGINE LUBRICANTS CRITICAL LUBRICATION REQUIREMENTS AND NORMAL SERVICE APPLICATIONS

API Letter Designation	Critical Lubrication Requirements	Normal Engine Service Applications
TA	• Piston Scuffing • Exhaust System Blocking	Mopeds and other Extremely Small Engines (Typically <50 cc)
TB	• Piston Scuffing • Deposit-Induced Preignition • Power Loss due to Combustion Chamber Deposits	Motorscooters and other Highly Loaded Small Engines (Typically 50 cc to 200 cc)
TC	• Ring Sticking • Deposit-Induced Preignition • Piston Scuffing	Various High-Performance Engines (Not Outboards) (Typically 20 cc to 500 cc)
TD	• Piston Scuffing • Ring Sticking • Deposit-Induced Preignition	Outboard Engines

(R) TABLE 2—PERFORMANCE AND SERVICE CLASSIFICATION TWO-STROKE-CYCLE GASOLINE ENGINE LUBRICANTS TEST METHODS AND PRIMARY CANDIDATE OIL PERFORMANCE CRITERIA

Letter Designation	Status	ASTM[1] Designation	Test Engine	Primary Performance Criteria
TA	Obsolete[4] as of MAR93	Not yet assigned	Yamaha CE50S	Tightening—Method in preparation Exhaust Blocking—Method in preparation
TB	Obsolete[5] as of MAR93	Not applicable	Vespa 125TS	Tightening—Method never developed Preignition—Method never developed Power Loss—Method never developed

SAE J2116 Revised JUN93

(R) TABLE 2—PERFORMANCE AND SERVICE CLASSIFICATION TWO-STROKE-CYCLE GASOLINE ENGINE LUBRICANTS TEST METHODS AND PRIMARY CANDIDATE OIL PERFORMANCE CRITERIA (CONTINUED)

Letter Designation	Status	ASTM[1] Designation	Test Engine	Primary Performance Criteria	
TC		D 4859-88	Covers Category TC comprehensively, including Primary Performance Criteria		
		D 4857-88	Yamaha RD 350B	Ring Sticking/Deposits In two (crossover) test runs	
				Second Ring Sticking, Avg	0.5 Max below Reference Oil
				Piston Skirt Varnish, Avg.	0.5 Max below Reference Oil
				Plug Fouls	2 Max above Reference Oil
				Preignition (major)	1 Max per run
				Exhaust Blocking	10% Max above Reference Oil
				Scuff/Seizure	None
				In one (without crossover) test run	
				Second Ring Sticking, Avg.	9.0 Min
				Piston Skirt Varnish, Avg.	Absolutely equal or better than Reference Oil
				Plug Fouls	1 Max
				Preignition	None
				Exhaust Blocking	5% Max above Reference Oil
				Scuff or other Lube-related damage	None
		D 4858-88	Yamaha CE50S	Preignition	
				Preignitions (major)	1 Max
				Other	See D 4859, Paras 6.4.2 and 6.4.3
		D 4863-88	Yamaha CE50S	Lubricity	
				Torque Drop	No more than reference oil within 90% confidence limit

SAE J2116 Revised JUN93

(R) TABLE 2—PERFORMANCE AND SERVICE CLASSIFICATION
TWO-STROKE-CYCLE GASOLINE ENGINE LUBRICANTS
TEST METHODS AND PRIMARY CANDIDATE OIL PERFORMANCE CRITERIA (CONTINUED)

Letter Designation	Status	ASTM[1] Designation	Test Engine	Primary Performance Criteria	
TD	Obsolete[3] as of MAR83	D 4681-87[2]	OMC 90 HP	Outboard Lubrication	
				Accelerated Lubricity	No piston scuff or significant bore damage
				Top Ring Sticking, Avg.	Not more than 1.0 points below reference oil engine
				Piston Varnish, Avg.	Not more than 0.5 points below reference oil engine
				Preignition	No more in reference oil engine
				Plug Fouling	Max of one more than in reference oil engine
				Exhaust Port Blocking	Max of 10% more than in reference oil engine

[1] Latest version of the ASTM designation should be used.
[2] The engine test in this Standard Specification is identical to that in National Marine Manufacturers (NMMA) TC-W (312-84).
[3] This category has been superseded and is no longer recommended by the National Marine Manufacturers Association.
[4] CEC withdrew support for this category.
[5] Test sponsor no longer desires this oil category.

(b) The JASO and Proposed ISO Two-Stroke Classifications

The API classifications listed above are generally recognized as obsolete, although the TC level is considered to define the minimum performance requirement for small land-based two-stroke engines.

Smoke emission from two-strokes has become a major concern, particularly in cities in Asia where many mopeds and passenger-carrying tricycles have produced major air-pollution problems.

The Japanese Standards Organization (JASO) has developed a reference oil, JATRE 1, that combines acceptable smoke levels with adequate detergency and other performance requirements. They have also developed engine tests that are used to define three quality levels using JATRE 1 as reference. These quality levels are:

FA - for use in less-developed countries
FB - a minimum domestic standard
FC - the recommended level for Japan

This classification system has provoked considerable interest and has been proposed as the basis of a new global system to be developed and administered by the International Organization for Standardization (ISO). European evaluation of JATRE 1 and the JASO classification indicated that the FC level was inadequate for general use and a new higher level of quality (but still based on JATRE 1 and the Japanese engine tests) was proposed. The ISO global classification would then be:

GB - corresponds to FB
GC - corresponds to FC
GD - recommended level for Europe

The engine tests and the limits for JASO and proposed ISO qualities are summarized below:

JASO Engine Tests

Test	Engine	Fuel/Oil Ratio	Parameters Measured
M340	Honda DIO AF27	50/1	Starting torque. Lubricity/ seizure (torque drop @ 300°C engine temperature)
M341	Honda DIO AF27	100/1	Detergency (ring sticking, piston cleanliness, and combustion deposits)
M342	Suzuki SX-800R	10/1	Exhaust smoke density (by light absorption)
M343	Suzuki SX-800R	5/1	Exhaust blocking (time to back pressure rise)

The Suzuki SX-800R is a small generator engine which is operated with a resistive load.

The Honda DIO is a small moped engine operated on a dynamometer/brake.

Current Pass/Fail Limits for JASO and Proposed ISO Specifications
(Minimum merit requirements versus JATRE 1 as reference)

		Quality Level			
Parameter	**Engine**	**FA**	**FB/GB**	**FC/GC**	**GD**
Smoke	Suzuki	30	45	85	85
Exhaust blocking	Suzuki	40	45	90	90
Initial torque	Honda	98	98	98	98
Lubricity	Honda	90	95	95	95
Detergency	Honda	80	85	95 (1-hour test)	125 (3-hour test)

The European industry is developing a new Piaggio engine test which they will propose for inclusion in GD or a higher quality level.

Other requirements
(Japan only, global not defined)

Viscosity	6.5 cSt min @ 100°C
Flash Point (P-M)	70°C min
Sulfated Ash	0.25% max

(c) Outboard Oil Qualification

The only effective specification is TC-W3, administered by the NMMA. This is a development of the earlier TC-WII, which is now obsolete. From Jan. 1996, new more severe limits are being introduced for the OMC 70HP and Mercury 15HP tests. Two passing results are required in the latter test.

TC-W3 Test Requirements from 1996:

Property	**Engine**	**Requirement**
General Performance	OMC 40HP	Average piston and ring varnish ratings better than/equal to "reference rating less 0.6."
Lubricity	Yamaha CE50S	Torque drop equivalent to reference at 90% confidence level.
Pre-ignition	Yamaha CE50S	Less than/equal to reference
Detergency	OMC 70HP	Second ring sticking and piston deposits: better than/ equal to reference.
Detergency	Mercury 15HP	No compression loss at 100 hours. Piston/ring scuffing better than/equal to limits. Needle bearing test must pass.

Other Tests

Miscibility	Less than/equal to 110% of reference
Rust	Better than/equal to reference
Filterability	20% maximum flow rate change
Fluidity	7500 cP @ –25°C maximum
Compatibility	Pass with TC-W11 reference oils

Appendix 15

The Engineering Society For Advancing Mobility Land Sea Air and Space®
INTERNATIONAL
400 Commonwealth Drive, Warrendale, PA 15096-0001

SURFACE VEHICLE INFORMATION REPORT

SAE J357 REV. FEB95

Issued 1969-08
Revised 1995-02

Superseding J357 JUN91

Submitted for recognition as an American National Standard

(R) PHYSICAL AND CHEMICAL PROPERTIES OF ENGINE OILS

Foreword—This document discusses a number of the physical and chemical properties of new and used engine oils. Where appropriate, standardized methods of test for these properties are indicated and a detailed listing included in the references section. This document provides those concerned with the design and maintenance of internal combustion engines with information relative to the terms used to describe engine lubricants.

This document may be used as a general guide to engine oil properties and as an outline for establishing oil quality inspection and maintenance programs.

1. ***Scope***—This SAE Information Report reviews the various physical and chemical properties of engine oils and provides references to test methods and standards used to measure these properties. It also includes general references on the subject of engine oils, base stocks, and additives.

2. *References*

2.1 Applicable Documents—The following publications form a part of this specification to the extent specified herein. The latest issue of SAE, ASTM, API, and CEC publications shall apply.

2.1.1 SAE PUBLICATIONS—Available from SAE, 400 Commonwealth Drive, Warrendale, PA 15096-0001.

SAE J183—Engine Oil Performance and Engine Service Classification (Other than "Energy Conserving")
SAE J300—Engine Oil Viscosity Classification
SAE J304—Engine Oil Tests
SAE J1423—Classification of Energy-Conserving Engine Oil for Passenger Cars, Vans, and Light Duty Trucks
SAE J2227—International Tests and Specifications for Automotive Engine Oils

2.1.2 ASTM PUBLICATIONS—Available from ASTM, 1916 Race Street, Philadelphia, PA 19103-1187.

ASTM DS 39b—Viscosity Index Tables for Celsius Temperatures
ASTM D 56—Test Method for Flash Point by Tag Closed Tester
ASTM D 91—Test Method for Precipitation Number of Lubricating Oils
ASTM D 92—Test Method for Flash and Fire Points by Cleveland Open Cup
ASTM D 93—Test Methods for Flash Point by Pensky-Martens Closed Tester
ASTM D 95—Test Method for Water in Petroleum Products and Bituminous Materials by Distillation
ASTM D 97—Test Methods for Pour Point of Petroleum Oils
ASTM D 156—Test Method for Saybolt Color of Petroleum Products (Saybolt Chromometer Method)

SAE J357 Revised FEB95

ASTM D 189—Test Method for Conradson Carbon Residue of Petroleum Products
ASTM D 287—Test Method for API Gravity of Crude Petroleum and Petroleum Products (Hydrometer Method)
ASTM D 322—Test Method for Gasoline Diluent in Used Gasoline Engine Oils by Distillation
ASTM D 341—Viscosity-Temperature Charts for Liquid Petroleum Products
ASTM D 445—Test Method for Kinematic Viscosity of Transparent and Opaque Liquids (and the Calculation of Dynamic Viscosity)
ASTM D 482—Test Method for Ash from Petroleum Products
ASTM D 524—Test Method for Ramsbottom Carbon Residue of Petroleum Products
ASTM D 664—Test Method for Acid Number by Potentiometric Titration
ASTM D 874—Test Method for Sulfated Ash from Lubricating Oils and Additives
ASTM D 892—Test Method for Foaming Characteristics of Lubricating Oils
ASTM D 893—Test Method for Insolubles in Used Lubricating Oils
ASTM D 974—Test Method for Acid and Base Number by Color Indicator Titration
ASTM D 1160—Method for Distillation of Petroleum Products at Reduced Pressure
ASTM D 1298—Test Method for Density, Relative Density (Specific Gravity) or API Gravity of Crude Petroleum and Liquid Petroleum Products by Hydrometer Method
ASTM D 1310—Test Method for Flash Point and Fire Points of Liquids by Tag Open Cup Apparatus
ASTM D 1500—Test Method for ASTM Color of Petroleum Products (ASTM Color Scale)
ASTM D 2270—Method for Calculating Viscosity Index from Kinematic Viscosity at 40 and 100 °C
ASTM D 2500—Standard Test Method for Cloud Point of Petroleum Oils
ASTM D 2887—Test Method for Boiling Range Distribution of Petroleum Fractions by Gas Chromatography
ASTM D 2896—Test Method for Total Base Number of Petroleum Products by Potentiometric Perchloric Acid Titration
ASTM D 2982—Test Method for Detecting Glycol-Base Antifreeze in used Lubricating Oils
ASTM D 3244—Standard Practice for Utilization of Test Data to Determine Conformance with Specifications
ASTM D 3524—Test Method for Diesel Fuel Diluent in Used Diesel Engine Oil by Gas Chromatography
ASTM D 3525—Test Method for Gasoline Diluent in Used Gasoline Engine Oils by Gas Chromatography
ASTM D 3607—Method for Removing Volatile Contaminants from Used Engine Oils by Stripping
ASTM D 3828—Test Method for Flash Point by Setaflash Closed Tester
ASTM D 3829—Test Method for Predicting the Borderline Pumping Temperature of Engine Oil
ASTM D 3945—Standard Test Methods for Shear Stability of Polymer-Containing Fluids Using a Diesel Injector Nozzle
ASTM D 4055—Test Method for Pentane Insolubles by Membrane Filtration
ASTM D 4485—Performance Specification for Automotive Engine Oils
ASTM D 4530—Test Method for Micro-Carbon Residue of Petroleum Products
ASTM D 4628—Test Method for Analysis of Barium, Calcium, Magnesium, and Zinc in Unused Lubricating Oils by Atomic Absorption Spectrometry
ASTM D 4683—Test Method for Measuring Viscosity at High Temperature and High Shear Rate by Tapered Bearing Simulator
ASTM D 4684—Test Method for Determination of Yield Stress and Apparent Viscosity of Engine Oils at Low Temperature
ASTM D 4739—Test Method for Base Number Determination by Potentiometric Titration
ASTM D 4741—Test Method for the Measurement of Viscosity at High Temperature and High Shear Rate by Tapered Plug Viscometer
ASTM D 4927—Test Methods for Elemental Analysis of Lubricant and Additive Components—Barium, Calcium, Phosphorus, Sulfur, and Zinc by Wavelength-Dispersive X-Ray Fluorescence Spectroscopy
ASTM D 4951—Determination of Additive Elements in Lubricating Oils by Inductively Coupled Plasma Atomic Emission Spectrometry

SAE J357 Revised FEB95

ASTM D 5002—Standard Test Method for Density and Relative Density of Crude Oil by Digital Density Analyzer
ASTM D 5119—Test Method for Evaluation of Automotive Engine Oils in the CRC L-38 Spark Ignition Engine
ASTM D 5133—Standard Test Method for Low Temperature, Low Shear Rate, Viscosity/Temperature Dependence of Lubricating Oils Using a Temperature Scanning Technique
ASTM D 5185—Standard Test Method for Determination of Additive Elements, Wear Metals, and Contaminants in Used Lubricating Oils by Inductively Coupled Plasma Atomic Emission Spectrometry
ASTM D 5293—Standard Test Method for Apparent Viscosity of Engine Oils Between -5 °C and -30 °C Using the Cold Cranking Simulator
ASTM E 1131—Test Method for Compositional Analysis by Thermogravimetry
ASTM MNL 1—Manual on the Significance of Tests for Petroleum Products: 5th Edition
ASTM STP 1068—High Temperature/High Shear (HTHS) Oil Viscosity: Measurement and Relation to Engine Operation
ASTM STP 1143—Low Temperature Lubricant Rheology Measurement Relevance to Engine Operations

2.1.3 OTHER PUBLICATIONS

API Publication 1509, latest edition, plus revisions
Federal Test Method Standard No. 791C, Method 203.1, "Pour Stability of Lubricating Oils"
Federal Test Method Standard No. 791C, Method 3470.1, "Homogeneity and Miscibility of Oils"
American Society of Lubrication Engineers, 1951, "The Physical Properties of Lubricants"
General Motors Engineering Standard GM 9099P, "Engine Oil Filterability Test (EOFT)," May 1980
CEC L-40-T-87, "Evaporative Loss," NOACK (DIN 51581) Method
CEC L-14-A-88 "Evaluation of the Mechanical Shear Stability of Lubricating Oils Containing Polymers" Method
C.M. Georgi, "Motor Oils and Engine Lubrication," New York: Reinhold Publishing Corporation, 1950
William A. Gruse, "Motor Oils, Performance and Evaluation," New York: Reinhold Publishing Corporation, 1967
A. Schilling, "Automotive Engine Lubrication," Broseley, England: Scientific Publications (G.B.) Ltd., 1972
R.C. Gunderson and A.W. Hart, "Synthetic Lubricants," New York: Reinhold Publishing Corporation, 1962
Dieter Klamann, "Lubricants and Related Products," Weinheim: Verlag Chemie (F.R.G.) GmbH, 1984
M. Campen, D. Kendrick, and A. Markin, "Growing Use of Synlubes," Hydrocarbon Processing, February 1982
G.J. Schilling and G.S. Bright, "Fuel and Lubricant Additives—II. Lubricant Additives," Lubrication Vol 63, No. 2, 1977
N. Benfaremo and C.S. Liu, "Crankcase Engine Oil Additives," Lubrication Vol 76, No. 1, 1990

3. General Description of Engine Oil Components—Modern engine oils consist of (1) base stocks and (2) the additives that are necessary to produce the required finished product performance. These engine oil components will be described in the following sections.

3.1 Base Stocks—Base stock properties have assumed increased importance with the addition of Base Oil Interchangeability Guidelines to the API Engine Oil Licensing and Certification System. These guidelines are intended to determine engine sequence tests required for variations in base stock compositions. See API Publication 1509.

3.1.1 REFINED PETROLEUM BASE STOCKS—Crude petroleum oil as it comes from the ground is a mixture of literally hundreds of hydrocarbon molecules of three basic types—paraffinic, naphthenic, and aromatic. Crude oils are classified according to the predominant type of hydrocarbon molecules they contain.

SAE J357 Revised FEB95

The first step in refining crude oil into useful products is the separation according to boiling range by atmospheric and/or vacuum distillation. The various fractions are then further processed into gaseous products, gasoline, diesel and burner fuels, lubricating oil stock, asphalt, etc. The lubricating oil stock is vacuum distilled, providing a series of base stocks of various levels of volatility and viscosity. The less viscous distillate fractions are called "neutrals" and the higher viscosity residual fractions are called "bright stocks." These fractions generally require further treatment to make them suitable for use as engine oils.

Historically, a nomenclature has evolved to identify neutral and bright stock fractions by their Saybolt viscosities. Neutral fractions are referred to by their nominal viscosity at 100 °F in Saybolt Universal Seconds (SUS). For example, a 150 neutral is a distillate fraction with a nominal viscosity of 29 mm^2/s (cSt) at 40 °C (150 SUS at 100 °F). Generally speaking, the viscosity for the distillate (neutral) fractions range from about 11 to 150 mm^2/s (cSt) at 40 °C (60 to 700 SUS).

In similar fashion, bright stocks are referred to by their nominal SUS viscosity at 210 °F rather than 100 °F. For example, a 150 bright stock is a residual fraction with a nominal viscosity of 30.6 mm^2/s (cSt) at 100 °C (150 SUS at 210 °F). For comparison purposes, if the viscosity of the bright stocks were measured at 40 °C rather than 100 °C, they would range from 140 to about 1600 mm^2/s (cSt) (650 to 7400 SUS).

The as-distilled base stock fractions may contain nitrogen- and sulfur-containing compounds, metal-containing compounds, and aromatic hydrocarbons of various structures. Many of these compounds can adversely affect the stability and performance properties of base stocks and the ability of various additives to enhance these properties for engine oil applications. These compounds are usually removed through extraction processes, using solvents such as phenol, furfural, or N-methyl pyrrolidone, or are modified by hydrotreating or hydrocracking.

Both hydrotreated and hydrocracked base stocks are typically composed of higher percentages of saturates and reduced sulfur contents relative to solvent refined petroleum base stocks. Hydrocracking can also increase the proportion of iso-paraffins at the expense of the less desirable hydrocarbons.

Waxy materials present in the base stock fractions may crystallize and agglomerate or congeal at low temperatures and thereby impede low temperature flow. These materials may be removed by solvent dewaxing processes employing solvents such as methylethyl ketone or propane, or by catalytic dewaxing.

These extraction and modification processes can be carried out either before or after final distillation into viscosity fractions. The choice depends on the processes employed.

The physical and chemical properties of the finished base stocks (often referred to as "virgin") will not be solely a function of crude source, but also will be dependent on the processes employed and the extent of refining employed.

3.1.2 RE-REFINED OR RECYCLED PETROLEUM BASE STOCKS—Used lubricating oils have been involved in recycling processes for over 65 years. Recycling normally involves the removal of volatile components produced in use as well as water, insolubles, and dirt. Little, if any, additional processing is involved. The resulting oil is not normally considered to be suitable for use in modern engines. However, it is often blended with other materials and burned as a fuel.

SAE J357 Revised FEB95

Recently, legislation and technical improvements in re-refining processes have increased the interest in re-refined oils. Re-refined petroleum base stocks may be manufactured from used oil by re-refining processes. Re-refined stock shall be substantially free from additives and from contaminants introduced from the re-refining process or from previous use. Re-refined oils can undergo one or more of the following processes: water separation, additive separation, solvent extraction, hydrotreating, and re-fractionation. The resulting finished, re-refined oil is often virtually indistinguishable from good quality virgin base stocks. These re-refined oils may be suitable for use in modern engines when treated with appropriate additives.

3.1.3 SYNTHETIC BASE STOCKS—Certain chemical compounds have been found to be suitable as base stocks for engine oil. These are referred to as synthetic lubricants and are defined as having been produced by chemical synthesis. These are manufactured by organic reactions such as alkylation, condensation, esterification, polymerization, etc. Starting materials may be one or more relatively pure organic compounds. Generally of simple composition, these compounds are obtained by chemically processing fractions from petroleum, natural gas, vegetable, or animal oil components. When vegetable or animal oil base lubricants are derived from natural, nonpetroleum sources rather than from synthesis, they are not considered synthetic lubricants unless the naturally occurring product has been chemically changed.

Classes of chemical compounds that might be used as synthetic base stocks after processing are shown in Table 1 along with distinct generic identification of the resulting fluids. A synthetic lubricant base stock may consist of any of the fluids shown in Table 1, or a mixture of compatible base fluids. This blending is usually practiced to enhance physical properties.

TABLE 1—CLASSIFICATION OF SYNTHETIC BASE FLUIDS

Class	Synthetic Fluids (Examples)
Synthetic Hydrocarbons	
Alkylated Aromatics	Alkylbenzenes
Polyolefins	Polyalphaolefins (Hydrogenated)
	Polybutenes
Organic Esters	
Dibasic Acid Esters (Diesters)	Adipates, Azelates, Dodecanedioates
Polyol Esters	Neopentyl or Hindered Esters
Polyesters	Dimer Acid Esters
Others	
Halogenated Hydrocarbons	Chlorofluorocarbon Polymers
	Fluoroesters, Fluoroethers
Phosphate Esters	Phosphate Esters of Isopropyl Phenol and Cresylic Acids
Polyglycols	Polyalkylene Glycols
Polyphenyl Esters	Meta bis (m-Phenoxyphenyl) benzene
Silicate Esters	Disiloxane Derivatives
Silicones	Phenyl, Methyl, and Alkylmethyl Silicones

Some synthetic base stocks are compatible with petroleum base stocks, and the two types may be blended to obtain desired physical and chemical properties. Such combinations are referred to as "partial synthetic base stocks."

Some synthetic base stocks are not compatible with either other synthetics or with petroleum base stocks. Therefore, lubricants containing synthetic base stocks should not be indiscriminately mixed.

SAE J357 Revised FEB95

The additive agents necessary in petroleum base stocks, synthetic base stocks, or partial synthetic blends intended for engine oils are also synthesized materials. However, even though these materials are synthesized, they should be referred to as additives and not included in the base stock description.

3.2 Additive Agents—A lubricant additive agent is defined as a material designed to enhance the performance properties of the base stock or to impart to the base stock properties that do not naturally exist with the base stock. These additive agents are used at concentration levels ranging from several parts per million to greater than 10 volume percent. Generally, additives are materials that have been chemically synthesized to provide the desired performance features, and they frequently contain an oil-solubilizing hydrocarbon portion as part of the molecule. Some additive agents are naturally occurring materials that have undergone only minor modifications to obtain the desired property. Additives can carry out their task of enhancing or imparting new properties to an oil in one of three ways—protection of engine surfaces, modification of oil properties, protection of base stocks. Engine protectors include seal swell agents, antiwear agents, extreme pressure (EP) agents, antirust agents, corrosion inhibitors, detergents, dispersants, and friction modifiers. Oil modifiers include pour point depressants, antifoam agents, and viscosity index (VI) improvers. Base stock protectors include antioxidants and metal deactivators. Some additives possess multifunctional properties.

Additive combinations contribute performance features which are required to satisfy the lubrication needs of modern engines under the most severe conditions and currently recommended oil change intervals. If additives have been put into the base stock to increase its commercial value to those who will use it for formulating engine oil, care should be taken to identify such additives so that further additive treatments will be compatible.

4. Physical and Chemical Properties—Understanding and agreeing on the methods of measurement of the physical and chemical properties of base oils and formulated engine oils can assist the user, the oil refiner, and the formulator to define a consistently uniform product. These properties are often used to establish acceptable levels of additive components in finished oils. Although oil performance in the engine is related to base stock and additive composition, it is often difficult to assign a specific aspect of such performance totally to the use of a specific additive or base stock. Part of the reason is that some of the physical and chemical properties of the oil overlap in their influences on engine performance and durability and it is presently difficult to directly and unambiguously attribute such effects to either the chemical or the physical properties of the oil. Progress in developing this level of understanding is being made. Some of these performance characteristics of engine oils are discussed in SAE J183, SAE J300, SAE J1423, and SAE J2227. At the present time, oil physical and chemical properties can be related to engine performance and durability only with the guidance of engine manufacturers and with appropriate and jointly accepted engine and/or field tests successfully completed on that oil.

At low temperatures where cranking/starting and engine oil pumpability are matters of concern, the physical properties of the engine oil can be directly related to its effects on any particular engine. Engine cranking and starting and oil pumpability are also related to a variety of other factors including engine response to oil rheology, as well as to nonrheological factors such as battery power and fuel volatility.

While the physical and chemical properties of an oil at operating temperatures are not related to oil performance in a simple way, these individual properties are meaningful and are related to the oil's ability to fulfill its function as a lubricant. These and the low temperature properties will be discussed in the following sections.

SAE J357 Revised FEB95

4.1 Viscosity—Viscosity of the engine oil is one of its most important and most evident properties. If sufficiently high, it is the source of the phenomenon of hydrodynamic lubrication in which the viscosity of the oil forces the bearing surface to ride on a thin film of oil and, thus, protect the lubricated surface from wear. Chemical additives, fuel dilution, contaminants from within and outside of the engine, wax in the oil, oil oxidation, volatilization, and many other materials found in or added to the oil affect the viscosity in advantageous or disadvantageous ways.

4.1.1 DEFINITION—Viscosity is defined as the internal resistance to flow of any fluid. It is expressed as follows in Equation 1:

$$\text{Dynamic Viscosity} = \frac{\text{Force / Sheared Area}}{\text{Velocity / Film Thickness}} = \frac{\text{Shear Stress}}{\text{Shear Rate}} \qquad \text{(Eq.1)}$$

The unit of measure for dynamic viscosity is the millipascal second (mPa·s), although the centipoise (cP) is also commonly used. One mPa·s equals 1 cP. Oils that exhibit a constant viscosity at all shear rates in this equation are known as "Newtonian" oils. In the absence of polymeric additives, most single grade oils are in this category at temperatures above their cloud point.

Oils that exhibit a viscosity which varies with changing shear rates in this equation are known as "non-Newtonian" oils. Multiviscosity graded oils formulated with polymeric additives are generally in this category.

Another form of viscometric expression involves the use of kinematic viscometers in which the liquid is driven by its own hydrodynamic head. This head varies directly with the density of the oil at the temperature of measurement. The relationship between kinematic and dynamic viscosity is as follows in Equation 2:

$$\text{Kinematic Viscosity} = \text{Dynamic Viscosity/Density of Liquid} \qquad \text{(Eq.2)}$$

The unit of measurement for kinematic viscosity is the millimeter squared per second (mm^2/s), although the centistoke (cSt) is commonly used. One mm^2/s equals 1 cSt. Density effects should be eliminated either by measuring the dynamic viscosity or by measuring kinematic viscosity and density at the temperature of interest and converting the values to dynamic viscosity.

4.1.2 VISCOSITY INDEX (VI)—Viscosity decreases rapidly with increasing temperature. For most oils, the relationship between viscosity and temperature can be approximated by the following empirical relationship in Equation 3:

$$\text{loglog(kinematic viscosity + 0.7)} = A + B\ \text{log(absolute temperature)} \qquad \text{(Eq.3)}$$

where:

A and B are constants, specific for each oil

This relationship, which is an approximation of the MacCoull, Walther, Wright equation, forms the basis for special viscosity temperature charts published in ASTM D 341. These charts permit the plotting of viscosity-temperature data as straight lines over the temperature range in which the oils are homogeneous liquids. The slope of these lines is a measure of the change in viscosity with temperature. It is dependent on the chemical composition of the oil and is described by an empirical relationship called VI. The higher the VI, the smaller the change in viscosity with temperature (slope). ASTM D 2270 is used to determine VI values. ASTM DS 39b is based on ASTM D 2270 and allows for more convenient determination of VI.

SAE J357 Revised FEB95

For engine oils, a relatively smaller change in viscosity with temperature (high VI) is desirable to provide a wider range of operating temperatures over which a given oil will provide satisfactory lubrication. At low temperatures, a relatively low viscosity oil is desirable to permit adequate cranking speed during starting, and then adequate flow to the oil pump and the entire engine oiling system after starting.

At high temperatures in a running engine, the oil viscosity must be high enough to maintain adequate film thickness between rotating or rubbing parts to minimize wear. Using a higher viscosity oil generally reduces oil consumption (past piston rings and valve guides) and blowby, but increases friction associated with oil film shearing in the piston/piston ring cylinder wall interface and bearings.

4.1.3 VISCOSITY INDEX IMPROVERS—To extend the upper temperature limit at which an oil will still provide satisfactory lubrication, polymeric additives, called Viscosity Index (VI) improvers, are widely used. Engine oils properly formulated with VI improvers generally contain lower viscosity base stocks which provide better low temperature cranking/starting and pumpability properties. As the oil temperature increases, the viscosity of the oil containing a VI improver decreases more slowly than the same oil without a VI improver, thus increasing the VI. The result is an oil that can give good starting/pumping response and is also effective in providing a more viscous oil film at operating temperatures than could be obtained with a single grade oil providing equivalent startability at low temperatures.

Oils containing a polymeric VI improver exhibit a decrease in viscosity as the shear rate or stress is increased. Because the viscosity of such oils depends on shear stress, they are called "non-Newtonian oils." Such change generally lasts only as long as the oil is operated under such high shear stress. When the shear stress is relieved, the oil reverts to its previous viscosity. This reversible decrease in viscosity due to shear is called "temporary shear (or viscosity) loss." When certain critically high shear stresses are imposed on a VI improver in oil solution, the viscosity contribution of the VI improver to both low and high shear rate viscosity can be permanently reduced. This nonreversible reduction in viscosity is called "permanent shear (or viscosity) loss." The magnitude of these temporary and permanent losses is dependent on the type and molecular weight of the VI improver used, as well as the actual service conditions.

The permanent shear stability characteristics of engine oils are evaluated by comparing the stripped viscosity of an engine oil after 10 h in the CRC L-38 engine test to the new oil viscosity. Shear stability can also be measured using CEC L-14-A-88.

4.1.4 VISCOSITY MEASUREMENT—The SAE J300 standard classifies oils into grades according to their kinematic viscosities measured at low shear rates and high temperature (100 °C), and their viscosities at high shear rate and high temperature (150 °C), and at both low and high shear rates at low temperatures (–5 to –35 °C).

Low shear rate kinematic viscosity is measured using ASTM D 445 and is reported in millimeters squared per second (mm^2/s), although the centistoke (cSt) is also commonly used. Kinematic viscosity is measured most commonly at 100 °C, and also at 40 °C if VI is to be determined.

At low temperature, the high shear rate viscosity is measured by ASTM D 5293. This is a multitemperature cold cranking simulator method. The low temperature, low shear rate viscosity is measured by ASTM D 4684. Both ASTM D 5293 and ASTM D 4684 report viscosities in millipascal seconds (mPa·s), although the centipoise (cP) is also commonly used. Results of both tests have been shown to correlate with engine starting and engine oil pumpability at low temperatures, although the precise correlation to modern engines has been questioned and is under active investigation with ASTM.

SAE J357 Revised FEB95

Oil viscosity at very high shear rates/stresses and at high temperatures (150 °C) is measured using ASTM 4683 or ASTM 4741 and is reported in millipascal seconds (mPa·s), although the centipoise (cP) is also commonly used. A comparable Capillary Viscometer method is currently being developed by ASTM. These methods are intended primarily to simulate operating conditions occurring in engine bearings.

4.2 Other Tests Pertinent to New and Used Oils

4.2.1 CLOUD POINT AND POUR POINT—The cloud point of a moisture-free oil is defined as the temperature at which a cloud or haze appears in the lower portion of the test oil when tested (i.e., cooled) by ASTM D 2500. The haze indicates the presence of some insoluble fractions, such as wax, at the temperature noted. In most applications, this haze will have little practical significance.

The pour point of an oil is defined as the lowest temperature at which the oil can be poured when tested by ASTM D 97. The pour point can be directly related to whether or not the oil can be poured from a container at low temperatures. Although pour point is a simple measure of wax crystal structure and low temperature viscosity, more precise and correlatable viscometric methods, such as ASTM D 3829 and ASTM D 4684, have been developed which better predict the ability of an oil to flow to the oil pump and throughout the system at low temperature. In actual practice, the oil in the crankcase will be a mixture of oil and small amounts of fuel fractions, the composition depending on several factors (see 4.3.4).

Some oils display an increase in pour point when exposed to repeated cycling at temperatures below and above the pour point. Appendix B of SAE J300 (taken from Federal Test Method, Standard No. 791C, Method 203.1) describes a procedure, commonly called the stable pour point, for evaluating the tendency of the pour point to so increase. While no longer a mandatory procedure in the low temperature classification requirements of SAE J300, the measurement of the stable pour point continues to be recommended when significant changes in formulation or base stock sources are made.

4.2.2 FLASH POINT AND FIRE POINT—The flash point of a petroleum product is the lowest temperature to which the product must be heated under specific conditions to give off sufficient vapor to form a mixture with air that can be ignited momentarily by a specified flame. Fire point is the lowest temperature to which a product must be heated under prescribed conditions to burn continuously when the mixture of vapor and air is ignited by a specified flame.

Flash and fire points are significant from the viewpoint of safety and should be related to the temperatures to which petroleum products will be subjected in storage, transportation, and use. Normally, engine oils will present no hazards in this respect but the minimum flash point that can be tolerated must be determined in each application. Flash point may also be used to indicate gross contamination of used oil by a volatile product such as gasoline or diesel fuel. Methods of obtaining this type of information are ASTM D 56, ASTM D 92, ASTM D 93, ASTM D 1310, and ASTM D 3828. ASTM D 92 and ASTM D 93 are the preferred methods for unused engine oils.

4.2.3 DISTILLATION DATA—The volatility characteristics of engine oils can be defined by distillation procedures. Because engine oils are comprised of relatively high boiling point fractions, which would thermally crack in an atmospheric distillation, a reduced-pressure (vacuum) distillation method, i.e., ASTM D 1160, must be used. ASTM D 2887, which gives boiling-range distribution data by gas chromatography, has gained acceptance and is often used in place of ASTM D 1160. ASTM D 2887 and CEC L-40-T-87 are both currently used to measure the evaporative loss or volatility of engine oils. Neither of these methods is entirely satisfactory and new methods are under development by ASTM. Correlations between performance characteristics, such as oil consumption, and the volatility characteristics of the oil in use need to be developed with actual engine tests.

SAE J357 Revised FEB95

4.2.4 ALKALINITY AND ACIDITY—The alkalinity or acidity characteristics of petroleum products can be measured by any one of several standardized methods. Methods currently used include ASTM D 664, ASTM D 974, ASTM D 2896, and ASTM D 4739. Changes in alkalinity or acidity with use give some indication of the nature of the changes taking place in the engine oil. For example, a reduction in base number can be ascribed to neutralization of basic additive components such as metal containing detergents as well as certain ashless dispersants. An increase in acid number may be ascribed to engine oil oxidation and/or contamination by products of combustion. Base number of a new oil is an indication of an oil's ability to resist the deleterious effects associated with high sulfur levels in diesel fuels. Different titration methods may yield different base numbers on the same oil. Therefore, caution is necessary in applying base number—oil performance relationships. For diesel engines, relationships have been published between base number of the new oil, change in base number during service, fuel sulfur content, and desired engine oil drain interval. The change in base number in service can be used under certain conditions to evaluate engine oil change interval practices. Both ASTM D 2896 and ASTM D 4739 methods are commonly used in these instances.

4.2.5 CARBON RESIDUE—The base stock components of engine oils are mixtures of many compounds that differ widely in their physical and chemical properties. Some vaporize at atmospheric pressure without leaving an appreciable residue. When destructively distilled, the nonvolatile compounds may leave a carbonaceous material known as carbon residue. Two methods used for evaluating base stocks in this respect are ASTM D 189 and ASTM D 524. ASTM D 4530 is essentially equivalent to ASTM D 189 while minimizing sample size. Engine oils containing ash-forming constituents, such as the additives commonly used in formulating oils, may give misleading high carbon residues by either method. Carbon residue has little value as a guide for predicting deposit-forming tendencies in automotive engines, but may relate to intake port deposits in certain large, two stroke cycle, natural gas fueled stationary engines.

4.2.6 ASH CONTENT—The amount of ash formed from burning engine oils may be obtained by ASTM D 482. However, ASTM D 874 is now the method most commonly used because it is a more accurate measure of ash-forming constituents. When tested by ASTM D 482, some metals are partially volatilized and lost, giving erroneously low values. The ash produced from burning new engine oils is principally related to the concentration of ash-producing additives in the oil. In addition to the additive contribution, the ash produced by used oils will also be a function of the amount of contaminants, such as lead compounds present in the engine oil if the engine is operated on a leaded gasoline. High values can also result from other contaminants, such as dirt, iron oxide, wear metals, and corrosion products. Ash forming substances in an oil may contribute to deposits on combustion chamber surfaces, spark plugs, and intake or exhaust valves, which can influence the combustion characteristics, exhaust valve sealing and certain driveability characteristics of an engine. However, the mechanism for the buildup of deposits in these areas is very complex and depends on many variables in addition to the ash content of the oil.

4.2.7 COMPATIBILITY—Engine oils are expected to be homogeneous and completely miscible with all other engine oils with which they might be mixed in service. When oils are mixed in any proportion, there should be no evidence of separation either of the additives or of the oils when the mixed oils are heated to a temperature as high as 232 °C and cooled to a temperature as low as the pour point of the mixture. The homogeneity and miscibility test currently used to evaluate engine oils is Federal Test Method Standard No. 791C, Method 3470.1.

4.2.8 FOAMING—Oils with poor antifoaming characteristics have been shown to result in decreased oxidation resistance and reduced lubricant efficacy. A bench test for determining this quality is ASTM D 892.

SAE J357 Revised FEB95

4.2.9 GRAVITY, COLOR, ODOR—Gravity (density) may be used to characterize the basic hydrocarbon type of the base stocks. Gravity and color are factors generally associated with the quality control of manufactured products rather than with performance characteristics. ASTM D 287 and ASTM D 1298 may be used to determine the gravity and density characteristics of oil. The color of engine oils may be specified by using ASTM D 156 or more commonly ASTM D 1500.

It is expected that engine oils will not produce offensive odors due to the nature of the base stocks or the additive agents with which the oil is compounded; nor should offensive odors or toxic vapors be generated during use of, or prolonged storage of, an engine oil. There are no standardized odor tests suitable for engine oils.

4.2.10 ELEMENTAL ANALYSIS—Elemental analysis of engine oils is often used as a means of quality control. Instrumental analytical techniques, such as emission spectroscopy, ICP, atomic absorption spectroscopy, and X-ray emission spectroscopy, are useful in this respect. Similar analyses of used oils will provide information relative to the changes in the elemental content of the engine oil. These data can also give a measure of contamination by materials such as ingested dirt, coolant, or products of combustion, especially with engines using leaded gasoline. They also can provide information relative to the extent of wear in the engine. Concentrations of the following elements are commonly determined:

a. Additive elements such as barium, boron, calcium, copper, magnesium, molybdenum, nitrogen, phosphorus, silicon, sodium, sulfur, and zinc.
b. Contaminants such as lead, silicon, chlorine, bromine, and potassium.
c. Wear metals such as aluminum, chromium, copper, iron, lead, molybdenum, and tin.

4.2.11 INFRARED ANALYSIS—Infrared spectrophotometry techniques are valuable in identifying the chemical structures found in base stocks and additives. Changes in these structures can be determined by comparing results of analyses of new and used oils. In used oils, it is also possible to measure oxidation and/or nitration, and to identify the presence of contaminants (e.g., fuel dilution and fuel soot), water, antifreeze, and similar materials.

4.2.12 FILTERABILITY—The tendency of an oil to form gels or other filter plugging material in the presence of water can be evaluated by use of the General Motors Engineering Standard GM 9099P. This method is being considered for standardization by ASTM in the near future.

4.3 Tests Pertinent to Used Oils

4.3.1 USED OIL PROPERTIES—The analysis of a used engine oil may be of value in establishing the condition of the engine and may be helpful in estimating the remaining useful life of the oil. To be of most value, used oil analyses must be taken periodically during the drain interval and a trend line established. The conditions of usage also must be considered in evaluating used oil analyses.

4.3.2 INSOLUBLE CONTENT—Insoluble materials found in both new and used engine oils may be determined using ASTM D 91, or the more frequently used ASTM D 893. Use of these methods permits an evaluation of the contaminant content and buildup of insoluble materials through oxidation, etc. However, the results must be judged with care, because minor changes in the analytical procedure can produce different results. For example, the age and purity of the coagulant solutions specified in ASTM D 893 can affect the results obtained.

With modern highly dispersant oils, the determination of insolubles has become increasingly difficult. Use of coagulant in ASTM D 893 may be required to make accurate determinations. ASTM D 4055 measures all insolubles greater in size than 0.8 mm. However, current interlaboratory precision (reproducibility) is poor.

SAE J357 Revised FEB95

4.3.3 COOLANT (MOISTURE) CONTENT—Small quantities of water will frequently be found in used engine oil as contamination from products of combustion, leakage from the cooling system, or condensation from atmospheric moisture. ASTM D 95 defines a process for determining the water content of used oil. For a qualitative determination, a commonly used simple test is to heat a drop of oil on aluminum foil. A snapping or crackling sound indicates free or suspended water in the oil. Cooling system leakage can be suspected when water is found in the oil on cool down after operation for several hours under high temperature conditions, such as interstate highway driving. The presence of glycol can be a more definitive indication of leakage. Glycol is detected by distillation of the aqueous material, followed by chemical analyses or infrared spectrophotometry on the distillate. A less complicated procedure which is adaptable to field kit use and gives positive, trace, or negative results is ASTM D 2982. Some additives commonly used in formulating engine oils contain glycol at a level that will give a positive result. If the new oil gives a positive result, the test in its simple form will be inadequate for detecting coolant glycol in used oils and the oil supplier should be consulted for advice.

4.3.4 FUEL DILUTION—Engines in good mechanical condition and operated at normal temperatures will usually show a small amount of fuel dilution in the used engine oil. Low operating temperatures, rich mixtures of fuel and air, and low ambient temperatures will promote fuel dilution, particularly if the engine is in poor mechanical condition or crankcase ventilation is inadequate. High dilution reduces oil viscosity and pour point. The presence of such dilution can cause accelerated wear and promote the formation of sludge, varnish, and rust. The presence of a high dilution level may indicate a need for engine maintenance. Dilution may be determined by ASTM D 322 or by ASTM D 3607. This latter method also produces a dilution-free sample for subsequent analyses. These methods are useful only with gasoline engines since the distillation range for diesel fuels in many cases overlaps that for the engine oils used in diesel engines. Procedures applicable to both diesel and gasoline engine oils are ASTM D 3524 and ASTM D 3525. For laboratories so equipped, ASTM D 2887 is also a suitable method for measuring fuel dilution.

The flash point may also be used to approximate fuel dilution. If the flash point test is utilized, it is extremely important to measure the flash point of the new (reference) oil to establish a correct baseline (see 4.2.2). The infrared spectrophotometer may also be utilized as a test tool to approximate fuel dilution in used oil. Viscosity decrease may also indicate the presence of fuel dilution; however, high levels of soot contamination can mask such reductions.

4.3.5 SOOT CONTENT—Soot content of used diesel engine oils can cause a number of problems related to additive complexing and abnormal viscosity increases. The soot is formed by the combustion process and is especially serious when combustion is incomplete. Currently soot can be detected by ASTM E 1131. A qualitative indication of soot content can be made by viscosity measurement.

5. ***Performance Characteristics***—In the operation of an internal combustion engine, engine oils are expected to lubricate, cool, seal, maintain cleanliness, and protect against wear and corrosion. An oil's ability to perform these functions depends on the combined effectiveness of its base stock and additives, as well as operating conditions, fuel quality, and the design and the mechanical condition of the engine. Although the physical and chemical tests described in the preceding sections can be used for quality control to insure manufacturing uniformity, they are not effective for accurately defining performance characteristics at operating loads and temperatures. Only actual performance evaluations in special laboratory engines and in field tests will define the capabilities of an engine oil. Laboratory diesel and gasoline engine tests that have become industry recognized for evaluating engine oils are described in SAE J304. SAE J183 and SAE J1423 classify oils according to performance criteria based on results from engine dynamometer tests. These criteria are generally correlated with field test results. Where other operational properties are of interest, specific tests must be developed using the equipment and conditions most relevant to a given situation.

SAE J357 Revised FEB95

Although the laboratory engine tests are necessary and valuable aids to engine oil development and evaluation, they have limitations. In many instances, the final proof-of-performance is established by field tests of the oil in actual vehicle service. While no industry standardized field test procedures are currently available, the SAE Lubricants Review Institute has furnished some guidelines in their procedures manual. The most meaningful results on a given oil are obtained by evaluation in the most severe type of service expected to be encountered by a particular engine oil.

6. ***Handling and Disposal of Used Oils***—Continuous, long-term contact with used engine oil has caused skin cancer in laboratory animal tests. Proper protective clothing and equipment (gloves, etc.) should be used when handling used engine oil. Exposed skin areas should be washed thoroughly after exposure to used engine oil.

Clothing should be laundered regularly, especially after contact with used engine oil. In addition, proper disposal of used engine oil is very important. For more information on handling and disposal of used engine oil, contact your oil supplier, the American Petroleum Institute, or appropriate federal, state, or local government agencies.

7. ***Conclusions***—The lubrication requirements for modern engines are extremely complex. Current engine oils are the result of extensive research and development aimed at meeting these requirements. It is not the objective of this document to treat the subject in detail. Rather, the purpose of this document is to define very briefly the terms frequently encountered in discussions of engine oils and engine oil performance for those technical people not directly associated with lubricants and lubricant development. For more detailed information on these matters, the reader is referred to the technical services offered by lubricant and additive manufacturers, appropriate engine manufacturer organizations, and pertinent literature available through the Society of Automotive Engineers, Society of Tribologists and Lubrication Engineers, American Society of Mechanical Engineers, American Petroleum Institute, American Society for Testing and Materials, etc. Information directly related to this document may be found in the listed references (see Section 2).

8. Notes

8.1 Marginal Indicia—The (R) is for the convenience of the user in locating areas where technical revisions have been made to the previous issue of the report. If the symbol is next to the report title, it indcates a complete revision of the report.

PREPARED BY THE SAE FUELS AND LUBRICANTS TECHNICAL COMMITTEE 1—
PASSENGER-CAR TYPE ENGINE OILS

Index

A

AAMA, 267, 308
Abbreviations, 489-492
ABOT test, 554
Accelerated mechanical tests, 10
Acceptable biodegradation, 397
Accuracy, of chemical analysis, 121-122
    in statistics, 139
    of test results, 147-148
ACEA, 115, 232, 260, 262, 263, 412
    classifications, 431
    diesel specifications, 241
    sequences, 238, 264, 581-588
Acetylenes, 497
Acid catalysis prevention, 85
Acid corrosion, causing engine wear, 179
    prevention, 85
Acid formation, problems, 39
Acid number in used oils, 166
Acidity, 116
Acryloid polymers, 71
Action of dispersants, 84
Acute toxicity of additives, 395
Add-on devices, 422
Additive concentrates, control of, 119
Additive manufacturing plant, 70
Additive supplier, role in production of oils, 354
Additive technology, future of, 418-419
Additive treatment, development of, 181
Additives, 69-94
    acute toxicity, 395
    alkaline, 40
    anti-foam, 93
    anti-rust, 80
    anti-wear, 85-88
    ATC classification of, 393

A

Additives (continued)
- boundary lubrication, 19
- definition, 69
- delivery of, 362
- detergent, 3, 182
- dispersing, in lubrication, 38
- effect on emissions, 400
- element determination, 167-168
- extreme pressure, 20, 88
- in gear oils, 279-281
- generic classification of, 393
- in greases, 313
- interchangeability, 94
- market, future, 434
- metal content, quality control, 119
- overbased, 80
- potent ashless dispersant, 182
- sludge-reducing, 3
- specifications for, 358
- toxicology of, 393-394
- traction, 89
- types, new, 188

Africa, EMD locomotives, 339
Afterburners, 422
Agip, 177
AICS, 389
Air binding, 107
Air-conditioner lubricants, 342-343
Air Jet test, 114
Air pollution, electric vehicles, 420
Alcohol, 498
Aldehyde, 498
Alkaline additives, 40
Alkalinity, 116, 338, 399
Alkanes, 496
Alkanolamides, 349
Alkanolamines, 349
Alkenes, 497
Alkyl benzenes, 65, 328
Alkylated aromatics, 65-66
Alkylates in greases, 312
Allison, 300, 607
Allison C-4/Cat.TO-2, 309

A

Aluminum, in used oils, 171
Aluminum greases, 311, 314, 316
Aluminum silicates in greases, 315
American Automobile Manufacturers Association, AAMA, 267
American Petroleum Institute, *see* API
American Society of Lubrication Engineers, 28
American Society for Testing and Materials, *see* ASTM
Amine dithiophosphates, 280
Amines, 85, 304, 305, 500
Amoco, 78
Amontons friction study, 14
Analytical systems, 425
Animal testing studies of toxicity, 396
Anionic emulsifiers, 348
Annular disc clutches, 297
Anti-corrosion, 4
  additives, 283
  properties of emulsifiers, 348
Anti-foam additives, 93, 185, 283
Anti-oxidants, 75-77, 194, 283
  in new formulations, 202
  in two-stroke oils, 337
Anti-rust, additives, 80
  in gear oils, 281
  performance, 207
  properties of emulsifiers, 348
Anti-wear additives, 82, 85-88, 176, 208
Anti-wear agent, ZDDP, 182
  in ATF formulation, 305
API,
  1952 service classification, 187
  "C" series classification, 266
  Certification Mark, 7, 258
  classification, 183, 194, 249, 431
    new, 62
    open-ended, 184
    performance, 187-188
  doughnut, 7
  Engine Oil Classification system 182
  Engine Oil Licensing and Certification System (EOLCS), 257-259, 263
  Engine Service Classification, 231, 250
  performance levels, 185, 265-266
  "S" series classification, 205, 236, 256266

A

API (continued)
- SC classification, 237
- Service Symbol, 250, 356
- SF classification, 226, 237
- starburst, 7
- symbol, 258
- two-stroke tests, 331-332

Apparent viscosity, ASTM D 1092, 317
Approval(s), 426-431
- for automatic transmission fluids, 599-608
- choice of base stocks, 355, 358
- of engine lubricants, 231-273
- of greases, 320-321
- of lubricants, 6-11
  - definition, 232

Approval procedures, outside U.S., 605-606
Approval tests, problems with, 220
Arctic oils, crankcase, 206
Aromatic amines as anti-oxidants, 199
Aromatics, 52, 497
- alkylated, *see* Alkylated aromatics

Ash, 118
Ash level of lubricants, 186
Ashless dispersants, 83, 182, 194
- thermally stable, 188

Ashless formulations, 335
- in outboard motor oils, 330

Ashless oils, 423
Asia/Pacific, 412, 431
- lubricant consumption, 242
- quality control, 429
- two-stroke oils in, 331

ASLE, 28
Association des Constructeurs Européens d'Automobiles, 260
Association Technique de l'Industrie Européene des Lubrifiants, 260
ASTM, 97, 191, 249, 257
- development of classification system, 184
- fuel economy test, 551
- global specifications, 333
- PG-2, 289
- precision data, 148
- Test Methods
  - D 92, 113-114
  - D 93, 113

A

ASTM (continued)
- Test Methods (continued)
  - D 97, 111-113
  - D 130, 126, 288
  - D 217, 317
  - D322, 165
  - D 445, 109
  - D 664, 116-118, 166, 167
  - D 665, 127
  - D 874, 118-119
  - D 892, 115-116, 288
  - D 893, 166
  - D 972, 114
  - D 1298, 116
  - D 2266, 320
  - D 2270, 103
  - D 2602, 109
  - D 2603, 110
  - D 2882, 303
  - D 2896, 116-118, 166-167
  - D 2923, 109
  - D 2983, 104
  - D 3340, 119-120
  - D 3524, 165
  - D 3525, 165
  - D 3828, 114
  - D 3945, 111
  - D 4624, 108, 109
  - D 4628, 120
  - D 4683, 108, 109
  - D 4684, 107, 109
  - D 4739, 116-118, 166-167
  - D 4741, 108, 109
  - D 4951, 121
  - D 5133, 104, 107, 109
  - D 5185, 121
  - D 5293, 107
- Test Monitoring Center, 256
- test programs, 517-520

ATC, 260, 263, 394, 402
- classification additives, 393
- Code of Practice, 263

ATF, 282, 291-309
- approvals and specifications, 304-308

A

ATF (continued)
- formulation, 304-306
- testing, 302-304

ATIEL, 260, 263
Atmospheric distillation, 55
Atomic absorption spectroscopy, 120, 519
Atomic absorption, to measure elemental content of additives, 362
Atomic number, 494
Atomic weight, 494
Atoms, 493
Australia, lubricant sales, 242
- notification laws, 386

Australian National Industrial Chemicals Notification and Assessment Scheme, 388-391
Austria, disposal of used lubricants, 401
- emissions legislation, 328

Austrian engineers AVL, 247
Autoflash test, 117
Automatic transmission, 341
- characteristics of, 294-299
- development of, 292-294
- in Europe, 291
- fluids, 282, 291-309
  - requirements of, 299-301
  - specifications and approvals, 599-608
- Ford Motor Co., 298
- oils for, 432

Automobile plants, industrial lubricants, 343-351
Autopour apparatus, 106
Aviation oil analysis, 170
AVL, 247
Axle and manual transmission lubricant viscosity classification, SAE J306, 589-597
Axle tests for hypoid gear wear, 286-287
Azeleic acids in greases, 314

B

B1-B3 specifications, ACEA, 241
Babbit, bearings from, 86
Bacterial growth in cutting oils, 349
Bactericides in cutting oils, 349
Bands, automatic transmision, 299

B

Barium,
additives, toxicity problems, 207
as detergents, 81
greases, 311
sufonates as detergents, 80
toxicity, 392
Barrels, storage of lubricants in, 374-375
Base Number, 116-118, 518
in used oils, 166-167
Base oil manufacturing plant, 51
Base oil stocks, hydrofinishing, 57
Base stocks, 49-69
alternative, 416-418
checks on synthetic, 363
choice of, 197-199
composition changes, 255
description of, 358
for hypoid gear oils, 290
reclaimed, 62-63
supplier, role in production of oils, 354
synthetic, 49-50
toxicity of, 392-393
unconventional petroleum, 198
Base units in S.I. system, 501
Batch blending, of oils, 364-369
Battery electric vehicles, 420-421
Bearing,
corrosion, 235, 236
tests for, 236
failure, 225
lubrication, in gear oils, 281
wear, 235
wedge generation, 29
"whirl", 29
Beauchamp Tower, 18
Belts, metal, for automatic transmission, 293
Bench performance tests, 124-127
Bench rust test, 127
Bench transmission tests, 303
Bench-rig tests, 519
Benzene ring, compounds based on, 497-498
Benzo-α-pyrene, 403
Berc alcohol/Ketone precipitation, 63

B

Bevel gears, 275, 276, 277
BIA, 331, 332
Big-end bearing lubrication in single cylinder engines, 328
Biodegradability, 329, 330, 335, 403, 412, 424, 431
  of lubricants, 397-398
  of polyalphaolefins, 198
  of used oils, 273
Biodegradable diesel fuel, 417
Biodegradable formulations, 329
Biodegradation tests, 398
Black sludge build up in rocker covers, 169
  problems, 216-219
Blend, making, for new formulation, 205
Blending, 353, 364-370
  in-line, automated, 369-370
Blind testing, 133, 160
Blotter spot test, 168-169, 218
Blowby, 37, 500
Blue Angel system, 403
Boating Industries Association, 331
Borate esters, 67, 340
Borax, 349
Bore polishing, 224, 235
  problems with, 240
Bore wear, 214-216, 240
Borg-Warner, 293
Boric amides and esters, 349
Boron, in used oils, 171
Boundary lubrication, 19-20
  in gearboxes, 279
BP, 177
Brake bands, 296
Brake fluids, 68
Brake, for engine on test bed, 127
Break-in oils, 208-209
Bright stocks, 55
British Leyland, 293
British Railways engines, 339
Broaching, 350
Brookfield Viscometer, 104, 105, 109, 112
Brookfield viscosity, 288
  tests for, 303, 517
Buick, 292

B

Bulk oil temperatures, 379, 380
Bulk storage, in tanks, 373-374
Bunding of tanks, 373
Burning of used lubricants, 400
Buttery greases, 313
Butyne, 497

C

C series classification, *see* API "C" series
C-4 requirements, Allison, 300
C-4 tests, 553-554
CA, API classification, 188
CAFE, 410
Calcium greases, 310, 313
Calcium long chain alkaryl sulfonate, 395
    irritancy, 394
Calcium long chain alkyl phenate, 395
    sulfide, 395
Calcium sulfonates as detergents, 80
Calibration, use of oils for, 159
Cam and tappet rig tests, 125-126
Canada, notification laws, 386
Cans, for storage of lubricants, 375
Capillary viscometer, 103, 109
Caravan tests, 137-138
Carbamates, 85
Carbon, 6
    and varnish, in engine deposits, 179, 235
Carbon black greases, 312, 315
Carbon formation, in lacquer deposits, 239
Carbon monoxide in exhaust emissions, 399
Carboxylic acid, 498
Carcinogenicity, 396
    of mineral oils, 393
CARE, 25
CAS numbers, 389
Case hardening, 24
Case/IH, 196, 605
Castor oil, hydrogenated, in greases, 311, 314
Catalyst converter, 220-221
Catalyst systems, 422
Catalysts, 500

C

Catalytic converter, blocking, 220
  on 2T exhausts, 335
Catalytic dewaxing, 49, 61-62, 416
Catalytic systems, 422
Catalytically dewaxed base stocks, 198, 341, 432, 434
Caterpillar Tractor Company, 7, 77, 78, 175, 186, 187, 270, 607
  engine tests, 239
    1-A, 189, 266
    1D, 240
    1G, 243, 428
    1G/1G2, 240
    1G2, 241, 256, 552
    1H, 190, 243
    1H/1H2 test, 240
    1H2 test, 241
    1K test, 241, 247, 552
    1M-PC, 256
    1MPC test, 552
    1N test, 241, 247, 552
    1N, 256
    L-1, 186, 189, 240
  micro oxidation test, 552
  pistons, 131
  Series 2 quality, 187
  Series 3 quality level, 187, 190, 191, 430
  single-cylinder test engine, 129
  TO-2 and TO-4 specifications, 300
Cationic emulsifiers, 348
Cavitation, 110
CB, API classification, 188
CC, API, 190, 194
CCMC, 193, 228, 232, 238, 260-261, 267
  D-1 to D-5, 241
  D-3, 224
  D-5, 340
  G series classifications, 270
  G1 to G3 specifications, 248
  G4 classification, 219, 267
  G5 classification, 219, 267
  PD2, 240, 241
  seals test, 554
  sequences, 252, 261, 577-580
CCS, *see* Cold Cranking Simulator

C

CD, API classification, 188
CD/SE quality oils, 191
CEC, 160, 161, 219, 261, 262-263
  cam and tappet rig test, 125
  global specifications, 333
  M102E black sludge test, 218
  two-stroke tests, 331
CEC-L-33-A-94 biodegradation test, 398, 404
CEC-L-33-T-82, 398
Cement kilns, for oil disposal, 402
Ceramic Applications in Reciprocating Engines, *see* CARE
Ceramics, 24-25
Certificates of analysis, 360
Certification Marks, 356
  API, 7
  EOLCS, 258
CF-2, API classification, 189
CF-4 oil costs, 428-429
CFCs, 342
CG-2, API classification, 189
CG-4, API classification, 189
Chainsaws, 328-329
Chariots, prehistoric, 310
Chassis classification, 320
Chemical analysis, precision, 121-122
Chemical reactions, 494-495
Chemical stability, 5
Chemical structures, 74, 79, 81, 495-498
Chemical tests,
  ASTM, 517-518
  on grease, 319
  and properties, 116-123
Chemistry, petroleum, 493-500
Chevrolet, engine in L-4 test, 129
CHFC, 343
China, 434
  lubricant sales, 242
Chloracne, 401
Chlorinated dibenzo-p-dioxins, *see* Dioxins
Chlorinated hydrocarbons, 280, 283
Chlorine compounds, in gear oils, 280
Chlorine, environmental effects, 403
Chlorine-free fluorocarbon, 343
Chlorofluorocarbons, 68, 342

C

Chromium, in used oils, 171
Chrysler, 292, 326
  automatic transmission requirements, 600
  MW 7176, 307
CID, 252
Citroën, 293, 340
Classification,
  based on toxicology, 392
  developments, 426-431
  difference between Europe and U.S.A, 263-264
  future, in Europe, 263-264
  in Germany, 262
  and labeling of new products, 387-392
  of lubricants, 231-273
    definition of, 231
    in Europe, 259-264
  open-ended, 184
  performance, 249-250
  SAE, 248-249
  in U.S., 248-249
Clay in greases, 312
Cleaning action, 4-5
Cleanliness in automatic transmissions, 299
Cleveland Open Cup tester, 113-114
Cloud point, 71
Clutches, 296-297, 299
  friction, 290
CMA Code of Practice, 157, 205, 254-257, 356, 429
  tests, 257
CNG, 423
CO emissions, 399
  reduction of, 327
Codes of Practice, ATC, ATIEL, 263
  US Chemical Manufacturers Association, 256
Coefficient of friction, definition of, 30
Coefficient of variation, 144
Cold Cranking Simulator, 104, 105, 106, 109, 249, 371
  in quality control, 371
Cold forming, lubricants for, 351
Cold sludge, 82
Collection of used oils, 400
Colza, 417
Commercial item description, 252

C

Commuter car, 419
Compatibility tests, in outboard oils, 334
Complaints, causes, 382-383
 documenting, 381
 about lubricants, 380-383
 procedure, 380-381
Complex greases, 312
Complex soaps, 312
Complexity of tests, 247
Component addition, in batch blending of oils, 366
Components, purchasing, 357-363
Composition of oils, marketing considerations, 354-357
Compressed natural gas, 423
Computer control in blending, 370
CONCAWE report, 392
Confidentiality for chemical substances, 391
Conformance audits, EOLCS, 258
Consistency of greases, 317
Constituents of lubricants, 49-94
Consumer buying habits, 425-426
Consumers, oil quality and, 408-409
Consumption, Asia Pacific region, 242
 problems with, 221-222
Contamination, 37, 165
 avoiding, in blending, 368
 complaints, 382
Continuously variable transmissions, 421
Conversion factors, in S.I. system, 504
Cook-out of greases, 313
Cooling, 4
 systems, specialized, 428
Cooperation, oil company and OEM, 228-229
Coordinating European Council for the Development of Performance Tests for Transportation Fuels, Lubricants and Other Fluids, CEC, 260-261
Copper, 350
 as anti-oxidant, 76, 229
 compounds, 85
 corrosion, in ATF testing, 303
 strip corrosion test, 126
 in used oils, 171
Co-polymers as viscosity modifier, 73, 74
Coriolis meters, 367
Corporate Average Fuel Economy, CAFE, 410

C

Corrosion, in engine, 39
in gearboxes, 279
inhibitors, 90-91, 283
in ATF formulation, 305
performance assessment, 236
protection, in gear oils, 281
tests, 126-127, 208
in greases, 319
salt water, 207
Cortina high-temperature deposit test, 248
Costs,
of developing a PMN, 387
pressures, 433
of tests, 243, 247, 427-429
see also Prices
Coulomb, friction study, 14
Coupling agents in cutting oils, 349
Covalent bond, 495
Cracked/isomerized base stocks, 434
Cracking, 500
reactions, 61
Crackle test for water, 361
Crankcase lubricants, future of, 407
SAE viscosity classification, 266
Crankcase oil, *see also* Engine oil
formulating, 196-205
future, 431-432
performance of, 233
quality levels and formulations, 175-209
specialized, 206-209
technology, 94
testing, 97-192
Crankcase ventilation problems, 219
Crankshaft problems, 225
CRC,
L-33 test, 287, 288
L-37 high torque axle test, 286, 288
L-42 high-speed shock loading test, 286, 288
L-60 test, 287, 288
Cross-head engine, 43
Crude oil distillation, 54
Crude oils, 51
Cultivators, 278

C

Cummins L10 high soot test, 552
Cummins NTC 400 test, 224, 241, 551
Cutting oils, 347-351
CVT, 421, 432
Cyclic compounds, 497-498
Cyclo-hexane, 498
Cycloparaffins, 52

D

D series classification, API, 187
D series tests, *see* specific test numbers under ASTM
D3 diesel classification, 224
DAF, 293
Daimler Benz seals test, 554
Damper oils, 341
Dangerous Substances Directive, 387
DEF 2101D Supplement 1, 215
Demulsifiers, 92-93
Denitrogenation reactions, 61
Density tests, 116
Deposit precursors, 39
Deposits, 5,
    see also Varnish, Sludge, Carbon, Gum
Dermatitis caused by lubricants, 393
Design of equipment, 10
Desulfurization reactions, 61
Detergents, and detergent inhibitors, 77-82, 203
    additives, 3, 7, 182
    barium compounds, 81
    formulations, 84-85
    gasolines, 180
    magnesium, 81-82
    treat levels, varying, for formulations, 188
Detroit Diesel 6V-53T test, 240
Detroit Diesel 6V92TA test, 256, 551
Detroit Diesel truck engines, 44
Deutscher Industrie-Normen classifications, 114, 262
Deutscher Normenausschuss, 262
Dewaxing, 58-59
    catalytic, 61-62

D

DEXRON™, 292, 605, 606
  II, 308
  IIE, 306
  III, 306, 308, 554, 600
Dicarboxylic acids and esters, 349
Dichlorodifluoromethane, 343
Diesel engines,
  classification for, 249
  lubrication requirements, 35, 41-42
  oils, 186-189
  tests, 189, 239-243
Diesel injector maintenance, 383
Diesel injector pump, 110
Diesel market, heavy duty, 243
Diesel particulates in exhaust emissions, 399
Diesters, 66
Difluorochloromethane, 343
DIN, 262
  51581 test, 114
Dioxins, 69, 401, 402, 412
Direct fuel injection, 326
Discrimination in test results, 155-156
Dispersants, 38, 82-83, 85, 283
  action, 84
  ashless, 83
  in ATF formulation, 305
  in automatic transmission fluids, 299
  in crankcase oils, 180
  in gear oils, 281
  in new formulations, 200
Dispersion of test results, 143
Disposal, 412, 413
Disposal requirements, 425
Disposal of used lubricants, 400-402
Distillation, 54-56
  test, 165
Distribution in statistics, 139
Distributions of test results, complex situations, 146-147
Disulfides, in gear oils, 279
DIY market, 425
DKA clutch pack rig, 308
DKA machine test, 304
Doughnut, API, 7
Drain interval, *see* Oil, change interval

D

Drawing, lubricants for, 351
Drilling, 347, 350
Drop point in greases, ASTM D 2265, 318
Drum storage, 363, 374-375
Dry friction and lubrication, 15-16
Du Pont, 82
Dumbbell blend, 114
Duplication of tests, 246
Dynaflow, 292
Dynamic friction, 15
Dynamic viscosity, 101, 104
Dynamometers, 127, 234, 428

E

E1-E3 specifications, ACEA, 241
Earthmovers, 42, 342
Eastern Europe, test procedures, 265
Eaton Axle, 290
Ecotoxicity, 394
EELQMS, 263
EGR, 413
EINECS, 389
Elasto-hydrodynamic lubrication, 21-22
Elastomer seal compatibility, in ATF testing, 303
Elastomeric seals, compatibility of, 223
Electric vehicles, 415, 420
Electro-Motive Division of General Motors, 126
Electrons, 494
Elemental contents, specification limits, 359
Elements, 493
    analysis test, 119-122
    in quality control, 371
ELTC, 262
EMD, 126
    locomotives, 337, 339
Emission spectroscopy, 120
    Inductively Coupled Plasma, 121
Emissions, to atmosphere, 412
    effects, 423-424
    legislation, 328, 413
    limits, 430

E

Emissions, to atmosphere (continued)
    lubricant effects on, 398-400
Emulsifiable oils, 347
Emulsifiers, 91-92, 348
Emulsions, 351
    of oil in water, 347
    in rocker cover, 219
    sludge, 92
EN 45001 laboratory accreditation, 264
End users, oil quality and, 408-409
Engine(s),
    cross-head, 43
    deposits, performance assessment, 235
    gasoline, oils for, 179-186
    four-stroke, 35
    gasoline-type, 37
    Otto-cycle, 34
    parts and rating, 131
    rig-test prices, 551-556
    sectional view, 34
    spark-ignition, 34
    two-stroke, 35-36
    see also Crankcase, Diesel engines, Marine engines
Engine Lubricants Technical Committee, 262
Engine Oil Licensing and Certification System, 250, 356, 373
Engine oils,
    classification, 111
    performance and service classification, SAE J183, 521-549
    properties required, 4-5
    stress and usage, 177
    tests, SAE J304, 507-515
    viscosity classification, SAE J300, 557-566
    see also Crankcase oil
Engine tests, 7-10
    ASTM, 520
    deposit tests, 186
    laboratory, 127-133
    Motorbecane, 332
    for new formulations, 204
    Piaggio, 332
    procedures, problems, 243-248
    requirements for, 130-133
    statistics in, 153-158
    Yamaha, 332

E

Engine wear, due to acid corrosion, 179
  performance assessment, 235
Engler viscometer, 101
Environment, 385-404
  aspects in marketing, 403
  concerns, use of chlorine compounds, 280
  effects of exhaust emissions, 398
Environmental factors, in defining quality, 273
  in future of lubricants, 407, 408
Environmental pressures, effects on future of lubricants, 412-414
Environmental Protection Agency, 401
Environmental regulations, 63
EOLCS, 250, 257-259, 356, 372
EP additives, 283
  in cutting oils, 349
EP requirements, 278
EP testers, 284-286
EP tests on greases, 319-320
EPA, US, 401
EPA (Gledhill) Shake Flask Test, 398
Epicyclic gears, 294-295
Equipment condition, testing for, 170-172
Equipment design, 10
Esso, France, 177
Esters, 66-67, 328, 329
  base stocks, 197-198, 416-417, 431
  based oils for two-stroke engines, 329
  in greases, 312
Ethane, 496
Ethene, 497
Ether, 498
Ethyl acetylene, 497
Ethylene, 497
Ethylene-propylene copolymers, 199
Ethyne, 497
EU, use of low-sulfur diesel fuel, 399
Euler, friction study, 15
Europe, automatic transmissions, 291, 293
  bus, heavy-duty and off-road transmissions, 607-608
  CCMC, 411
  future for classifications, 263-264
  future market, 431
  lubricant classification in, 259-264

E

Europe automatic transmissions (continued)
OEMs in, 259-260
oil quality, 216
passenger car transmissions, 606-607
railroad engines, 339
rig-test prices, 552-553
specification, 237
European Engine Lubricants Quality Management System, 263
European Union Dangerous Substances Directive, 387-391
Evaporation in greases, ASTM D 972/D 2595, 318
Exhaust emissions, hazards of, 398-400
targets, meeting, 193
Existing Chemical Substance List, 389
Explosive properties, classification based on, 387
External scavenging system, 326
Extrapolation in statistics, 140
Extreme pressure additives, 20, 88
Extreme pressure lubrication, 20-21
Exxon, 69, 78, 80, 82
Eye irritancy, 395

F

FA quality, 333, 631
Factories, lubricants in, 378-380
Falcon rust test, 184
Falex tester, 285
Fats, in gear oils, 279
in greases, 311
Fatty acids, 306
esters, in gear oils, 279
in greases, 311
sodium salts, 348
Fatty amides, 306
Fatty oils, 306, 351
FB quality, JASO, 333, 631
FC quality, JASO, 333, 631
Federal Trade Commission, 397
Fiat, 196, 259, 293, 326
Fiat Lubrificante, 259
Field testing, 133-138
Fill-for-life, 413, 432

F

Filter blocking, 383
Filtration tests, in outboard oils, 334
Filtration, use to lengthen oil life, 180
Fingerprints, infrared spectra as, 357
Finishing processes, 57-58
Finland, 224
  disposal of used lubricants, 401
Fire hazards, 342
Fire-resistant hydraulic fluids, 345
Fischer-Tropsch process, 64
Flame photometry, 119-120
Flammable properties, classification based on, 387
Flash point, 113-114
  test, 303, 518
    on used oils, 165
Fleet operators, 409
Fleet trials, 135-137
Fluidity test, in outboard oils, 334
Fluon™, 23
Fluoro elastomers, 223
Foam reduction, 93
Foam tests, in ATF testing, 303
Foaming, 5, 518
  and stability, 115-116
Ford, 196, 270, 292, 293, 300, 326
  15, 000 cycle clutch friction test, 554
  20, 000 cycle test, 554
  automatic transmissions, 89, 298
  Cortina test, 270
  engine testing, 237
  ESP-M2C138-CJ, 307
  ESP-M2C166-H, 307
  ESP-M2C185A, 307
  ESP-M2C33F and G, 307
  of Europe, 169, 218, 238, 259
  Falcon rust test, 184
  frictional requirements, 301
  M2C 101-B, 184
  M2C134D, 605
  MERCON™, 300, 307, 600, 605
  Model T, epicyclic gears, 294
  New Holland, 196
  shift feel test, 554
  SQM-2C9010A, 307

F

Ford (continued)
- Tornado, 224
- tractor businesses, 196
- UK, 238
- WSP-M2C185A, 307

Forklift trucks, safety, 374
Formulation, changes in, 255
- crankcase oil, 175-209
- finalizing, 203-205
- future changes, 416-419
- new, order of testing in, 204

Formulator, role in oil production, 354
Foul air venting, 217
Four-ball EP test, 285, 290, 320, 371, 519
Four-stroke engine cycle, 35
France, disposal of used lubricants, 401
- IFP, 326
- Motorbecane, 331, 332
- railways, 340

Frequency diagram of test results, 141
Fretting wear, ASTM D 4170, 320
Friction, 11-21
- in ATF, 282
- clutches, 290
- enhancers, 88-89
- in journal bearings, 30-31
- modifier, 20, 32, 88-89
  - in ATF formulation, 305-306
  - in gear oils, 281
- in unlubricated conditions, 27-28

Friction reducers, 4, 88-89
- in gear oils, 281

Friction test, 308
- in ATF testing, 303

Friction, mechanics of, 11-15
Frictional characteristics, in automatic transmissions, 299
- tests for, 304

Frictional heat, 347
FTIR fingerprinting, 357
Fuel(s),
- alternative, 423
- dilution problems, 222
- dilution tests on used oils, 165-166

F

Fuel(s) (continued)
economy, 431
injected Peugeot TU3 100 hr test, 552
injection engines, 326
oil, used oil disposal, 401
oil ratio, for two-stroke engines, 327, 329
quality in engine testing, 129, 130
soot, 40
sulfur, 338
Fuels and Lubricants Working Group, CCMC, 260
Fuller's earth, 57
Future of lubricant sales, 407-434
FZG, ASTM 4998 test, 552
gear machine, 308
gear test, 125, 290
machine ASTM 5182, 286

G

G-5 test, 238
G1-G5 test sequences *see* CCMC
GA, GB, GC wheel bearing classifications, NLGI, 320
Gas chromatography, 123
tests, 165
Gas turbines, 421
oils, 335-337
analysis, 170
Gasoline engine oils, 179-186
performance requirements, 236-239
Gasoline-type engines, 37
Gaussian curve, 143
GB, CEC, 333, 631
GC, CEC, 333, 631
GD, CEC, 333, 631
Gear(s), epicyclic, 294-295
forms, 275
Gear oils, 275-291, 432
additives, 279-281
formulation, 281-284
hypoid, base stocks for, 290
quality levels, 287-290
specifications, 287-290

G

Gear oils (continued)
  testing, 284-287
  sulfur compounds in, 279-280
  viscosity classification for, 266, 289
Gear test, FZG, 125
Gearbox, manual, 281-282
  corrosion, 279
  greases in, 346
Geiger counter, 121
General Electric, 339
General Motors, *see* GM
General Motors DEXRON™, *see* DEXRON™
German standardization organization, 262
Germany, 607
  "Blue Angel" system, 403
  disposal of used lubricants, 401
  gear research institute, 286
  used oil collection, 62
GF-1, ILSAC performance standard, 267, 567-576
Glassware tests, 223
Gledhill Shake Flask Test, 398
Global specifications for two-stroke tests, 333
Glossary, 437-487
GM, 270, 300, 326
  6.2 litre wear test, 241, 256, 552
  6041-M, 184
  6137-M, 306
  6297-M, 306
  20, 000 cycle test, 554
  Allison C-4, 307
  band clutch friction test, 554
  clutch plate test, 554
  Electro-Motive Division, 126
    railroad engines, 44, 337
  engine testing, 237
  fluid, 293
  oxidation test, 554
  performance standards, 183
  Research, new transmission, 292
  specification for automatic transmissions, 599, 600
  "Type A" specification, 292
  type frictional requirements, 302
  vehicle test, 554

G

Grand Prix racing engine, 272
Graphite greases, 26
Grease kettle, 316
Greases, 2, 3, 309-321, 433
    advantages, 309
    alternative fluids in, 312
    carbon black, 315
    characteristics of, 313-315
    complex, 312
    copper-containing, 26
    definition of, 26, 309
    disadvantages, 310
    high temperature flow in, ASTM D 3232, 318
    manufacture, 315-316
    metals in, 311
    oil separation in, ASTM D 1742, 318
    penetration test, 519
    in small gearboxes, 346
    soap based, 311
    solid-thickened, 315
    specifications and approvals, 320-321
    synthetic-based, 321
    testing 316-321
    use in motor vehicles, 321
Greenhouse gas emission, 415
Grinding, 347, 350
Guillaume Amontons, friction study, 13
Gums, 39

H

Halogenated hydrocarbons, 68
Haltermann, 163
Hard fluids, 305
Hardware, new types of, 419-421
    problems, lubricant sensitive, 421-422
Hazards, in drum storage, 363
    fire, 342
HC in exhaust emissions, 399
Health, 385-404
Health Aspects of Lubricants, CONCAWE report, 392

H

Health hazards of exhaust emissions, 398
    tricresylphosphate, 280
Heat removal, in gear oils, 281
    in metal working, 350
Heat soak-back in gas turbine oils, 337
Heavy duty oils, 189, 428, 430
    with detergents, 182
Heavy-duty transmissions, specifications for, 601
Helical gears, 275, 277, 281
Heterocyclic compounds, 496
High dispersancy oils, 177
High temperature/high shear viscosity, 108, 249
Hindered phenols, as anti-oxidants, 199, 304
Hindered-ester fluids, 337
Histogram of test results, 141, 142
Holland, Van Doorne, 293
Honda, 608
    engine tests, 632
        2-T, costs of, 555
Hooke, elasticity study, 14
HSE, 385-404
HT/HS viscosity, 108, 249
Humidity cabinet panel rust test, 207
Husqvarna chainsaw test, 555
Hybrid vehicles, 420-421
Hydraulic drive, 421
Hydraulic fluids, tractor, 308-309
Hydraulic oils, 340-342, 432
    industrial, 344-345
Hydraulic suspensions, 340, 421
Hydro-isomerization, 49, 60
Hydro-isomerized base stocks, 416
Hydrocarbon emissions, 327, 399
Hydrocracking, 49, 60
    base stocks, 198, 416
Hydrodynamic lubrication, 17, 29-33
Hydrofinishing, 57
Hydrogen reforming, 60, 198
Hydrogenation, 57, 500
Hydrophilic/lipophilic balance, 348
Hydrostatic lubrication, 33
Hydrotreating, 58
12-hydroxystearic acid, in greases, 311, 313, 314

H

Hypoid gears, 275, 277, 278, 346
Hypoid rear axle oils, 283-284

I

IAPAC engine, 326
IARC, 393
ICP, 121
    spectroscopic analysis, 519
        to measure elemental content of additives, 362
IFP 326
    acid/clay, 63
    propane precipitation, 63
ILSAC, 267, 411
    ATF standard, 308
    Certification Mark, 258, 426
    GF-1, 567-576
    performance standard, 567-576
    specifications, 265
    standard GF-1, 267
    worldwide, 434
IMO, 402
In-town runabout car, 419
Incineration of used lubricants, 401
India, 434
    lubricant sales, 242
    use of two-stroke oils, 331
Inductively Coupled Plasma, *see* ICP
Industrial gear lubricants, 346
Industrial hydraulic oils, 344-345
Industrial lubricants in automobile plants, 343-351
Infrared absorption, 123
    test on additives, 362
Infrared fingerprinting of oils, 123
Infrared spectra, use for monitoring samples, 357
Infrared spectrophotometry, 168, 371
Inhalation, toxicity by, 393
Inhibited ester-based lubricants, 337
Inhibited mineral oils, 340
Inhibitors, *see* Additives
Injection of oil, 334
Insolubles, levels of, in used oils, 166

I

Inspection of materials, 360-363
Institut Français du Pétrole, IFP, 326
Institute of Petroleum, 262
Intermediate oils, 51
Internal-combustion engines, lubrication requirements, 34-35
International Agency for Research on Cancer, IARC, 393
International Lubricant Standardization and Approval Committee, ILSAC, 267
International Maritime Organization, IMO, 402
International Organization for Standardization, (ISO), *see* ISO
International specifications, 306
International Standards Organization, *see* ISO
Interpolation in statistics, 140
Ionic bonds, 495
IP, 262
Iron, in used oils, 171
Irritancy of additives, 394, 395
ISO, 266, 344, 631
  9000 quality systems, 264
    series accreditation, 255
  9000/EN45001 accreditation, 429
  classification, 344
    global, 631
  standards, 248
    worldwide, 266
  two-stroke classifications, 631
Iso-butylene, 497
Isocyanates for polyurea greases, 316
Iso-octane, 496
Isotopes, 494
Isuzu, 25
Italy, disposal of used lubricants, 401
  Piaggio, 331, 332, 333, 632
IUPAC chemical names, 496

J

J 20 D, John Deere, 605
J I Case/International Harvester MS 1207, 605
Japan, ash level of lubricants, 186
  ATF approval requirements, 608
  Automobile Manufacturers Association, JAMA, 265, 267, 308
    performance standard, 567-576

J

Japan, ash level of lubricants (continued)
classifications in, 264-265
future market, 431
HD diesel engines, 413
industry standards, 264
lubricant sales, 242
manufacturers, 293, 308
motorcycle oils, 329
notification laws, 386
OEMs, 272
oil evaluation tests, 130
quality control 429
SAE meeting, 1994, 326
sequences, 412
tests, 243
two-stroke market, 331
use of low-sulfur diesel fuel, 399
valve wear tests, 126
Japanese Automobile Standards Organization, *see* JASO
Japanese Chemical Substances Control Law, MITI, 388-391
Japanese Standards Organization, JASO, 264, 333, 631
2T Engine Oil Standards, 265
engine tests, 632
two-stroke classifications, 631
JATRE-1, 333, 631
JD 27, John Deere, 196
JIC 187, CASE/IH, 196
JIS K2215, 264
John Deere, 196
J 20 D, 605
Journal bearings, friction in, 30-31
lubrication requirements, 29

K

Kerosene, 280
Ketone, 498
Kinematic viscosity, 101-103
tests for, 517
Korea, notification laws, 386
KTI thin film evaporation, 63
Kurt Orbahn, 111

L

L-1 test, Caterpillar, 129, 239, 240
L-4 test, Chevrolet, 129
L-38 test, Labeco, 111, 130, 184, 266, 251
  stay in grade test, 552
L-60, CRC, 289
LA-LB chassis classification, 320
Labeco L-38 test, 111, 130, 184, 266, 551
Labeling of new products, 387-392
  deliberate mislabeling, 212-213
Laboratory tests, 7-10, 99-123
  examination of samples, following complaints, 381-382
  glassware, 223
  length of, 136
  performance, 125
  on petroleum products, 148-152
  precision of, 517-520
Lacquer, 6, 239
Lagging of storage tanks, 373
Laminar flow and viscosity, 100
Lawnmowers, 327-328
Laws, notification, for new substances, 386-387
LD50 of additives, 395
Lead naphthenate, 346
Lead, in used oils, 171
  soaps, 279, 283
Leakage tendency in greases, ASTM D 1263, 319
Leaks, in storage tanks, 373, 374
Legislation, for new substances, 387
Leonardo de Vinci, friction experiments, 13
Licensing procedure, EOLCS, 258
Limitations of oils, 5-6
Limited-slip differentials, 290-291
Limits, specification, 150-152
Line flushing, in batch blending of oils, 364
Liner wear, 235
Lip seal compatibility problems, 223-224
Liquefied petroleum gas, 423
Liquid chromatography, 123
Liquid flow, 99
Liquid polymers, 351
Lithium greases, 311, 314
Long chain polymers, as viscosity modifiers, 199
Long-distance vehicle, future, 420
Long-life motor oils, 183

L

Low ash formulation, 335
Low-potency gear oils, 421
Low-pour point oils, 432
Low-sulfur diesel fuel, 399, 434
Low-temperature flow problems, 383
Low-temperature specification testing, 371
Low-temperature torque in greases, ASTM D 1478, 318
Low-viscosity oils, 267, 431
LPG, 423
Lubri-graph, 78
Lubricant Test Monitoring System, 256
Lubricants, consumption, Asia Pacific region, 242
  definition, 1
  design interactions, problems with, 214-227
  functions of, 4-6
  history of, 1-3
  problems, experience with, 211-229
  sales, future of, 407-434
  tests, 7-10
  used, toxicity of, 396-397
Lubricants Review Committee, 252
Lubricants Review Institute, 252, 255
Lubricated sliding, 16-18
Lubricating greases, SAE J310, 609-623
Lubricating oils, general purpose, for machinery, 344
Lubrication, boundary, 19-20
  and dry friction, 15-16
  elasto-hydrodynamic, 21-22
  extreme pressure, 20-21
  full-fluid, 16-18
  fundamentals of, 11-21
  hydrodynamic, 17
  requirements, 29-46
    of 2-T engines, 45
    of diesel engines, 41-42
    of internal-combustion engines, 34-35
    of marine engines, 42-44
    of railroad locomotives, 42
    simple systems, 29-33
    of two-stroke engines, 44-45
  of tractors, 194
Lubricity, in metal working oils, 350
  testing, 303
Lubrizol, 78, 80, 83, 86, 88, 417

M

M series classification, 187
M102E test, 219, 229, 238, 245, 552
M1139, Massey-Ferguson, 196
M1143, Massey-Ferguson, 605
M2C 159-C, Ford, 196
M2C134D, Ford, 605
M2C138-CJ, Ford, 300
M2C166-H, Ford, 300
M340 to M343 tests, JASO, 632
Machine capability, 151, 152
Mack, 270, 271, 290
    GO-H/S, 290
    spalling test, 289
    T-6 and T-7 tests, 241, 551
    T-8 test, 200, 241, 256, 551
Magnesium additives, 81-82
Maintenance, 382, 383
    DIY, 425
MAN cylinder head components, 247
MAN D-2866 KF (QC 13-017), 552
MAN QC approval, 248
MAN seals test, 554
MAN single cylinder test, 553
Mannich formaldehyde process, 83
Manual gearboxes, 281-282, 432
    motor oil in, 282
Manual transmissions, classifications for, 289
Manufacture of grease, 316
Manufacturing, base oil, 51
Marine engines, lubrication requirements, 42-44
Marketing initiatives, 228-229
Marketing of lubricants, 353,
    SHE issues, 403-404
Mass flow meters, 367
Mass spectroscopy, 123
Massey Ferguson, 196
    M1139, 196
    M1143, 605
Matthys Garap continuous process, 63
Mayonnaise, in rocker cover, 219
Mazda, 608
Mechanical failure, 382
Mercaptan, 499

M

Mercedes-Benz, 238, 241, 259, 270, 271, 290, 293, 308, 606, 607
  Betriebstoff - Vorschriften/Specifications for Service Products, 259
  environmental evaluation of oils, 403
  M102E engine test, 200, 218
  OM tests, 241
MERCON™, Ford, 307, 554
  M2C-185A, 300
Mercury 15HP test, 556, 633
Metal belts, for automatic transmission, 293
Metal content, 119
Metal deactivators, in ATF formulation, 304
  in gear oils, 281
Metal containing detergents, 335
  inhibitors, 193
  reduction of, 413
  in two-stroke oils, 328
Metal passivation, see Metal deactivators
Metal phenates, 79, 84, 85
Metal salicylates, 78, 79, 85, 201
Metal soaps, 85
Metal sulfonates, in new formulations, 201
  structure of, 80
Metal surfaces, in air, 27
  in vacuum, 28
Metal-working oils, 347-351
Metal-working operations, 349
Metals, used in greases, 311
  as solid lubricants, 24
Methane, 496
Methyl acetylene, 497
2-methyl propene, 497
Middle East, EMD locomotives, 339
  test procedures, 265
MIII sludge test, 553
MIL-2-104, 186
MIL-2-104B, 187
MIL-2-105, 287
MIL-B, 190
MIL-Board, 252
MIL-G-23549, 320
MIL-G-244139, 320
MIL-G-819371, 321
MIL-L-2104, 251, 252

M

MIL-L-2104-A, 187
MIL-L-2104B, 190, 215, 216
MIL-L-2104C, 191, 192, 215, 216
MIL-L-2104F, 252
MIL-L-2105, 287
MIL-L-15719, 321
MIL-L-20260, 207
MIL-L-45199, 187, 190
MIL-L-46152, 191, 192, 251, 252
MIL-L-46167B, 206
Military Specifications, 272
Military, Supplement 1 and 2 specifications, 187
Milling, 347, 350
  grease, 316
Mineral oils, 351
Mini-Rotary Viscometer, MRV, 104, 106, 107, 108, 109, 227, 249, 518
  in quality control, 111, 371
Ministry of Defence, UK, 254, 270
  approvals, 157
  requirements, 248
Mirrlees corrosion test, 126
Miscibility tests, in outboard oils, 334
MITI, 388-391
Mitsubishi, 608
Mixed soap greases, 311
Mixing equipment, 367
Mixing oil blends, time required, 367
ML, light duty oil, API, 183
MM, moderate duty oil, API, 183
Mobil, 78, 80, 177, 292
Modified Sturm Test, 398
Molecules, 493
Moly greases, 26
Molybdenum disulfide, 33, 346-347
Monitoring program, 212
Monitoring systems, external, 372
Mono-cylinder engine tests, 247
Monsanto, 71, 88
Mopeds, 327-328
Motor industry, and technical societies, 427
  and user quality levels, 271-273
Motor Vehicle Manufacturers Association, MVMA, 253, 255, 265, 267
  performance standard, 567-576

M

Motorbecane, France, 331, 332
Motorcycles, 329-330
MRV, *see* Mini-Rotary Viscometer
MRV-TP-1, 228
    procedure, 107, 228
MS 1207, API, 605
MS quality of oils, API, 183, 184, 191
    Sequence tests, 234
MS VE test, 126
MSE IIIE test, 126
MTAC, 256
Multigrade oils, 33
    in gearboxes, 282
    shear stability, 109-111
    see also Viscosity modifiers
Multiple Test Acceptance Criteria, MTAC, 256
Multiple tests, 254
Multi-purpose oils, 192-193, 411
    gasoline/diesel oils, 189-194
MVMA, 253, 255, 267, 567-576
MW 7176, Chrysler, 307
MWMB (KD 12E) test, 552

N

N-methyl-2-pyrrolidone, 56
n-octane, 496
Naphthenic oils, 51
Napier "Deltic" engines, 339
National Lubricating Grease Institute number, 317
National Marine Manufacturers Association, 334
National oil companies, test procedures, 265
Neopentyl esters, 66-67
Neutralization number, 116, 518
New Zealand, disposal of used lubricants, 401
Newton, fluid flow, 14
Newtonian fluids, 101
NICNAS, 388-391
Nissan, 608
  detergency test, 555
    KA24E low-temperature valve train wear test, 265
    SD22 detergency test, 555

N

Nitriding, 24
Nitrogen, chemistry of, 499-500
    oxides in exhaust emissions, 399, 413
NLGI grease classification, 317, 320
NMMA, 334, 633
NO_x in exhaust emissions, 399
    reduction in, 413
Noack test, 114, 115, 125
Nonionic emulsifiers, 348
Non-Newtonian fluids, 101
Normal distribution of test results, 143, 145
North Africa, test procedures, 265
North America, EMD locomotives, 339
Notification laws, 388-391
    for new substances, 386-387
Nuclear magnetic resonance, 123
Nucleus, 494
Nylon, 24

O

OECD Modified Sturm Test, 398
OEM, approval, 136
    definition, 11
    future developments, 410
    role in oil reformulation, 408
    specifications, Europe, 259
Off-road transmissions, specifications for, 601
Oil(s),
    change interval, 177, 179, 180, 218
    classification of, 51
    companies, future developments, 409
        role in oil reformulation, 408
        role in production of oils, 353
    condition, tests for evaluation of, 164-170
    consumption, problems with, 221-222
    formulator, role in production of oils, 354
    to fuel ratio, for two-stroke engines, 327, 329
    injection, 334
    Labeling Assessment Program, OLAP, 212
    marketer, role in production of oils, 353

O

Oil(s) (continued)
- performance, *see* Performance
- pump screen clogging, 235
- quality,
  - development, 430-431
  - improvement, 180
  - *see also* Quality
- ring clogging, 235
- specialized, 325-351
- supply, 425-426
- and water emulsions, 347
  - in hydraulic fluids, 342

OL-1 engine test, Caterpillar, 229
OLAP, 212
Oldsmobile Hydromatic, 292
Oldsmobile Safety Transmission, 292
Olefin oligomers, 64-65
Olefin/alkyl ester co-polymer, 395
Olefins, 497
Oleic acids in greases, 311
Oleophilic surfaces, 312
Oligomerization, 497, 500
OM364A test, 552
OM441LA test, 553
OM602A test, 552
OM616 Combi test, 552
OMC tests, 633
- 2-T, costs of, 556
- 40HP, 334
- 70HP test, 633

On-board diagnostics (OBD), 410, 425
Orbital Engine Company, 326
Organic esters, 66-67
Organic phosphates, 305, 306
Organizations involved in classifications, 248-267
Organo-clay greases, 315
Original Equipment Manufacturer, see OEM
Oronite, 78, 83
Otto-cycle engines, 34
Outboard, ashless oils, 335
- Marine Corporation, U.S., 331
- motors, 330-331
- oil qualification, 633-634

O

Overbased additives, 80, 85
Overbased phenate, 194
Overbased sulfonate, 194
    and phenates, 188
Overheating, problem of, in batch blending of oils, 366
Oxidation inhibiting, 84
    in ATF formulation, 304
    in gear oils, 279
Oxidation, 235
    resistance of polyurea greases, 314
    stability, 5, 338
        in greases, ASTM D 942, 318
        in hydraulic fluids, 341
    tests, 124, 130
Oxide layer, 28
Oxidizing properties, classification based on, 387
Oxygen sensor, 220-221, 422
Oxygenated compounds, 498-499

P

Packages, for storage of lubricants, 375
Packard, 88, 283
PAGs, 340, 343
PAHs, 393, 396
Palmitic acid in greases, 311
PAO, 65, 198, 416
Paper chromatography tests, 168-170
Paraffinic oils, 51
Paraffins, 496
Paraflow, 71
Parallel shafts in gears, 275
Paratone, 72
Particulate traps, 422
Particulates, diesel, 399
Passenger car gasoline oil sequences, 238-239
PCA, carcinogenicity, 393
PCBs, 402
PCV (Positive Crankcase Ventilation), 38
Pensky-Martens flash point tester, 113, 114
Performance, classifications, 249-250
    engine , 235

P

Performance, classifications (continued)
levels, 231-273
measurement, 233-248
parameters, 233-236
requirements for gasoline engine oils, 236-239
standards, 183, 249
definition of, 257
worldwide, 266
targets, in testing, 203
testing, 97, 124-138
Perkins Engines, 272
Personal Rapid Transport, 419
Petroleum Additives Panel, 394
Petroleum chemistry, 493-500
Petroleum derived base stocks, 416
Petter, AV-1 tests, 266
AVB tests, 552
engines, 129-130
W-1 test, 266, 553
Peugeot, 241, 259, 270, 290
TU3 high temperature deposit test, 553
XUD 7TE endurance test, 553
XUD II ATE soot loading test, 553
Phenates, 84, 85, 305
in new formulations, 201
overbased, 188
Phenolic compounds, 84
Phenothiazine, 337
Phenylamines, 337
Philippines, notification laws, 386
Phillips, 163
ammonium sulfate process, 63
Phosphate esters, 68
Phosphonates, 85
Phosphorus, in gear oils, 280
pressure to reduce level of, 221
Phthalate esters, 66
Phthalocyanines, 312
Physical and chemical properties of engine oils, SAE J357, 635-647
Physical properties, in evaluation of formulations, 203
tests for, 99-116
Physical tests, ASTM, 517-518
Piaggio, Italy, 331, 332, 333, 632

P

Pigging, in batch blending of oils, 364
Piston, cleanliness, 9
  cylinder lubrication in single cylinder engines, 328
  deposits, 40, 189
    tests for, 236
  groove deposits, 42, 235
  to liner clearances, 193
  rings, in diesel engine tests, 239
Planetary gears, 294
Planing, 350
Plant, for additive manufacturing, 70
  base oil manufacturing, 51
  for batch blending of oils, 364
PM in exhaust emissions, 399
PMN, 387
PNAs, 396, 424
Poly alkyl methacrylate, 395
Polyacrylates as viscosity modifier, 73
Polyalkylene glycols, 340, 343
Polyalkylstyrenes as viscosity modifier, 73
Polyalphaolefins, 65, 416
  as base stocks, 198
Polyaromatic hydrocarbons, 396
Polybutene, 64, 182, 208, 329
  lubricants, 329
  thickener, 328
  as viscosity modifier, 73
Polychloro-biphenyls, 402
Polycyclic aromatic hydrocarbons, 393, 396
Polycyclic aromatics, carcinogenicity, 393
Polyethers, 67
  as base stocks, 198
Polyfumarate co-polymers as viscosity modifier, 73
Polyglycols, 67
  as base stocks, 198
  in greases, 312
Polyhalocarbons in greases, 312
Polyisobutylene, 208
  viscosity modifiers, 182
Polymer rupture, viscosity loss, 110
Polymer solutions, 351
Polymerization, 497, 500
Polymers, notification of, 390
  as viscosity modifiers, 199

P

Polymethacrylates, 199
  as viscosity modifier, 73, 74
Polynuclear aromatics, 396, 424
Polyol esters, 66, 343
Polyolefin, 395
  amide alkeneamine, 395
    borate, 395
  in greases, 312
Polyphenyl ethers in greases, 312
Polysulfide, 499
Polytetrafluoroethylene, 23, 24, 26
Polyurea greases, 314-316
Positive crankcase ventilation, PCV, 38, 184
Potentiometric titration, 118
Pour point, 111-113
  depressants, 71-72, 181
    checking, 362
  in quality control, 371
  test, 518
Power generation, 42
Power take-off, 194, 195
Powershift transmission fluid classification, SAE J1285, 301, 602-604
Poyaud, 340
Pre-ignition problems, 227
Pre-manufacturing notification, 387
Pre-marketing notification, 387
Precision, of chemical analysis, 121-122
  data, published by ASTM, 148
  of laboratory tests, 517-520
  lack of, in engine testing, 243-244
  in statistics, 139
Predicting the future, 414-416
Prefixes in S.I. system, 503
Premium oils, 182
Preservative oil, 207
Pressing, lubricants for, 351
Pressure profile in journal bearing, 18
Prices, of lubricants, 335
  rig-tests, 551-556
  see also Costs
Primary amine, 500
Primary biodegradation, 397
Pro-friction additives, 89

P

Problems, black sludge, 216-219
  bore wear, 214-216
  crankcase ventilation, 219
  crankshaft, 225
  in engine test procedures, 243-248
  fuel dilution, 222
  inadequate test procedures, 227-228
  inappropriate lubricant, 212
  inappropriate quality of lubricants, 213-214
  lip seal compatibility, 223-224
  lubricant design interactions, 214-227
  with lubricants, practical experience, 211-229
  mayonnaise, 219
  misblended lubricant, 212
  mislabeling (deliberate), 212-213
  oil consumption, 221-222
  pre-delivery, 222
  pre-ignition, 227
  process changes, 226
  production, 225
  pumpability, 227
  sludge, 212
  tappet pitting, 225-226
  valve deposits, 222
Process capability, 151, 152
Process changes, problems with, 226
Production of lubricants, 353-383
Production problems, 225
Propane, 496
Propene, 497
Properties,
  physical, tests for, 99-116
  required of engine oils, 4-5
  tested in new formulations, 204
  see also under specific properties
Proportional blender, 369
Propylene, 497
Protocol for engine testing, EOLCS, 257
PRT, 419
PSA TU3M hot test pistons, 131
PTFE, 23, 24, 26
PTO, 194, 195
Pulsair system, 367

P

Pumpability problems, 227
Pumps, in hydraulic systems, 344
  high-performance, 345
Purchase specifications, 357
Purchasing components, 357-363
Purchasing lubricants, 375-377
Pyridine, 500
Pyrrole, 500

Q

Quality, approval procedures, 429-430
  considerations, in purchasing lubricants, 376-377
  control, over additive metal content, 119
    in blending, 370-372
    of purchased products, 360
    of samples, in blending, 368
  in crankcase oils, 175-209
  levels, in gear oils, 287-290
    of oils, 184-185
  of lubricants, 271-273
    inappropriate, 213-214
    in tractors, 195
  of oils, external factors affecting, 419-420
  of parts in testing, inconsistent, 245
  seals, 7
Quaternary amines in greases, 312

R

R series refrigerants, 343
Racing motorcycles, 330
Racing Oil, Esso, 177
Rack-and-pinion mechanism, 276
Railroad locomotives, lubrication requirements, 42
Railroad oils, 337-340
Random variation in statistics, 139
Rapeseed oil, 417
Rating booth, 131, 132
Rating of wear, 132
Ravenfield Viscometer, 108-109

R

Raw materials, supply of, 353-354
Reactions in hydrogen reforming and hydrocracking, 60-61
Reactions, chemical, 494-495
Reading-across of prior test results, 203
Rear axle, limited-slip differentials, 290-291
Reclamation and recycling, 50, 62-63, 400, 412
Redwood viscometer, 101
Reference fuels, in testing, 162-163
Reference lubricants, 158-162
Reference samples, storage of, 357
Refining processes, conventional, 54-59
Reforming reactions, 61
Refrigerants, 68
Refrigerator lubricants, 342-343
Regular oils, 182
Renault, 259, 290, 293, 308, 606, 607
Renk, 607, 608
Repeatability, of spectrographic methods, 122
  in test results, 149, 154
  of tests, 517
Reproducibility, in test results, 149, 154
  of tests, 517
Re-refining of used lubricants, 400
  see also Reclamation
Retail crankcase oil, requirements for, 354
Re-use of used lubricants, 400
  see also Reclamation
Reynolds, Osborne, 18
Ricardo Consulting Engineers, 414
Rickshaws, three-wheel motorized 325
Rig-tests, 7, 97
  limitations of, 226-227
  prices, 551-556
Ring sides and rear, 235
Ring sticking test, 248
Ring wear, 235
Ring-opening reactions, 60
Road testing, 304
Rocker cover, emulsion and sludge in, 39, 219, 235
Rohm and Haas, 71, 82
Roll stability, D 1831, 317
Root-mean square deviation, 144
Rotational viscometer, 104, 109
  see also MRV

R

Round-robins, 160
Rover, 293
Run-in oils, 208-209
Running in, 21
Rust and corrosion, performance assessment, 236
Rust inhibitors, 90-91, 344
 in ATF formulation, 305
Rust protection, in ATF testing, 303
Rust tests, 236
 in outboard oils, 334
Ryder gear machines, 286

S

S series, API Engine Service Classification, 250
SA oil quality, API, 185
Saab turbo deposit test, 553
Saab-Scania, 259
SAE, 191, 228, 249, 252
 development of classification system, 184
 J183, 184, 236, 250, 521-549
 J300, 107, 248, 249, 557-566
 J304, 507-515
 J306, 289, 589-597
 J308, 287
 J310, 321, 609-623
 J357, 49, 635-647
 J1285, 301, 602-604
 J2116, 331, 626-630
 J2227, 259, 265
 label monitoring program, 212
 Lubricants Review Institute, 251-252
 No. 2 Machine test, 304
 Oil Labeling and Assessment Program, 372
 Surface Vehicle Information Report, *see* SAE information report number
 Surface Vehicle Recommended Practice, *see* SAE recommended practice number
 Surface Vehicle Standard, see SAE standard number
 viscosity classification, 103, 227, 248-249
  for crankcase and gear lubricants, 266
Safety, forklift trucks, 374
Safety, Health and Environment, 385-404, 424-425

S

Salicylates, 78, 79, 85, 201
Salt water corrosion protection in outboard motors, 330
Salt water corrosion test, 207
Santopour, 71
Saturated compounds, 496
Saturation reactions, 60
Sawing, 350
Saybolt viscometer, 101
SB oil quality, API, 185
SC oil quality, API, 185
Scania, 607
Scooters, 327-328
Scuffing, 225
SD oil quality, API, 185
SE oil quality, API, 185
Seal compatibility, 283
Seal degradation, 235
Seal swell agents, in ATF formulation, 306
Sealed-for-life, crankcases, 425
  joints, 321
Sealing, 5
  gland, 43-44
Seals of quality, 7
Sebacic acids in greases, 314
Secondary amine, 500
Sensitization potential of additives, 394
Sequence, definition of, 232
Sequence tests,
  Sequence I tests, 183
  Sequence II low-temperature rust test, 237
  Sequence II-B/III-B, 184
  Sequence II-D, 207
  Sequence III high-temperature test, 237
  Sequence III tests, 183
  Sequence III-E test, 202
  Sequence III-E viscosity requirements, 238
  Sequence IV tests, 183
  Sequence V test, 131, 183, 237, 245
  Sequence V-B, 184
  Sequence V-C test, 237
  Sequence V-D test, 237
  Sequence V-E test, 200, 219, 238, 520
  Sequence 2D rust screener test, 551
  Sequence 2D rust test, 551

S

Sequence tests (continued)
- Sequence 3E test, 551
- Sequence 5E sludge test, 551

Series 2 quality, 187
Series 3 quality level, Caterpillar, 187
Series 3 type oils, 215
Service duty, oil classifications based on, 183
Service Symbol, API, 250
Setaflash apparatus, 114
Seveso, 401
SF oil quality, API, 185
SG oil quality, API, 185
SG oils, description when not approved, 356
SH oil quality, API, 185
Shake Flask Test, 398
SHE, 385-404
Shear rate, 100
Shear stability, 125, 161, 199
- in greases, 314
- Index, 110, 111
- of multigrade oils, 109-111
- tests, 124-126

Shear stress, 101
Sheet polymers, 351
Shelf-life of products, 374
Shell, 78, 80, 83
- Four-ball EP tester, 285, 290, 317, 320, 519
- viscosity modifiers, 75

Shift performance, in ATF testing, 303
Shock absorber oils, 341
Short testing, in quality control, 377
SI system of units, 98, 501-505
Silica aerogel in greases, 312
Silicon, in used oils, 171
Silicones, 68
- in greases, 312
- to reduce foam, 93

Silver corrosion test, EMD, 126
Silver plated bearings, 338
Silver plated turbocharger parts, 338
Single cylinder engine for tests, 130
SISU, Finland, 224
Skewed distribution of results, 145, 146

S

Skin absorption, toxicity by, 393
Skin irritancy, 393-395
Slideway lubricants, 345
Sliding between gears, 278
Sliding systems, lubrication requirements, 29
Sludge, deposits, 235
    problems, 211, 216-219
    production of rocker cover, 39
    tests for, 236
    see also Deposits, Gums
Sludge-handling properties, 192
Sludge-reducing additives, 3
Smoke emission, in outboard motors, 330
    test, 328
    from two-strokes, 631
Snowmobiles, 335
SOCAL, 78
Society of Automotive Engineers *see* SAE
Sodium greases, 311, 314
Sodium nitrite, 349
Sodium sulfonates, 92, 348
Soft fluids, 306
Solid lubricants, 22-26
Soluble oils, 347
Solvent extraction of distillates, 56
Solvents, in dewaxing, 58
Soot, 40
    formation test, 200
South America, EMD locomotives, 339
Soviet Union, test procedures, 265
Spain, disposal of used lubricants, 401
Spark-ignition engines, 34
    classification for, 249
Specialized oils, 325-351
Specifications,
    for automatic transmission fluids, 599-608
    of crankcase oil, 197
    for gear oils, 287-290
    of greases, 320-321
    in Japan, 264
    limits, 150-152
    of lubricants, 231-273
        definition, 232
    new, advantages of, 267-271

S

Specifications (continued)
of oils, standard, 426
see also under individual specification numbers
Spectroscopy, atomic absorption, 120
emission, 120
X-ray fluorescence, 121
Spent oil, 400-402
Spillage and drips, safety hazards, 378
Spiral bevel gears, 276, 277
Spur gears, 275, 277
SSI, *see* Shear Stability Index
Stable Pour Point, 113
Stamping, lubricants for, 351
Standard deviation, 144
Standard Oil of California, SOCAL, 78
Standard Oil Development Laboratories, 69
Standard Oil of Indiana, 78
Starburst, API, 7
Static friction, 15, 300
Statistics, principles, 139-148
sampling, for quality control, 370, 377
testing, 133, 139-163, 153-158, 203
Stearic acid in greases, 311
Stick-slip, 32
Storage of additives, 363
Storage of lubricants, 373-375
Storage oils, 206-208
Straight spur gears, 281
Stress of engine oil, and usage, 177
Stress on lubricant, 177-178
Stribeck curve, 31
Structural analysis, 123
Structural chemistry, 74, 79, 81, 495-498
Stuck rings, 40
STUO, Super Tractor Universal Oils, 194-196, 308
specifications, 196
Styrene isoprene copolymers, 199
Subaru, 326
Substances, new, notification laws for, 386-387
Succinic acid derivatives, 207
Succinimides, 85, 182
as dispersants, 83
Sulfated ash test, 118-119

S

Sulfonates, 85, 305
 overbased, 188
 skin irritation, 394
Sulfonation, 500
Sulfur, chemistry of, 499-500
 in gear oils, 279-280
 in fuel, 338, 340
 levels, reduction in, 413
Sulfuric acid, 39
Sulfurized compounds, 304
Sulfurized fats, 283
Sulfurized oils, 85
Sulfurized phenates, 79, 84
Sulfurous acid, 39
Sump, sludge deposits in, 235
Sunflower oil, 417
Super Tractor Universal Oils, 194-196, 308
Supercharged 1-D test engine, 187
Superchargers, 422
Supplement 1 and 2 specifications, US military, 187
Surface protection, 23
Suspension of acids and fuel residues, 85
Suzuki engine tests, 632
 2-T, costs of, 555
Sweden, 607
Switzerland, emissions legislation, 328
Synchronized gearbox, lubrication, 282
Synthesized hydrocarbons, 64
Synthetic base stocks, 49-50, 197
 checks on, 363
Synthetic chemical fluids, 340
Synthetic oils, 177
Système International of units, 501-505

T

TA test, SAE, 332
Taiwan, emissions legislation, 328
Tall oil acids in greases, 311
Tanks, for bulk storage of lubricants, 373-374
Tapered Bearing Simulator, 104, 108, 109
Tappet pitting, 225-226

T

Tapping, 350
TB test, SAE, 332
TBN, 201, 207
additives, 413
formulations, 415
of railroad engines, 338
TBS, Tapered Bearing Simulator, 104, 108, 109
viscometer test, 518
TC test, SAE, 332
TCDD, 401
TCP, tricresyl phosphate, 280, 345
TCW tests, 556
TC-WII, 332, 334, 556, 633
TC-W3, 332, 334, 556, 633
Technical Committee of Petroleum Additive Manufacturers, ATC, 260, 393, 395
Technical societies, future role, 411-412
and motor industry, 427
role in oil reformulation, 408
Teflon™, PTFE, 23, 24, 26
Temperatures of bulk oils, 379, 380
Tertiary amine, 500
Tests, blind, 160
caravan, 137-138
costs, 427-429
developments, 426-431
new, justification for, 246-248
engine, 127-133
statistics used in, 153-158
equipment condition, 170-172
for evaluating oil condition, 164-170
field, 133-138
gear oils, 284-287
greases, 316-321
irrelevant to the market, 245
laboratory, 7-10, 99-123, 148-152
for measuring performance, 236
need for new types, 240
for physical properties, 99-116
precision measurement, 150
procedures, inadequate, 227-228
requirements for, 130-133
results, frequency diagram, 140-141
repeatability, 149
reproducibility, 149

T

Tests, blind (continued)
  rig, 7, 9
  sequences for MS quality, 183
  statistics of lubricants, 139-163
  time to run, 136
  use of reference fuels, 162-163
  used oils, 164-172
Tetrafluoromethane, 343
Texaco, 80
Thailand, emissions legislation, 328
Thermally stable ashless dispersant, 188
Thermally stable ZDDP, 188
THF, Tractor Hydraulic Fluids, 194, 300, 308-309, 341, 605
Thickeners, non-soap, in greases, 312
Thickening, 75, 235
  test, 130
Thiodiazoles, 283
  as corrosion inhibitors, 90
Thiol, 499
Thiophene, 499
Thiophosphates, 85
Thiophosphonate, 194
THOT test, 4L60, 303
Timing case, 235
Timken OK Load, 285
Timken test, ASTM D 2509, 319-320
Timken tester, D 2782, 285
Tin, in used oils, 171
Titration method D 664, 167
Titrimeter, 117
TLTC, 262
TO-2, Caterpillar, 300
TO-4, Caterpillar, 300
TO4-Bronze friction retention test, 551
TO4-Sequence 1220 test, 551
Torque converter, 292, 296, 297
Total base number, 201, 413, 415
Toxicity of mixtures, 394
  of unused lubricants, 394-396
  of ZDDP, 396
Toxicology of lubricants, 392-397
Toyota, 293, 308, 326, 608
  high temperature oxidation test, 555
  valve train wear test, 555

T

TP-1 procedure, 107, 228
Trabant, 325
Trace metal analysis of used oils, 171
Traction additives, 89
Tractor hydraulic fluids, 194, 300, 308-309, 341
  specifications, 605
Tractor, lubrication areas, 194
Trading standards, 372
Transatlantic differences in testing, 238
Transforming data (statistics), 145
Translucent fluids, 347
Transmission cycling test, in ATF testing, 303
Transmission Lubricants Technical Committee, 262
Transmission oxidation test, in ATF testing, 303
Transmission tests, costs, 553-555
Transmissions, new, 421
Transportation of lubricants, 402-403
Tray-type distillation, 55
Triazoles, 283
  as corrosion inhibitors, 90
Tribology, definition of, 12
  fundamentals of, 11-21
  testers, 127
Tributyl phosphate, 280
Tricresylphosphate, 280, 345
2, 2, 4-trimethyl pentane, 496
Tripartite of SAE, ASTM and API, 249-250
Trouble shooting, 380-383
True value in statistics, 139
TSC-1/2/3/4, API, 332
Turbine oil, 345-346
Turbochargers, 422
Turning in metal working, 347, 350
Two-stroke,
  -cycle gasoline engine lubricants performance and service classification, SAE J2116, 626-630
  cycles, for passenger car engines, 415
  design for passenger cars, 420
  engine, 35-36
    lubrication requirements, 44-45
  oils, 325-335
    general purpose, 334-335
    tests and specifications, 331-334
  tests, 555-556
Types of oils, future varieties, 416

U

Ultimate biodegradation, 397
Union Oil, 78, 80
Units, in S.I. system, 501-505
    used in laboratory practice, 505
Universal oils, 191
    gear, 283
Unleaded fuels, 413
Unsaturated compounds, 496
Ureas in greases, 312
U.S.,
    Air Force, aviation oil analysis, 170
    army, 212
        specifications 2-104, 7
    Chemical Manufacturers Association, 256
        ecotoxicity testing, 394
    classifications, 261, 248-259
    disposal of used lubricants, 401
    Environmental Protection Agency, 398, 401
    Federal Trade Commission, 397
    future market, 431
    lubricant consumption, 242
    military, 111, 113, 176, 190, 207, 269
        and SAE Lubricants Review Institute, 251-252
    military specifications, 237, 287, 290
        grease, 320
        MIL-2-104, 186
        MIL-L-46167B, 206
    Navy specification 14-0-13, 7
    notification laws, 386
    passenger car transmissions, 600-601
    rig-tests, costs, 551-552
    Sequence tests, 130
        V-E test, 219
    Tripartite arrangement, 411
    TSCA, 388-391
    use of low-sulfur diesel fuel, 399
Used lubricants,
    collection in Germany, 62
    disposal of, 378, 400-402
    tests on, 164-172
    toxicity, 396-397
    see also Recycling
Users, role in oil reformulation, 408
U-tube determination of kinematic viscosity, 103, 106

V

Vacuum distillation, 3, 55
Valency, 495
Valve body, for operating clutches and bands, 297
Valve deposits, problems with, 222
Valve lifter rusting/corrosion, 236
Valve train, 87
  wear, 240
    pitting, 235
    test, 186, 236
Van Doorne, Holland, 293
Vane-type pumps, 344-345
Variance, 144
Varnish, 6, 39
  formation, standards for, 239
Vegetable oil base stocks, 417
Vehicle hoists, 342
Vehicle manufacturers, future developments, 410
V.I., 53, 72
V.I. improver, *see* Viscosity modifier
Vickers Vane Pump Test, 303, 307, 554
Viscometers, 101, 103, 104, 106, 518
Viscosity, 5, 16-17, 99-108, 235
  in ATF testing, 303
  classification, 344
    for Gear Oils, 289
    SAE J300, 107
    SAE, 248-249
    system, modifications to, 107
  effect on emissions, 400
  future, 431
  in gear oils, 288
  high-temperature/high shear, 108, 186, 228
  of hydraulic oils, ISO classification, 266
  Improver, *see* Viscosity modifier
  Index, 33, 53, 72
    to test base oils, 358, 361
  loss, 109, 282, 383
  in lubrication, 32
  modifiers, 72-75, 181, 185, 203
    in ATF formulation, 305
    in automatic transmission fluids, 299
    checking, 362
    choice of, 199-200

V

Viscosity (continued)
    modifiers (continued)
        polyisobutylene, 182
        to reduce wear, 180
        specification of, 358
        structure of, 74
    in quality control, 371
    requirements, 238
    SAE 40, 339
    tests, 517
        on used oils, 165
Voith, 607, 608
Volatility, 5
    of base stock, 114-115
    effect on emissions, 400
Volkswagen, 208, 241, 259, 270, 290, 606, 607
    1.6 TC Diesel test, 241
    1302 wear test, 553
    cam & tappet rig test, 553
    intercooled turbo diesel test, 553
    naturally aspirated diesel test, 553
    quality control, 429
    seals test, 554
Volvo, 293, 607, 608
    Drain approval, 136
    TD 120, 224
    Truck, 259
VW, *see* Volkswagen

W

W1 Cortina test, 248
W2 tests, 2-T, costs of, 556
Warner Gear, 293
Warranty conditions, 410
Wartburg, 325
Waste disposal, 412, 413
Water, crackle test for, 361
    in oil emulsions, 347
    off-spray test, ASTM D 4049, 318
    for removing heat, 347

W

Water, crackle test for (continued)
sensitivity in greases, 314
wash-out test, ASTM D 1264, 318
Water-based hydraulic fluids, 345
Water-based lubricants, future of, 417-418
Water/glycol mixtures, as lubricant, 417-418
Wax hydro-isomerization, 60
Waxes, 351
Wear, 11-21
reduction, 4
in gear oils, 281
using viscosity modifiers, 180
test, in ATF testing, 303
Wedge generation in bearing, 29
Wet brakes, 308
Wheel bearing classification, 320
Wheel bearing test, 319
high-speed, ASTM D 1741 and D 3336, 319
Worldwide system for specifications and qualities, 434
Worm gears, 275, 277, 278, 346

X

X-ray analyzer, 117
X-ray fluorescence analysis, 121, 519
to measure elemental content of additives, 362

Y

Yamaha engines, 332
Yamaha tests, 334, 633
2-T, costs of, 555

Z

ZDDP, 76, 87, 180, 182, 194, 208, 225, 226, 280, 282, 304, 305, 344, 394, 424, 432
anti-wear properties, 418
irritancy, 394
in new formulations, 202
reduction of levels, 413
structure of, 86

Z

ZDDP (continued)
  thermally stable, 188
  toxicity, 394,
Zero ash oils, 432
ZF, 290, 293, 308, 606, 607, 608
Zinc alkaryl dithiophosphate, 395
Zinc alkyl dithiophosphate, 395
Zinc dialkyldithiophosphates, *see* ZDDP
Zinc, environmental effects, 403